THE FLORA

OF

WEST LANCASHIRE:

HELL CRAG, TARNBROOK FELL,

Showing fallen blocks of Millstone Grit. The dark stain on the face of the cliff at the extreme left is a moss, *Andreæa Rothii*.

Frontispiece.

THE FLORA

OF

WEST LANCASHIRE:

*(Vice-county 60 of " Watson's Topographical Botany,"—the portion of
Lancashire north of the Ribble and south of Morecambe Bay).*

BY

J. A. WHELDON, F.L.S.,

VICE-PRESIDENT OF THE LIVERPOOL BOTANICAL SOCIETY;

AND

ALBERT WILSON, F.L.S., F.R. MET. SOC.

(FORMERLY OF GARSTANG).

ILLUSTRATED WITH MAP and

FIFTEEN REPRODUCTIONS FROM PHOTOGRAPHS

of some of the most Interesting and Characteristic Scenery.

EP Publishing Limited
1978

This is a reprint of the 1907 edition, originally privately
published by the authors.

Republished 1978 by
EP Publishing Limited
East Ardsley, Wakefield
West Yorkshire, England

ISBN 0 7158 1361 7

British Library Cataloguing in Publication Data
Wheldon, J A
  The flora of West Lancashire.
  1. Botany – England – Lancashire
  I. Title II. Wilson, Albert
  581'.9'4276   Q K306
  ISBN 0-7158-1361-7

Please address all enquiries to EP Publishing Limited
(address as above)

Printed in Great Britain by
The Scolar Press Limited
Ilkley, West Yorkshire

CONTENTS.

LIST OF ILLUSTRATIONS.

(From photographs by Albert Wilson.)

These plates originally found throughout the volume are, in this reprint, to be found at the end of the volume.

Lancashire has always been remarkable for the number of its botanists, and it is, therefore, a singular fact that, with the exception of Westmorland, it stands alone amongst the northern counties of England in having no complete Flora of either of its vice-counties. In attempting to supply this deficiency in the case of West Lancashire, the Authors have followed the conventional models in the plan of the Flora. No innovations have been attempted, the work consisting of the usual descriptive introductory chapters, followed by a list of the species with their range and distribution within the county, and localities for the rarer ones. The idea of including what is known as a "Botanical Survey" on lines recently much advocated was considered and abandoned. It seemed desirable to the writers that the "Flora of West Lancashire" should follow the usual course of evolution and commence with a list of plants on the usual lines. To have added a "Botanical Survey" would not only have delayed publication, but might possibly have entailed the omission of the cryptogamic section of the list in order to keep the size of the book within reasonable limits. The elementary facts are here marshalled and arranged, and it is hoped they will be of assistance in furthering any future research into the problems of plant œcology that may be undertaken in West Lancashire.

It would be contrary to general experience of similar lists of plants to anticipate that the present one is free from errors. They seem to be inevitable in a work of this nature. An endeavour has been made to reduce them to a minimum by the personal verification of as many facts as possible and by the submission of critical or doubtful specimens to those best able to determine them. Such help has been freely and generously accorded, and it is owing to assistance received from the specialists named elsewhere that the Authors rely with some confidence on the general accuracy of the work.

In extenuation of certain shortcomings, of which the Authors are very sensible (amongst others the few references to public libraries and herbaria, which might possibly have yielded information had opportunities of consulting them been available), they would point out that the time spent upon the work cannot be

B

accurately measured by the number of years it has taken to bring it to its present stage. Both Authors reside at a considerable distance from the vice-county, and this has entailed, in wearisome railway journeys, a serious waste of the brief holidays snatched from business as opportunity offered. It was found impossible to obtain leisure for both field work and extensive library research, and the latter had to be sacrificed to some extent to the former. The Flora must be regarded therefore, chiefly as an account of the existing state of the vegetation of the vice-county rather than as a compilation of historical records. At the same time, all information to which we have had access, left by former workers, has been fully utilized, and in many instances has received recent verification.

During the progress of the work we have been brought into communication with various botanists who have made observations of local plants. We gratefully record our thanks to these and to many friends and correspondents who have kindly aided us with notes or specimens. Their names will be found in the list of authorities consulted. We have also to gratefully acknowledge help of another and equally important nature from the following botanists, who have obligingly placed their special critical knowledge of certain groups of plants at our service :—For Phanerogamic plants : Messrs. E. G. Baker, F.L.S., Arthur Bennett, F.L.S., Rev. G. R. Bullock-Webster, G. Claridge Druce, F.L.S., Alfred Fryer, A.L.S., James Groves, F.L.S., F. J. Hanbury, F.L.S., Rev. E. F. Linton, Rev. W. R. Linton, Rev. E. S. Marshall, F.L.S., Rev. W. M. Rogers, F.L.S. For Mosses : Messrs. J. E. Bagnall, A.L.S., H. N. Dixon, F.L.S., E. C. Horrell, F.L.S., W. E. Nicholson, and Mons. F. Renauld and Herr G. Roth. For Hepatics : Messrs. H. W. Arnell, Symers M. Macvicar, and M. B. Slater, F.L.S. For Lichens : Miss A. Lorraine-Smith and Messrs. E. M. Holmes, F.L.S., and J. A. Martindale.

WEST LANCASHIRE.

H. C. Watson in "Topographical Botany" divided Lancashire into three portions or vice-counties, viz., Lake Lancashire, West (or Mid) Lancashire, and South Lancashire.

The northern division, or Lake Lancashire, comprises the portions of the county north of Morecambe Bay, which he united with Westmorland, forming V.C. 69. The middle portion, or West Lancashire, V.C. 60, of which this work treats, consists of the portion south of Westmorland and north of the Ribble. The large area south of the Ribble forms South Lancashire, V.C. 59.

Compared with most other vice-counties, West Lancashire is rather small, the area being only 492 square miles.*

Its greatest length from the summit ridge of Greygarth Fell, in the extreme north-east corner, to Lytham in the west, is about 41 miles, and its greatest breadth, from Mitton in the east to Blackpool in the west, is 25 miles. There is, however, in this small area a great diversity of surface, and consequently, as would be expected, a rich and varied flora. The coast line is long and irregular, measuring about 102 miles from Silverdale to the estuary of the Ribble near Preston. It presents alternations of sandy shore, muddy salt-marshes, banks of shingle, and rocky cliffs of limestone and grit. In the north of the vice-county there is a varied series of geological strata, the principal of which, as affecting the flora, are Millstone Grit, Scar Limestone and Upper Silurian.

As in the neighbouring parts of Westmorland and Lake Lancashire, the scar limestone covers a considerable area near the shore of Morecambe Bay, and forms a rich combination of wild scar or cliff, flat creviced "pavement" and rocky hills, often beautifully clothed with thickety, aboriginal woodland, in which the yew, hazel and juniper form conspicuous features. Eastwards, across the Vale of Lune, Greygarth Fell rises to nearly 2,100 ft., and on the lower slopes of this the scar limestone appears again, forming the plateau of Leck Fell with its wonderful vertical chasms or "pot-holes," resembling those of Ingleborough and Penyghent.

* Lake Lancashire with Westmorland contains about 1,050 square miles and South Lancashire 1,130 square miles.

Here also the Leck Beck, a tributary of the Lune, which rises on Greygarth Fell, cuts a romantic glen called Ease Gill, first through Yoredale limestone and shales, and afterwards through scar limestone and upper silurian.

The principal rivers of West Lancashire are the Lune, the Keer, the Cocker, the Wyre, and the Ribble. The first and last belong as much to Westmorland and Yorkshire respectively as to Lancashire, for they only flow through the latter county in the lower part of their courses. The beauty of the Vale of Lune throughout almost its entire length is well known, but owing to the land being mostly highly cultivated, and the geological formation mainly Millstone Grit in the Lancashire portion, this is less interesting botanically than the country higher up the valley in Westmorland. The same remark applies, to some extent, to the valley of the Ribble from Stonyhurst to Preston. The Keer, and Cocker, and the lower portion of the Wyre draining the Fylde country, flow mostly through a low-lying fertile plain, now generally highly cultivated, but formerly covered by large areas of swampy woodland and peat-moss. Small remnants of the latter still remain and yield numerous interesting plants which, with the large quantities of fallen tree trunks dug out from the land, tell us something of the character of the vegetation in bygone times.

A large portion of the east of the vice-county, especially along the Yorkshire boundary, consists of broad elevated moorlands or fells, formed of Millstone Grit, which rise generally to an altitude of about 1,500ft., but reach 1,836ft. on Wardstone near the head of Wyresdale. On the north slopes the fells are drained by tributaries of the Lune, and on the west and south by feeders of the Wyre and Ribble. These streams cut deeply into the moors, forming steep narrow glens which are known as "gills" in the drainage area of the Lune, but further south as "cloughs." The upper parts of these are wild stony gullies, but lower down they are generally thickly wooded with Oak, Mountain Ash, Birch, etc., and they open out into beautiful dales. The higher parts of the moors are generally thickly peat-covered and boggy, with extensive areas of cotton grass and sphagnum, alternating with heather and bilberry in the drier situations. The lower slopes are clothed with a similar vegetation with the addition of bracken, and the character of the surface gradually changes to more or less cultivated upland pasture, which latter forms a large part of the remainder of the fell-country.

Further and more detailed particulars as to the geology and other physical features will be found in the descriptions which follow of the different districts.

Although the more easily accessible parts of West Lancashire have for many years been visited from time to time by various botanists, the more remote dale-districts and higher fells remained almost entirely neglected. So much was this the case that up to about the year 1880 West Lancashire was, as a whole, perhaps the least worked vice-county botanically in the north of England. The rich flora of the limestone district bordering Westmorland had long been fairly well investigated, as had also that of parts of the Fylde and the district about Preston, but as regards the flora of the upper dales little or nothing was known.

For botanical purposes the vice-county may be very naturally and conveniently divided into three main divisions, North, East, and West, each presenting marked differences as to surface, elevation, geological formation and climate.

The North Division is separated from the remainder of the vice-county by the river Lune from its mouth upwards to the point where it is joined by the Wenning, beyond which this tributary forms the line of demarcation to the Yorkshire boundary.

The West Division consists of the nearly level plain lying between the estuaries of the rivers Lune and Ribble, and intersected about midway by the river Wyre. The high road running south from Lancaster to Preston forms a convenient means of separating this division from the East Division.

The East Division comprises the country east of the high road from Lancaster to Preston, as far as the Yorkshire boundary, and consists largely of elevated moorlands and dales.

We have further divided the above three main divisions into a total of eight minor divisions or districts. These we have endeavoured to make as natural as possible, being guided by the configuration of the country and varying character of its surface. The physical features of West Lancashire do not always lend themselves to the plan of dividing it into districts according to the drainage areas of the river-systems, nor yet in all cases to making the rivers themselves form the dividing lines. We have therefore been obliged to resort to a combination of the two. In most cases the latter method appeared more applicable, but in that of the districts of the East Division we have adopted the former, taking the dividing watersheds as boundaries. In separating two

of the North Division districts, viz., "Kellet" from "Silverdale and Heysham" we have taken a more artificial boundary, making the main high road from Lancaster to Burton serve as a dividing line. This forms a good demarcation, separating the maritime and mostly low-lying area of "Silverdale and Heysham" from the uplands of "Kellet." The following are the eight sub-divisions of the vice-county :—

NORTH DIVISION—3 districts.
1. "Leck."
2. "Kellet."
3. "Silverdale and Heysham."

WEST DIVISION—2 districts.
4. "Pilling."
5. "Kirkham."

EAST DIVISION—3 districts.
6. "Wray and Quernmore."
7. "Wyresdale and Bleasdale."
8. "Longridge."

NORTH DIVISION.

DISTRICT No. 1—"LECK."

This district contains a greater variety of surface than any other in the vice-county. It consists of a long, narrow strip of land extending from near the source of the Leck Beck on Greygarth Fell to the junction of the Wenning with the Lune below Hornby, a distance of twelve miles. On the north it is bounded by Westmorland and on the east by Yorkshire; the river Wenning separates it on the south from District 6, and the Lune on the west from District 2. Geologically it contains no less than six distinct strata, viz. :—Upper Scar Silurian, Carboniferous or Great Scar Limestone, Yoredale Rocks, Millstone Grit, Coal Measures, and Permian Sandstone. Towards the north-eastern corner the bare mountain ridge of Greygarth Fell or "Gragreth" reaches an altitude of 2,080ft., the highest ground in West Lancashire. The portion of the fell on the Lancashire side that rises above 1,800ft., which is the lower limit in the North of England of the Infer-Arctic zone of Watson, measures over three miles in length by from half to a quarter-of-a-mile in width. This is the only ground in the vice-county coming within this zone except a very small area on Wardstone near the head of Wyresdale. A wall which serves as the county boundary runs along the summit-ridge following the watershed. The ground slopes away rapidly on the east, or Yorkshire, side, into the bare upland valley of Kingsdale, but on the west the slope is more gradual. Geologically the ridge is mostly composed of Yoredale Grit, with a little Yoredale Limestone. The ground overlying the grit consists of peaty moorland, very boggy in places, but that over the limestone is dryer and covered with the fine grassy turf so characteristic of this formation. The surface is bare and wind-swept, with very little exposure of limestone rock. but the soil is favourable to the growth of a few plants of a xerophilous character, and the total number of species found on the limestone and grit combined is considerable, though fewer than those which occur at a similar altitude on the neighbouring mountain of Ingleborough, where patches of scar afford suitable situations and shelter. The following is a complete list, so far as yet observed, of the plants which occur on Greygarth above the 1,800ft. contour line. Those marked with an asterisk are, we believe, peculiar to the small area of Yoredale Limestone, and are found within a mile of the north-east extremity of the district :—

Infer-Arctic Zone in West Lancashire.

FLORA OF GREYGARTH FELL (summit ridge only).

1,800–2,080 feet.

Ranunculus Flammula, L.
*   ,,   acris, L.
   ,,   repens, L.
Cardamine hirsuta, L.
* Draba verna, L.
Cochlearia alpina, Wats.
Viola palustris, L.
Cerastium vulgatum, L.
Stellaria media, Vill.
   ,,   uliginosa, Murr.
Sagina procumbens, L.
Montia fontana, L.
Oxalis Acetosella, L.
* Trifolium repens, L.
Rubus Chamæmorus, L.
Potentilla sylvestris, Neck.
* Alchemilla vulgaris, L.
Saxifraga hypnoides, L.
Chrysosplenium oppositifolium, L.
Epilobium palustre L. f. minor simplex
Galium saxatile, L.
   ,,   umbellatum, Lam.
Bellis perennis, L.
* Achillea Millefolium, L.
Carduus lanceolatus, L.
Leontodon autumnalis, L.
Taraxacum officinale, Web.
Vaccinium Vitis-idæa, L.
   ,,   Myrtillus, L.
Calluna vulgaris, Hull.
* Veronica arvensis, L.
*   ,,   serpyllifolia, L.
* Veronica Chamædrys, L.

* Euphrasia borealis, Townsend.
* Thymus Serpyllum, L.
* Prunella vulgaris, L.
Rumex Acetosa, L.
   ,,   Acetosella, L.
Urtica dioica, L.
Empetrum nigrum, L.
Juncus squarrosus, L.
   ,,   conglomeratus, L.
Luzula campestris, D.C.
   ,,   multiflora. Lej.
Scirpus cæspitosus, L.
Eriophorum vaginatum, L.
   ,,   polystachion, L.
Carex echinata, Murr.
   ,,   leporina, L.
   ,,   rigida, Good.
   ,,   Goodenowii, Gay.
   ,,   flacca, Schreb.
Anthoxanthum odoratum, L.
Agrostis tenuis, Sib.
Aira cæspitosa, L.
   ,,   flexuosa, L.
Holcus mollis, L.
Poa annua, L.
   ,,   pratensis, L.
Festuca ovina, L.
Nardus stricta, L.
Cryptogramme crispa, Br.
Lastrea aristata, Britt. & Rendle
         (L. dilatata Presl.) v. nana.
Lycopodium Selago, L.

Sphagnum rubellum, Wils., v. versicolor
   ,,   ,,   ,,   v. rubrum
   ,,   ,,   v.purpurascens
   ,,   ,,   ,,   v. viride
   ,,   acutifolium, Russ. & Warnst.,
         v. purpurascens
   ,,   quinquefarium, Warnst.,
         v. roseum
   ,,   ,,   v. fusco-flavum
   ,,   subnitens, Russ. & Warnst.,
         v. obscurum
   ,,   cuspidatum, Russ. & Warnst.,
         v. submersum
   ,,   ,,   v. falcatum
   ,,   recurvum, Russ. & Warnst.,
         v. mucronatum
   ,,   subsecundum, Limpr.
   ,,   rufescens, Warnst.
   ,,   papillosum, Lindb.
         v. normale
Andreæa petrophila, Ehrh.
Oligotrichum hercynicum, Lum.

Polytrichum alpinum, L
   ,,   piliferum, Schreb.
   ,,   strictum, Banks.
   ,,   commune, L.
Ditrichum homomallum, Hpe.
Ceratodon purpureus, Brid.
Dichodontium pellucidum, Schp.
   ,,   v. fagimontanum, Schp.
Dicranella heteromalla, Schp.
Dicranoweisia cirrata, Lindb.
Blindia acuta, B. & S.
Campylopus flexuosus, Brid.
Dicranodontium longirostre, B. & S.
   ,,   v. alpinum, Schp.
Dicranum scoparium, Hedw.,
         v. spadiceum, Boul.
   ,,   fuscescens, Turn.
Fissidens adiantoides, Hedw.
Grimmia apocarpa, Hedw.
   ,,   pulvinata, Sm.

Grimmia Doniana, Sm.
Rhacomitrium fasciculare, Brid.
   ,,   heterostichum, Brid.
   ,,   ,,   v.gracilescens, B. & S.
   ,,   ,,   v. alopecurum, Hübn.
   ,,   lanuginosum, Brid.
   ,,   canescens, Brid.
Ptychomitrium polyphyllum, Fürnr.
Tortula intermedia, Berk.
Barbula rubella, Mitt.
   ,,   rigidula, Mitt.
Leptodontium flexifolium, Hpe.
Trichostomum tortuosum, Dixon.,
   ,,   fragilifolium, Dixon
Encalypta streptocarpa, Hedw.
Orthotrichum saxatile, Milde.
   ,,   cupulatum, Hoffm.
Splachnum sphæricum, L.
Funaria hygrometrica, Sibth.
Aulacomnium palustre, Schwaeg.
Philonotis fontana, Brid.
Webera nutans, Hedw.
   ,,   v. longiseta, B. & S.
   ,,   v. bicolor, Schp.
   ,,   albicans, Schp.
Bryum inclinatum, Bland.
   ,,   pallens, Sw.
   ,,   affine, Ldb.
   ,,   cæspiticium, L.
   ,,   capillare, L.
   ,,   obconicum, Hornsch (?)
   ,,   argenteum, L.
Mnium affine, Bland.
   ,,   serratum, Schrad.
   ,,   rostratum, Schrad.
   ,,   hornum, L.
Thuidium tamariscinum, B. & S.
Climacium dendroides, W. & M.
Camptothecium sericeum, Kindb.
Brachythecium populeum, B. & S.

Brachythecium plumosum, B. & S.
Eurhynchium confertum, Milde.
   ,,   murale, Milde.
Plagiothecium elegans, Sull.
   ,,   undulatum, B. & S.
Hypnum fluitans, L.
   ,,   v. atlanticum, Ren.
   ,,   v. ovale, Ren.
   ,,   v. gracile, Boul.
   ,,   v. Jeanbernati, Ren.
   ,,   uncinatum, Hedw.,
   ,,   v. plumosum, Schp.
   ,,   cupressiforme, L.,
   ,,   v. ericetorum, B. & S.
   ,,   molluscum, Hedw.
   ,,   cuspidatum, L.
   ,,   Schreberi, Willd.
Hylocomium splendens, B. & S.
   ,,   loreum, B. & S.
   ,,   squarrosum, B. & S.
Lejeunea serphyllifolia Dicks.
Blepharozia ciliaris, L.
Lepidozia reptans, L.
Kantia trichomanis, L.
   ,,   Sprengelii Mart.
Scapania undulata, L.
   ,,   irrigua, Nees.
Diplophyllum albicans, L.
Mylia Taylori, Hook.
Plagiochila asplenioides, L.
   ,,   spinulosa, Dicks.
Jungermania inflata, Huds.
   ,,   Flœrkii, Web. & Mohr.
   ,,   var. heterostipa
   ,,   Lyoni, Tayl.
   ,,   gracilis, Schleich.
   ,,   ventricosa, Dicks.
   ,,   minuta, Crantz.
Nardia scalaris, Schrad.
Pellia Neesiana, Gottsche.

Cetraria islandica var. crispa Nyl. and other interesting lichens also occur.

The ridge of Greygarth extends half a mile northwards beyond the place at the north-east extremity of the district where Westmorland, Yorkshire, and Lancashire meet. and culminates in Dent Crag, 2,250ft. On the south-west or Westmorland side of this summit a stream rises which, after a course of three-quarters of a mile in Westmorland, forms the boundary between Lancashire and Westmorland for about three miles. The stream is called the Leck Beck, and flows down a rough, stony ravine known as Ease Gill. We will describe the course of this beck down to its junction with the Lune, 7½ miles below, and afterwards return to the description of Greygarth Fell. At the point

where the beck becomes the county boundary the elevation is 1,570ft.; a little below this it flows over ledges of limestone and grit belonging to the Yoredale Series, and forms several small but romantic waterfalls. The following are the rarer plants which adorn the stream-side and rock crevices :—

Cochlearia alpina, Wats.
Alchemilla vulgaris, L.,
         v. alpestris (Schmidt)
Saxifraga hypnoides, L.
Hieracium rigidum, Hartm.,
         v. tridentatum (Fr.)
   ,,   murorum, L.
   ,,   v. micracladium Dahlst.
   ,,   duriceps,
         v. cravoniense, F.J.H.
Asplenium viride, Huds.

Lycopodium alpinum, L.
Selaginella selaginoides, Link.
Andreæa Rothii, W. & M.
Seligeria pusilla, B. & S.
Weisia rupestris, C. M.
Bartramia Œderi, Sw.
Breutelia arcuata, Schp.
Plagiobryum Zierii, Lindb.
Orthothecium intricatum, B. & S.
Hypnum stramineum, Dicks.

The gill then widens a little and falls less steeply, the bed and banks of the stream being strewn with large numbers of boulders brought down during floods. The banks rise steeply on either side, with here and there a sparkling rill from the moorland above. Half-a-mile lower, at an altitude of 1,150ft., the Yoredale rocks are left behind and the Great Scar Limestone is reached. The stream, after flowing a short distance over a smooth, sloping floor of marble and through a narrow rock-bound channel, plunges 27ft. into a deep basin or hollow in the limestone with water-worn sides. The rocks rise perpendicularly on both sides and the water is very deep. This waterfall is known as Ease Gill Foss. There is no surface outflow from the basin except during rainy weather, the water escaping underground, and below this point for the next mile or so there is frequently little or no water in the bed of the gill. Bare limestone scars rise at intervals on either hand, but especially on the Lancashire side, and the whole gill is exceedingly rough and rocky. "As rough as Leck Beck" is an old proverb of the district and a very true one, for a rougher beck is hardly to be found in Lancashire or Westmorland. About three-quarters-of-a-mile below the Foss another very interesting place called the Dry Foss is reached. The course of the beck here is between overhanging rocks of limestone, water-worn and polished, and hollowed out beneath in a most curious manner. The Dry Foss, as its name implies, is perfectly dry except after heavy rain, the water finding its way underground. A short distance lower there is another steep fall in the gill, and below this the limestone rocks rise over 50ft. above the stream, forming a fairly flat, rather gloomy enclosure beneath, which has been compared to the inside of a church, and is locally known as "Ease Gill Kirk." Beyond

this are some curious holes in the rocks called "The Witches' Caves," and the stream again flows through a deep, narrow gully in the limestone, its sides beautifully clothed with shrubs, ferns, and mosses. The water-worn rocks in the bed of the gill must be climbed over to explore it properly, and this can only be done in dry weather, when the water is low. About 250 yards below the Kirk the beck receives a stream on the right from "Bull Pot of the Witches," and soon afterwards ceases to form the boundary of the counties, a wall then marking this over Casterton Fell. Still following the stream, the limestone is left behind and we reach the Upper Silurian formation (Coniston Grit). The following are the rarer plants found on the Scar Limestone of Middle Ease Gill from the Foss downwards :—

Actæa spicata, L.
Draba incana, L.
Saxifraga hypnoides, L.
Galium umbellatum, Lam.
Scabiosa Columbaria, L.
Hieracium murorum, L.
   ,,   duriceps,
         v. cravoniense, F.J.H.
Myosotis sylvatica, Hoffm.
Carex sylvatica, Huds.
Festuca sylvatica, Vill.
Polypodium Robertianum, Hoff.
Asplenium viride, Huds.
Lycopodium Selago, L.

Fissidens decipiens, De Not.
Trichostomum nitidum, Schp.
Ulota Bruchii, Hornsch.
   ,,   crispa, Brid.
Bartramia Œderi, Sw.
Philonotis calcarea, Schp.
Lepidozia Pearsoni, Spruce.
Scapania aspera, Müll. & Bern.
Metzgeria pubescens Schrank.
   ,,   conjugata, Lindb.
Lejeunia calcarea, Lib.
Saccogyna viticulosa Mich.
Peltidea aphthosa, Ach.
Lecidea cupularis, Ehrh.

After half-a-mile of gradual fall in a rough stony bed, the stream once more cuts a gorge, this time in the Coniston Grit. The contrast between this glen and the limestone glen just described is, of course, very great ; each has its own peculiar features dependent on its geological formation, and each its characteristic flora. The high banks are wooded and heather-clad, and form a charming background to the beautiful rock scenery and waterfalls along the course of the stream. Casterton Fell rises steeply on the right, mostly grassy and bracken-covered, but with patches of Silurian rock on the upper part. Amongst these the Parsley Fern grows luxuriantly, much more so than elsewhere in the vice-county. *Andreœa petrophila* also occurs here, as on the rocks near the stream, but it is rather scarce. The beck next bends sharply to the left, and the glen opens out with a stretch of fairly level, grassy ground on the left bank. The Silurian rocks give place to Drift deposits overlying Coal Measures, which latter belong to the Ingleton coal-field. The steep hillside on the right is clothed with a thick wood, in which *Orthotrichum pulchellum* grows, and soon the whole valley becomes wooded and continues so almost to the village of Leck, which is pleasantly situated on the left bank of the stream. Half-a-mile lower is Cowan Bridge, on the main road from Kirkby Lonsdale to Ingleton, and here the comparatively level valley of the Lune is reached. After a further course of about two miles through field and woodland the Leck Beck joins the Lune near Nether Burrow. The following are some of the rarer plants of the Silurian rocks in Lower Ease Gill, the boggy ground near the stream, and the wooded valley about Leck :—

Geranium sylvaticum, L.	Orthotrichum pulchellum, Sm.
Chrysosplenium alternifolium, L.	,, stramineum, Hornsch.
Galium uliginosum, L.	Ulota Bruchii, Hornsch.
Primula farinosa, L.	Bartramia ithyphylla, Brid.
Cryptogramme crispa, Br.	Webera elongata, Schwgr.
Asplenium Adiantum nigrum, L.	Bryum filiforme, Dicks.
Polypodium Phegopteris, L.	,, alpinum, Huds.
Andræa petrophila, Ehrh.	Heterocladium heteropterum, B. & S.
Diphyscium foliosum, Mohr.	Thuidium delicatulum, Mitt.
Swartzia montana, Lindb.	Amblystegium fluviatile, B. & S.
Fissidens osmundoides, Hedw.	Frullania fragilifolia, Tayl.
Barbula spadicea, Mill.	Parmelia conspersa, Ach.
Zygodon Mougeotii, B. & S.	Lecanora ventosa, Ach.

Returning to the ridge of Greygarth, we find the southern summit nearest Ingleton is broader and more boggy than the northern part already described. Only Yoredale Grit occurs here, and this is mostly covered with peat. A few grit rocks near the summit-wall at an altitude of 2,000ft. yield *Andreæa petrophila, Oligotrichum hercynicum,* and small starved plants of *Cryptogramme crispa.* On the boggy ground of the fell top *Blepharozia ciliaris* is very fine and abundant, and amongst the heather abundance of *Cetraria islandica.* The views from the south slope are magnificent and extensive. To the north-east the bleak ridge of Whernside is seen across Kingsdale, whilst Ingleborough rises steeply to the south-east. The whole of the country south of Ingleton stretching away to the fells of Bowland is spread out like a map, as is also the valley of the Lune down to Lancaster, with the sea beyond. If the day be sunny, the river itself is seen shining like silver as it flows through the meadows about Arkholme. Proceeding westwards from the summit, a gradual slope is traversed for half-a-mile over peaty ground and abundant Cloudberry plants ; the fell side then becomes steeper, with scree-slopes of broken grit rock, in the crevices of which *Lycopodium Selago* grows to a large size, some stems being about a foot in length. *Andreæa Rothii, A. petrophila,* and *Tetraplodon mnioides* also occur, and in boggy ground on the slope towards Ease Gill *Cinclidium stygium* grows. A cart-track which leads down the

fell to the village of Leck is next seen, and if this be followed for about a mile the limestone plateau of Leck Fell is reached. This is well worth a visit, not only because of its botanical interest, but also on account of the wonderful vertical chasms, or "Pot Holes," in the limestone which occur here. These compare favourably in size and general interest with those found on Ingleborough and Penyghent and on the Yorkshire side of Greygarth. The principal are Death's Head Hole, Rumbling Hole, and Navel Pot, their depths being, according to Balderston,* 200ft., 140ft., and 118ft. respectively. There are also two curious pairs of holes called the Upper Eye Holes and the Lower Eye Holes. The left eye of the upper pair is 45ft. deep, and the others, though shallow, are interesting. Death's Head Hole and Rumbling Hole are very dangerous chasms, but their sides are beautifully adorned with vegetation, and the latter has been called by Balderston the "Fairies' Workshop." The ground round the mouth of the pots is generally funnel-shaped and often clothed with bushes, whilst the inaccessible rock-ledges are covered with plants, ferns, and mosses for many yards below the level of the moor.

Navel Pot is so called from its shape, and although less deep, is much more capacious than the others, measuring 40 yards long by about 20 wide in the centre. Its sides are not so uniformly vertical, having more rock-ledges and grass slopes with numerous shrubs. *Actæa spicata* is found in one of the pot holes, and with the exception of the neighbouring locality in Ease Gill, does not occur, so far as we know, elsewhere in Lancashire. Other interesting plants of Leck Fell are *Draba incana, Chrysosplenium alternifolium, Hieracium gothicum, Polypodium Robertianum,* Hoff., *Lastrea rigida, Asplenium viride,* and *Sphagnum quinquefarium.* Scar Limestone rocks crop out at intervals near the pot-holes and the surface is mostly dry and grassy, but on the north-east part towards Ease Gill it is rather boggy and deeply heather-clad.

From the pot-holes the easiest way to Leck is by the road traversed in the descent of Greygarth Fell, but the most interesting is by way of Ease Gill Kirk (half-a-mile north of the pots) and through Lower Ease Gill, already described. In the field or allotment on the south side of the road, and at about the same elevation on the fell as the pots, is situated Lost John's Cave. This extends, according to Balderston, for 222 yards.

---

* See "Ingleton, Bygone and Present," by R. R. and M. Balderston.

The lower slope of Greygarth, south of Leck Fell, is known as Ireby Fell, from the village of Ireby. It is mostly grassy and heathery and yields the rare *Amblyodon dealbatus,* but does not appear to possess any special botanical interest. The geological strata represented are Yoredale rocks, Coal Measures, and Permian Conglomerate.

Turning now to the low country we find that this is very beautiful, fine old trees being numerous, but being generally well cultivated it is not very interesting botanically except along the banks of the rivers. Just below where the Lune enters the vice-county, a quarter-of-a-mile below Kirkby Lonsdale Bridge, the left bank rises steeply and is wooded for a short distance, but soon afterwards it becomes flat and so continues most of the way to where the district terminates below Hornby. The fall in the river is gradual in this part of its course, being only 60ft. in about seven miles. On the banks near Tunstall and Melling there are considerable tracts of shingle, occasionally submerged, and much waste, marshy ground with abundance of *Carex acuta* and thickets of *Salix purpurea.* At Nether Burrow the Lune receives the Leck Beck, and near Tunstall the Cant Beck and the river Greeta. The banks of the latter stream, from the Yorkshire boundary below Black Burton to its junction with the Lune, are beautifully wooded, and for some distance in the upper part the river flows through a narrow glen with rocky sides. Flat mossy rocks also occur in the bed of the stream, and there are several small water-falls. *Trollius europæus* and *Sesleria cærulea* are locally abundant. The following rarer plants also occur :—

Cochlearia alpina, Wats.	Tetraphis Browniana, Grev.
Geranium sylvaticum, L.	Dichodontium flavescens, Lindb.
Rubus saxatilis, L.	Fissidens crassipes, Wils.
Circæa intermedia Ehrh.	Campylostelium saxicola, B. & S.
Gymnadenia conopsea, Br.	Weisia tenuis, C.M.
Allium oleraceum, v. complanatum, Fr.	,, rupestris, C.M.
Blysmus compressus, Panz.	,, verticillata, Brid.
Carex pendula, Huds.	Hylocomium brevirostre, B. & S.
Avena pratensis, L.	Blepharostoma trichophylla Dill.
Polypodium Dryopteris, L.	Metzgeria conjugata, Lindb.
,, Phegopteris, L.	

The Allium and Sesleria have probably been brought down by floods from the limestone scar district about Ingleton. The Vale of Lune, from Nether Burrow to Tunstall and Melling, contains much rich meadow and grazing land. About four miles below where it receives the Greeta, the Lune is joined by the Wenning. Except for the land near the Lune, which consists of

flat fields and marshes, the country between the Greeta and Wenning is hilly and very picturesque. The geological formation is Millstone Grit overlaid with Drift. The following are some of the rarer plants found on the banks of the Lune and Wenning :—

Thalictrum collinum, Wallr.	Salix purpurea, L.
Cochlearia alpina, Wats.	Potamogeton densus, L.
Hypericum maculatum, Crantz.	Scirpus sylvaticus, L.
Agrimonia odorata, Mill.	Blysmus compressus, Panz.
Hippuris vulgaris, L.	Carex acuta, L.
Myriophyllum alterniflorum, DC.	Festuca elatior, L.
Senecio saracenicus, L.	Polytrichum nanum, Neck.
Myosotis sylvatica, Hoffm.	Tortula mutica, Lindb.
Salix phylicifolia, L.	Orthotrichum rivulare, Turn.
,, nigricans, Sm.	Amblystegium fluviatile, B. & S.

DISTRICT No. 2—"KELLET."

This district lies immediately to the west of District 1. It is bounded on the north by Westmorland, on the east and south by the Lune, and on the west by the high road leading from Burton to Lancaster. With the exception of the low ground in the valley of the Lune, and along the course of the Keer, the land consists mostly of upland pasture, having an elevation of about 400ft. There are three principal geological strata represented, viz., Scar Limestone, Yoredale Rocks and Millstone Grit. The flora is rather rich considering that the district possesses no coast line and the highest elevation is only 860ft. The limestone in the west largely accounts for this, but the banks of the Lune yield a number of interesting plants, as do also the small areas of aboriginal swamp, boggy moor and woodland which still remain in certain parts of the district. The only stream of importance apart from the Lune is the Keer, which, rising on Docker Moor just over the Westmorland boundary, flows south-westwards and enters the sea near Carnforth, the last mile of its course being through District 3.

In our description we will take the country north of the Keer first, beginning with the township of Dalton. This is part of the parish of Burton in Westmorland, and was added to Westmorland for civil purposes in 1896. Above the hamlet of Dalton Houses, and one mile north-east of Dalton Hall is the richly wooded limestone scar called Dalton Crag, the highest ground in District 2. It is really the south-west portion or Lancashire side of Hutton Roof Crag, the actual summit of which is in Westmorland, a few yards beyond our limits. Farlton Knott, Hutton Roof Crag and

Dalton Crag form a fine range of limestone scars, and are highly interesting botanically. The surface of Dalton Crag consists principally of rocky and thickety woodland, with extensive areas of limestone pavement. This latter is much indented with long, narrow and deep crevices or "clints" in which *Polypodium Robertianum* and *Lastrea rigida* are abundant. On the higher portion of the hill there are fewer trees and some extent of grassy and bracken-covered moorland interspersed with patches of pavement and fine limestone terraces. The following are the rarer plants of this beautiful tract of country :—

Thalictrum montanum, Wallr.	Pottia recta, Mitt.
Asperula cynanchica, L.	Barbula recurvifolia, Schp.
Inula vulgaris, Trev.	Trichostomum crispulum, Bruch.
Cynoglossum officinale, L.	„ nitidum, Schp.
Juniperus communis, L.	Ulota Bruchii, Hornsch.
Taxus baccata, L.	„ crispa, Brid.
Polygonatum officinale, All.	Funaria calcarea, Wahl.
Convallaria majalis, L.	Thuidium recognitum, Lindb.
Sesleria cœrulea, Ard.	Cylindrothecium concinum, Schp.
Melica nutans, L.	Hylocomium rugosum, De Not.
Polypodium Robertianum, Hoff.	Riccia Lescuriana, Aust.
Lastrea rigida, Presl.	Reboulia hemisphærica (L.)
Fissidens decipiens, De Not.	Scapania aspera, Bern.

On a piece of heathery ground, on the Westmorland side of the hill, about 400yds. beyond the boundary, *Listera cordata* and *Lycopodium alpinum* were found by one of the authors in 1878, at the unusually low elevation of 700ft.

To the south-west of Dalton Crag the country falls gradually towards the high road leading southwards from Burton, which, as before stated, forms the boundary of the district on its western side. The land immediately to the left of the road and extending by way of Priest Hutton to Borwick and Carnforth consists principally of rich pasture and meadow, the Scar Limestone not appearing much on the surface. Further east this rock is succeeded by Yoredale Grit, which occurs as a band parallel to the limestone, and composes the range of hills which extends from Henridden, on the south side of Dalton Crag, to near Borwick. The surface of these hills consists of grass pasture with a considerable proportion of woodland, and the country generally is very beautiful. The Yoredale Grit also continues in a north-easterly direction to the upland country between Docker Moor, Hutton Roof and Whittington, and in a southerly direction to near Halton. In the neighbourhood of Hutton Roof village, near the county boundary (which here is rather irregular in outline), the Yoredale Grit has been extensively quarried for a long period,

yielding excellent flag stones. Some parts only of the Upper Keer district are well cultivated, the remainder consisting of fir plantations and moorland. A heathy swamp on North's Lots, Whittington Moor, is rich in bog plants especially Sphagna, and an adjoining marshy meadow yields *Scutellaria minor*. Near Borwick the strip of Yoredale rock is replaced in an easterly direction by Millstone Grit, and this formation extends across the Keer and over almost the whole of the hill country eastwards and southwards to the Lune. As before stated, the Keer rises near the Westmorland boundary on Docker Moor. It flows southwestwards, and a mile from its source cuts a narrow gulley through Millstone Grit strata, forming at one point a waterfall about 20ft. high. The steep banks of the ravine are thickly clothed with trees, the lower portion being known as Wash Dub Wood. Several interesting mosses occur here :—

Sphagnum inundatum, Warnst.	Antitrichia curtipendula, Brid.
Tetraphis Browniana, Grev.	Heterocladium heteropterum, B. & S.
Weisia verticillata, Brid.	Eurhynchium pumilum, Schp.
Zygodon conoideus, H. & T.	Hylocomium brevirostre, B. & S.
Ulota Drummondii, Brid.	Chomiocarpon quadratus Scop.
„ Bruchii, Hornsch.	Scapania compacta, Dum.

The stream then flows for two miles through low flat meadows at the foot of Arkholme Moor and through the beautiful country about Capernwray Park. The remainder of its course across the low ground from Borwick to Carnforth does not call for special remark, but several interesting plants are found upon its banks or in the neighbourhood. Mid-way between Borwick and Whittington and one mile south-east of Wash Dub Wood is the little hamlet of Docker. Near this, at the foot of Docker Moor, there is a small deep bog or swamp surrounded by dry hilly pastures. It probably occupies the site of an ancient tarn, and although only a few acres in extent, it is exceedingly rich in interesting plants, as shown by the following list. Those marked with an asterisk are, so far as we know, not found elsewhere in the vice-county :—

*Genista anglica, L.	Carex disticha, Huds.
„ tinctoria, L.	* „ diandra, Schrank.
Comarum palustre, L.	„ canescens, L.
Parnassia palustris, L.	Lastrea spinulosa, Presl.
Galium uliginosum, L.	Ulota Bruchii, Hornsch.
Andromeda Polifolia, L.	*Hypnum vernicosum, Lindb.
Menyanthes trifoliata, L.	* „ v. majus, Lindb.
Veronica scutellata, L.	„ scorpioides, L.
Salix pentandra, L.	„ giganteum, Schp. (fruiting)
„ repens, L.	Sphagnum Warnstorfii, Russ.
Orchis incarnata, L.	„ contortum (Schultz.) Limpr.
Carex dioica, L.	

C

The surface of the bog as seen about the end of May is gay with a wealth of *Menyanthes*, *Salix pentandra*, and *Genista anglica*. The country between here and Kirkby Lonsdale is very picturesque and generally well wooded. Whittington is a pretty village, sheltered by fine trees and high ground to the north and west. One mile south-west of Kirkby Lonsdale, near the Westmorland boundary, there is an outcrop of Scar Limestone which extends westwards into Westmorland to the north of Hutton Roof. This yields numerous calcicole plants, including *Draba muralis*. Coming now to the Lune, we find that the right bank from Whittington to Arkholme and Gressingham is mostly low and bordered by flat fields. The country is thus very similar to that on the opposite side about Nether Burrow and Tunstall, already described. A little above Arkholme the river divides into two streams for about a mile, forming an island. Near this there is a good deal of waste, marshy and shingly ground, liable to be flooded.

From Arkholme downwards to Gressingham, Aughton, and Caton the river generally flows near to the hill country on its right bank, the slopes being often steep and well wooded. The fall in the river is slight, and much low, marshy, meadow land extends along the opposite bank. (See District 6.) At Caton and the Crook of Lune the country is very picturesque, the banks of the stream being generally richly wooded to the water's-edge. The bed of the river from the Crook to the village of Halton is rocky in places, and several interesting plants occur, which we shall mention when describing District 6. About two miles below Halton the Lune becomes tidal, and near this point, at Skerton Bridge, Lancaster, the district under consideration terminates.

It now only remains to describe the western and central portions south of the Keer. The geological formation of the former consists mainly of Scar Limestone and of the latter of Millstone Grit, with a narrow band of Yoredale Grit between them, the same as already mentioned as occurring further north near Borwick. The limestone, a portion of the large area which extends northwards for 20 miles, has its southern limit near the village of Slyne. About a mile south-east of Carnforth it forms a ridge of high ground called Kellet Seeds, below which, to the north and south respectively, are the villages of Over Kellet and Nether Kellet. The surface of Kellet Seeds consists partly of upland pasture, but the remainder is beautiful and rocky woodland similar in character to that on Dalton Crag, but the limestone pavement is much less extensive.

A mile to the east-south-east of Nether Kellet a small stream, which rises on boggy ground at the foot of some limestone scars, falls into a large cave or pot-hole called Dunnald Mill Hole. In dry weather the cavern can be followed for many yards, when the stream disappears amongst the rocks and further progress is impossible. The following rarer plants are found on Kellet Seeds and in the neighbourhood of Over Kellet and Nether Kellet :—

Aquilegia vulgaris, L.	Convallaria majalis, L.
Minuartia verna, Hiern.	Paris quadrifolia, L.
Hypericum montanum, L.	Carex muricata, L.
Euonymus europœus, L.	Sesleria cœrulea, Ard.
Rhamnus catharticus, L.	Kœleria cristata, Pers.
Hippocrepis comosa, L.	Campylopus fragilis, v. densus, B. & S.
Sedum purpureum, Tausch.	Pottia recta, Mitt.
Scabiosa Columbaria, L.	Ulota phyllantha, Brid.
Ligustrum vulgare, L.	Orthotrichum stramineum, Hornsch
Verbascum Thapsus, L.	Amblystegium compactum, Aust.
Epipactis atrorubens, Schultz.	Anthoceros lævis, L.
Orchis ustulata, L.	Reboulia hemisphærica, L.
„ morio, L.	Scapania aspera, Bern.
Ophrys muscifera, Huds.	Lejeunea Mackaii, Hook.
Habenaria bifolia, R. Br.	„ Rossettiana, Mas.
„ montana, Dur. & Schinz.	

The country to the west of Nether Kellet about Bolton-le-Sands, and southwards towards Halton and Skerton, is generally well cultivated, whilst that in an easterly direction to Aughton and Gressingham consists mostly of rather rough pasture with a few marshes and woods. A small swamp between Nether Kellet and Carnforth yields *Myrica Gale* and several interesting bog plants. South of Capernwray and east of Over Kellet the country is very diversified and picturesque. About a mile from the latter village is situated the Carnforth Waterworks reservoir, below which towards Capernwray the valley is clothed with a thick wood known as Kellet Park Wood. *Rubus suberectus* grows luxuriantly here, some stems being over 10ft. high. Eastwards, again, there is a rather extensive tract of woodland called Lord's Lot Wood, in the neighbourhood of which are small patches of boggy ground and swamp. Beyond this the surface for about two miles consists of marshy moorland and rough furze-covered pasture, and is known as Arkholme Moor and Gressingham Moor. The ground then slopes away quickly to the flat meadows by the Lune. The following are some of the rarer plants of this upland district :—

Millegrana Radiola, Druce.	Carex canescens, L.
Rubus fissus, Lindb.	Sphagnum compactum, DC.,
Anagallis tenella, L.	v. subsquarrosum, Warnst.
Centunculus minimus, L.	„ inundatum, Warnst.
Scutellaria minor, Huds.	„ obesum, Warnst.
Myrica Gale, L.	„ turfaceum, Warnst.
Narthecium ossifragum, Huds.	Scapania compacta, Dum.

DISTRICT No. 3—"SILVERDALE & HEYSHAM."

We now come to the western or maritime portion of the North Division. This may be conveniently divided into two sub-districts—a northern one, consisting of the country about Silverdale and Carnforth, in which the geological formation is almost entirely Mountain Limestone, and a southern one, comprising the Heysham Peninsula, which is principally composed of Millstone Grit (usually overlaid with Drift) and a little Permian Sandstone. The whole district has a rich flora, but the northern part especially so. The phanerogams and ferns of Silverdale have been treated of in a paper to "The Naturalist" by Mr. S. Lister Petty.* In an area of about two square miles he enumerates 403 species. Additional plants we have observed since bring up the total to about 500, whilst in the larger area represented by the sub-district north of Hest Bank, including about 15 square miles, we have records of 619 species. The number observed in the whole of District 3 amounts to about 710, and more are likely to be found. The country about Silverdale is beautifully diversified, and within a mile of the village presents a rich combination of rocky and thickety woodland, open hill-pasture, limestone scar and pavement, sea cliff and muddy seashore, with cornfield, meadow, and garden. In common with the adjoining districts of South Westmorland and Lake Lancashire, and as the figures already given indicate, the flora is exceedingly varied, and vegetation generally is luxuriant. Beginning at the Westmorland boundary, half-a-mile north-west of Silverdale, and following the coast-line southwards, we find that this is generally marked by cliffs of scar limestone, which continue round the promontory known as Jenny Brown's Point, with a wide area of mud-flats beneath them stretching away to the sea. Further south the level salt-marsh is bordered by the Furness Railway for some distance, the coast being mostly low to the mouth of the Keer. On these flats *Statice Armeria* is so abundant that the ground is pink with its flowers in early June. The principal limestone hills about Silverdale, west of the railway, are Castlebarrow to the north, Heald Brow to the south, and the high ground between the village and the station. Castlebarrow is the southern or Lancashire side of Middlebarrow, the wooded hill which rises from the low ground near Arnside Tower and opposite to the fine scree-covered slope of Arnside Knott. The Westmorland boundary runs from the

* "Some Plants of Silverdale, West Lancashire," by S. Lister Petty.— "The Naturalist," Feb., 1902.

shore near Silverdale Cove over the top of Middlebarrow and across the low ground eastwards to Leighton Beck. The surface of the Silverdale hills consists in parts of rough rocky pasture and limestone pavement, with numerous juniper bushes, but more generally it is clothed with a thick natural wood. This consists largely of Hazel, Oak, and Ash, with other common trees and shrubs, whilst many rarer species as Yew, Juniper, the Buckthorns *(Rhamnus catharticus* and *R. Frangula)*, Spindle Tree and Privet are here thoroughly native and abundant. White Beam *(Pyrus Aria v. rupicola)*, Small-leaved Lime *(Tilia cordata)*, Cornel and Spurge Laurel are also found, though less commonly. The introduced Conifers, Larch, Spruce, and Scotch Fir have been much planted in the woods and add greatly to the beauty of the district. In spring and early summer the country abounds with a wealth of flowers. Cowslips are especially fine and quite a feature of the district ; the rocky pastures and banks are gay with Rock Rose, Wild Thyme, and Salad Burnet, whilst *Orchis morio, Gymnadenia conopsea, Habenaria bifolia,* and other Orchidaceæ adorn the grassy hills and woodland glades.* In some of the woods the Daffodil and Lily of the Valley are plentiful. About the limestone scars *Sesleria cærulea* is the prevailing grass, and many lime-loving mosses as *Hylocomium rugosum, Funaria calcarea,* and *Weisia crispata* are here abundant. The following are the rarer plants of Silverdale, including only the district west of the Furness Railway, from the Westmorland boundary at the top of Eaves Wood and Castlebarrow to the shore marsh south of Heald Brow, and as far as the point where the combined streams or ditches from Hawes Water and Storrs Moss flow under the railway :—

Aquilegia vulgaris, L.	Triglochin maritimum, L.
Helleborus viridis, L.	Zannichellia palustris, L.
Cardamine impatiens, L.	Scirpus maritimus, L.
Viola hirta, L.	Blysmus rufus, Link.
,, silvestris, Reich.	Schœnus nigricans, L.
Hypericum montanum, L.	Carex arenaria, L.
Tilia cordata, Mill.	,, digitata, L.
Geranium sanguineum, L.	,, distans, L.
,, columbinum, L.	,, extensa, Good.
Euonymus europæus, L.	,, Pseudo-cyperus, L.
Rhamnus catharticus, L.	Sesleria cœrulea, Ard.
,, Frangula, L.	Melica nutans, L.
Hippocrepis comosa, L.	Sclerochloa maritima, Lind.

* These plants were formerly more abundant than now. *Ophrys muscifera,* for example, which was frequent 50 years ago, is now rare owing to careless eradication. *Orchis ustulata* and *O. pyramidalis* also occurred, but they have not been seen of late years. Altogether there are records of about 17 species.

Spiræa Filipendula, L.	Lepturus filiformis, Trin.
Potentilla verna, L. (?)	Asplenium marinum, L. (extinct)
Poterium Sanguisorba, L.	,, Adiantum-nigrum, L.
Rosa spinosissima, L.	Ceterach officinarum, DC.
Pyrus Aria, v. rupicola, Syme.	Lastrea rigida, Presl.
Crithmum maritimum, L.	Fissidens decipiens, De Not.
Cornus sanguinea, L.	Pottia recta, Mitt.
Galium umbellatum, Lam.	,, Heimii, Fürnr.
Asperula cynanchica, L.	,, littoralis, M t.
Scabiosa Columbaria, L.	,, intermedia, Fürnr.
Aster Tripolium, L.	,, minutula, Fürnr.
Erigeron acre, L.	,, lanceolata, C.M.
Antennaria dioica, Gaert.	Tortula ambigua, Angstr.
Inula Conyza, DC.	,, aloides, De Not.
Carduus pycnocephalus, L.	Barbula recurvifolia, Schp.
Serratula tinctoria, L.	,, sinuosa, Braith.
Centaurea Scabiosa, L.	Weisia microstoma, C.M.
Hieracium murorum, L.	,, crispata, C.M.
Lysimachia Nummularia, L.	,, verticillata, Brid.
Gentiana Amarella, L.	Trichostomum crispulum, Bruch.
Cynoglossum officinale, L.	,, mutabile, Bruch.
Lithospermum officinale, L.	,, ,, v. littorale, Dixon
Atropa Belladonna, L.	,, nitidum, Schp.
Verbascum Thapsus, L.	Pleurochæte squarrosa, Lindb.
Linaria repens, Mill.	Orthotrichum leiocarpum, B. & S.
Veronica hybrida, L. (extinct ?)	,, straminevm, Hornsch.
Nepeta Cataria, L.	Funaria calcarea, Wahl.
Salvia Verbenaca, L.	Breutelia arcuata, Schp.
Verbena officinalis, L.	Bryum murale, Wils.
Atriplex laciniata, L.	,, roseum, Schrad.
Suæda maritima, Dum.	Mnium cuspidatum, Hedw.
Daphne Laureola, L.	,, stellare, Reich.
Parietaria ramiflora, Moench.	Thuidium recognitum, Lindb.
Taxus baccata, L.	Cylindrothecium concinnum, Schp.
Spiranthes spiralis, C. Koch.	Brachythecium glareosum, B & S.
Epipactis atrorubens, Schultz.	,, albicans, B. & S.
Orchis pyramidalis, Linn. (extinct ?)	Hypnum chrysophyllum, Brid.
,, morio, L.	Porella lævigata, Schrad.
Ophrys muscifera, Huds.	Lejeunea Mackaii, Hook.
Narcissus Pseudo-narcissus, L.	,, Rosettiana, Mass.
Polygonatum officinale, All.	Scapania aspera, Mull. & Bern.
Convallaria majalis, L.	Riccia sorocarpa, Bisch.
Paris quadrifolia, L.	Lichina confinis, Ag.
Sparganium minimum, Fr.	

East of the railway and north of the road leading from Silverdale Station to Burton is another tract of country similar to the preceding and exceedingly rich botanically. Like the country further north and west it consists largely of rough limestone pasture and woodland. The Leighton Beck, a small stream rising on Hale Moss and entering the Kent estuary at Arnside, marks the Westmorland boundary for some distance on the north. The principal hills are Trowbarrow, Gatebarrow, and Thrang End. These rise from low flat ground, formerly consisting of peat-moss but now mostly reclaimed. One mile north-east of Silverdale Station and between Trowbarrow and the western end of the

Gatebarrow woods, is situated a tarn called Hawes Water, covering about 15 acres. This is almost the only tarn West Lancashire possesses. The ground near it is boggy and marshy, and yields the following plants :—

Nymphæa lutea, L.	Schœnus nigricans, L.
Castalia alba, Greene.	Cladium Mariscus, Brown.
Viola canina, L.	Carex dioica, L.
Primula farinosa, L.	,, Pseudo-cyperus, L.
Myrica Gale, L.	,, vesicaria, L.
Epipactis longifolia, All.	Selaginella selaginoides, Link.
Sparganium minimum, Fr.	Hypnum elodes, Spr.
Juncus obtusiflorus, Ehrh.	

A small stream, rising 150yds. away in a deep pool called Little Hawes Water, enters the eastern end. A local tradition says : " Hawes Water is bottomless, but Little Hawes Water is still deeper !" Between the Hawes Waters and the Leighton Beck is the somewhat extensive rocky woodland and limestone pavement of Gatebarrow. Here occur *Helleborus fœtidus, Serratula tinctoria, Ophrys muscifera, Polygonatum officinale, Convallaria majalis, Carex digitata, Polytrichum nanum, Hylocomium brevirostre, H. rugosum, Riccia Lescuriana,* and other interesting plants. The ground by the Leighton Beck for some distance above and below the old Leighton Beck furnace— now long unworked—consists mostly of cultivated arable land, but half-a-mile or more lower down at Cold Well, near the place where the stream is crossed by the road from Arnside to Yealand, it flows through a swampy wood bordering the north end of a small limestone scar. A number of interesting plants occur about here, including *Primula farinosa, Schœnus nigricans, Amblystegium irriguum, Hypnum scorpioides,* and *Hypnum intermedium.*

South of Gatebarrow is Thrang End, the higher parts of which are bare rocky pasture, but the ground on the eastern side is thickly wooded and descends in steep limestone slopes and scars to Thrang Moss. This and Hale Moss formerly extended over the whole of the low ground eastwards towards the villages of Holme and Burton. Although the bog has been drained a part of it still remains uncultivated, and the following plants lingered on up to 1901 and probably may yet occur :—*Drosera rotundifolia, Parnassia palustris, Rubus fissus, Myrica Gale, Polytrichum gracile,* and three or four species of *Sphagnum.*

Proceeding southwards and crossing the road from Silverdale to Burton, already mentioned, we come to Yealand Storrs, Yealand Redmayne and Yealand Conyers, three little villages beautifully situated and sheltered by the range of limestone hills extending south from near Thrang End to Cringlebarrow and Warton Crag. On the west of Cringlebarrow is the low ground about Storrs Moss, now nearly all drained and cultivated. A few bog and marsh plants still remain, and the ditches yield luxuriant *Rumex Hydrolapathum*. Cringlebarrow is a richly wooded hill with limestone pavement similar to Gatebarrow. Below it on the south-west is Leighton Hall Park, and southwards again the terraced limestone hill called Warton Crag, overlooking Warton and Carnforth. These hills yield nearly all the rarer calcicole plants enumerated as occurring near Silverdale, and also some others. Eastwards from the Yealands to the boundary of District 2 the country is mostly grass-pasture and corn-field. A small swampy pool near the locks on the Kendal canal, three-quarters of-a-mile north-west of Borwick, contains several interesting plants, including *Sparganium minimum* and *Carex elata v. turfosa*.

Crossing the Keer at Carnforth we find hills of gravel composed entirely of waterworn and rounded limestone pebbles. *Melilotus officinalis, Centaurea Scabiosa,* and many xerophilous plants are here plentiful. At the mouth of the Keer estuary and towards Bolton-le-Sands the coast line is marshy, but near the latter place there is a stretch of shingle on which *Glaucium flavum* grows abundantly. The following plants also occur on this coast :

Cerastium tetrandrum, Curtis.	Blysmus rufus, Link.
Arenaria Lloydii, Jord.	Schoenus nigricans, L.
Matricaria inodora, v. salina, Bab.	Carex distans, L.
Artemisia maritima, L.	,, extensa, Good.
Limosella aquatica, L.	Lepturus filiformis, Trin.
Obione portulacoides, Moquin.	Bryum Marratii, Wils.
Eleocharis uniglumis, Schultes.	Amblystegium salinum, Carr.
Scirpus panciflorus, Light.	

The country for half-a-mile inland consists principally of flat meadow with numerous ditches, in which *Potamogeton pectinatus* and *Ruppia rostellata* are plentiful. Between Bolton-le-Sands and Hest Bank the ground is similar, and the ditches yield various water-plants, including *Scirpus Tabernæmontani* ; the coast-line, however, is less interesting botanically than further north.

We now come to the Heysham Peninsula, or land on the north of the Lune estuary, which stretches southwards opposite to Glasson Dock and terminates in Sunderland Point. It presents a long extent of coast line ; the distance across the neck of the peninsula from Hest Bank to the Lune at Lancaster is only 2¼ miles, whereas along the shore by way of Heysham, Sunderland

Point and Overton, it is 19 miles. The land generally is low and flat, the highest elevation being about 100ft. on the gritstone hills near Heysham. A peat moss, now mostly reclaimed, formerly covered about a square mile of country to the east of this village. *Rynchospora alba, Myrica Gale, Narthecium ossifragum,* and several interesting mosses, including *Polytrichum gracile,* still occur.

With the exception of the peat moss, the ground near the coast presents a greater interest to the botanist than that inland, but the extensive growth of Morecambe along the shore from Bare almost to Heysham, a distance of over three miles, and the construction of the harbour or dock a mile south-west of the latter village, has destroyed many plants that formerly flourished. *Glaucium flavum* was abundant near Morecambe up to about the year 1880, but is now gone. In making the dock, the Far Naze sea-cliffs, on which grew *Asplenium marinum* and other interesting plants, were entirely removed. The remainder of the coast has happily seen less change, and many parts remain but little altered, especially along the estuary of the Lune. But the coast there is low, flat and marshy, yielding more particularly salt-marsh plants, and the flora characteristic of sea-cliffs is wanting. The Heysham cliffs are almost the only rocks remaining, now that those at the Far Naze are destroyed, and the district being much over-run by excursionists, very few interesting plants are left except such as are either inconspicuous or out of reach. Between Heysham Dock and Sunderland Point the coast is mostly sandy, but there are some grit-rocks and banks of Permian Marl. The following are the rarer plants :—

Cochlearia danica, L.	Calamagrostis epigejos, Roth.
Cakile maritima, Scop.	Triticum junceum, L.
Sagina maritima, Don.	Elymus arenarius, L.
Rosa spinosissima, L.	Asplenium marinum, L.
Sedum anglicum, Huds.	Grimmia maritima, Turn.
Eryngium maritimum, L.	Tortula ruraliformis, Dixon.
Carduus pycnocephalus, L.	Trichostomum flavo-virens, Bruch.
Limonium vulgare, Mill.	Bryum Marratii, Wils.
,, humile, Mill.	Brachythecium albicans, B. & S.
,, binervosum, C.E.S.	Frullania Tamarisci, v. cornubica, Carr.
Centaurion vulgare, Rafn.	Lichina confinis, Ag.
Obione portulacoides, Moquin.	Ramalina polymorpha, Ach.
Salsola Kali, L.	,, scopulorum, Ach.
Polygonum Roberti, Lois.	,, cuspidata, Nyl.
Parietaria ramiflora, Moench.	Physcia aquila, Nyl.
Eleocharis uniglumis, Schultes.	Opegrapha saxicola, Ach.
Carex arenaria, L.	

To the east of Sunderland Point and almost all the way up the Lune estuary the coast, as before stated, is bordered by salt

marshes. Near Overton, *Limonium vulgare* and *L. humile* are very abundant and the muddy pools yield *Limosella aquatica.* The following are the rarer plants found on or near the banks of the estuary between Sunderland Point and Skerton, near Lancaster :—

Ranunculus Baudotii, Godr.	Suæda maritima, Dum.
Cochlearia anglica, Linn.	Juncus maritimus, Lam.
Alsine marginata, Reich.	Triglochin maritimum, L.
,, media, Crantz.	Ruppia rostellata, Koch.
Trifolium fragiferum, L.	Zannichellia palustris, L.
Apium graveolens, L.	Scirpus Tabernæmontani, Gmel.
Œnanthe Lachenalii, C. Gmel.	Calamagrostis epigejos, Roth.
Aster Tripolium, L.	Lepturus filiformis, Trin.
Limonium vulgare, Mill.	Pottia Heinii, Fürn.
,, humile, Mill.	,, intermedia, Fürn.
,, binervosum, C.E.S.	,, Starkeana, C.M.
Samolus Valerandi, L.	Ulota phyllantha, Brid.
Limosella aquatica, L.	

In meadows near Torrisholme *Crocus vernus* appears to be thoroughly naturalised, and is so plentiful that the blossoms give a rich purple tint to the ground in early spring. Ditches are frequent throughout the peninsula, and produce a number of interesting hygrophilous plants, including *Scirpus fluitans* and *Juncus obtusifolius.* We have now described the whole of the district except the narrow strip of country east of the railway and west of the high road between Skerton and Bolton-le-Sands. This is mostly pasture land, rising towards the east, and does not appear to possess any special botanical interest.

WEST DIVISION.

DISTRICT No. 4—" PILLING."

The portion of the " West Division " north of the river Wyre and extending from the Lune at Lancaster to the Wyre at Shard Bridge, a distance of 15 miles, forms District 4. It is principally low and flat, a large part being only 15 to 50ft. above sea level. Almost the only hilly country, except the small area in the west near the Wyre estuary (where Preesall Hill reaches 100ft.), is the strip of rising ground which begins about two miles from the the eastern boundary, and extends from near Lancaster to a mile-and-a-half south-west of Garstang ; this forms the base of the uplands of the East Division. , The highest elevation is found a little to the south-west of Scotforth and attains 175ft. Geologically the district consists of Triassic Marl and Permian Sandstone with a little Millstone Grit, but almost the whole is deeply overlaid with Drift and in some parts also with Peat. A large portion, covering an area of about 25 square miles, and stretching from near Stalmine, Preesall and Pilling on the west to Tarnacre and Nateby on the east, and also northwards to Winmarleigh and the river Cocker, formerly consisted almost entirely of peat moss, but by the year 1900 this had been nearly all reclaimed. The only remnant is a part of Cockerham Moss covering about half a square mile. In the place of bog we now see fields of corn, potatoes and other crops, and the whole district is highly cultivated. The chief botanical interest, apart from the patch of bog, now lies along the coast, on the banks of the Wyre and of the Preston and Kendal Canal, and in the numerous ditches which drain the low reclaimed land. Small ponds are also numerous, and these, together with the above, yield quite a number of interesting hygrophilous plants. The district, taken generally, is rather bare ; woods are few and of small extent, and the larger forest trees are mostly planted. In contrast to this, there is abundant evidence from the immense number of prostrate tree-trunks and stumps found in the peat and disclosed from time to time in making ditches, ploughing, etc., that the country was once clothed with extensive forests.

Beginning in the north at the ancient and now rapidly extending county-town of Lancaster and following the south side of the Lune estuary down to its mouth below Glasson Dock, we find a similar type of ground to that already described on the northern or opposite bank. (See District 3.)   Level stretches of muddy salt-marsh border the river most of the way, especially near Conder Green and Glasson. The following are the rarer plants :—

Trifolium fragiferum, L.	Ruppia rostellata, Koch.
Œnanthe Lachenalii, Gmel.	Zannichellia pedunculata, Reich.
Aster Tripolium, L.	Carex distans, L.
Artemisia maritima, L.	,, extensa, Good.
Suæda maritima, Dum.	Sclerochloa rigida, Link.
Juncus maritimus, Lam.	Poa compressa, L.
Triglochin maritimum, L.	Triticum pungens, v. littorale, Host.

Near Cockersand Abbey—very little of the ruins of which are now standing—the coast becomes drier, with flat rocks of Permian sandstone on the shore, but further south, about the estuary of the Cocker, it is again bordered by muddy salt-marshes. These continue for the next five miles, or to one mile west of Pilling, where the coast changes its character and becomes dry and sandy with tracts of shingle. The following are characteristic plants of the salt-marshes about Cockerham and Pilling :—

Cochlearia officinalis, L.	Atriplex littoralis, L.
,, anglica, L.	Salicornia europæa, L.
Alsine media, Crantz.	Suæda maritima, Dum.
,, marginata, Reich.	Juncus maritimus, Lam.
Apium graveolens, L.	,, Gerardi, Lois.
Œnanthe Lachenalii, Gmel.	Triglochin maritimum, L.
Aster Tripolium, L.	Scirpus maritimus, L.
Artemisia maritima, L.	Carex distans, L.
Statice Armeria, L.	Sclerochloa maritima, Lindl.
Glaux maritima, L.	Lepturus filiformis, Trin.
Plantago maritima, L.	

To the west of Fluke Hall, near Pilling, a sea-wall has been built, which has tended to reduce the number of interesting plants, but the following rarer species may still be found between Pilling and Knott End. *Limonium binervosum*, C.E.S., is especially fine and abundant :—

Cakile maritima, Scop.	Obione portulacoides, Moq.
Trifolium arvense, L.	Atriplex littoralis, L.
,, fragiferum, L.	,, Babingtonii, Woods.
Eryngium maritimum, L.	,, laciniata, L.
Chærophyllum Anthriscus, Lam.	Salsola Kali, L.
Limonium vulgare, Mill.	Polygonum Roberti, Lois.
,, humile, Mill.	Carex arenaria, L.
,, binervosum, C.E.S.	Triticum junceum, L.

The country inland between Lancaster and Cockerham, about Aldborough, Stodday, and Thurnham, is rather better wooded, more hilly, and generally more picturesque than that further south. Half-a-mile south-west of Galgate there is an outcrop of

Millstone Grit rock, the surface of which is rounded and striated by glacial action. Several plants of a heath-loving character, which for want of a suitable surface are very rare in most parts of the district, occur here. West of Thurnham towards Glasson and Cockersand Abbey, and southwards to the Cocker below Cockerham, the ground is mostly low and flat and drained by ditches. The following are some of the more interesting plants :—

Ranunculus Baudotii, Godr.	Callitriche autumnalis, L.
Œnanthe fistulosa, L.	Hydrocharis Morsus-ranæ, L.
Bidens cernua, L.	Carex disticha, Huds.
Hottonia palustris, L.	Nitella opaca, Agardh.
Samolus Valerandi, L.	

Crossing the Cocker and proceeding rather over a mile southwestwards we come to Cockerham Moss. The elevation of this bog is 25ft. above the sea, and the surface where undisturbed is almost perfectly flat. As in the case of the great red bogs of Ireland, its level is 1oft. or more above that of the surrounding country, and previous to about 25 years ago, when the bog was much more extensive than now, the distant view as seen on a clear day from near the centre was very striking. The effect produced by the bog being slightly raised, added to that due to the rotundity of the earth, made trees and farmhouses on the flat plain at the edge of the moss appear partly submerged, and the moss itself seemed to stretch away to the foot of the Wyresdale fells, although in reality a tract of country five miles wide intervened. The part of the bog least affected by drainage, and where the surface approaches most nearly to its aboriginal condition, is towards the north side. This has been preserved by the owner for game purposes. The south side has recently been much cut up and spoiled by a " Moss-litter " Company, and many years ago a drain was cut across the centre of the moss. On the north-west side the Black-headed Gull breeds in large numbers, the part called " Gull Island " being almost covered with nests during the breeding season. The birds are preserved, and in order to visit the moss it is necessary to obtain a permit. Now that lowland peatbogs of any considerable extent are becoming very rare in England, it may be well to indicate the characteristic flora of Cockerham Moss, and give a list of the plants which occur where the surface has been but little affected by drainage. In the wettest parts the bog is generally covered, as would be expected, with a thick carpet of Sphagnum, amongst which the cranberry is very abundant, forming a close trailing network of wiry stems. The phanerogamic flora of this wet Sphagnum ground is peculiar and

almost restricted to the following 11 species. Those marked with an asterisk are becoming scarce :—

Drosera rotundifolia, L.	Eriophorum vaginatum, L.
* ,, anglica, Huds.	*Rynchospora alba, Vahl.
Vaccinium Oxycoccus, L.	Carex canescens, L.
Andromeda Polifolia, L.	* ,, limosa, L.
Erica Tetralix, L.	Molinia cœrulea, Mœnch.
Narthecium ossifragum, Huds.	

Other plants of the moss, most of them growing where the surface is rather drier and more uneven, are :—

Rubus plicatus, W. & N.	Empetrum nigrum, L.
Calluna vulgaris, Hull.	Orchis latifolia, L.
Melampyrum pratense, L.	Juncus bulbosus, L.
Betula pubescens, Ehrh.	Eriophorum polystachion, L.
Salix cinerea, L.	Lastrea spinulosa, Presl.
Myrica Gale, L.	

The Mosses and Hepatics found on the bog include some interesting species :—

Sphagnum fimbriatum, Wils., v. tenue, Grav.	
,, Warnstorfii, Russ., v. flavescens, Warnst.	
,, rubellum, Wils., vs. pallescens, purpurascens, flavum, and versicolor.	
,, fuscum, Klinggr., v. fuscescens, Warnst	
,, subnitens, Russ. & Warnst., v. flavescens, Warnst.	
,, teres, Angstr., v. squarrosulum, Warnst.	
,, riparium, Angstr.	
,, cuspidatum, Russ. & Warnst., v. plumosum, N. & H.	
,, ,, ,, v. serratum, Lesq. & James.	
,, pulchrum, Warnst.	
,, obtusum, Warnst.	
,, recurvum, Russ. & Warnst., v. mucronatum, Warnst.	
,, molluscum, Bruch.	
,, medium, Limpr., vs. roseum, roseo-pallescens, and purpurascens.	
,, cymbifolium, Limpr.	
,, turfaceum, Warnst.	
,, papillosum, Lindb., vs. normale and sublæve.	

Polytrichum juniperinum, Willd.	Hypnum fluitans,
,, strictum, Banks.	,, v. elatum, Ren. & Arnell.
,, gracile, Dicks.	,, cupressiforme,
,, commune, L.	,, v. ericetorum, B. & S.
Dicranella cerviculata, Schp.	Kantia trichomanis, L.
Campylopus flexuosus, Brid.	,, submersa, Arnell.
,, pyriformis, Brid.	Cephalozia lunulæfolia, Dum.
Dicranum scoparium, Hedw.,	,, bicuspidata, L.
,, v. ericetorum, Corb.	,, Lammersiana, Hub.
Leucobryum glaucum, Schp.	,, Sphagni, Dicks.
Tetraplodon mnioides, B. & S.	,, fluitans, Nees.
Aulacomnium palustre, Schwgr.	,, divaricata, Sm.
Webera nutans, v. longiseta, B. & S.	Mylia anomala, Hook.
Hypnum fluitans, L., v. gracile, Boul.	Jungermania ventricosa, Dicks.
	Nardia scalaris, Schrad.

The following are the principal lichens :—

Cladonia pyxidata, Fr.	Cetraria aculeata, Fr.
,, digitata, Hoffm.	Parmelia sulcata, Tayl.
,, coccifera, Schær.	,, physodes, Ach.
Cladina sylvatica, Nyl.	Lecanora varia, Ach.
,, uncialis, Nyl.	Lecidea decolorans, Florke.

On one part of Cockerham Moss the Birch forms dense thickets and has increased during recent years, the ground having become drier and more favourable to its growth. In ditches on the borders of the moss *Epilobium obscurum*, Schrad., is very abundant, and on land recently reclaimed *Viola Lloydii*, Jord., and *V. carpatica*, Borbas.

The country between Cockerham Moss and Garstang, by way of Winmarleigh or Nateby, is similar to that about Cockerham. Ditches are frequent and yield numerous interesting plants. *Osmunda* occurred near Nateby up to about 25 years ago and at Cogie Hill up to 10 years ago, but is now apparently extinct. *Thalictrum flavum*. L., still grows near Winmarleigh. In a marshy pond at Cabus, *Typha angustifolia*, L., grows in great abundance, and the canal in the same neighbourhood contains *Nitella prolifera*, Kutz. At Garstang we come to the river Wyre, which divides the district under consideration from District 5. From this place to Cart Ford Bridge, below St. Michaels, the Wyre flows through a low alluvial plain and winds considerably, the distance being 10 miles by the river and five miles in a straight line. Its channel is generally rather deep and narrow, and being liable to floods from the heavy rain which often falls on the Wyresdale and Bleasdale Fells, the banks have been artificially raised by dykes to protect the surrounding country. This is mostly grass land. The soil exposed on the banks is in some parts boulder-clay and in others sandy and loamy alluvium. A little below St. Michaels the river becomes tidal, and one-and-a-half miles below Cart Ford Bridge it widens into an estuary. The distance from Cart Ford to the Shard bridge is three-and-a-half miles ; near the latter the estuary contracts somewhat, the distance from bank to bank at the bridge being less than 300 yards, but a little further down it widens to half-a-mile. From Shard to Staynall and Knott End the banks are marshy in some places and yield the following plants :—

Cochlearia anglica, L.	Atriplex deltoidea, Barb.
Brassica monensis, Huds.	,, Babingtonii, Woods.
Alsine marginata, Reich.	,, laciniata, L.
,, media, Crantz.	Obione portulacoides, Moq.
Aster Tripolium, L.	Suæda maritima, Dum.

Tanacetum vulgare, L.
Artemisia maritima, L
Limonium vulgare, Mill.
   ,, humile, Mill.
   ,, binervosum, C.E.S.
Atriplex littoralis, L.

Triglochin maritimum, L.
Carex distans, I..
   ,, extensa, Good.
Triticum pungens, v. littorale, Host.
Pottia littoralis, Mitt.
Tortula aloides, De Not.

The more central parts of the district, including the country about Preesall, Stalmine, Hambleton, Rawcliffe, and Pilling, are almost all now under cultivation and present no special interest to the botanist. Little remnants of partly drained moss-land still exist, and there are also some small woods. About these a few bog and heath-loving plants still continue, survivals of the aboriginal flora. The geological formation of this area is Permian Marl, generally overlaid with deposits of Glacial Drift and Peat. Below the marl in the neighbourhood of Preesall, at a depth of about 250ft., there is an extensive deposit of Rock Salt. This is the same age geologically as the Rock Salt of Cheshire.

DISTRICT No. 5—"KIRKHAM."

Although this district is much larger than any of the others— it has an area of about 130 square miles—a large proportion of its surface presents a remarkable degree of sameness, and if it were not for the interesting plants of its varied coast line, its flora would be only common-place in interest. It consists principally of the highly-cultivated country known as the Fylde, and contains, besides numerous large villages, the towns of Blackpool and Fleetwood and the western half of Preston. The ground generally is more undulating and not quite so flat as that in District 4, but the highest elevation is only 140ft. The district has few woods and practically no bogs, almost all the latter having been reclaimed. Somewhat extensive bogs and marshes formerly existed, especially about Catforth and Inskip and in the country generally along the course of the Wood Plumpton Brook, but these were drained about 50 years ago by cutting a large drain to the river Wyre below St. Michaels. The geological formation is mostly Triassic Pebble Beds and Triassic Marl, much overlaid with deep deposits of Glacial D ft and Boulder Clay, and along the coast from Blackpool to Lytham with blown sand from the sea-shore. The salt marshes bordering the Ribble and Wyre estuaries produce some interesting plants, but it is the line of coast sand-hills or dunes between Lytham and Blackpool that have yielded the

richest harvest to the botanist. But these, alas! are fast disappearing before the builder, and the coast is now much spoiled from a botanical point of view. We shall refer to this again further on.

Starting in the north-eastern corner of the district at the bridge over the Wyre near Garstang, and proceeding southwards over pasture and meadow land, between the river and the high road, which here form the boundaries of the district, we soon reach the west side of Bowgrave Hill (123ft.). From this there is an extensive view of the Fylde plain and the low country about Pilling and Cockerham. We then come to the river Calder, which joins the Wyre near Catterall. The ground about here is low and flat, and there are several old mill-dams and ponds, which yield numerous water plants. Two miles further south we reach the river Brock, another tributary of the Wyre. The land between the Calder and Brock consists of river alluvium, and is only 25ft. above sea level. The rarer plants are *Potamogeton obtusifolius, Sparganium simplex,* and *Scirpus sylvaticus.* Following the course of the Wyre from the point near St. Michaels where it is joined by the Brock, to near Poulton-le-Fylde, we find the country very similar to that on the opposite bank already described in District 4. It is nearly all well-cultivated ground, mostly under grass, and presents but little interest to the botanist. The r ver banks, ditches, and ponds are, however, deserving of further work. Amongst the rarer plants so far observed *Thalictrum flavum, Œnanthe aquatica,* and *Bidens tripartita* may be mentioned. From the Shard Bridge down to Fleetwood Docks an extensive area of salt-marsh and waste ground borders the estuary. This yields several interesting plants, and in the neighbourhood of the docks many foreign plants of a casual character have been found, including the following :—

Lepidium Draba, L.
Sisymbrium pannonicum, Jacq.
Medicago sativa, L.
Vicia villosa, Roth.
Lactuca virosa, L.

Chenopodium opulifolium, Schrad.
   ,, serotinum, L.
   ,, murale, L.
   ,, rubrum, l..
Serrafalcus arvensis, Gour.

Rounding the promontory upon which the town of Fleetwood stands, the coast becomes dry and sandy, and so continues southwards to Rossall and Cleveleys. The flora is very different from that of the muddy Wyre estuary, as shown by the following list of rarer plants :—

D

Cochlearia danica, L.
Cakile maritima, Scop.
Eryngium maritimum, L.
Carduus pycnocephalus, L.
Limonium binervosum, C.E.S.
Convolvulus Soldanella, L.
Obione portulacoides, Moq.
Atriplex laciniata, L.

Atriplex deltoidea, Bab.
   ,, littoralis, L.
Salsola Kali, L.
Polygonum Roberti, Lois.
Carex arenaria, L.
Triticum pungens, v. littorale, Host.
   ,, junceum, L.
Tortula ruraliformis, Dixon.

Between Cleveleys and Blackpool there are fine cliffs of reddish-brown Boulder Clay rising near Norbreck to a height of 113ft. They are continually being undermined by the sea, and large pieces of earth frequently become detached and fall. The clay contains many boulders composed of Lake District rocks, which are striated and rounded by ice and water. The coast here yields *Plantago Coronopus v. ceratophyllum* Rap., which appears in West Lancashire to be confined to this boulder clay. We now come to Blackpool, which extends along the coast for over three miles, and it seems likely that before long houses will reach all the way to St. Annes and Lytham. The change during the last 25 or 30 years is remarkable. Up to about the year 1875 the coast from Blackpool to Lytham was truly a botanist's paradise and presented a plant association scarcely to be met with elsewhere in England, except on the similar and perhaps even richer coast of South Lancashire between Southport and Formby. The sand dunes were then practically undisturbed and, in the neighbourhood of where St. Annes now stands, extended a considerable distance inland. They formed a high belt rising along the shore, the surface consisting in some parts of bare drifting sand and in others partially clothed with tufts of sea marram *(Ammophila arenaria)*. On the land side of these dunes were flat stretches of boggy and sandy marsh, whilst inland again were more sand-hills and marshes succeeding one another and intermixed. The boggy hollows formed beautiful natural gardens, filled during the summer with a wealth of flowers delightful to behold ; and in them also grew many rare plants and mosses, unnoticed by the ordinary passer-by, but of fascinating interest to the botanist. It is painful to think that many of these have now gone and none of the old marshy ground remains in its original condition ! Where not already built upon it is " improved " and drained and given over to crowds of excursionists or converted into golf links. Here amongst a silvery carpet of dwarf willow *(Salix repens var. argentea)* grew a rich profusion of *Epipactis longifolia, Orchis incarnata, Pyrola rotundifolia v. arenaria, Parnassia palustris,* and the curious *Monotropa Hypopitys.* In July the damper spots

were pink with *Anagallis tenella* and with it flourished *Carex Œderi, Selaginella selaginoides, Equisetum variegatum, Bryum Warneum, Bryum lacustre, Hypnum Wilsoni v. hamatum, Hypnum lycopodioides,* and other interesting species. On the drier ground about the foot of some of the dunes grew *Convolvulus Soldanella, Brassica monensis,* and *Vicia lathyroides,* whilst in many parts the sand was gay with bright patches of Rest Harrow, Yellow Bedstraw, Skullcap, and Centaury, of which latter three species were abundant. The following is a list of the rarer plants of this interesting coast line ; those marked with an asterisk are very scarce and some of them, we fear, are nearly or quite extinct :—

*Thalictrum dunense, Dum.
Ranunculus Baudotii, Godr.
Cochlearia danica, L.
Sisymbrium Thalianum, Gay.
Brassica monensis, Huds.
Diplotaxis muralis, DC.,
   v. Babingtonii, Syme.
Cakile maritima, Scop.
Reseda lutea, L.
Viola canina, L.
   ,, Curtisii, Forst.
   ,, Pesneaui, Lloyd.
Polygala oxyptera, Reich.
Silene maritima, With.
Cerastium tetrandum, Curt.
   ,, semidecandrum, L.
Arenaria Lloydii, Jord.
Sagina nodosa, Fenzl.
Trifolium arvense, L.
*   ,, suffocatum, L.
   ,, fragiferum, L.
Vicia lathyroides, L.
Rubus cæsius, L.
Rosa spinosissima, L.
Saxifraga tridactylites, L.
Parnassia palustris, L.
Eryngium maritimum, L.
Erigeron acre, L.
Inula vulgaris, Trev.
Carlina vulgaris, L.
Cichorium Intybus, L.
Hieracium umbellatum,
   v. coronopifolium, Bernh.
Taraxacum officinale,
   v. erythrospermum, DC.

*Pyrola rotundifolia,
   v. arenaria, Koch.
*Monotropa Hypopitys, L.
Glaux maritima, L.
Anagallis tenella, L.
Samolus Valerandi, L.
Erythræa vulgaris, Rafn.
   ,, pulchella, Fr.
Gentiana baltica, Murb.
Cynoglossum officinale, L.
Myosotis collina, Hoffm.
Echium vulgare, L.
*Convolvulus Soldanella, L.
*Cuscuta Epithymum, Murr.
*Lasiopera viscosa, Hoff.
Calamintha vulgare, L.
Scutellaria galericulata, L.
Salsola Kali, L.
*Euphorbia Paralias, L.
   ,, portlandica, L.
Epipactis longifolia, All.
   ,, Helleborine, Crantz.
Orchis incarnata, L.
Eleocharis uniglumis, Schult.
*Blysmus rufus, Link.
Carex arenaria, L.
   ,, Œderi, Retz.
Phleum arenarium, L.
Ammophila arenaria, Link.
Festuca fasciculata, Forsk.
Triticum junceum, L.
Elymus arenarius, L.
Equisetum variegatum, Schleich.
Selaginella selaginoides, Link.
Chara vulgaris, v. longibracteata, Kütz.
Nitella glomerata, Chev.

Pleuridium alternifolium, Rabenh.
Pottia Heimii, Fürnr.
   ,, littoralis, Mitt.
Tortula ruraliformis, Dixon.
Barbula Hornschuchiana, Schultz.

Trichostomum flavo-virens, Bruch.
Encalypta vulgaris, Hedw.
*Amblyodon dealbatus, P. Beauv.
*Meesia trichoides, Spruce.
Leptobryum pyriforme, Wils.

Bryum pendulum, Schp.
* ,, Warneum, Bland.
,, inclinatum, Bland.
* ,, lacustre, Brid.
* ,, uliginosum, B. & S.
,, intermedium, Brid.
,, atropurpureum, Web. & Mohr.
,, roseum, Schreb.
Mnium cuspidatum, Hedw.
Camptothecium lutescens, B. & S.
Brachythecium albicans, B. & S.
,, Mildeanum, Schp.
Eurhynchium megapolitanum, Milde.
Amblystegium filicinum,
,, v. Whiteheadii, Wheldon.
Hypnum elodes, Spr.

Hypnum polygamum, Schp.
,, aduncum, v. paternum, Sanio.
,, Sendtneri, Schp.
* ,, Wilsoni, Schp.
* ,, ,, v. hamatum, Lindb.
* ,, lycopodioides, Schwgr.
,, revolvens, Sw.
,, cupressiforme,
v. tectorum, Brid.
* ,, cordifolium, Hedw.
* ,, giganteum, Schp.
Kantia trichomanis, L.
Nardia scalaris, Schrad.
Pellia calycina, Tayl.
Chomiocarpon quadratus, Scop.

East of Lytham we come to the mouth of the Ribble estuary, and the character of the coast again changes. Instead of sand-hills the shore is low and marshy in places, with quantities of *Carex extensa* and other salt-marsh species. Near Freckleton there are banks of boulder clay, which yield *Helminthia echioides*, a plant so far unknown elsewhere in the vice-county. From Freckleton to Preston the estuary was formerly bordered by extensive salt-marshes, but these were largely reclaimed when the Ribble Docks were made. *Ranunculus Baudotii, Cochlearia anglica, Œnanthe Lachenalii, Ruppia rostellata, Carex distans, Lepturus filiformis,* and other halophytes still occur, however. About the docks a large number of casuals have been found, many of the species being the same as those observed at Fleetwood. As before stated, the more central parts of the district are generally well cultivated and do not present many interesting botanical features. Between St. Annes and Poulton-le-Fylde is situated a little tarn called Marton Mere, and the neighbouring country contains numerous small ponds and ditches. The following are the more interesting plants of the district about Marton, Weeton, Singleton, and Poulton-le-Fylde :—

Castalia alba, Greene.
Pimpinella major, Huds.
Œnanthe fistulosa, L.
Bidens cernua, L.
,, tripartita, L.
Hottonia palustris, L.

Typha angustifolia, L.
Sparganium neglectum, Beeby.
Lemna gibba, L.
Potamogeton obtusifolius, M. & K.
Scirpus lacustris, L.
Glyceria aquatica, Wahlb.

A fair proportion of corn has been grown near Treales, Wharles, and Inskip and in some other parts, but the area devoted to this has diminished during recent years. In the marshy district about Catforth and Inskip, *Osmunda* formerly occurred in at least two or three localities, but it is now, we believe, extinct.

EAST DIVISION.

DISTRICT No. 6—WRAY & QUERNMORE.

This district consists of the portion of the East Division which comes within the drainage area of the River Lune, and comprises the moorland and dale districts of Hindburn, Roeburn-dale, Foxdale, Udale, and Conder.

It is bounded on the east by Yorkshire, on the north by the Wenning and the Lune, on the west by the high road from Lancaster to a little south of Galgate, and on the south by the water-shed between the Wyre and Lune.* The district contains the eastern half of the town of Lancaster and the villages of Wray, Quernmore, Caton, and Galgate, but apart from these it is very thinly populated, and no less than 33 square miles, out of a total area of 73, consist of wild and uninhabited moor. The only ground below 200ft. is in the valley of the Lune and in the district between Lancaster and Galgate, including the lower part of the Conder valley. Much of the country, especially that in Hindburn, Roe-burndale, and Foxdale, is very picturesque, a land of purple fells and breezy upland pastures, with many sparkling moorland streams and richly-wooded glens. Geologically the district is almost entirely composed of Millstone Grit rocks (shales and grits). From the Yorkshire boundary near Low Bentham to the Lune below Hornby the country is drained by the Wenning and its important tributary, the Hindburn. The latter stream and its feeder, the Roeburn, rise on the summit-ridges of the fells which here divide Yorkshire from Lancashire and separate the district under consideration from the Forest of Bowland. Starting from the boundary of the counties between Low Bentham and Wennington and proceeding south-eastwards near the River Hindburn, we first traverse well-cultivated meadow and pasture land nicely diversified with woodland, and then come to the more elevated district above the hamlet of Low Gill. The country here is dotted over with farm-houses, but the ground is rather rough, with a good deal of wet pasture. It is drained by the beck which

* The small area drained by the upper Cocker to the south-east of Galgate is included in that of the Wyre.

rises on Tatham Fells and which flows down a deep wooded valley to the Hindburn. Higher up, again, we come to a wide stretch of heathery moorland and sphagnum-swamp, called Loftshaw Moss and Redmire, which extends eastwards into Yorkshire, where it is known as Burn Moor. To the south is the upper portion of the Tatham valley, containing a few green fields, but these are principally on the north or sunny side of the beck. Steep banks of dark shale border the stream in some parts, and on these the curious moss *Discelium nudum* is very abundant. Clothing large surfaces with its green persistant protonema, from which the numerous erect setæ extend like the bristles of a brush, it forms a striking object. In other parts the slopes above the beck are covered with sphagnum and various mosses, amongst which may be particularly mentioned *Dicranella squarrosa*. This, although rarely found in fruit, here fruits abundantly. In the stream itself grows *Hypnum ochraceum* var. *complanatum*. Two miles southwards we come to Lythe Fell and Whiteray Fell, which are drained by the Whiteray Beck. This flows in a north-westerly direction, almost parallel to the Tatham Beck, and joins the Hindburn below Lythe. The moors near its source reach an altitude of about 1,500ft. A few patches of broken grit rocks occur near the summit-ridge, amongst which *Lycopodium Selago* grows, but, generally speaking, the surface is thickly peat-covered and the vegetation principally Cotton-grass, Sphagnum, and *Campylopus flexuosus*. The road from Bentham to Slaidburn crosses these fells, attaining a height of 1,350ft. Near the road, at a point a little below the bridge over the Little Moor Beck (a feeder of the Whiteray Beck), are two pretty waterfalls. The rocks here yield *Tetradontium Brownianum* and *Andreæa Rothii*. Still proceeding southwards, the country becomes more and more bleak as we ascend Botton Head Fell, the third highest elevation in the vice-county. Although the summit is only 1,784ft. above sea level, it is surrounded by a large expanse of moor and is one of the most secluded spots in the North of England. The view is very extensive ; to the south are the wild fells of Bowland, with Pendle Hill beyond ; westwards rise Mallowdale Fell and the dark slopes of Wardstone and Wolf-hole Crag ; to the north the distant Vale of Lune is spread out like a map, whilst to the north-east, on the horizon, are the fell summits of Greygarth, Whernside, and Ingleborough. The north side of Botton Head Fell is drained by two branches of the Hind-burn, the Middle Gill stream to the east and the Dale Beck to the west. These flow down deep rugged cloughs with steep heathery

sides. A number of interesting plants, especially ferns and mosses, occur here, and the locality deserves further investigation. The plant-association in Dale Gill is rather peculiar. The geological formation consists of Millstone Grit rocks (shales and grits), but we have here such plants as *Juniperus communis, Cystopteris fragilis, Asplenium Trichomanes,* and *Trichostomum tortuosum* growing along with or near to *Cryptogramme crispa, Polypodium Dryopteris,* and other common calcifuge species. The rocks near the stream yield *Zygodon Mougeotii* and *Breutelia arcuata,* whilst higher up the fell *Oligotrichum incurvum* is abundant. Over a large area of Botton Head Fell *Rubus Chamæmorus* is common, and on one portion near the head of Middle Gill it forms almost the only vegetation. The surface is almost everywhere thickly peat-covered and well suited to its growth, and on the west side of Dale Gill it descends as low as 1,280ft.! Adjoining Botton Head Fell, on the north-west of Dale Gill, are Greenbank Fell and Thrushgill Fell. These rise to a height of about 1,300ft., and form wide stretches of heather-land and wet peat-bog. The latter fell is a locality for *Dicranodontium longirostre* var. *alpinum*. Following the Hindburn downwards, past the farms of Greenbank and Botton Head, the banks become wooded and very beautiful. Near Ivah the stream is joined by an important tributary, which rises on Thrushgill Fell and Summers-gill Fell. Near Botton Mill this flows through a richly-wooded ravine, in which are two or three small but very pretty waterfalls. The country here is perhaps the most interesting in Hindburn, but owing to being difficult of access it is little known. West of Botton Mill is an extensive area of boggy moorland known as Goodber Common and White Moss, extending westwards to near Salter, in Roeburndale. The banks of the Hindburn from Ivah to near Cowkins, a distance of two-and-a-half miles, generally rise steeply and are beautifully wooded. The river then takes a more westerly course for two miles to the village of Wray, where it is joined by the Roeburn. Near Millhouses, a hamlet about a mile above Wray, are the falls of the Hindburn. The bed of the stream here is rocky and picturesque and well worth a visit. After heavy rain the falls are very fine, but the rocks and banks can only be explored when the river is low. One-and-a-half miles below Wray, at the foot of Hornby Park Wood, the Hindburn unites with the Wenning. The following are some of the rarer plants of the Hindburn valley and the extensive fells which rise above it :—

Ranunculus Lenormandi, F. Schultz.
Corydalis claviculata, DC.
Stellaria nemorum, L.
Rubus suberectus, Anders.
  „ saxatilis, L.
  „ Chamæmorus, L.
Circæa alpina, v. intermedia, Ehrh.
Solidago Virgaurea, v. cambrica, Huds.
Jasione montana, L.
Vaccinium Vitis-Idæa, L.
Andromeda polifolia, L.
Myosotis repens, G. Don.
Juniperus communis, L.
Carex canescens, L.
Cryptogramme crispa, R. Br.
Polypodium Dryopteris, L.
  „ Phegopteris, L.
Lycopodium Selago, L.
Sphagnum fimbriatum, Wils.
  „ Russowii, Warnst.
  „ compactum, D.C.
  „ Gravetii, Warnst.
  „ medium, Limpr.
Andreæa petrophila, Ehrh.
  „ Rothii, W. & M.
Tetraphis Browniana, Grev.
Catharinea crispa, James.
Oligotrichum hercynicum, Lam.
Polytrichum alpinum, L.
Brachyodus trichodes, Fürnr.
Dicranella rufescens, Schp.
Campylopus fragilis, B. & S., v. densus.
  „ paradoxus, Wils.
Campylostelium saxicola, B. & S.
Leptodontium flexifolium, Hpe.
Weisia rupestris, C.M.
Zygodon Mougeotii, B. & S.
Ulota Bruchii, Hornsch.
Splachnum sphæricum, L.
Tetraplodon mnioides, B. & S.

Discelium nudum, Brid.
Breutelia arcuata, Schp.
Mnium serratum, Schrad.
Pterygophyllum lucens, Brid.
Heterocladium heteropterum, B. & S.
Eurhynchium rusciforme,
  „ v.atlanticum, Brid.
  „ v.alopecurum, Brid.
Hypnum ochraceum,
  „ v. flaccidum, Milde.
  „ „ v.complanatum, Milde.
  „ stramineum, Dicks.
  „ Patientiæ, Ldb.
Blepharozia ciliaris, L.
Trichocolea tomentella, Ehrh.
Cephalozia sphagni, Dicks.
Scapania purpurea, Dill.
Jungermania cordifolia, Hook.
Mylia Taylori, Hook.
Nardia hyalina, Carr.
Blasia pusilla, L.
Pellia Neesiana, Limpr
  „ calycina, Tayl.
Racodium rupestre, Pers.
Bæomyces rufus, D.C.
Cladonia chlorophæa, Flk.
  „ bacillaris, Nyl.
Platysma ulophyllum, Nyl.
Feltidea aphthosa, Ach.
Gyrophora torrefacta, Cromb.
Pertusaria dealbata, Nyl.
  „ Wulfenii, D.C.
Lecidea crustulata, Ach.
  „ lucida, Ach.
  „ flexuosa, Fr.
  „ disciformis, Fr.
Opegrapha atra, Pers.
Verrucaria æthiobola, Wahl.
  „ epidermidis, Ach.

The country near the Wenning, between Low Bentham and Wray, and downwards to its junction with the Lune, consists principally of rich pasture and meadow, and is similar to that about Wennington, Melling, and Wrayton in District 1. We will now return to Botton Head Fell and proceed westwards, across a mile or more of fell, to the summit of the pass between Croasdale and Roeburndale, over which runs the track from Slaidburn to Hornby. The ground here is rather flat and consists of peat-bog. The moors slope away to the south into Whitendale, whilst the summit of Wolf-hole Crag (1,731ft.) rises about a mile away to the south-west. Proceeding north-westwards and following the infant Roeburn, a small peaty stream in the midst of the swamp, the narrow valley of Roeburndale is seen ahead. About a mile from its source the stream descends quickly over a boulder-strewn

bed, with numerous small waterfalls. The grit rocks here yield *Andreæa Rothii* and *A. crassinervia*. The ground on the right is generally boggy, whilst on the left there rises a steep heathery rock-bank, dripping wet and Sphagnum-clothed. This is, indeed, a place to delight a Sphagnologist or, in fact, any one who loves wild and beautiful country. Nowhere have we seen such a richly-coloured carpet of Sphagnum forms. Here *Sphagnum rubellum* var. *rubrum* grows in fine crimson patches, contrasting beautifully with the deep green and yellow species. Every tint is represented, from purple, red, and pink, to yellow, brown, and green. Little rills come tumbling down the banks at intervals, and the rocks and heathery slopes are adorned with Mountain Ash and Birch and a few stunted shrubs of Oak and Juniper. The latter is abundant near at hand, by the Lary Syke, a stream which descends a stony gully from the slopes of Wolf-hole Crag. From this point a hard, rough walk of about half-an-hour through deep heather and Cowberry—the latter being in some places about 2ft. high—will bring us to the top of the crag. The view from this is very wild and bleak, and similar to that from Botton Head Fell. Below the junction of the Lary Syke with the Roeburn the valley deepens considerably and continues exceedingly boggy for two miles, especially on the left bank of the stream. Mallowdale Fell, which is continuous with Wolf-hole Crag, now rises on the left, and Salter Fell on the right. Mallowdale Bridge is next reached. This is close to Mallowdale Farm, the highest habitation in the dale, and here some green fields begin and the banks of the river become beautifully wooded. A little below the bridge a tributary comes down from Mallowdale Fell. The upper course of this is down a steep stony gully on the west side of Mallowdale Pike; but lower down it flows through a narrow, thickly-wooded glen, the slopes of which are beautifully clothed with *Polypodium Dryopteris* and *P. Phegopteris*. *Epilobium angustifolium* is also exceedingly fine and abundant. From Mallowdale to Lower Salter Bridge the course of the Roeburn is bordered by green meadows and woods and the dale becomes very beautiful, the views looking back to Mallowdale Pike being especially fine. The river next cuts a gorge through dark shale, the banks being high and precipitous for some distance. On the right it is joined by the Harter Beck, a wild moorland stream from Goodber Common. The lower course of this is over a rocky bed in a deep wooded gully with several charming little waterfalls. The valley of Roeburndale from this point to Wray is richly wooded. The country to the east consists of upland pasture, whilst to the west rise the

slopes of Whitmoor. The following are some of the rarer plants of Roeburndale and the adjoining fells. The district is still imperfectly worked and will probably yield other plants of interest :—

Stellaria nemorum, L.
Rubus saxatilis, L.
Cratægus oxyacanthoides, Thuill.
Carduus heterophyllus, L.
Vaccinium Vitis-Idæa, L.
Andromeda polifolia, L.
Andreæa petrophila, Ehrh.
  „ Rothii, Web. & Mohr.
  „ crassinervia, Bruch.
Tetraphis Browniana, Grev.
Catharinea crispa, James.
Oligotrichum hercynicum, Lam.
Dicranodontium longirostre, B. & S.,
  „ v. alpinum, Schp.
Weisia rupestris, C.M.
Ulota Bruchii, Hornsch.
Splachnum sphæricum, L. fil.
Bartramia ithyphylla, Brid.
Breutelia arcuata, Schp.
Myosotis repens, G. Don.
Juniperus communis, L.
Carex pendula, Huds.

Cryptogramme crispa, R. Br.
Polypodium Dryopteris, L.
  „ Phegopteris, L.
Lycopodium Selago, L.
  „ clavatum, L.
Mnium subglobosum, B. & S.
Pterygophyllum lucens, Brid.
Heterocladium heteropterum, B. & S.
Sphagnum quinquefarium, Warnst.
  „ molluscum, Bruch.
  „ compactum, D.C.
  „ contortum, Schultz.
  „ aquatile, Warnst.
  „ pulchrum, Warnst.
  „ Torreyanum, Sulliv.
  „ medium, Limpr.
Lepidozia Pearsoni, Spruce
Kantia Sprengelii, Mart.
Mylia Taylori, Hook.
Nardia compressa, Hook.
Alectoria bicolor, Nyl.
Gyrophora polyphylla, Turn.

The dale west of Roeburndale and south-west of Whitmoor is called Littledale. Included in this are two branch dales running parallel to one another in a north-westerly direction, from the northern slopes of Wardstone. These are Foxdale and Udale. The former is a wild, rugged glen in the upper part, but with considerable wood lower down, and about Littledale Hall and Cross Gill it is beautifully clothed with Spruce and Scotch Fir. Udale is similar in character but less rocky. Both dales contain many dripping moss-banks and rocky rills, which drain the boggy fells above. The Udale beck rises a little to the north of the summit of Wardstone (1,836ft.), the second highest elevation in our area. We shall have to speak more particularly of this fell when describing District 7. Below Cross Gill, where the Foxdale and Udale becks unite, the stream enters a deep, narrow wooded glen, and is henceforward known as the Artle Beck. Three miles lower it joins the Lune near Caton. The following are some of the rarer plants of Littledale, the northern slopes of Wardstone, and Littledale Fell :—

Rubus Chamæmorus L.
Vaccinium Vitis-Idæa, L.
Populus tremula, L.
Juniperus communis, L.
Listera cordata, R. Br.

Carex dioica, L.
Hymenophyllum peltatum, Desv.
Asplenium Adiantum-nigrum, L.
Andreæa petrophila, Ehrh.
  „ Rothii, W. & M.

Catharinea crispa, James.
Diphyscium foliosum,
  „ v. acutifolium, Ldb.
Rhabdoweisia denticulata, B. & S.
Weisia rupestris, C.M.
Breutelia arcuata, Schp.
Hypnum revolvens, v. Cossoni, Ren.
Sphagnum Russowii, Warnst.
  „ quinquefarium, Warnst.
  „ crassicladum, Warnst.

Frullania Tamarisci, L.
Blepharozia ciliaris, L.
Bæomyces roseus, Pers.
Gyrophora torrefacta, Cromb.
  „ polyrhiza, Krb.
  „ polyphylla, Turn.
  „ flocculosa, Turn.
Lecanora glaucoma, Ach.
Lecidea sanguinaria, L.

North of Littledale rises the bare, flat-topped mass of moorland called Caton Moor (1,184ft.). The north side is known as Claughton Moor and slopes down to the level valley of the Lune, whilst the eastern portion is named Whitmoor, and has already been mentioned. It is drained by several small streams, some of which have cut deep gullies in the dark shale of which the moor is composed ; their banks are of considerable interest botanically, whilst the uplands above are rich in Sphagna. From Hornby Holmes to Caton the Lune flows through low flat fields, which are often partly flooded in the autumn and winter. The bottom of the valley is rather bare of trees, but the lower slopes of Claughton Moor are pleasantly wooded. South and west of Caton the country is richly wooded, especially about Quernmore Park and at the Crook of Lune. On a damp wall by the road-side near the latter grow *Seligeria recurvata* and *S. pusilla*, whilst the river banks and land adjoining between Hornby and Lancaster yield the following :—

Trollius europæus, L.
Stellaria nemorum, L.
Hypericum maculatum, Crantz.
Geranium sylvaticum, L.
Galium boreale, L.
Hieracium Orarium, Lind.,
  „ v. ravusculum, Dahlst.
  „ duriceps,
  „ v. Cravoniense, Hanb.
  „ diaphanoides, Lindb.
Lysimachia Nummularia, L.
Myosotis sylvatica, Hoffm.
Salix purpurea, L.
Orchis Morio, L.
Allium vineale, L.

Scirpus sylvaticus, L.
Carex acuta, L.
  „ pendula, Huds.
Equisetum hyemale, L.
Distichium capillaceum, B. & S.
Fissidens crassipes, Wils.
Tortula mutica, Ldb.
Barbula lurida, Ldb.
  „ spadicea, Mitt.
Orthotrichum rivulare, Turn.
Fontinalis gracilis, Lindb.
Chomiocarpon quadratus, Scop.
Parmelia scortea, Ach.
Lecidea albo-atra, Hoffm.
Lecanora expallens, v. lutescens, Nyl.

On the west of Quernmore Park rises Lancaster Moor. This is thickly wooded above the park, but nearer Lancaster it consists principally of upland pasture, and possesses no particular botanical interest. The remainder of District 6, still to be described, is comprised in the drainage area of the river Conder, above the village of Galgate. The Conder rises on Black Fell and Clougha, the most westerly part of the Wardstone range of fells. Although

Clougha Pike rises to an elevation of only 1,325ft., it is a splendid view-point, the prospect being unbroken for many miles, especially over the low country and the sea, and it includes the whole of Morecambe Bay. In clear weather the range of Lake mountains from Black Combe to High Street, and further to the east the Lunedale fells, stand out boldly on the horizon. The western end of Clougha is indented with two deep cloughs or ravines in the Millstone Grit, from which the hill takes its name. These are very rocky and mossy, and form a "happy hunting ground" for the bryologist and lichenologist. Such plants as *Dicranum fuscescens, Tetraphis pellucida, Mylia Taylori,* and *Bazzania trilobata* grow to a luxuriant size, whilst other interesting mosses and lichens are the following :—

Tetraphis Browniana, Grev.	Lophocolea cuspidata, Limpr.
Polytrichum alpinum, L.	Kantia arguta, Mart.
Campylopus atrovirens, De Not.	Plagiochila spinulosa, Dicks.
,, brevipilus, De Not.	Jungermania gracilis, Schleich.
Leptodontium flexifolium, Hpe.	,, minuta, Crantz.
Tetraplodon mnioides, B. & S.	,, incisa, Schrad.
Amblyodon dealbatus, P.B.	Coniocybe pallida, Fr.
Bryum alpinum, Huds.	Cladonia squamosa, Hoffm.
Webera annotina, Schwgr.	,, bacillaris, Nyl.
Heterocladium heteropterum, B & S.	Usnea ceratina, v. scabrosa, Ach.
Plagiothecium pulchellum, B. & S.	Alectoria bicolor, Nyl.
Hypnum fluitans, L.,	Platysma diffusum, Nyl.
,, v. molluscum, Sanio.	,, Fahlunense, Nyl.
Sphag ım turfaceum, Warnst.	Parmeliopsis ambigua, Nyl.
,, quinquefarium, Warnst.	Gyrophora flocculosa, T. & B.
Blepharostoma trichophylla, L.	Lecanora elegans, Ach.
Lepidozia cupressina, Sw.	Lecidea crustulata, Ach.
,, Pearsoni, Spruce.	,, lithophila, Ach.
Scapania resupinata, Dill.	,, plana, Lahm.

The south side of Clougha is drained by the Rowton Brook, which rises on Red Moss and joins the Conder near the scattered little village of Quernmore. The country above Quernmore is fairly well wooded, but lower down it is rather bare and has no features that call for special remark. The surface is almost entirely grass-land.

DISTRICT No. 7—WYRESDALE AND BLEASDALE.

This district lies immediately to the south of the one just described, and is very similar to it in general character of surface and physical features. It forms one of the most natural of our districts, containing the drainage area of the river Wyre, and its tributaries the Calder and Brock, so far as their courses in the "East Division" are concerned. It is bounded on the north by

District 6 and on the west by the high road dividing the "East" and "West Divisions"; Yorkshire bounds it on the east, whilst on the south it is separated from District 8 first by the watershed between the Brock and the Loud, and second—further west—by the rising ground between the Brock and the Barton Brook.

The eastern portion of the district consists of high, bleak moors, the principal summits of which are Wardstone, 1,836, Tarnbrook Fell (White side), 1,556, Marshaw Fell, 1,568, Bleasdale Fell, 1.505, and Fairsnape Fell, 1,701 feet respectively. The area of unenclosed fell within the district amounts to about 26 square miles out of a total area of about 70. The remainder consists largely of upland pasture with a fair proportion of woodland, especially along the banks of the principal streams. It includes also some low ground below 200ft. in the valley of the Wyre and in the strip of country along the base of the hills between Bay Horse and Brock. The lowest elevation, 50ft., is found near Garstang. Geologically a large part of the district, comprising the whole of Wyresdale and the area drained by the Calder, consists of Millstone Grit. A small area south-east of Garstang is composed of Permian Sandstone, much overlaid with Drift, whilst in the greater part of Bleasdale and the valley of the Brock the formation is Bowland Shale, which is considered to belong to the Yoredale series of Carboniferous rocks. The district generally is rather thinly populated and contains no large villages, Scorton and Calder Vale being the most important. We will begin our description at the head of Wyresdale and proceed downwards. The upper portion of the valley, or Over Wyresdale, as it is called, is divided into two parts by a tongue of high ground named Greenside. The stream of the northern branch-dale is known as the Tarnbrook Wyre and that of the southern as the Marshaw Wyre. The former rises on Tarnbrook Fell, an extensive tract of high moorland, which forms a plateau 1,500 to 1,750 feet elevation on the south-east side of Wardstone. This fell and the summit of Wardstone are mostly covered with thick peat, but a few masses of Millstone Grit rock appear on the surface in some parts. A large portion is exceedingly boggy, and there are extensive areas where the vegetation consists almost entirely of Sphagnum and Cotton grass. In others the peat is either deeply indented by weathered trenches or has been completely denuded. In the lower layers, a foot or more above the gritstone rock, as shown in section on the sides of the gullies, are great numbers of trunks, branches, and roots of Birch trees, the remains of an ancient forest. The fact that these are found on an exposed wind-swept fell, at a

height of 1,600 to 1,700 feet, an altitude much higher than that at which trees will grow in such a situation at the present day, is evidence that the climate at the time of their growth was considerably warmer than now. Bearing on the same subject, Mr. R. Lloyd Praeger, in a paper on peat-bogs,[*] says :—"The lower layers of our peat-bogs are usually full of the stumps and roots of trees. . . . . The distribution of these trees, consisting chiefly of Scotch Fir and Oak buried in bogs, deserves special consideration. They are found to extend in Scotland far north of the present limit of the species, while in Ireland, where the Scotch Fir is now extinct as a native, the remains of this tree occur in enormous quantity throughout the land. In both countries the old forest appears at elevations in the mountains far beyond the present upper limit of the species. Along the exposed western coasts of Scotland and Ireland enormous stumps may be dug out at various levels where now, even with artificial protection, trees will not grow." The plateau of Tarnbrook Fell, as already stated, is for the most part bare and wind-swept, and affords little protection for plants. The flora is poor in variety of species, as shown by the following list of phanerogams and vascular cryptogams observed above the 1,500ft. contour line ; below this the slopes of the fell are craggy in places and the flora is much richer :—

Rubus Chamæmorus, L.	Luzula sylvatica, Gaud.
Potentilla silvestris, Neck.	Scirpus cæspitosus, L.
Drosera rotundifolia, L.	Eriophorum vaginatum, L.
Galium saxatile, L.	Carex pulicaris, L.
Andromeda polifolia, L.	,, echinata, Murr.
Vaccinium Vitis-Idæa, L.	Agrostis tenuis, Sibth.
,, Myrtillus, L.	Aira flexuosa, L.
,, Oxycoccos, Roth.	Festuca ovina, L.
Calluna vulgaris, Hull.	Nardus stricta, L
Erica Tetralix, L.	Hymenophyllum peltatum, Desv.
Rumex Acetosella, L.	(Pteris aquilina, ceases at 1,500ft.)
Empetrum nigrum, L.	Blechnum Spicant, With.
Listera cordata, R. Br.	Lastræa aristata, Britt & Rendle,
Juncus squarrosus, L.	v. nana, Newm.
,, effusus, L.	

The above plants are characteristic of the gritstone fell-summits of the East Division, rising above 1,500ft. After draining the upland peat-bog above described, the Tarnbrook Wyre descends rapidly into a wild ravine known as the Great Clough. There are numerous falls along the course of the stream, and its bed is strewn with large boulders. To follow it is rather difficult and requires some hard scrambling, especially in the upper part of the gill. The slopes are boggy and precipitous,

[*] "Knowledge," October, 1902.

and in some parts are clothed with bracken six to seven feet high, whilst a few trees of mountain ash and birch adorn the rocky sides of the stream. About half-way down the clough it receives on its left bank an important tributary, which flows down a stony gill called Gavell's Clough. This rises in the peat-bog on the west side of Wolfhole Crag. The fell-slopes about the Great Clough and Gavell's Clough yield the following interesting plants :—

Caltha palustris, v. minor, Syme.	Philonotis calcarea, Schp.
Vaccinium Vitis-Idæa, L.	Breutelia arcuata, Schp.
Myosotis repens, G. Don.	Hypnum stramineum, Dicks.
Listera cordata, R. Br.	Marsupella emarginata, Ehrh.
Carex dioica, L.	Jungermania minuta, Crantz.
Cryptogramme crispa, R. Br.	Cephalozia fluitans, Nees.
Polypodium Phegopteris, L.	Racodium rupestre, Pers.
Lycopodium Selago, L.	Sphaerophorus compressus, Ach.
Andræa Rothii, W.M.	,, coralloides, Pers.
,, crassinervia, Bruch.	Stereocaulon evolutum, Graewe.
Oligotrichum hercynicum, Lam.	,, denudatum, Flk.
Polytrichum alpinum, L.	Gyrophora torrefacta, Cromb.
Brachyodus trichodes, Fürn.	,, polyphylla, Turn.
Blindia acuta, B. & S.	,, flocculosa, Turn.
Campylopus flexuosus,	Parmelia prolixa, Nyl.
v. zonatus, Limpr.	Lecanora Dicksonii, Nyl.
Fissidens osmundoides, Hedw.	Pertusaria dealbata, Nyl.
Leptodontium flexifolium, Hampe.	Lecidea confluens, Web
Amblyodon dealbatus, P. Beauv.	

The moor on the left bank of the stream below Gavell's Clough is called the White side and that on the right bank the Black side of Tarnbrook Fell, probably because the former is clothed with cotton-grass, sedges, and grass (principally *Molinia cœrulea*) and has a greyish-white appearance, whilst the latter is more heathery and rocky and has areas of dark peat-bog. West of the Great Clough, the slope of Tarnbrook Fell is girdled at a height of about 1,450ft. by a line of broken gritstone scars known as Long Crag, Hell Grag, and Thorn Crag. Immense blocks of fallen rocks are here piled upon one another in the roughest confusion. In the deep shaded hollows amongst these, various mosses and hepatics such as *Dicranum fuscescens, Dicranodontium longirostre, Bazzania trilobata,* and *Mylia Taylori* grow to a luxuriant size, whilst other rarer plants of the fell-side are the following :—

Vaccinium Vitis-Idæa, L.	Sphagnum aquatile, Warnst.
Rynchospora alba, Vahl.	,, medium, Limpr.
Hymenophyllum peltatum, Desf.	Lepidozia Pearsoni, Spruce.
Lycopodium Selago, L.	,, cupressina, Sw.
Campylopus atrovirens, De Not.	Parmelia lanata, Wallr.
Bryum alpinum, Huds.	,, tristis, Nyl.
Hypnum scorpioides, L.	Gyrophora torrefacta, Cromb.
Sphagnum rubellum, Wils.	,, polyphylla, Turn.
,, pulchrum, Warnst.	Lecidea sanguinaria, Nyl.
,, molluscum, Bruch.	

Hell Crag is the finest of all the three scars above-named, and, although of small extent, is exceedingly wild and rugged. At the foot of the Great Clough, half-a-mile above the hamlet of Tarnbrook, the Black side of Tarnbrook Fell is especially boggy and yields a rich variety of Sphagna and other interesting plants. Near Tarnbrook green fields begin and the brown fells are left behind. The valley soon afterwards becomes increasingly beautiful and the banks of the stream are clothed with trees, shrubs, and luxuriant ferns. *Rosa mollis* also is particularly fine and abundant. Rather over a mile below Tarnbrook the river receives a large feeder on the right bank called the Grizedale River. This drains the western flank of Wardstone and Abbeystead Fell. Half-a-mile lower the Tarnbrook Wyre is joined by the Marshaw Wyre, just before it enters the Abbeystead reservoir, which was constructed some years ago by the Lancaster Corporation. It is a fine sheet of water, and adds considerably to the beauty of the valley. The slopes about the reservoir are well wooded, and, from a footpath on the south side, fine views of the surrounding country are obtainable. The following are some of the more interesting plants of the Tarnbrook branch of Wyresdale additional to those mentioned in the three preceding lists :—

Rubus fissus, Lindb.	Catharinea crispa, James.
,, Scheutzii, Lindeb.	Discelium nudum, Brid.
Carduus heterophyllus, L.	Cladonia pityrea, Flk.
Myrica Gale, L.	Lecanora sympagea, Nyl.
Salix decipiens, Hoffm.	,, calva, Nyl.
Habenaria montana, Durand & Schinz.	Pertusaria leioplaca, Schaer.
Andreæa petrophila, Ehrh.	

Turning now to the Marshaw Wyre and proceeding to its source on the moors to the south-east of Tarnbrook, we find a similar fell-country to that already described. The moors here are called Threaphaw Fell and Winfold Fell, and lie immediately to the west of Brennand, a secluded valley in the Forest of Bowland. The Trough of Bowland, through which runs the road from Lancaster to Clitheroe, at a height of about 1,000ft., is situated to the south of Winfold Fell. It forms a narrow nick or trough amongst the moors which rise 400 to 500 feet above the level of the road. The slopes are steep on the east or Yorkshire side, but are more gradual on the west. South-west of the Trough is Blaze Moss, another extensive area of boggy fell and sphagnum-swamp, on some parts of which *Andromeda polifolia* is extremely abundant. The Marshaw Wyre rises near the Yorkshire boundary, west of

the pool called Brennand Tarn,* and is formed of various streams which drain the deep and tortuous gullies in the peat so numerous on these moors. The road from the Trough of Bowland descends by the edge of Blaze Moss, following the Wyre for some distance through old plantations of Scotch Fir, which form a pleasing variety from the bareness of the fells. Near Marshaw the stream is joined by a tributary from Black Clough and the moors on the south side of Blaze Moss. Black Clough in its upper part is a deep narrow ravine with heathery sides, and a stream at the bottom flowing over a bed of smooth rock. Lower down, the gully is beautifully adorned with Mountain Ash and Birch, and forms one of the most charming of our moorland cloughs. There are numerous small waterfalls, and the banks are rocky and clothed with ferns. The steep fell slopes above are covered in some parts with deep bracken, in others with bilberry, amongst which *Trientalis europæa* grows plentifully. From the foot of the clough across to Blaze Moss, and also along the north side of Marshaw Fell, the ground is very boggy and yields a rich variety of sphagna and other mosses. Amongst these may be particularly mentioned *Hypnum sarmentosum* and *Philonotis calcarea*. The former grows at an altitude of only 600ft., a very low one for this moss, but the locality is cold and exposed, with a northerly aspect. The occurrence of *Philonotis calcarea* here, as elsewhere on the gritstone fells, is interesting; it is apparently restricted to small patches of boggy ground surrounding springs where it is associated with other calcicole species, as *Hypnum commutatum*. It is probable that these springs well up from underlying strata somewhat calcareous in character (dark-coloured impervious shales with which the millstone grit is interpolated) and are sufficiently charged with lime to form a suitable habitat for these plants.

The land on the north or sunny side of the dale has been drained and converted into green fields, or planted with pine woods, for nearly two miles above the farm of Marshaw, the cultivation extending to the summit of Greenside; but on the south side, the heathery slopes of Marshaw Fell come down almost to the banks of the river. This latter fell extends some distance in a westerly direction, where it is known as Hawthornwaite Fell, and rises steeply from the valley to an altitude of 1,568ft. It is a prominent object from almost all parts of Wyresdale. The summit forms part of a nearly level plateau extending southwestwards along the Yorkshire boundary to a place called Johnny

<hr>

* Brennand Tarn is just in Yorkshire and the stream from it flows eastwards
into the Brennand.

E

Pye's Clough Top, and to White Moss at the head of the Langden. An extensive area on the summit has been almost denuded of peat and is entirely bare of vegetation. The banks of the Wyre from Marshaw downwards are well wooded, and about the Abbeystead reservoir, near which, as before stated, the two branches of the Wyre unite, the country is especially beautiful. The following are the rarer plants of the Marshaw branch of the dale :—

Rubus fissus, Lindl.	Fontinalis squamosa, L.
Vaccinium Vitis-Idæa, L.	Hypnum scorpioides, L.
Andromeda polifolia, L.	,, falcatum, Brid.
Trientalis europæa, L.	,, stellatum, Schreb.
Myosotis repens, G. Don.	,, exannulatum, Gümb.
Salix phylicifolia, L.	,, stramineum, Dicks.
Narthecium Ossifragum, Huds.	,, sarmentosum, Wahl.
Carex dioica, L.	Sphagnum Warnstorfii, Russ.
,, canescens, L.	,, rubellum, Wils.
Catharinea crispa, James.	Cladonia cæspititia, Flk.
Polytrichum alpinum, L.	Alectoria jubata, Nyl.
Tetraplodon mnioides, B. & S.	Platysma ulophyllum, Nyl.
Philonotis calcarea, Schp.	Parmelia lævigata, Ach.
Bryum alpinum, Huds.	Lecanora chlarona, Nyl.

Below Abbeystead the dale becomes much wider and the fells recede a long distance from the river. On the north is Abbeystead Fell, a large rather flat stretch of moorland below the southeast extremity of Clougha. This is drained by the Grizedale River on the east, and further west by the Damas Gill stream. The latter flows down a long wooded clough and joins the Wyre near Dolphinholme. Abbeystead is hardly to be called a village, but consists of farmhouses scattered over the north side of the dale. From the foot of the reservoir to Dolphinholme, a distance of two-and-a-half miles, the banks of the Wyre are very picturesque. On the left bank it receives the streams from Catshaw Greave and Hall Gill, each of which descends a narrow wooded ravine. A little below the foot of Hall Gill the Wyre flows through a deep glen or gorge, and is bordered by cliffs of sandstone and shale overhanging the river in places. The banks are thickly wooded for some distance and yield the following rarer plants :—

Rubus fissus, Lindl.	Heterocladium heteropterum, B. & S.
,, saxatilis, L.	Chomiocarpon quadratus, Scop.
Hieracium murorum, L.	Lecidea crustulata, Ach.
Festuca sylvatica, Vill.	,, lucida, Ach.
Polypodium Robertianum, Hoffm.	,, rivulosa, Ach.
Tetraphis Browniana, Grev.	Opegrapha herpetica, Ach.
Weisia verticillata, Brid. (fruiting).	,, atra, Pers.
Zygodon Mougeotii, B. & S.	Arthonia pruinosa, Ach.

The Catshaw Beck is formed by the union of two streams, one rising on Hawthornthwaite Fell and the other on White Moss.

Both of these descend stony gullies, which are known as the Hawthornthwaite and Catshaw Greaves respectively. The latter is the principal ravine and has two branches, both very steep and rugged. The gritstone rocks in each of the greaves yield *Andreæa Rothii* var. *falcata,* and in the Catshaw Greave there is an abundance of *Rhacomitrium fasciculare*. South-westwards from Catshaw the range of moors is known as Catshaw Fell and Harrisend Fell. These are very wet and similar in character to those already described, but are less high, the summit of the latter being only about 1,000ft. From Dolphinholme to Garstang, where the Wyre leaves the district, the distance is about six miles. The river winds considerably and the banks are pleasantly wooded in parts. The land adjoining, as a so the country northwards to the boundary of District 6, is almost all under grass and well cultivated. The remainder is woodland.

On the south side of Catshaw and Harrisend Fells we come to another tributary of the Wyre called the Grizedale Brook. This must not be confused with the stream of similar name which joins the Tarnbrook Wyre above Abbeystead. The Grizedale Brook takes a nearly straight south-westerly course, joining the Wyre above Garstang. For the first two miles it flows over very wet moors, and its banks are rich in Sphagna and other bog-plants. *Anagallis tenella* and *Vaccinium Oxycoccos* are here especially abundant, the latter fruiting freely, and on the steep bank to the left of the stream *Breutelia arcuata* grows more luxuriantly than we have seen it elsewhere in the vice-county. Below Harrisend Fell the Grizedale valley becomes deep and narrow, and is known as Nicky Nook. The wooded slope of Barnacre Moor rises steeply on the left, and on the right, but rather lower down, Nicky Nook Hill. A reservoir was constructed here many years ago for the supply of Blackpool, Fleetwood, etc., with water. This adds to the beauty of the valley, but its construction has tended to reduce the number of interesting plants. The "Nook" has long been a place of picnic for parties from Preston, etc., and several flowering plants and ferns formerly found here have become scarce or extinct. Amongst these are *Hypericum Androsæmum*, *Wahlenbergia hederacea*, *Polypodium Phegopteris*, and *Cryptogramme crispa*. The north slopes of the valley are clothed with bracken and heather, *Erica cinerea* being especially fine and plentiful. Below Nicky Nook the Grizedale valley opens out into the plain and the stream soon afterwards joins the Wyre.

On the higher part of Barnacre Moor there was formerly a tarn called Sconce Tarn, and the ground in the neighbourhood was boggy, yielding various interesting plants, but some years ago, when the water supply in Grizedale became insufficient, a reservoir was made on the site of the tarn, into which water is now conveyed by a culvert from the river Calder above Oakenclough. The Calder, which must not be confused with the two rivers of the same name in East Lancashire and the West Riding of Yorkshire, rises on White Moss and Luddocks Fell at the north side of the Bleasdale moors. For the first three miles its course is westerly and entirely amongst the fells. These generally rise steeply from its banks, the ground in some parts being very boggy and similar to that by the Grizedale Brook on Harrisend Fell. Near Oakenclough green fields begin and the stream takes a more southerly course, with Barnacre Moor on the right. From Oakenclough to the village of Calder Vale, and thence to about a mile above Catterall, the valley is deep and narrow and the banks of the stream thickly wooded. The low country is then reached and the river soon afterwards enters the West Division. Near Catterall it is joined by a small stream called the Little Calder, which drains a part of Barnacre and the low ground south of Greenhalgh Castle, between Sullam and Bowgrave Hill. This low ground was formerly covered by a small peat-moss, but it is now highly cultivated. The following are the rarer plants found in the drainage area of the Calder from the moors above Oakenclough to the low country near Garstang :—

Rubus fissus, Lindl.	Splachnum sphæricum, L.
,, Chamæmorus, L.	Philonotis calcarea, Schp.
Vaccinium Vitis-Idæa, L.	Mnium subglobosum, B. & S.
Andromeda polifolia, L.	Pterygophyllum lucens, Brid.
Utricularia vulgaris, L.	Heterocladium heteropterum, B. & S.
Habenaria viridis, R. Br.	Hypnum riparium, v. longifolium, Brid.
Carex pendula, Huds.	,, exannulatum, Gümb.
,, lævigata, Sm.	,, stramineum, Dicks.
Catharinea crispa, James.	

We now come to Bleasdale, the broad upland valley to the south-west of Fairsnape Fell and Parlick Pike. It extends from the Bleasdale Fells on the north, to Beacon Fell on the south, and forms a fairly level plain indented by the narrow valley of the Brock and its branches. It formerly contained a good deal of moorland and peat moss, but most of this has been reclaimed, and the surface now consists principally of rough pasture with numerous plantations of Scotch Fir. Fairsnape Fell rises boldly from the dale to an elevation of 1,701ft., and commands an extensive

prospect westwards to the sea. The view from Parlick Pike is also very fine, especially over the country in District 8 and eastwards into Yorkshire. Except for a few broken grit rocks the sides of these fells are mostly smooth and grass-covered, and are too dry to yield much of botanical interest. In the deep clough between Fairsnape Fell and Hazelhurst Fell, however, the ground is more boggy, and yields *Primula farinosa, Parnassia*, etc. On Parlick also there are too small but steep shaley gullies, in one of which *Brachyodus trichodes* grows. The slopes of Hazelhurst Fell and the moors further west are mostly clothed with a carpet of bilberry, heather being comparatively scarce. The summits are wet, and like that of Fairsnape Fell are covered with thick peat, which is weathered and indented by numerous boggy gullies. The river Brock has two main branches, one rising on Hazelhurst and Fairsnape Fells and draining the north side of Bleasdale, and the other flowing from Parlick and the country further south. They are about equal in size, and unite about three-quarters-of-a mile south-west of Admarsh. Below this point the valley of the Brock is narrow and in most parts beautifully sylvan and picturesque. Near Claughton Park it flows through a rocky channel between steep banks of shale upon which grow several interesting mosses, including *Weisia verticillata* and exceedingly fine *Hypnum commutatum*. Rather over a mile lower down it enters District 5. A large part of Bleasdale and the drainage area of the Brock is still somewhat imperfectly worked, and may possibly yield further plants of interest. The following are the rarer species at present known ; of these, *Gentiana pneumonanthe* is the most worthy of note :—

Stellaria nemorum, L.	Casex pendula, Huds.
Rubus fissus, Lindb.	(Osmunda regalis, L., extinct.)
,, criniger, Linton.	Equisetum maximum, Lam.
Chrysosplenium alternifolium, L.	Brachyodus trichodes, Fürnr.
Vaccinium Vitis-Idæa, L.	Barbula spadicea, Mitt.
Andromeda polifolia, L.	Weisia verticillata, Brid.
Primula farinosa, L.	Mnium stellare, Reich.
Gentiana Pneumonanthe, L.	,, subglobosum, B. & S.
Myosotis repens, G. Don.	Brachythecium albicans, B. & S.
Habenaria montana, Dur. & Schinz.	Hypnum Patientiæ, Lindb.
Narcissus Pseudo-narcissus, L.	,, stramineum, Dicks.
Carex canescens, L.	Lecidea crustulata, Ach.

DISTRICT No. 8—LONGRIDGE.

This district is bounded on the north, from the village of Barton to the summit of Parlick Pike, by District 7, and afterwards, along the ridge of the Leagram Fells to the river Hodder, by Yorkshire. On the east and south the Hodder and Ribble separate it from Yorkshire and South Lancashire respectively, whilst on the west the main road from Preston to Barton divides it from District 5.

The general character and physical features of the district are in many respects similar to those of Wyresdale and Bleasdale, but there is much less high fell and moorland, the greater part of its surface consisting of pasture. Some portions are rather bare and uninteresting, but others, especially in the valleys of the Hodder and Ribble, are richly wooded and very picturesque. The principal elevations are Longridge Fell (1,016ft.), Parlick Pike (1,416ft.), and the ridge of the Fairsnape and Leagram Fells reaching a height of from 1,500 to 1,700 feet. The lowest ground is in the valley of the Ribble near Preston, where the river becomes tidal. Geologically, the district consists principally of Carboniferous rocks—Millstone Grit series and Bowland Shales, with small areas of Scar Limestone—but about Preston and Barton the formation is Triassic Pebble Beds, generally overlaid with Drift. The Scar Limestone occurs near Whitewell, Greystoneley, and Thornley, and has an important bearing on the flora. The district generally is rather thinly populated, but it contains the villages of Longridge, Grimsargh, and Chipping, and about half of the large town of Preston. The drainage areas divide it naturally into three parts as follows :—

1st.—The north-eastern portion, drained by the Hodder and its large tributary the Loud, and comprising the country east of Beacon Fell between the Leagram Fells and Longridge Fell.

2nd.—The southern or Ribble portion, containing the land to the south of Longridge Fell and the Blundel Brook.

3rd.—The area south and south-west of Beacon Fell. This slopes away gradually to the Fylde plain and is drained by the Barton and Blundel brooks.

Beginning our description with the 1st, or Loud, area we find that the river of this name rises to the east of Beacon Fell, and after first taking a very circuitous course flows north-eastwards

into the Hodder. The Leagram Fells rise boldly on the north side of the Loud valley, some two miles from the stream, their summits thickly peat-covered and their sides covered with heather and bracken. They form the southern extremity of the wild moors already described, which stretch away for many miles into Bleasdale and Wyresdale and the Forest of Bowland. Parlick Pike, their most south-western extremity, rises steeply about two miles from Chipping, and is well worth ascending on account of the fine view it commands,* but the hill is rather wanting in botanical interest. The country west and south-west of Chipping about the source of the Loud consists mostly of bare upland pasture. Eastwards it becomes more wooded and it increases in interest as the Hodder is approached. Near Thornley the Loud is joined by two important tributaries from the Leagram Fells, the Chipping Brook on the west from Parlick and Wolf Fell and, further east, the stream from Burnslack and Leagram. The chief feeder of the former is the beck from White Stone Clough, a wild gully with steep heathery sides on the south-east side of Fairsnape Fell. The Burnslack stream comes from Dewhurst Clough, another of the deep-cut, heather-clad ravines so characteristic of the gritstone fells of West Lancashire and Bowland. Along the boggy summit-ridge near the sources of these becks the Cloudberry *(Rubus Chamæmorus)* is remarkably abundant, particularly between White Stone Clough and the moor on the Langden side called Brown Berry Plain. The country known as Little Bowland to the north-east of Leagram is drained by a beck with three principal branches rising on Burnslack Fell, Fair Oak Fell and Totridge Fell, and which, after a course of three miles, joins the Hodder a little above where the latter is joined by the Loud. The rocks exposed along the upper reaches of the beck are Millstone Grit and Bowland Shales, and the surface generally is swampy moorland and wet pasture ; but near Greystoneley the country becomes richer with many well cultivated fields, and the stream cuts a glen in the Scar Limestone. The ravine is beautifully wooded and contains a fine waterfall in wet weather but in dry times the rocky bed of the beck has only pools at intervals, the water flowing under ground. Several interesting plants grow here, including *Festuca sylvatica* and *Mimulus Langsdorffii*. North of Greystoneley and east of the beck the Scar Limestone reaches a height of 800ft. It extends to the valley of the Hodder opposite Whitewell, and appears also on the Yorkshire side of the river. The surface of the high ground is mostly dry grassy pasture, with here and there patches of limestone scar, but the slopes above the Hodder are thickly wooded and very steep, the river flowing through a beautiful gorge for over two miles. The wood on the Lancashire side below Whitewell is rocky in some parts with limestone cliffs, and in others it is very swampy owing to numerous calcareous springs which issue from the limestone. In one spot there is a dripping rock-bank composed of tufa, overhanging the river and most charmingly clothed with moss and ferns. *Hypnum commutatum* is here in great luxuriance and *Weisia verticillata* fruits freely, whilst near at hand grow abundance of *Festuca sylvatica, Lathræa squamaria* and *Equisetum maximum.* Except for a small patch on the wooded hill-slope known as Fence Wood, the limestone does not appear far north-eastward from the farm called Dinkling Green (a mile west of Whitewell) being overlaid with Bowland Shales. These shales extend northwards across the county boundary and form the lower slopes of Totridge Fell. About two miles below Whitewell, and a mile above, where it is joined by the Loud, the Hodder leaves the limestone and flows through shales. It continues to do so along the remainder of its course to the Ribble except for a small area of limestone near Harmer Ridding. The country all the way is well wooded and very picturesque.

The following are the rarer plants of the right bank, or Lancashire side, of the Hodder (including the adjacent land within about a mile of the river) from the county boundary, one mile north-west of Whitewell, to the junction of the Hodder with the Loud :—

Arabis hirsuta, Scop.
Draba muralis, L.
Stellaria nemorum, L.
Viola sylvestris, Reich.
Rubus saxatilis, L.
Inula Conyza, DC.
Myosotis sylvatica, Hoffm.
Mimulus Langsdorffii, Donn.
Lathræa squamaria, L.
Scirpus sylvaticus, L.
Carex sylvatica, Huds.
    ,,   pendula, Huds.
Festuca sylvatica, Vill.
Triticum caninum, L.
Weisia microstoma, C.M.
Trichostomum crispulum, Bruch.
    ,,   nitidum, Schp.
Zygodon conoideus, H. & T.

Zygodon Stirtoni, Schp.
Orthotrichum cupulatum,
    ,,   v. nudum, Braith.
Aulacomnium androgynum, Schwgr.
Mnium serratum, Schrad.
Pterygophyllum lucens, Brid.
Lejeunea ulicina, Tayl.
Choniocarpon quadratus, Scop.
Reb022lia hemispherica, L.
Collema multipartitum, Sm.
Collemodium fluviatile, Nyl.
Parmelia cetrarioides, Nyl.
    ,,   subaurifera, Nyl.
Physcia lychnea, Nyl.
Thelotrema lepadinum, Ach.
Lecidea concentrica, Dav.
Graphis sophistica, Nyl.
Endocarpon miniatum, L.

The following are the more interesting plants of the Leagram Fells and the north side of the Loud valley :—

Corydalis claviculata, DC.
Rubus Chamæmorus, L.
Pimpinella major, Huds.
Carduus heterophyllus, L.
Andromeda polifolia, L.
Primula farinosa, L.
Myosotis repens, G. Don.
Mimulus Langsdorffii, Donn.
Listera cordata, R. Br.
Epipactis longifolia, Allion.
Gymnadenia conopsea, R. Br.
Carex Hornschuchiana, Hoppe.
Festuca sylvatica, Vill.
Asplenium viride, Huds.
Ceterach officinarum, Willd.

Polypodium Phegopteris, L.
Andreæa petrophila, Ehrh.
    ,,   Rothii, W. & M.
Philonotis calcarea, Schp.
Mnium subglobosum, B. & S.
Brachythecium glareosum, B. & S.
Mylia Taylori, Hook.
Jungermania gracilis, Schleich.
Racodium rupestre, Pers.
Lecanora ventosa, Ach.
    ,,   fuscata, Nyl.
Urceolaria bryophila, Nyl.
Lecidea mollis, Wahl.
    ,,   rivulosa, Ach.

The banks of the Loud near where it joins the Hodder are prettily wooded, but the country higher up about Thornley and on the north slopes of Longridge Fell is somewhat bare. In the outcrop of Scar Limestone near Thornley there are old quarries containing pools and marshy ground, which yield several interesting plants. Longridge Fell, as its name implies, forms a long, rather narrow ridge, rising gradually towards the east from the village of Longridge and reaching a height of 1,016ft. On the higher parts of the fell the surface consists of heathery moor, but a considerable area has been planted with Scots Pine, etc. The hill terminates somewhat abruptly at the eastern end (which is known as Kemple end), and descends in beautifully wooded slopes to the banks of the Hodder.

The following are the more interesting species found south of the Loud and on the north and east sides of Longridge Fell, including the Lancashire side of the Hodder valley from the Loud downwards to near Mitton :—

Corydalis claviculata, L.
Pimpinella major, Huds.
Ceratophyllum demersum, L.
Lathræa squamaria, L.
Orchis Morio, L.
Eleocharis multicaulis, Sm.

Catharinea crispa, James.
Oligotrichum hercynicum, Lam.
Barbula lurida, Ldb.
    ,,   spadicea, Mitt.
Mnium stellare, Reich.
Blepharozia ciliaris, L.

We now come to the portion of the district drained directly by the Ribble. This consists of the south side of Longridge Fell and the land from Stonyhurst and Ribchester westwards to Preston. It forms a gradual slope facing towards the south, but falling steeply near the banks of the river in some parts. Generally speaking it is well wooded and was formerly a very rich tract of country, but it is now somewhat spoiled by smoke from Preston, Blackburn, and other towns. The geological formation about Longridge Fell and Ribchester is principally Millstone Grit, and

further west it is Triassic Pebble Beds, generally overlaid with thick deposits of glacial Boulder-clay and Gravel. There are some fine sections of the latter deposits on the bank of the Ribble between Ribchester and Redscar. The river flows with many windings through the low alluvial valley. The distance to Preston from the point where the Hodder joins the Ribble, following the river, is about 22 miles, and in a straight line 11½ miles. The plants of the eastern portion of the district have been investigated by the Rev. John Gerard and the Rev. C. A. Newdigate and others from Stonyhurst College and recorded in the "Flora of the Stonyhurst District." The western portion also has been worked by members of the Preston Naturalists' Society. The following are the rarer plants found between Stonyhurst and Longridge and on and near the banks of the Ribble between where it is joined by the Hodder and Preston :—

Corydalis claviculata, L.
Radicula sylvestris, Druce.
Hypericum maculatum, Crantz.
    ,,   Androsæmum, L.
Potentilla verna, L.
Saxifraga granulata, L.
Pimpinella major, Huds.
Senecio saracenicus, L.
Carduus heterophyllus, L.
Lysimachia Nummularia, L.
Myosotis sylvatica, Hoffm.
Lamium Galeobdolon, Crantz.
Rumex aquaticus, L.

Salix phylicifolia, L.
Neottia Nidus-avis, Rich.
Gagea lutea, Gawl.
Paris quadrifolia, L.
Carex sylvatica, Huds.
    ,,   Pseudo-cyperus, L.
Glyceria plicata, Fr.
Fissidens incurvus, Starke.
Tortula mutica, Ldb.
Barbula lurida, Ldb.
    ,,   spadicea, Mitt.
Orthotrichum rivulare, Turn.
Hylocomium brevirostre, B. & S.

Coming now to the area drained by the Barton and Blundel brooks, we find a country consisting almost entirely of pasture, and sloping towards the west from a height of about 850ft. on Beacon Fell to 100ft. at Barton. Although a pleasant district, it contains but little woodland or rough ground, and is therefore not of much interest to the botanist. The geological formation of the upper portion is Carboniferous Shale, and of the lower Triassic Pebble Beds, often overlaid with Drift. The most interesting ground is along the banks of the streams.

METEOROLOGY AND CLIMATE.

In common with a large part of the North-west of England, West Lancashire has, during most of the year,* a fairly mild and humid climate, intermediate in character, as would be expected, between that of the West of Ireland and the portion of the North of England east of the Pennine range of hills.

Different parts of the vice-county, of course, vary considerably according to their altitude and their proximity or otherwise to the sea.

Of the five factors, Sunshine, Temperature, Rainfall, Humidity and Wind-force, which go to determine Climate, the most important, in its influence on Plant-life, is Temperature. Treating first of this, we find that, although it is subject at certain times to nearly as great extremes as occur further east, it is generally more equable, and it is, on an average much milder in West Lancashire in Winter and Spring than it is in Yorkshire and the Eastern Counties generally. This greater mildness is largely due to the warming effect of the Gulf Stream as brought more directly to our coasts by the Irish Sea. Snow seldom lies long over the plain near the coast, or in fact anywhere, in West Lancashire, except on the higher fells near the Yorkshire boundary, and the amount that falls is generally much less than in the West Riding of Yorkshire.

The table shown on page 60 gives particulars of Temperature for the 10 years 1891—1900 at Eruna Hill, near Garstang, and is compiled from observations made there by Mr. Sydney Wilson, a brother of one of the authors. The station is 100ft. above sea level and seven miles from the nearest sea-coast, and may be taken as fairly representative of the whole of the vice-county.

The extremes of temperature on the hills are, of course, less great than in the low country, the maxima being lower and the minima higher. Along the valleys of the Lune and Ribble, for example, and especially in the Fylde plain, frost during calm, clear weather is much more severe than on the hills. It frequently occurs in late spring or early autumn, cutting down delicate plants, as Potatoes, Dahlias, etc., when the upland districts entirely escape. Speaking particularly of the Garstang district,

---

* The Spring in West Lancashire is frequently dry, much more so than in Yorkshire, particularly when there is a prevalence of northerly or easterly winds.

TEMPERATURE AT BRUNA HILL, GARSTANG—1891-1900.

	Mean temperature.	Extremes.				Number of days with max. 70° or above.			Number of days with min. 32° or below.		
		Maximum.	Date.	Minimum.	Date.	Average.	Greatest number.	Least number.	Average.	Greatest number.	Least number.
January	37·3	53·4	19th, 1898	12°·0	4th, 1893	0	0	0	13·6	26	2
February	38·5	58·5	25th, 1900	7·5	19th, 1892 8th, 1892	0	0	0	11·7	26	6
March	40·7	68·5	27th, 1894	16·0	12th, 1892	0	0	0	12·9	24	5
April	46·2	76·0	25th, 1893	22·0	16th, 1892	0·9	7	0	5·2	10	2
May	50·6	81·5	30th, 1895	24·7	18th, 1891	2·3	7	0	2·0	5	2
June	57·5	85·5	18th, 1893	30·8	14th, 1895	7·7	15	4	0·1	1	0
July	59·3	81·3	7th, 1893	38·0	27th, 1896	8·3	16	2	0	0	0
August	59·0	84·0	25th, 1899	35·0	10th, 1892	6·7	22	0	0	0	0
September	54·9	83·0	28th, 1895	29·8	27th, 1894	3·5	13	0	0·2	1	0
October	47·4	71·5	1st, 1895	17·0	28th, 1895	0·1	1	0	4·9	12	0
November	44·0	61·0	2nd, 1894	21·0	3rd, 4th, 5th, 1896	0	0	0	5·1	12	0
December	39·7	58·3	5th, 1898	14·5	28th, 1899	0	0	0	8·7	21	0
Averages and extremes of 10 years	47·9	85·5		7·5		29·5			64·4		

where observations have been made for a number of years, the daily minimum temperatures are frequently from 8° to 10° lower in the plain than on the adjacent hills of Barnacre or Bleasdale. The direction of the wind, as in other parts of England, has a very great effect in determining temperature. During summer, E., S.E., and S. winds are the warmest about Garstang, and the same probably applies to the whole of the vice-county. Blowing, as they do, from the inland districts of England, they have become warm by passing over the land, whilst those from the S.W., W., and N.W. are comparatively cool. Winds from the last quarter are particularly cool about Garstang in summer and, indeed, all over the portion of the vice-county south of Lancaster. Further north they are less so, their temperature having been raised by passing over Cumberland and Lake Lancashire. During a warm day in settled summer weather, the coming on of a sea-breeze in the early afternoon (produced by the temperature having become high inland) will frequently cause the thermometer to fall about 10°. This often takes place before the usual time of highest temperature—viz., 1.30 to 3 p.m.—has been reached, rendering the maximum on such days much lower than would otherwise have been the case, and it is only when the gradients for E. and S.E. winds are sufficiently pronounced to overcome the tendency to sea-breeze formation that very high temperatures are recorded. Such a case occurred on July 20th, 1901, when the thermometer rose to 89·4° in the screen at Garstang. The lowest winter temperatures usually occur with light N. or N.E. winds or during anti-cyclonic calms. Winds from the S.W. or W. during winter, blowing direct from the Irish sea, bring, as would be expected, very mild weather ; but it is particularly in the spring, when there is a prevalence of easterly winds, that West Lancashire is proportionately most favoured as compared with the country east of the Pennine range of hills. For these winds generally bring bright sunshine and dry atmosphere to the west, whilst on the eastern side they are usually accompanied by dull skies and frequently by snow or cold rain. Now, bright sunshine greatly raises the day temperature and has a marked effect on plant life, forcing into bloom the early spring flowers, and we find that vegetation generally is usually from a week to a fortnight forwarder than in Yorkshire. This advance continues until July, when the higher summer maximum temperatures in-land begin to tell in the opposite direction. In West Lancashire, spring flowers are often to be found very early in the season, especially on sheltered banks amongst the hills. This is particularly noticeable in the Silverdale

district, where the dry limestone soil, proximity to the sea, and shelter of the rocks, all tend to produce this result. In some years the forwardness of vegetation would be much more pronounced than it is, if it were not for others factors in the climate which tend in an opposite direction. The following are one or two of the most important. When, for example, a "northerly type" of weather prevails in winter or early spring the night temperatures occasionally fall much lower in Lancashire than in Yorkshire and other parts of the North of England east of the Pennine range of hills. In such weather the barometer is high over Ireland and the Atlantic, and an area of low pressure occupies North Germany and the region to the east of the North Sea. With these conditions there is a tendency for small secondary depressions to develop over and travel down the North Sea, and these commonly produce squally, unsettled weather, with considerable cloud and showers of sleet or snow on the *east* side of Great Britain, and consequently a moderately low temperature there with but little range. In Lancashire, on the contrary, the weather is quiet and fine, and the air being dry, radiation is excessive and the bright sunshine prevalent during the day gives place to severe frost at night, especially in the low country. Minima 10° below those registered at the same time in Yorkshire have sometimes been observed at Garstang. This greater range of temperature has, of course, an injurious effect on early flowering plants. The smaller snowfall in Lancashire is also at times an indirect cause of injury to vegetation, as when severe frost occurs there is not so much protection as in those districts where the snow-fall is greater.

Turning now to Bright Sunshine, we find that with the exception of the sea-coast and the western portion of the Fylde plain, which are very sunny, West Lancashire is, during a large part of the year, not particularly favoured. When westerly winds are blowing, moisture laden from the sea, clouds generally form rapidly as the uplands are approached. This is, of course, due to the cooling (caused by reduced pressure) of ascending currents and the consequent condensation of moisture. The dale districts are thus often without sunshine when the sky near the coast and over the sea is almost or perfectly cloudless. But when easterly winds prevail, as so frequently happens in Spring, the converse is the case. These winds are *descending* currents in West Lancashire, blowing down from the Pennine range of hills, and being thus warmed by compression often give us bright sunshine and a relatively warm temperature, as compared with the West Riding of Yorkshire, which is then covered with cloud. This tendency to

bright weather with easterly winds is much more pronounced than it is in South Lancashire and is apparently due to two causes. Firstly, the Pennine range in that part of Yorkshire to the east of West Lancashire is generally much broader, and in its northern part more lofty, than it is further south,[*] and the condensation and precipitation of moisture with easterly winds in the former region more complete than it is in the latter, rendering the air relatively drier under these conditions in West Lancashire than it is in South Lancashire ; and, secondly, the part of Yorkshire east of West Lancashire contains no large smoke-producing towns, whereas that east of South Lancashire contains the great manufacturing districts of the West Riding. These produce an enormous amount of smoke, and this, blowing westwards over the similarly smoky districts of South Lancashire, increases the dimness of the atmosphere and tendency to cloud formation there to no inconsiderable extent.

The following table shows the average number of hours of bright sunshine in each month at Blackpool during the five years 1901-1905 :—

Jan.	Feb.	Mar.	April.	May.	June.	July.	Aug.	Sept.	Oct.	Nov.	Dec.
43·1	63·4	111·9	178·4	218·6	225·2	211·4	197·4	150·5	82·2	60·9	37·3

Blackpool, though a very sunny place, is faifly representative of the coast line generally. At Lancaster the amount is only about three-quarters of that at Blackpool, and further inland amongst the hills the sunshine rapidly diminishes, the upper dale-districts of Bleasdale, Wyresdale, Roeburndale, etc., getting very much less than Lancaster. Under certain conditions of stormy and unsettled weather (belonging to what is known as Easterly type), when the centre of an area of low pressure exists to the southward, and strong east winds are blowing over the North of England, the following interesting phenomenon of cloud distribution has been noticed in the Garstang district by one of the authors. The sky to the east over Yorkshire is then usually quite overcast and low cloud-fog often covers the higher fells of Bleasdale and Wyresdale. Over the western slopes of the hills

---

[*] The breadth of the Pennine range from the hills near Masham on the east to Greygarth on the west is 35 miles, and from the neighbourhood of Dewsbury to the moors above Rochdale only *half* that distance. In the former district many of the fell summits rise to over 2,200ft., in the latter they only reach about 1,500ft. In addition to this, the drying effect of the northern part of the Pennine range is augmented when N.E. winds are blowing by the North Yorkshire moors between Thirsk and Whitby, whilst further south between Dewsbury and Bridlington these winds blow over a comparatively level country.

the clouds melt away, giving place to clear blue sky, which extends westward for some miles in a long, broad strip parallel to the fells ; but exactly at the point where the low plain is reached the clouds rapidly re-form and the whole western sky is thickly overcast. The clear blue strip is, of course, produced by the compression and consequent liberation of latent heat due to the descending current from the fells, as before mentioned, and the reformation of cloud is, no doubt, partly accounted for by the wind striking the level ground below and rebounding upwards, as in the case of the cloud produced by the Helm wind on the western slope of Cross Fell ; but it is also due to the general tendency to the production of cloud, during such conditions of weather, which reasserts itself as soon as the local influence of the Pennine range has been left behind and eliminated. This phenomenon of cloud absorption and reproduction, with easterly winds, along the base of the fells in West Lancashire, although only evident to the eye in certain states of the weather, is, no doubt, always more or less existent, and the fact that it is so is confirmed by rainfall measurements in the district. There is a marked increase of rain with easterly winds in the western part of the Fylde plain as compared with the eastern part, and we believe this is almost as great as the converse result with winds from the opposite quarter. This brings us to the consideration of rainfall generally throughout the vice-county. As will be seen from the following table of observations, it varies considerably in different parts. As would be expected, there is a marked increase towards the hills, the amount being moderate on the sea coast, but very large amongst the fells near the Yorkshire boundary.

RAINFALL IN WEST LANCASHIRE.

NORTH DIVISION.

		Annual fall. Inches.
Overton,	average of 4 years 1891-1894...	33·19
Hest Bank,	,, 4 ,, 1891-1894...	35·05
Heysham,	,, 6 ,, 1892-1897...	38·76
Hornby Castle,	,, 10 ,, 1891-1900...	43·36
Arkholme,	,, 10 ,, 1891-1900...	46·55
Silverdale,	,, 3 ,, 1898-1900...	46·55

WEST DIVISION.

Blackpool,	average of 7 years 1894-1900...	33·39
St. Michaels-on-Wyre,	,, 10 ,, 1891-1900...	36·25
Pilling,	,, 3 ,, 1898-1900...	36·47
Weeton,	,, 10 ,, 1891-1900...	36·75
Elswick,	,, 9 ,, 1891-1899...	37·13

EAST DIVISION.		Annual Fall. Inches.
Garstang (Bruna Hill),	average of 10 years 1891-1900...	37·53
,,	,, 35 ,, 1872-1906...	40·25
Haighton (near Fulwood),	,, 10 ,, 1891-1900...	37·63
Beacon Fell,	,, 10 ,, 1891-1900...	38·51
Lancaster (Marton Street),	,, 10 ,, 1891-1900...	39·45
Caton,	,, 10 ,, 1891-1900...	40·97
Wray,	,, 10 ,, 1891-1900...	44·69
Abbeystead,	,, 10 ,, 1891-1900...	46·20
Longridge Fell (Jeffrey Hill),	,, 10 ,, 1891-1900...	46·47
Stonyhurst,	,, 10 ,, 1891-1900...	47·95
Barnacre Reservoir (near Garstang),	,, 10 ,, 1891-1900...	49·81
Fairsnape Clough,	,, 3 ,, 1898-1900...	62·87
Wardstone,	,, 7 ,, 1894-1900...	64·69

In most if not in all parts of the vice-county August is the average wettest month and April the driest. In many of the fell-districts no observations have yet been made, but it is probable, judging from the rainfall of stations just over the Yorkshire boundary, as Langden, Whitendale and Kingsdale, that the annual fall will equal, if not exceed, an average of 70 inches. A rain gauge placed by one of the authors on the summit plateau of Fairsnape Fell, at a height of 1,630ft., gave in 1906 a rainfall of 87·02 inches, distributed as follows :—

January ... 10·21	April ... 3·30	July ... 4·92	October ... 10·68
February .. 9·62	May ... 8·93	August ... 6·35	November . 6·47
March ... 7·42	June ... 3·86	September. 2·83	December . 5·43

The year 1906 had about an average rainfall in the North of England generally, but the fall was probably somewhat heavier than usual in the locality named.

The humidity in the fell-districts is also great and conditions generally favourable to the growth of moisture-loving mosses and hepatics which are correspondingly abundant and luxuriant. Of the relative humidity there we have no observations, but the following averages at 9 a.m. during the years 1901-1905 at Blackpool, may be quoted :—

	Jan.	Feb.	Mar.	April.	May.	June.	July.	Aug.	Sept.	Oct.	Nov.	Dec.
Per cent.	90	89	85	78	77	75	76	80	81	84	88	90

In the greater part of the North and East Divisions the humidity is, we believe, considerably greater than the above figures would indicate. The dryest month of the year generally is June and the dampest probably December. We now come to the last of the five elements which determine climate, viz., Wind. As elsewhere on our western coasts, westerly winds are the most prevalent and blow at times with great force. They have an injurious effect on trees, and prevent them from growing to a large size except in the sheltered dale-districts and amongst the hills of

F

the North Division. Some parts of the Fylde plain and the western slopes of the hills rising from it on the east are especially wind-swept, and the trees on the coast are much blown out of shape and distorted in an easterly direction. Sometimes the sea-spray is carried far inland. The day after the great gale from the west and north-west of December 22nd, 1894, the branches of trees and leaves of evergreen shrubs in the Garstang district, seven miles from the sea, were seen to be coated with salt. Next day, soon after the commencement of a drizzly rain, one of the authors collected into a cup the drops hanging from tree-twigs, until a few ounces of liquid were obtained. This, when evaporated to dryness, yielded sea-salt equal to 6,080 grains, or nearly 14oz., to the gallon.* On this occasion the amount of salt deposited over the country must have been enormous, and much damage was done to evergreen trees and shrubs. For example, the branches of Yew trees exposed to the west and north-west about Garstang were killed. During the storm the wind velocity, as shown by the anemometer at Fleetwood, equalled 107 miles per hour, and for four consecutive hours it was over 100 miles per hour. This was, of course, very exceptional, but salt-spray is often carried inland, and it, combined with the force of the wind, has a very adverse effect on the growth of trees, especially in spring when the leaves are young and delicate. Another wind-carried agent, deleterious to plant life, is the smoke from South and East Lancashire and the West Riding of Yorkshire. With south-east winds this commonly blurs the whole atmosphere, rendering the distant landscape grey and hazy, and on some days has a marked effect in lowering the temperature by partially obscuring the sun's rays, as shown by careful observations made near Garstang. A notable point in connection with the flora of West Lancashire is the singular rarity of bark-loving mosses belonging to the genus *Ulota*. These plants, growing on the trunks and branches of trees, are peculiarly exposed to the evil effects of smoke, and are apparently absent except in the north-east and north of the vice-county furthest from the smoke-producing districts. The same applies to lichens which are very scarce and poorly developed in the southern part of the area.

From what we have said it will be evident that the climate of West Lancashire varies greatly in different parts, and the effect of this in modifying vegetation is well marked. Considered broadly there are three distinct climates prevailing, the limits of which correspond very closely with those of our three main divisions North, West and East. In the North Division which, as we have

* See *Meteorological Magazine*, Jan., 1895.

seen, has a somewhat heavy rainfall, sunshine is rather deficient, but the country generally is more sheltered than either in West or East, and less exposed to strong winds from the Irish Sea. The average summer temperature is also rather higher, and it is here that we have some of the richest country in the vice-county. This is especially the case amongst the limestone hills and in the Lune valley. The greater portion of the arable land is well suited to the growth of oats, but owing to the heavy annual rainfall and usually wet Augusts, the climate most years is not so satisfactory for wheat culture. The part where most of the latter crop is grown is in the low-lying Heysham peninsula.

The climate of the West Division is marked by a smaller rainfall and more bright sunshine than either of the other divisions. Wheat generally forms a satisfactory crop and a good deal is grown, especially in the western districts. In the East Division, as we have before pointed out, the rainfall is heavy, the cloud excessive and the humidity great. Very little corn is grown except oats, and that only to advantage in the more favoured parts. The limit of cultivation, especially of arable land, is low as compared with that in North and West Yorkshire *east* of the Pennine range of hills, and there is very little ploughed land in West Lancashire at a higher altitude than 600ft. The reason for this is the generally excessive rainfall and humidity and want of sunshine during the summer and early autumn.

During the prevalence of westerly winds at these periods the western slopes of the fells are often bathed in cloud fog, whilst the summits further east, in Yorkshire, are clear and even in sunshine. This is, however, less marked at other times of the year, and in spring, with easterly winds, as before stated, the converse is the case. But at that season grain crops are only in a vegetative condition, and the effect is of less value than would appear, as it is during the flowering and seed maturing periods that sunshine is most needed. Different years, however, as in other parts of England, vary greatly, and it may be worth putting on record here that during the exceptionally fine and warm summer of 1893 the Spanish Chestnut (*Castanea sativa*) ripened its fruit in the neighbourhood of Garstang, just within the borders of the East Division, and young trees were afterwards raised by one of the authors from seed grown therefrom. But Garstang, it must be remembered, is in the low country and the climate there is much drier and more sunny than in the dales.

There are other points that might be mentioned with regard to Meteorology and its effect on vegetation, but our space is limited and we must now conclude this sketch.

VERTICAL AND HORIZONTAL RANGE OF THE SPECIES.

Before considering the distribution of the plants within our boundaries, the relation which the Flora of West Lancashire bears to that of the rest of Great Britain may be briefly indicated. The central position of the vice-county from a phyto-geographical point of view, its diversity of surface, both geological and altitudinal, and its extensive seaboard, together give its Flora peculiar interest, and the large number of species relative to its area is, no doubt, due to these diverse conditions. Owing to the first of them, its intermediate position, we find nearly all the Watsonian types of distribution represented, only plants of the Arctic type being absent. These types, as defined by Watson, are in general use in British Floras for indicating the geographical affinities of the species, and it is, therefore, unnecessary here to occupy space with definitions of them, which may be found in Watson's Topographical Botany. In our vice-county, as might have been anticipated, plants of the Germanic or most easterly type are but few, North Yorkshire having four times and East Yorkshire three times as many as are found in West Lancashire. With regard to the Atlantic type, the converse is found to be the case, West Lancashire having more than twice the number found in North Yorkshire and four times as many as in East Yorkshire. Owing to its situation near the extreme limits of the range of these two types, they form but a small proportion of the Flora of the district. The plants of the Intermediate and Scottish types are fairly well represented, but those of the Highland type are found in very limited numbers, as our hills are not lofty or craggy enough to afford suitable habitats for many of them. Plants of the English type, which presumably reached West Lancashire from a south-easterly direction, are in smaller proportion to those of the British type than is found to be the case in those counties situated in about the same parallel east of the Pennine Chain. While in North and East Yorkshire the ratio is approximately as 1 : 2, in West Lancashire the English element declines until it is about 1 : 3. It would seem reasonable to presume that the extension of the English and Germanic types in a westerly direction may have been retarded by this range of hills, but the chief cause is probably to be found in the character of the West Lancashire climate, the autumn of which is generally

cool and damp and unsuitable for the ripening of certain seeds. The following table shows how many species of the various Watsonian types fall under each of the three higher grades of citizenship, and also how the totals compare with those of the more eastern Yorkshire Floras :—

WATSONIAN TYPES of distribution.	Natives.	Colonists	Denizens	Total for West Lancashire.	Total for North Yorkshire.	Total for East Yorkshire.
Highland... ...	12	—	—	12	32	—
Scottish	37	—	1	38	47	19
British	526	29	5	560	527	531
Germanic ...	8	1	—	9	40	26
English	187	14	17	218	307	286
Atlantic	15	—	1	16	7	4
Intermediate ...	9	—	6	15	34	9
Local	3	—	1	4	11	3
Total ...	797	44	31	872	1005	878

Of the 797 species which may be regarded as probably genuine native plants, the North district yields 696, the West 609, and the East 591. The last named owes its inferiority partly to the absence of sea-coast, partly to its having no extensive areas of limestone, and partly to the large extent of heathland and upland pasture, yielding plant associations consisting of vast numbers of gregarious individuals, but of very few species. The relative richness of this small vice-county may be gathered from the following data. Although it has been probably far from completely investigated, and its total area amounts to only 492 square miles, or about 0·55 per cent. of that of Great Britain, nearly half the species recorded for the larger area have been found within its borders! The ninth edition of the London Catalogue of British plants enumerates 1,930 species of seed plants and higher cryptogams (excluding the *Characeæ*), and of these 958 or about 49 per cent. are recorded for our district. The British Mosses, including the *Sphagna* may be estimated at 650 species, and the Liverworts at 250 species ; and of these 330 and 94 respectively,

or about 47 per cent., occur within our boundaries. But the figures given for the Phanerogams include many species which have little or no claim to be regarded as natives. Eliminating all those either starred or printed in italics in the catalogue cited, as being clearly not indigenous in Britain, and also deducting a few plants found only in the Channel Islands, which belong to the Continental rather than to the British Flora, there remain 1,695 species which are presumably aboriginal inhabitants of Great Britain, or have become so thoroughly naturalised as to be hardly distinguishable from true natives. Restricting our species in a similar way, and accepting the London Catalogue standard of specific rank, the figures will be found to still retain about the same proportion, our list showing 872 species, or about 50 per cent. of the British Flora. In the following table the species of the higher grades of citizenship will be found compared with some other Floras of the Humber and Mersey Provinces, it being understood that the figures are of necessity not mathematically precise, owing to divergent views as to the status of some of the species :—

Phanerogams and Higher Cryptogams.	W. Lancs. area, 492 sq. miles.	Cheshire. area, 1026 sq. miles.	W. Yorks. area, 2760 sq. miles.	N. Yorks area, 2112 sq. miles.	E. Yorks, area, 1175 sq. miles.
Natives	797	872	877	884	764
Colonists	44	30	58	84	} 114
Denizens	31	20	78	37	
Total	872	922	1013	1005	878

Of a lower grade of citizenship than those already considered are the plants of casual occurrence and accidental introduction, ballast plants, and garden escapes. These, which number considerably over one hundred, are classed as aliens. Some are better established than others ; in fact, a few of them appear likely to attain sufficient permanency to entitle them to rank as Colonists or Denizens. In such cases it is often very difficult to assign them to the correct category. By paying due regard to their ascertained range elsewhere in Britain, and to the circumstances accompanying their appearance in each locality, we hope we have classed them as accurately as possible, but in such a matter there is great room for diversity of opinion.

Various species have been recorded from time to time on more or less good authority, which it seems desirable at present to exclude. It is possible that further research may result in the restoration of some of them to the West Lancashire list, and the Authors will be glad of such information as may lead to the removal of any of them from the class "Incognita" to which they are at present relegated. Some of them have crept into the list through errors of locality, arising from the similarity of names of places in our district to those of other counties, or to carelessness in labelling, or to the accidental transference of specimens from another county to sheets of West Lancashire plants in the herbarium. Some of the following species are certainly, and others probably erroneously, recorded through one or another of these accidents, viz. :—*Anemone Pulsatilla, Minuartia lepto-phylla, Cochlearia grænlandica,* and *Diplotaxis tenuifolia.* Other errors have arisen through closely allied plants being mistaken for each other, either because they were not well understood by British botanists at the time the records were made, or because the specimens were named from imperfect material, or because of the inexperience of the person making the record. Amongst such may be reckoned the following :—*Viola lutea, Trifolium suffocatum, Geranium rotundifolium, Rubus Salteri, Potentilla fruticosa, Cotoneaster integerrima, Bryonia dioica, Senecio palustris, Carduus nutans, C. eriophorus, Hieracium diaphanum,* and *Hordeum maritimum.* Still other records are now regarded with a certain amount of doubt because, however well authenticated originally, specimens have not been seen in recent years, and if they ever occurred they are now apparently extinct. It is amongst these especially that one may look to the ultimate restoration of some species to our list of plants. It is probable that some of them, however, were originally little more than casuals of the grain-refuse or garden-outcast types. The following are the species which we regard as probably extinct :—*Crambe maritima, Ranunculus sardous, R. Lingua, Hypericum elodes, Drosera longifolia, Isatis tinctoria, Smyrnium Olusatrum, Epilobium roseum, Bupleurum tenuissimum, Dipsacus sylvestris, Veronica hybrida, Utricularia minor, Allium Scorodoprasum, Tulipa sylvestris, Alopecurus myosuroides.*

This list of probable extinctions is, perhaps, too small, but we hesitate to add several species to it which may possibly still linger in localities which require further exploration, but for which there are no recent records.

(A.)—PLANT DISTRIBUTION AS AFFECTED BY ALTITUDE.

The surface of Great Britain was divided by H. C. Watson into six zones of temperature, three Arctic and three Agrarian. Of these, three only are represented in West Lancashire, viz., the Infer-Arctic, Super-Agrarian, and Mid-Agrarian. Our highest summit, Greygarth Fell (2080ft.) comes within the Infer-Arctic zone, which includes in the North of England any hill country between 2,700 and 1,800 feet in altitude, having a mean annual temperature of from about 39° to 42°. The Super-Agrarian zone includes the country between 1.800 and 900ft., with a temperature of from 42° to 45°, whilst the Mid-Agrarian zone consists of the remainder of the land down to sea level, with a temperature of from about 45° to 48°. The Infer-Agrarian zone is not represented in West Lancashire, as the most southerly part of the vice-county is in about lat. 53°, 45', whilst the most northerly limit of this zone is at *sea level* in lat. 53°, 30'.

These zones are well understood by students of plant distribution in Great Britain, and the relative extent of their areas has a marked effect in determining the character of the vegetation of the various British counties. Therefore, in spite of the fact that the boreal limits of plants do not depend entirely on temperature, it is of utility to retain them for the purpose of supplying a ready means of comparing one local Flora with another.

The study of the effects of altitude on the distribution of our species is attended by several drawbacks. We have practically but one plain and one range of hills of any size, and these latter are not greatly diversified in aspect or geological formation. Only a very small portion of our area enters the Infer-Arctic zone, *i e ,* the wind-swept and exposed summits of Greygarth Fell and Wardstone, and of the first-named we can claim only one side, the eastern slope being in Yorkshire. Consequently there is a greater dearth of Alpine species than is found to be the case at similar elevations in districts where the hills rise higher and are more craggy, in which case the zone in question is not only reinforced by descending species from above, but, owing to greater protection being afforded, more species are enabled to maintain a footing, and the ascending species also attain a greater altitude than with us.

The summit of Wardstone, the highest of our East group of Fells, is thickly capped with weathered peat, with little or no outcrop of rock except on the N.W. side, and it therefore affords little

shelter for plants. It appears to have been thickly wooded in former times, but with a change of climatic conditions the woods disappeared and were succeeded by peat-bogs. Remains of trees still abound, especially on the S E. side, where they are to be seen on the sides of gullies, deeply buried in the peat. In comparatively recent times the plateau no doubt presented a wide stretch of cotton-grass bog or moor, such as clothes many of our lower summits to-day. But the peat has been worn away by the elements into deep gullies, and the weathering process is now so rapid that considerable areas are devoid of vegetation, the surface consisting of bare peat. The deep trenches cause a certain amount of drainage to the portions that remain—islands or flat tables of peat—which have withstood the action of wind and rain, being protected by the roots of Bilberry, Heather, and Cotton-grass. These, with other subordinate species, form a somewhat transitional flora, an assemblage of peat-loving plants which cannot be classed under any of the recognised moorland associations. They consist, in fact, of isolated patches of nearly all of them, varying from yard to yard according to the exposure and amount of moisture. The summit of Fairsnape Fell (see illustration) and others of our eastern group of hills present similar features.

The following is a complete list of the plants observed within the Infer-Arctic zone on the summit of Wardstone. It will be noticed that the Mosses with Hepatics outnumber the Flowering plants and Ferns by about two to one :—

Galium saxatile	Rhacomitrium lanuginosum
Calluna vulgaris	Dicranella heteromalla
Erica Tetralix	Dicranum fuscescens
Vaccinium Myrtillus	,, majus
,, Vitis-idæa	,, scoparium
,, Oxycoccos	Dicranum scoparium, var. turfosum
Empetrum nigrum	Campylopus flexuosus
Eriophorum vaginatum	Plagiothecium elegans
Festuca ovina	,, undulatum
Agrostis tenuis	Hypnum Schreberi
Aira flexuosa	,, cupressiforme, var.
Juncus squarrosus	ericetorum
Scirpus cæspitosus	Hylocomium loreum
Lastrea aristata	Kantia trichomanis
	Lepidozia setacea
Sphagnum papillosum	,, reptans
,, submitens	,, Pearsoni
Tetraphis pellucida	Diplophyllum albicans
Polytrichum commune	Scapania resupinata, var. minor
,, piliferum	Mylia Taylori
,, strictum	Jungermannia Floerkii
Andreæa petrophila	,, ventricosa
,, Rothii	
Rhacomitrium fasciculare	Sphærophoron fragile

Platysma glaucum	Parmelia omphalodes
Cetraria aculeata	Gymphora flocculosa
Cladonia cervicornis	,, torrefacta
,, macilenta f. carcata	Lecidea contigua
Parmelia saxatilis	,, rivularis

The Infer-Arctic zone of Greygarth Fell is rather more extensive and more diversified in its lithological character. A complete list of the Florula of this summit will be found on page 8, of which the following is a summary :—

Infer-Arctic Florula of Greygarth Fell.			
Phanerogams 61	Sphagnaceæ 9
Vascular Cryptogams ... 3	Hepaticæ 20
Musci 70*	Lichens, about ... 50

* The list of Mosses on p. 8 contains 68 species, but since this was printed 2 others have been discovered, viz. : *Dicranella cerviculata* and *Hypnum falcatum.* The Hepatic, *Cephalozia fluitans,* has also been added.

The only species limited to this zone in West Lancashire are *Carex rigida, Cephalozia heterostipa, Cetraria islandica,* and *Cladonia rangiferina.*

Alpine and arctic conditions favour the production of perennial plants. In the hot coast region of Granada, Boissier found the annuals outnumbered the perennials. In Dauphinié, Bonnier and Flahault compared the proportions in three different zones of altitude. The results arrived at are shown in the annexed table, along with the figures we have obtained from the West Lancashire species. Judged by this standard, our general Flora comes out as slightly more boreal than that of the Pine Forest region of Dauphinié.

Percentage of	Perennials.	Biennials.	Annuals.
Granada	45	4	51
Dauphinié. Lower region	40	—	60
,, Pine Forest region ...	67	—	33
,, Upper Alpine region ...	94	—	6
West Lancashire...	75	5	20
,, ,, Infer-Arctic zone ...	85	11	4

and our Infer-Arctic zone as somewhat less so than the Upper Alpine Region.

In addition to affecting their duration in the manner described, alpine conditions tend also to a reduction in the size of plants. The Infer-Arctic Flora is, generally speaking, a dwarf one, trees being altogether, and many of the plants are of much smaller stature than they attain in the lower zones. For example, *Epilobium palustre,* the Marsh Willow Herb, occurs in a diminutive form with stems sometimes little more than an inch high, with its leaf area proportionately reduced, and frequently bearing but one flower. It then presents a not very distant resemblance to the more arctic *Epilobium alpinum. Lastrea aristata* and other plants are also found in very dwarf forms. The absence of other and more marked special adaptations to prevent the plant suffering from wind and sun exposure is not particularly remarkable when we remember that our Infer-Arctic summits are so extensively covered by thick beds of water-retaining peat and so abundantly supplied with moisture, and are also, owing to their geographical position, frequently swathed in cloud-fog.

In dealing with the altitudinal range of the species, we propose dividing them into two main groups, one of species having their headquarters amongst the hills, and principally of northern distribution ; the other chiefly occupying the lower country and usually less boreal in their horizontal range. The former will be termed *Descending* and the latter *Ascending* species. Although these terms express well the *general* tendency of the groups, it must be understood that there are some sub-groups whose range is not quite accurately denoted by them.

A critical survey of our Montane or Descending species at once discloses how materially they differ amongst themselves as to their vertical range. Some are limited to mountain summits, others descend to sea-level, and still others occupy an intermediate position, avoiding the two extremes. In the majority of cases, no doubt, the dominant factor affecting the altitudinal range of a species is temperature, for each plant has not only its specific zero, below which the seeds will not germinate, but each also varies in its power of resisting cold and ripening fruit. In addition, there are active modifying influences at work in wind, sunshine, soil, moisture, presence or absence of insects, and the competition of other plants.

Exposure to wind has a most decided effect, and many plants which would be cut down on fully exposed slopes reach considerable altitudes in cloughs and pot-holes and when sheltered by broken ground or rocks. An example may be cited in *Adoxa Moschatellina*, which in West Lancashire ascends to only 1,050ft., but in Yorkshire we have seen it in sheltered hollows of the rocks at 2,200ft., and in Perthshire, notwithstanding the higher latitude, it reaches 3,500ft.

As exemplifying the effect of soil on altitudinal range, the following facts are worthy of consideration :—In North Yorkshire, according to Mr. Baker, *Galium umbellatum* does not descend below 600ft., *Polypodium Robertianum* below 1,200ft., *Sesleria cærulea* below 600ft., and *Minuartia verna* below 450ft. These four species descend much lower, some of them to sea-level, in West Lancashire, where the scar limestone, to which they are so partial, outcrops at a lower elevation than in North Yorkshire. It is obvious that the position of the limestone rocks, and not simply temperature, largely determines the altitude of these species in the North of England.

In the Agrarian Zone the operations of man—drainage, cultivation, the smoke of manufacturing towns, etc.—materially interfere with Nature, and render it difficult to make deductions without falling into error. Viewing our species in their Lancashire habitats only, some of them might be placed in the Montane category quite erroneously, because at the present day they survive in the hilly districts only, having been exterminated at lower altitudes by the effects of human industry. Many plants, formerly plentiful in bogs or heathery tracts in the plains, are now either extinct there or rapidly becoming so. They still thrive amongst the hills, not because a lofty habitat is essential, but because there only the boggy and peaty soil they require is left unreclaimed. Keeping these and other facts in view, we find our Montane species fall naturally into the following four groups :—

1. Eu-Montane Species.
2. General Montane Species.
3. Sub-Montane Species.
4. Pseudo-Montane Species.

A brief definition of these groups follows, with an illustrative list of the West Lancashire species falling under each.

1. The Eu-Montane group consists of those plants which have their head-quarters in the Arctic regions, represented in England

by mountain summits ranging between 3,200 and 2,000ft. Many of the species of this group ascend still higher in Scotland, where the mountains are more lofty and afford a large amount of shelter. As a rule they do not descend much below the lower limit of the Inter-Arctic Zone, and they never establish themselves, even accidentally, in the low country. The group is poorly represented in West Lancashire, containing only the following :—

Rubus Chamæmorus	Dicranum scoparium, var.
Carex rigida	spadiceum
	Hypnum sarmentosum

Our list of two phanerogams appears insignificant beside the nineteen of Westmorland and the seven each credited to North and West Yorkshire.

2. General Montane Group.—This embraces plants having their headquarters, *i.e.*, attaining their average maximum abundance and luxuriance, between 1,000 and 2,000ft. Horizontally they are distributed principally in Scotland and the North of England, becoming rarer or disappearing altogether further south. They are represented in West Lancashire by the following species :

Draba incana	Diphyscium foliosum
Cochlearia alpina	Grimmia Doniana
Saxifraga hypnoides	Swartzia montana
Juniperus communis	Blindia acuta
Vaccinium Vitis-idæa	Campylopus atrovirens
,, Myrillus	Dicranodontium longirostre
Empetrum nigrum	Dicranum fuscescens
Listera cordata	,, scoparium, v. turfosum
Trientalis europæa	Rhacomitrium heterostichum vars.
Sesleria cœrulea	Bartramia Œderi
Cryptogramme crispa	Webera bicolor
Asplenium viride	Splachnum sphæricum
Lycopodium alpinum	Tetraplodon mnioides
,, Selago	Plagiobryum Zierii
Andreæa Rothii	Bryum alpinum
,, petrophila	Cinclidium stygium
,, crassinervia	Mylia Taylori
Oligotrichum incurvum	Scapania resupinata
Polytrichum alpinum	

Sesleria is placed doubtfully in this group. It has a very peculiar distribution. It is remarkable that it is absent from the scar-limestone of Wales, Cheddar, etc., and even from Derbyshire. According to Hooker it ascends to 2,500ft. in the Highlands of Scotland, and to nearly that height on Mickle Fell in North Yorkshire. It is exceedingly abundant on the West Yorkshire limestone, generally at about 1,400ft. In West Lancashire it grows luxuriantly at or about sea-level near Silverdale. On the whole it appears to be most at home with the General Montane plants, amongst which other species occasionally appear quite

abundantly at low altitudes in exceptional situations, as in the case of *Helianthemum marifolium* on Humphrey Head, *Sedum roseum* on maritime rocks in Scotland, species of *Andreæa* about sea-level near Barmouth, and *Catoscopium nigritum* on the coast at Formby.

3. The Sub-Montane group consists of a larger number of species than either of those already described. They flourish best about the moorland cloughs, and in the higher upland pastures, with us being most abundant between 500 and 1,000ft. They thin out in both an upward and a downward direction, rarely ascend into the Infer-Arctic zone, and only casually occur in the plains. Many of the general ascending species, although *less restricted* to the hill country, ascend far higher than do some members of the present group.

The following West Lancashire species come under the Sub-Montane category :—

Actæa spicata	Euphrasia borealis
Ranunculus Lenormandi	Rumex aquaticus
Trollius europæus	Salix nigricans
Stellaria nemorum	,, phylicifolia
Geranium sylvaticum	Luzula sylvatica
Prunus Padus	Carex binervis
Epilobium angustifolium (as native)	,, Hornschuchiana
	Hymenophyllum peltatum
Rubus Scheutzii	Lastrea montana
Pyrus Aucuparia (as native)	Polypodium Robertianum
Saxifraga umbrosa	,, Dryopteris
Myrrhis odorata	,, Phegopteris
Catharinea crispa	Cystopteris fragilis
Polytrichum urnigerum	Lycopodium clavatum
Dicranum majus	Weisia rupestris
Dicranodontium longirostre	Ulota Bruchii
Brachyodus trichodes	Zygodon Mougeottii
Fissidens osmundoides	Discelium nudum
Rhacomitrium aciculare	Mnium subglobosum
Barbula spadicea	Breutelia arcuata
Carduus heterophyllus	Heterocladium heteropterum
Crepis paludosa	Orthothecium intricatum
Hieracium duriceps	Hyocomium flagellare
,, sylvaticum	Hypnum ochraceum
,, gothicum	,, exannulatum
Wahlenbergia hederacea	Hylocomium loreum
Erica cinerea	,, brevirostre
Primula farinosa	Lejeunia calcarea
Myosotis repens	Saccogyna viticulosa
,, sylvatica	Lepidozia Pearsoni

A group of plants having both a montane and a littoral rôle of distribution occurs in Yorkshire and elsewhere, comprising such plants as *Silene maritima, Plantago maritima, Armeria maritima,* etc. The mosses *Amblyodon dealbatus* and *Catoscopium nigritum* should perhaps be grouped here. Of these only the

Amblyodon occurs in both montane and littoral stations in West Lancashire. The phanerogams are probably the descendants of halophytes which fringed the mountain peaks when they formed a scattered archipelago of islets in the pre-historic sea. Originally this islet flora would be more numerous, and it is now probably only represented by a few tolerant and hardy species able to survive the various changes which have supervened. Their absence on the hill summits of West Lancashire may be due to their extinction through competition of the swamp heatherland vegetation ; or possibly to the fact that when Ingleborough and Whernside were islands, our lower summits were submerged and unable to accommodate a land flora.

4. The Pseudo-Montane group forms an intermediate link between the ascending and descending species, and might with equal propriety be classed with the former. It consists of plants which not only flourish and fruit freely in the plains under suitable conditions, but which thrive equally well on the hills, sometimes growing luxuriantly as high as the middle limits of group 2. They are really general ascending species which are extremely intolerant of drainage and cultivation; their requirements being peat bogs or barren heathy ground. In consequence of the gradual disappearance of the conditions essential to them, they are diminishing species in the plains, both in West Lancashire and many other districts in England. When, as not unfrequently happens, they nearly or quite disappear from the lower levels through the effects of cultivation, they still continue to thrive amongst the hill moorlands, and then are hardly distinguishable from the preceding groups. The following are West Lancashire examples :—

Viola palustris	Eriophorum vaginatum
Drosera rotundifolia	,, polystachion
,, anglica	Carex canescens
Rubus Idæus	Molinia cœrulea
,, fissus	Sphagnum—many species
Montia fontana	Polytrichum commune
Chrysosplenium oppositifolium	,, strictum
Parnassia palustris	Campylopus flexuosus
Erica Tetralix	Leucobryum glaucum
Calluna vulgaris	Fissidens adiantoides
Melampyrum pratense	Aulacomnium palustre
Vaccinium Oxycoccos	Hypnum stellatum
Narthecium Ossifragum	,, fluitnas
Rynchospora alba	,, revolvens
Juncus squarrosus	,, scorpioides
Scirpus cæspitosus	

The remainder of the flora consists of ascending species, which may be divided into two well-marked groups, viz., a

General Ascending group, embracing the bulk of our flora, and a Sub-Ascending group, containing those plants which are more rigidly confined to the low country and have also frequently a more southern horizontal range. With us the members of this sub-group can hardly be said to ascend at all, many of them only occurring at or very near sea-level. Excluding the strictly maritime plants, which number about 50, the following is a list of the species which in West Lancashire do not ascend above 150ft., and in consequence may be placed in the Sub-Ascending group. It will be observed how few of them (about one-third, marked by an asterisk) are of the more northern types of distribution in Great Britain.

Thalictrum flavum	Lycopsis arvensis
Ranunculus fluitans	Cuscuta Epithymum
,, divaricatus	Solanum nigrum
• ,, heterophyllus	Hyoscyamus niger
,, Baudotii	Linaria repens
,, Lingua	• ,, vulgaris
• ,, sceleratus	•Euphrasia curta
•Castalia alba	•Lasiopera viscosa
Nymphæa lutea	•Orobanche minor
Fumaria purpurea	Scrophularia aquatica
Sisymbrium Sophia	Nepeta Cataria
Sinapis nigra	•Lamium amplexicaule
Diplotaxis muralis	• ,, mollucellifolium
•Viola canina	• ,, hybridum
Saponaria officinalis	•Littorella uniflora
•Cerastium tetrandrum	Chenopodium urbicum
Malva rotundifolia	,, murale
•Erodium cicutarium	Rumex Hydrolapathum
Ononis spinosa	Ceratophyllum demersum
Trifolium arvense	Elodea canadensis
•Ornithopus perpusillus	Hydrocharis Morsus-ranæ
Rubus plicatus	Juncus obtusiflorus
,, rhamnifolius	Sparganium neglectum
,, calvatus	Acorus Calamus
,, incurvatus	Lemna trisulca
•Hippuris vulgaris	,, polyrhiza
•Myriophyllum alterniflorum	Butomus umbellatus
,, verticillatum	•Potamogeton perfoliatus
Callitriche autumnalis	• ,, crispus
,, polymorpha	• ,, densus
•Peplis Portula	Zannichellia palustris
Apium graveolens	•Eleocharis uniglumis
Œnanthe fistulosa	Scirpus lacustris
Smyrnium Olusatrum	,, Tabernæmontani
Pastinaca sativa	Blysmus compressus
Bidens cernua	•Schœnus nigricans
,, tripartita	•Cladium Mariscus
Helminthia echioides	•Carex vulpina
Lactuca virosa	• ,, acuta
Pyrola rotundifolia	• ,, Œderi
Monotropa Hypopitys	Calamagrostis epigejos
Hottonia palustris	Aira caryophyllea
Samolus Valerandi	Sclerochloa loliacea
Blackstonia perfoliata	Festuca fasciculata
•Gentiana baltica	Hordeum murinum
Symphytum officinale	Equisetum hyemale

West Yorkshire, with its much greater area, presents suitable localities for some of the species named above at greater altitudes than is the case with our district. The following sub-ascending species of West Yorkshire are not found in West Lancashire :—*Barbarea stricta, Viola lactea, Ulex minor, Lathyrus palustris, Sium latifolium, Peucedanum palustre, Carduus pratensis, Campanula patula, Veronica triphyllos, Polygonum minus, Carex paradoxa, Sclerochloa distans, Lastrea cristata, L. Thelypteris, Pilularia globulifera.*

B.—PLANT DISTRIBUTION AS AFFECTED BY THE CHARACTER OF ROCKS AND SOILS.

In considering the lithological range of the components of the Flora, two prominent factors require attention—viz., the physical and the chemical characters of the soil. Viewing the species first as affected by the constitution and physical condition of the subjacent rocks, two principal groups become prominent, not always sharply separated, it is true, but sufficiently marked to give a decided character to the flora of a district according to the predominance of one or the other. These groups are termed the Xerophiles (dry-loving) and the Hygrophiles (damp-loving), the former preferring hard, non-absorbent, little detritus-bearing formations, and the latter those which are more porous and absorbent and more readily disintegrated. These two classes of rock are termed Dysgeogenous and Eugeogenous by Thurmann, and the characteristic features of their vegetation have been well described in Baker's "North Yorkshire," and with very slight alteration in the lists of species named, his account could be made to apply to similar tracts of country in West Lancashire. It has been recently advanced, however, that the separation of these two groups of plants depends less on the physical characteristics of the rocks and their comparative humidity than on their chemical composition ; and further, that many plants hitherto regarded as Hygrophiles in reality exhibit the characteristic structure of, and really are, Xerophiles. This will be referred to more fully in succeeding pages ; for the present the broad lines of Thurmann's idea will be adopted, without, however, attempting to describe all the minor sub-divisions of his groups. It is necessary to note that the term " Xerophile " is employed in its strictly literal meaning— *i.e.*, a plant loving a dry habitat and practically restricted to the drier rocks in the vice-county. A Xerophyte, on the contrary,

G

denotes a plant having xerophytic structure, its organs being modified to prevent rapid transpiration, which effect may be produced by other causes than growth in arid situations. Indeed, owing to the peculiar phenomenon of physiological drought (see page 91), many plants exhibiting xerophytic modifications actually grow in marshes and swamps.

The physical properties of the soil which principally affect the distribution of plants are its water capacity, its power of capillary conduction, and its permeability. The water capacity is greatest in those soils consisting largely of peat or clay, and least in those of limestone. The hard limestone rocks of the Silverdale district differ greatly in power of absorption from the blown sands of the dunes extending from Lytham to Blackpool, but both are remarkably permeable, so that they soon dry, and they each support a markedly xerophilous type of vegetation. The permeability of the limestone is due to drainage on a large scale from its numerous cracks and fissures ; the rock itself is, perhaps, the least permeable or absorbent of all our formations. The valley of the Ribble and a great part of District 5 are overlaid by glacial drift, which contains a good deal of clay, and is consequently damp and cold.

Owing to its humid climate and to the greater part of its solid geology consisting of distinctly eugeogenous strata, West Lancashire has a flora markedly hygrophilous in character, the xerophilous species forming only about 12½ per cent. of the whole. The members of the latter group become predominant in the rocky woods, crags, and fissured pavements of the Scar Limestone tracts of our North Division, and are much rarer, and the majority of them absent, from the Glacial Drifts, and from the peaty and sandstone areas of the East and West. In the North the hard limestone lies near the surface, and it not only retains little moisture, but it is also so impervious that it prevents the rising of sub-soil water by capillarity, thus causing at times excessive drought, which the natural humidity of our climate fails to entirely correct. The climate of West Lancashire is moister than that of North Yorkshire, with the result that many species which are only partially xerophilous in the latter county are strictly confined to dysgeogenous habitats in the former. This is quite in accordance with the axiom of phyto-geography laid down by Mr. Baker in "North Yorkshire," that " in proportion as the climate is damper, the characteristically dry-loving species are more and more rigidly restricted to the dry-soiled tracts of country." In the list of West

Lancashire Xerophiles given below, all of which exhibit a very marked preference for calcareous situations, those which are only sub-xerophilous in North Yorkshire are marked with an asterisk, and those which Mr. Baker does not indicate as having any xerophilous restriction whatever, with the obelisk sign. The number which we find it necessary to so mark affords a striking confirmation of the truth of the generalization quoted above :—

Thalictrum collinum	Scabiosa columbaria
Helleborus viridis	†Antennaria dioica
Aquilegia vulgaris (native)	Inula vulgaris
Actæa spicata	Taraxacum erythrospermum
*Arabis hirsuta	*Ligustrum vulgare
Cardamine impatiens	*Atropa Belladonna
Draba incana	†Verbascum Thapsus
,, muralis	†Clinopodium vulgare
,, præcox	• ,, Acinos
Sisymbrium Thalianum	Nepeta Cataria
Helianthemum Chamæcistus	Salvia verbenaca
*Viola hirta	*Lithospermum officinale
*Minuartia verna	Taxus baccata
Hypericum montanum	†Spiranthes autumnalis
†Tilia cordata	Epipactis atrorubens
Geranium sanguineum	Ophrys muscifera
† ,, columbinum	†Orchis ustulata
†Euonymus europæus	,, pyramidalis
Rhamnus catharticus	Polygonatum officinale
*Anthyllis Vulneraria	Convallaria majalis
Hippocrepis comosa	Carex digitata
Spiræa Filipendula	Sesleria cerulea
*Poterium Sanguisorba	†Koehleria cristata
Pyrus rupicola	*Melica nutans
Sedum purpureum	Polypodium Robertianum
*Galium umbellatum	Lastrea rigida
Asperula cynanchica	Ceterach officinarum

†Weisia crispa	Anomodon viticulosus
,, crispata	Cylindrothecium concinnum
Ditrichum flexicaule	†Camptothecium lutescens
†Trichostomum crispulum	Eurynchium tenellum
,, nitidum	†Hypnum chrysophyllum
,, tortuosum	Thuidium recognitum
Bryum murale	Hylocomium rugosum
Funaria calcarea	Neckera crispa
Pleurochæte squarrosa	Scapania aspera
Fissidens cristatus	

When they are not alternated with bands of clay or other impervious layers, some sands approach the limestone in permeability, and sandy tracts with good drainage support, even with us, a somewhat xerophilous type of flora. Plants which are less rigidly confined to the true dysgeogenous strata, occurring not unfrequently on dry sandy ground, and banks where water rapidly percolates away, and where the conditions approach in dryness to those of the limestone formations, are brought together in the following list of sub-xerophilous species :—

Hypericum hirsutum	Centaurea Scabiosa
Arenaria serpyllifolia	Galium Mollugo
„ leptoclados	„ verum
Malva moschata	Gentiana Amarella
Viola sylvestris	„ campestris
Vicia lathyroides	Cynoglossum officinale
Rubus saxatilis	Thymus Serpyllum
Daucus Carota	Myosotis collina
Torilis nodosa	Origanum vulgare
Pimpinella Saxifraga	
Valerianella olitoria	Eurynchium murale
Saxifraga tridactylites	„ crassinervium
Sedum acre	Tortula montana
Hieracium murorum	Encalypta streptocarpa
„ diaphanoides	„ vulgaris
Carduus crispus	Orthotrichum saxatile
Carlina vulgaris	Camptothecium sericeum
Erigeron acre	Bartramia Œderi

In addition to the preceding, and to be grouped with them, are a number of dry-loving plants which exhibit in West Lancashire a decided preference for maritime situations. They are restricted to the more arid portions of the sand dunes, dry sea banks, and sandy or shingly shores. Some of the true Xerophiles also occasionally find a congenial home in such localities, and especially is this the case where there is a considerable admixture of shell *débris* with the sand. Amongst these latter may be enumerated *Draba præcox, Geranium sanguineum, Anthyllis Vulneraria, Ditrichum flexicaule,* and *Camptothecium lutescens.* The following is a list of the more interesting maritime sub-Xerophiles :—

Thalictrum dunense	Euphorbia portlandica
Glaucium flavum	Plantago Coronopus
Brassica monensis	Convolvulus Soldanella
Diplotaxis muralis	Salsola Kali
Cakile maritima	Carex arenaria
Silene maritima	Festuca fasciculata
Erodium cicutarium	Phleum arenarium
Viola Curtisii	Ammophila arenaria
Polygala oxyptera	Triticum junceum
Trifolium arvense	„ pungens
Cerastium tetrandrum	„ acutum
„ semidecandrum	Elymus arenarius
„ arvense	
Matricaria maritima	Tortula ruraliformis
Rosa spinosissima	Barbula Hornschuchiana
Eryngium maritimum	Trichostomum flavo-virens
Atriplex laciniata	Bryum pendulum
Polygonum Roberti	Brachythecium albicans
Euphorbia Paralias	Eurynchium megapolitanum

It may be here remarked that several plants of this group

the cell sap varies with the conditions of growth, and it seems reasonable to assume that this aids them within certain limits in tiding over periods of extra stress induced by excessive dryness or moisture, even where no marked structural modifications are developed. To enter into further details of these mesophytes would be out of place here, and space does not admit of lists of species of the minor associations into which they have been divided being given. A few outstanding features, however, require notice, and it may be of some little interest to continue the comparison of the Flora with that of neighbouring counties.

Commencing with the weeds of cultivated ground, we find that several species which are frequent in Yorkshire, especially its south-eastern portion, are either absent from West Lancashire or of such very rare occurrence as to render it probable that they are little more than casual aliens. As examples may be instanced *Ranunculus sardous, R. arvensis, Silene noctiflora, S. anglica, Anthemis arvensis, Legousia hybrida, Vicia tetrasperma, Torilis arvensis, Galium tricorne,* and *Alopecurus myosuroides.*

The following species are also comparatively infrequent, and although we have ranked certain of them somewhat doubtfully as colonists, they might with almost equal propriety have been regarded as only of alien rank :—*Sinapis alba, S. sinapioides, Papaver Argemone, Thlaspi arvense, Lychnis Githago, Scandix Pecten-Veneris, Torilis nodosa, Cichorium Intybus, Filago germanica, Avena fatua, Centaurea Cyanus,* and *Orobanche minor.*

Better established and both more abundant and more widely spread, but still far less so than on the eastern side of the Pennines, are the following :—*Valerianella dentata, Raphanus Raphanistrum, Lepidium campestre, Lepidium heterophyllum, Melilotus officinalis, Vicia sativa, Brassica Rapa, B. Napus, Linaria minor, Lithospermum arvense, Veronica didyma, V. Buxbaumii,* and *Stachys arvensis.*

On the black soil farms of the Fylde district, consisting of dark peaty soil reclaimed from the moss-lands, which are largely devoted to potato cultivation, *Galeopsis speciosa* is a frequent and conspicuous ornament ; and a large blue-flowered pansy, which has been referred to *Viola carpatica Borbas,* is abundant about the borders of fields. But the common poppy *(P. Rhœas)* is absent, or of very casual occurrence ; and throughout the county, except, perhaps, in a very limited area in the north, it is sufficiently infrequent to render its localities worth recording. The

found to the east of the Pennine chain in Yorkshire do not occur in West Lancashire. As compared with North Yorkshire, the number is smaller than might have been anticipated when the relative areas of the two vice-counties are considered. The following is a list of North Yorkshire Xerophiles which appear to be absent from West Lancashire :—*Anemone Pulsatilla, Helianthemum marifolium, Linum perenne, Astragalus danicus, A. glycyphyllos, Galium erectum, Carduus eriophorus, C. acaulis, Thymus Chamædrys, Ophrys apifera, Bromus erectus, Brachypodium pinnatum, Hordeum europæum.*

If we may judge from their ascertained range in Great Britain, the majority of these are unlikely to occur spontaneously within our area ; but a few of them approach our borders so nearly that it is a singular fact that they have never crossed them. It seems impossible to find any satisfactory reason for the absence of *Helianthemum marifolium* (found to the north, east, and west of our district), except that given by Canon Kingsley to account for a similar anomaly of plant distribution—viz., " that it had not *happened* to get there !" There is not a solitary instance of the occurrence of *Clinopodium Calamintha* ; and *Cerastium arvense,* a not unfrequent member of this group in Yorkshire, is with us one of the rarest of plants.

On the other hand, our Scar-Limestone yields a few interesting species which have not been found in North Yorkshire—viz., *Cardamine impatiens, Draba muralis, Polygonatum officinale,* and *Lastrea rigida.* On comparing the xerophilous plants of the sand-dune tract with those of similar habitat to the south of our area in the Mersey Province, we find that there is no species peculiar to West Lancashire, but there are several absentees. These latter are as follows :—*Teesdalia nudicaulis, Filago minima, Hypochæris glabra, Onopordum Acanthium,* and *Ophrys apifera.*

From the sub-xerophilous plants there is a gradual transition to the decidedly hygrophilous species, through many minor groups of plants which are more or less indifferent as to the degree of humidity of the matrix in which they grow. Many species thrive equally well in either wet or dry localities, some assuming hydrophytic or xerophytic modifications of structure in order to accommodate themselves to their environment, others showing but little alteration in their external features. Dr. Drabble has recently shown* that in some of these species the osmotic strength of

* "The Relation Between the Osmotic Strength of Cell Sap in Plants and their Physical Environment." E. & H. Drabble.—*Bio-Chemical Journal,* Vol. 2, p. 117. 1907.

same would apply to *Chærophyllum Anthriscus,* and in a still greater degree to *Scleranthus annuus.*

Many wayside and hedgerow plants found in neighbouring counties will be sought for in vain in West Lancashire. *Ranunculus parviflorus, Sagina ciliata, Arctium majus, Carduus nutans, Bryonia dioica, Artemisia Absinthium, Picris hieracioides, Clinopodium Calamintha, Serrafalcus secalinus,* and *S. commutatus* are quite unknown, and *Acer campestre* and *Ballota nigra* nearly so. The following common plants are markedly infrequent in West Lancashire, and many miles might be traversed without any of them being seen :—

Geranium pusillum	Veronica hederifolium
Dipsacus sylvestris	Poa compressa
Lamium album	Scleterchloa rigida
„ amplexicaule	Festuca sciuroides
„ hybridum	Hordeum murinum
Marrubium vulgare	Polystichum angulare

In dealing with our hydrophilous species several extensive groups of maritime and moorland species will not be noticed here, as they are described in succeeding pages. Schimper, in treating of water-loving plants, which he terms Hydrophytes, divides them into two principal groups, viz. : Aquatic and Semi-aquatic species. He excludes those which display xerophytic structure, even though they always grow in wet localities. By adopting the term Hydrophiles for the group we may still use his chief divisions, regardless of whether the species display xerophytic modifications or not. The more interesting West Lancashire Hydrophiles may be enumerated in the following six groups.

I. Isoëtes Type :—Completely submerged rosette plants, with cylindrical leaves. These are mostly lacustrine, and our two tarns do not appear to yield any of them. The only species representing the group is *Littorella uniflora.*

II. Nymphæa-Hippuris Type :—Rooting plants which reach the surface by means of long-stalked leaves or long shoots. They are well represented in West Lancashire, as shown by the following list of species, many of which are abundant :—*Ranunculus heterophyllus, R. peltatus, R. Baudotii, R. Lenormandi, Nymphæa lutea, Castalia alba, Hippuris vulgaris, Callitriche vernalis, C. stagnalis, C. polymorpha, C. intermedia, C. autumnalis, Polygonum amphibium, Sparganium minimum,* Potamogeton natans, P. polygonifolius, P. alpinus, P. heterophyllus, and Catabrosa aquatica.*

III. Naias Type:—Rooting plants with long flowing shoots or leaves, usually entirely submerged. except the flowering shoots. The species are less abundant than the length of the list apparently indicates. several of them being very local, and most of them of rare occurrence:—*Ranunculus divaricatus, R. fluitans, R. Drouettii, Myriophyllum alterniflorum, M. verticillatum, M. spicatum, Hottonia palustris, Utricularia vulgaris, U. minor, Ceratophyllum demersum, Elodea canadensis, Potamogeton perfoliatus, P. crispus, P. densus, P. obtusifolius, P. Friesii, P. pectinatus, P. pusillus, Ruppia rostellata, Zannichellia palustris, Scirpus fluitans, Chara fragilis, C. vulgaris, C. aspera, Tolypella glomerata, T. prolifera, and Nitella opaca.*

IV. Hydrocharis Type.—Includes plants which float in the water unattached, or rooting plants which rise to the surface to flower, sinking again afterwards. The following six examples occur:—*Hydrocharis Morsus-ranæ, Stratiotes aloides, Lemna minor, L. gibba, L. trisulca,* and *L. polyrhiza.*

V. Podostemon Type.—Non-rooting lithophytes. which attach themselves to stones or rocks beneath the surface of the water. None of the higher orders of plants in this group occur in England, but examples amongst the mosses are numerous. The following are a few characteristic examples:—*Sphagnum crassicladum, S. obesum, Fissidens crassipes, Grimmia apocarpa var. rivularis, Rhacomitrum aciculare, Cinclidotus fontinaloides, Orthotrichum rivulare, Fontinalis antipyretica, F. squamosa, Brachythecium rivulare, B. plumosum, Hyocomium flagellare, Eurynchium rusciforme, Amblystegium irriguum, A. fluviatile, Hypnum palustre,* and *H. ochraceum.*

VI. Semi-Aquatic Type.—Includes a large number of hydrophilous species which, although rooting under water, have the greater part of their stems and leaves elevated above the surface. They are copiously represented in West Lancashire, and the following selections from a much more extensive list may be given as interesting examples: *Comarum palustre, Spiræa Ulmaria, Œnanthe fistulosa, Œ. Lachenalii, Œ. aquatica, Menyanthes trifoliata, Myosotis repens, Limosella aquatica, Iris Pseudacorus, Typha latifolia, T. angustifolia, Sparganium neglectum,*

Acorus Calamus, Alisma Plantago, A. ranunculoides, Butomus umbellatus, Eleocharis palustris, Scirpus lacustris, S. Tabernæmontani, S. maritimus. Osmunda regalis, and many sedges, grasses, and horsetails.

Many of these are marginal plants, growing by ditches and streams or in the shallower parts of ponds and swamps. Amongst such are *Caltha palustris, Spiræa Ulmaria, Sium erectum, Bidens spp, Carex acuta* and other sedges, *Juncus obtusiflorus* and other rushes, *Phragmites communis,* and *Equisetum palustre.* These plants root freely in either very wet or somewhat drier places, and by their ability to grow in either, bridge over the gap between aquatic and land plants. They act as pioneers in winning back portions of land from the water. Some, like *Molinia* and *Carex stricta* form tussocks; others send their roots from the shallow to the deeper water, forming peninsulas or marginal bands around ponds and lakes. In these, débris of all kinds becomes entangled, in which other plants obtain root-hold. Of these latter, mosses of various kinds are amongst the first to help in the process of reclamation. The various oases thus formed ultimately coalesce, and if not artificially cleared away, ponds, ditches and slow streams, as also more extensive bog areas, become gradually converted into a solid peaty soil. Professor Yapp, who has paid much attention to fen plants, in a lecture to the Liverpool Botanical Society, pointed out that the taller growing species frequently exhibit xerophytic modifications not found in those of more humble growth. The former are exposed to sun and drying winds; the latter are afforded shelter from both. The conditions of temperature and moisture to which each is exposed must consequently almost represent two quite different climates or climatic zones. In fact the difference between the reeds and the *Bidens* or *Caltha* growing amongst them present a similar contrast to that between a pine forest and its undergrowth. Professor Yapp finds that the Meadow Sweet, in wet reedy places, loses much of the tomentum from its leaves, especially the radical ones, but in drier situations they are always normal.

On comparing our list of aquatic plants with that of the neighbouring county of Yorkshire, we observe that a number are missing from our district. A list of these is given below, and those which have been found just beyond our borders in Westmorland, or in the more southern portions of the Mersey Province are distinguished by an asterisk.

*Ranunculus Lingua
*Barbarea stricta
*Stellaria palustris
Malachium aquaticum
Callitriche pedunculata
*Elatine hexandra
*Drosera longifolia
*Cicuta virosa
Sium latifolium
Lysimachia thyrsiflora
Polygonum mite
    ,,   minus

*Alisma natans
*Scheuchzeria palustris
Potamogeton flabellatus
    ,,   lucens
Eleocharis acicularis
Eriophorum paniculatum
Carex paradoxa
    ,,   filiformis
*Calamagrostis lanceolata
*Lastrea Thelypteris
*Lycopodium inundatum
*Pilularia globulifera

Only four hygrophilous plants appear in the West Lancashire list that are absent from North Yorkshire, viz., *Callitriche autumnalis, Wahlenbergia hederacea, Cladium Mariscus,* and *Tolypella intricata.* These, however, are all found in West Yorkshire. The following frequent West Lancashire species appear to occur much less frequently in Yorkshire than they do with us:—*Castalia alba, Œnanthe crocata, Carex Pseudo-cyperus,* and *Andromeda polifolia.*

So far we have considered the plants of the district as being divisible into three principal groups, Xerophiles, Mesophiles and Hydrophiles. The physical characters of the soil largely determine the degree of moisture, on which these groups depend. But it is very evident that another agent also has its influence. On examining the plant associations of habitats having about the same degree of humidity, we frequently find the species in them are totally dissimilar. Compare the groups of species found in the following examples of boggy ground; the calcareous shore of Haweswater Tarn, with its *Primula farinosa, Cladium, Schœnus, Juncus obtusiflorus* and *Hypnum falcatum;* the swampy margin of a pool on the glacial drift in Kirkham division, with *Ranunculus Flammula, Bidens tripartita, Alisma Plantago. Sparganium neglectum,* and *Hypnum aduncum;* a peat bog in the Pilling district, with *Drosera, Andromeda, Carex limosa, Myrica,* and *Sphagnum;* and the estuarine marshes of the Wyre, with *Alsine media, Limonium, Statice, Aster Tripolium,* and *Sclerochloa maritima.* In all these stations the water supply is fairly constant and really in excess of plant requirements, and the conditions of altitude and exposure practically identical. We are led, therefore, to the conclusion that some component of the soil must be the factor which determines the presence or absence of these various species. That the chemical composition of the soil has a potent influence was pointed out long ago by Nageli. Other observers have also worked in this field with interesting results, which have recently been summarised

in Schimper's great work on plant distribution. It is not always easy to determine the respective share of physical and chemical causes in the production of our various plant associations, and such recondite problems cannot be fully dealt with in this work and are beyond its scope. But as certain large groups of plants are obviously greatly influenced by the chemical constituents of the soil, they must be briefly noticed.

The chemical constituents which affect the flora on anything like a large scale are few in number, and of general occurrence although very variable in proportional quantity. Their action in forming plant associations may be described as follows. One of the elements existing in the soil on being increased much beyond the usual proportions becomes deleterious to certain plants, which then suffer in their ability to compete with more tolerant species. If the noxious ingredient of the soil be largely increased, a greater number of species suffer, and the growth of many plants may be totally prevented. A small group of tolerant species then becomes dominant, and large plant associations are recognisable which appear to have been formed in this manner. Some of the species, however, instead of merely tolerating the salts referred to, appear to have an absolute craving for them, and are unable to exist long without a regular supply. They might almost be compared with human victims of the morphia or arsenic habit. At first a certain resistive power to the poison has been established, which has become intensified in the course of time, until the originally injurious substance is a necessity. One result of growth in strong saline solutions is the production of xerophytic features, induced by what is called *physiological drought.* It has been shown that the roots of plants take up more water when it is pure than when it contains salts in solution, and that there is a fixed limit of concentration for each species, beyond which absorption ceases. In consequence of this, a plant may be reduced to a condition of drought although growing in a bog, as absolute with regard to its physiological requirements as if it grew on the driest rock. It must then assume the characteristic xerophytic devices to safeguard transpiration, otherwise it will be unable to withstand special seasons of stress due to excessive transpiration. Of the elements which appear to exert such an influence on plant life as that described, some are almost absent from the soils of West Lancashire, as for instance the salts of Magnesium, to which are due many of the interesting features of the flora of Dolomitic areas. But we have three groups of plants somewhat well marked off from the rest of our vegetation,

which are apparently dependent to a large extent on their predilection for, or their power of resisting, the salts of Sodium and
Calcium, and Humic acid. The more interesting features of these
groups must be very briefly noticed.

I.· *Halophilæ, or Salt-loving plants,* depend largely upon
the proportion of chloride of sodium present. They have the
peculiar property of being able to store salt in their tissues, and
some of them will do this even when growing under ordinary conditions, finding sufficient of this ubiquitous element in common
garden soil. They greedily absorb it and tenaciously retain it,
although it is said to have but little nutritive value. Although
this salt is not in the ordinary sense a plant food, it has been
found of practical utility by agriculturists, when applied to land
with other manures. Certain cereal crops grown on loamy soils
will sometimes be improved by it, and it has been used with this
object on farms distant from the sea. But although quantities of
not more than 5cwt. per acre have been found to have a beneficial
effect on corn lands, one drawback has been found to its use, viz.,
that it checked the subsequent growth of " seeds " (rye grass and
clover) on the same soil, so that the secondary crop was much
inferior. Mangolds and cabbages, as also celery, being sea plants,
are found to benefit by the addition of salt to the manures used
for them.

In our salt-marshes, fully exposed as they are to sun and wind,
a high state of concentration of the saline solution has often to be
endured, which only a limited number of plants can tolerate.
Some plants are able to withstand solutions of as great a strength
as 50 per cent. without injury, whereas a two or three per cent.
solution kills many ordinary plants in a short time. According
to Schimper those which survive this latter strength may be
regarded as halophytes. With us all the species are maritime or
estuarine. Nearly all are fleshy or succulent, having well
developed water storage systems, and they are often covered with
glaucous pruina, or display familiar xerophytic features. Dr.
Drabble has shown that their sap exhibits a very high osmotic
strength, without which their water reserve would be rapidly
extracted in solutions of high concentration. The group is well
represented in West Lancashire, attaining its greatest richness on
the muddy salt marshes bordering the principal estuaries. Here
there are extensive tracts in which only halophytes are found.

The majority of the species enumerated in the lists of plants
on page 28 belong to this salt marsh group, and to them may be

added *Sagina maritima, Pottia Heimii, Bryum Marrattii* and
Amblystegium salinum. In drier, sandy or shingly parts of the
coast, the following species occur :—

Glaucium flavum	Convolvulus Soldanella
Cochlearia danica	Carex arenaria
Cakile maritima	,, extensa
Brassica monensis	Triticum junceum
Honkeneja peploides	,, acutum
Eryngium maritimum	,, pungens
Salsola Kali	Psamma arenaria
Atriplex arenaria	

The species which affect rocky and stony places are of necessity but poorly represented, as the coast of West Lancashire
presents but few suitable stations for them. Some of them occur
on artifically-made stone embankments. Representative species
are *Crithmum maritimum, Silene maritima, Statice maritima,
Limonium binervosum, Asplenium marinum* and *Grimmia
maritima.* Several lichens could be added to this group. Some
of the species, as *Lichina confinis,* delight in being submerged by
every tide. Others grow where they are occasionally wave or
spray washed, amongst these being *Ramalina scopulorum,
R. cuspidata,* several species of *Placodium, Physcia aquila,
P. parietina, Verrucaria halophila, V. glaucina,* and *Opegrapha
rupestris.* Many other species attain special luxuriance, or are
most abundant near the sea coast.* It is probable, however, that
the purity of the atmosphere, as also its moistness, and its greater
freedom from smoke may partly account for this.

Plants of the halophilous group which do not occur in West
Lancashire, but are found a little further south in the Mersey
province are *Alsine rupicola* and *Zostera marina ;* and a slight
extension of our boundaries to the North would enable us to
include *Crambe maritima, Raphanus maritimus* and *Mertensia
maritima,* none of which now grow in West Lancashire.

Very few plants of this group seem able to compete with the
general vegetation in soils unimpregnated with salt. *Cochlearia
alpina,* a sub-species probably of *C. officinalis,* occurs on our
hills, and *C. danica* may be rarely found on banks some little
distance from the sea, but we have no instance of a maritime
species spreading as a weed over wide inland areas, as *Salsola
Kali* is said to have done in Canada. On the inland hills of the

* For an interesting note bearing on this subject see Journ. Bot., 1905, p. 31,
where Mr. West, of Bradford, attributes the partiality of certain lichens for
sea-rocks to the richness of the rock surface in nitrogenous matter, derived
from the droppings of sea-birds.

counties north and east of our district *Statice maritima, Silene
maritima* and *Plantago maritima* may be found, but only in
places where competition is strictly limited. On the other hand
a great many ordinary plants encroach upon the strictly maritime
species in situations where the amount of salt is somewhat
diminished, and as they recede from the sea or tidal influences, the
latter become only sub-dominant and then soon disappear. The
species which are not completely halophilous, but which are able
to resist occasional overhead spraying or root suffusion with sea
water, form a connecting link with the general flora. Some of
them, judged by Schimper's standard, appear from time to time as
true halophytes, but they occur just as often in ordinary situations.
In this group of what might be called potential halophytes may be
classed the following West Lancashire species :—

Thalictrum minus	Senecio Jacobœa
Ranunculus sceleratus	Carduus tenuiflorus
Sisymbrium Sophia	Crepis virens
Diplotaxis muralis	Thrincia nudicaulis
Viola Curtisii	Sonchus arvensis
Cerastium tetrandrum	Samolus Valerandi
Arenaria leptoclados	Cynoglossum officinale
Sagina nodosa	Hyoscyamus niger
Erodium cicutarium	Limosella aquatica
Ononis spinosa	Chenopodium rubrum
,, repens	Atriplex patula
Trifolium fragiferum	,, deltoidea
Lotus corniculatus	Polygonum aviculare
Vicia lathyroides	Rumex crispus
Erythræa umbellatum	Juncus obtusiflorus
Sedum acre	Potamogeton pusillus
Apium graveolens	,, pectinatus
Œnanthe fistulosa	Zannichellia palustris
Pastinaca sativa	Scirpus Tabernæmontani
Daucus Carota	Carex Œderi
Galium verum	Agrostis palustris
Matricaria inodora	Festuca rubra

II. *Calciphilæ or Lime-loving plants,* are those restricted to
soils containing a considerable proportion of calcium carbonate.
Like the salt previously dealt with, this also has a low nutritive
value, but it is supposed to be useful in fixing certain deleterious
products of plant digestion and assimilation, hence the formation
of raphides, etc., in their tissues. It is insoluble in pure distilled
water, but in the presence of rain water containing carbon dioxide
derived from the atmosphere, a soluble bi-carbonate is formed,
which can be absorbed by plants. It may thus act as does sodium
chloride in the production of a special plant association. The
flora of our calcareous districts exhibits strongly xerophytic
characters, but we should not like to affirm that these are entirely

due to the lime. It is not always easy to determine the respective
share of physical and chemical influences in producing the xerophytic features of some of our plant associations, and this is
especially true of the *Calciphilæ.* In West Lancashire there is a
marked difference between the vegetation of the calcareous hills
of North and the sandstone ones of East, but as this coincides
with a difference in physical structure and comparative dryness as
well as of chemical composition, it is difficult to assign to either a
preponderating influence on the vegetation. The Silverdale rocks
are in places very dry almost throughout the year, and the xerophytic flora may be due to this. The flora of the hills of East
is subject to very wet conditions during part of the year, and, in
certain parts to great drought and insolation at other times, and
some of the plants seem specially modified to meet these
varied conditions. Many indifferent species occur on both
types of hills. For instance, bracken, which is such a characteristic plant of the gritstone cloughs, seems able to adapt itself
to the limestone soil of Dalton and Warton Crags, on which it
may be seen growing out of bare limestone crevices, with its
roots closely applied to the rock itself. *Calluna vulgaris* has been
seen in similar situations in Yorkshire. Masclef grew the
bracken on both calcareous and clay soils, under equal conditions
as to moisture, and found it assumed xerophytic characters on
the former, which would appear to indicate that solution of lime
has the power of checking absorption and inducing physiological
drought as in the case of chloride of sodium. A list of species
which occur in the drier calcareous habitats is given on page 83.
To these may be added, as representing calciphilous species, the
following, which are usually found under somewhat moister
conditions.

Geranium pratense	Festuca sylvatica
Rhamnus Frangula	Asplenium viride
Saxifraga hypnoides	Seligeria pusilla
Chrysosplenium alternifolium	Fissidens decipiens
Ribes rubrum	Weisia rupestris
Parnassia palustris	,, verticillata
Hieracium sylvaticum	Bartramia Œderi
,, gothicum	Philonotis calcarea
Pyrola rotundifolia	Cinclidotus fontinaloides
Primula farinosa	Anomodon viticulosus
Plantago media	Orthothecium intricatum
Corylus Avellana	Hypnum commutatum
Epipactis longifolia	,, falcatum
Orchis morio	,, molluscum
Gymnadenia conopsea	Riccia Lescuriana
Juncus obtusiflorus	Lejunia calcarea
Schœnus nigricans	,, Rossettiana
Cladium Mariscus	,, Mackaii

Pteris, Ephemerum serratum and *Weisia crispa*, which have been regarded as strictly calciphobe species on the continent, all occasionally occur on limestone soils with us.

III. *Humus-loving plants.* — Humus, consisting of the organic products of the decomposition of animals and plants, is one of the most important constituents of soils. It is rich in carbonaceous and nitrogenous material, the former probably not directly assimilable, but the latter is liberated for plant food by the action of certain fungi *(Mycorhiza* and *Bacteria)* which abound in all humus soils. Earthworms, as shewn by Darwin, have an enormous influence on the fertility of soil, bringing up mineral matter from the deeper layers and incorporating it with humus. Humus is mostly acid in re-action, and is a very complex material, containing, amongst other compounds, humic acid and its derivatives. This acid forms with alkalies soluble, and with alkaline earths insoluble, salts. According to its degree of acidity, humus is known as *mild* or *acid,* and its quality has a considerable influence on plants and helps to determine the character of the flora. It is necessary, therefore, to regard each type of humus separately and consider its influence upon plant distribution.

A. *Mild Humus,* when thoroughly aerated and mixed with the mineral constituents of the subsoil by worms or by cultivation, is known as mould and forms the ordinary soil of fields, hedgebanks, and pastures. It largely increases the water-holding power of poor sandy ground. On river banks, and in woods of certain types, the soil usually contains a very large proportion of humus, which is mostly light and well aerated, and consequently of the mild order. This is especially the case in mixed deciduous woods. In pine woods the layers of fallen needles lie closer, the soil is more compact, and insect life is less varied and abundant, the result being a lower degree of areation and a greater tendency to develop acid humus. Consequently, we find a considerable reduction in the number of species of plants able to thrive in such a situation. A further contributing cause, however, must not be overlooked, viz., the great want of light in such a wood.

The following are characteristic woodland and river alluvium plants in West Lancashire, attaining their greatest abundance and luxuriance in the rich mild humus soils of such localities :—

Anemone nemorosa	Hypericum Androsæmum
Cardamine amara	Lychnis diurna
Stellaria nemorum	Geranium sylvaticum
Hypericum maculatum	,,　　Robertianum

Oxalis Acetosella	Ajuga reptans
Impatiens Noli-tangere	Lamium Galeobdolon
Geum rivale	Mercurialis perennis
Chrysosplenium oppositifolium	Ulmus montana
Sanicula europæa	Urtica dioica
Angelica sylvestris	Neottia Nidus-avis
Heracleum Sphondylium	Epipactis Helleborine
Lonicera Periclymenum	Orchis mascula
Circæa lutetiana	Narcissus Pseudo-narcissus
Asperula odorata	Allium ursinum
Petasites ovatus	Endymion non-scriptum
Crepis paludosa	Gagea lutea
Arctium nemorosum	Paris quadrifolia
Carduus heterophyllus	Arum maculatum
Lysimachia nemorum	Carex remota
Myosotis sylvatica	,,　　sylvatica
Solanum Dulcamara	Milium effusum
Scrophularia nodosa	Festuca gigantea
Veronica montana	Equisetum sylvaticum
Stachys sylvatica	,,　　maximum

It does not seem desirable to take up space by enumerating the various minor groups which connect this section with the Acid Humus series. They depend largely upon the amount of shade, moisture, and richness of the soil in humus. As a contrast to those just noticed, however, an extreme example at the opposite end of the series may be instanced in the florula of an upland pasture on the slopes of the gritstone fells. The soil consists largely of silica, and owing to exposure to frequent heavy rains and the inclination of the surface, the amount of humus retained is comparatively small. When we consider the extreme insolubility of quartz, it must be regarded mainly as a diluent or vehicle for more essential plant foods, and not as supporting a "silica" flora. In the upland sandy and stony pastures referred to, we find a group of plants thriving under conditions of soil and exposure which most plants cannot tolerate. The salt-lovers, lime-lovers, and woodland plants do not find their special requirements here, and the humus supply is too small, or the other conditions too austere, to enable many of the more indifferent species to eke out a living. The upland pasture association on the slopes of our gritstone fells is essentially, therefore, an assemblage of individuals of negative tendency, very hardy and requiring no special luxuries. The following are characteristic species of such a locality :—

Ranunculus acris	Achillea Millefolium
Polygala serpyllacea	Scabiosa Succisa
Ulex Gallii	Campanula rotundifolia
Galium saxatile	Euphrasia nemorosa
Hypericum humifusum	,,　　borealis
,,　　pulchrum	Prunella vulgaris
Potentilla silvestris	Rumex Acetosella
Leontodon autumnalis	Juncus conglomeratus

H

Luzula campestris	Agrostis tenuis
Carex pilulifera	Aira præcox
,,　　ovalis	Sieglingia decumbens
,,　　binervis	Cynosurus cristatus
Festuca ovina	Nardus stricta
Aira flexuosa	Botrychium Lunaria

Where the ground is not too steep or elevated for meadowland, and the conditions are somewhat improved by artificial aid from the farmer, some of the above species disappear. They are, however, replaced by others in much increased variety, and in the height of the blooming season some of these upland meadows present an exceedingly beautiful and flowery appearance. The display is enhanced by the grasses attaining a shorter stature than in the low country. On this account the grass is not so well able to crowd out the less desirable flowering weeds, nor does it hide them so much. Most of the following species were observed in meadows at elevations of from 550 to 850 feet in Roeburndale and Wyresdale :—

Lychnis Flos-cuculi	Rhinanthus Crista-galli
Trifolium pratense	Ajuga reptans
Lotus corniculatus	Primula veris
Lathyrus montanus	Plantago lanceolata
,,　　pratensis	Rumex obtusifolius
Alchemilla vulgaris	,,　　Acetosa
Sanguisorba officinalis	Polygonum Bistorta
Galium palustre	Orchis mascula
Carum majus	,,　　maculata
Angelica sylvestris	Habenaria viridis
Achillea Ptarmica	,,　　montana
Chrysanthemum Leucanthemum	Listera ovata
Senecio aquaticus	Authoxanthum odoratum
Centaurea nigra	Alopecurus pratensis
Carduus palustris	Phleum pratense
,,　　heterophyllus (rare)	Dactylis glomerata
Hypochœris radicata	Poa pratensis
Leontodon hispidus	Lolium perenne
,,　　autumnalis	Equisetum sylvaticum

In addition, various members of the upland pasture association are more or less abundant. Both these groups become poorer in species as the altitude increases, and at their extreme limits are with difficulty preserved from degenerating into the Heath or Grass Moors to be described later.

The coast sand-dunes exhibit a similar excess of silica and deficiency in humus to that of the upland pasture-lands, but the vegetative features are entirely different. These sand-hills are not subject to the frequent mists that keep up the supply of moisture on the fells, and the rainfall is also smaller and showers

less frequent. They are, therefore, far drier and probably also warmer. In addition, the finely divided sand is constantly shifting. This affects the vegetation, the most abundant plants being those which have adopted special habits to overcome this disadvantage and prevent themselves from being overwhelmed. Apart from the ordinary aerial xerophytic contrivances, which enable them to keep such moisture as they are able to obtain, they have further improved their chances by the development of a great power of underground growth. The subterranean parts of some species, such as *Salix repens.* often stretch for enormous distances under the sand-banks. When buried in the shifting sand, the stems at once accommodate themselves to the new conditions. They assume some of the functions of roots, fixing the plant, binding the soil around it, and issuing numerous adventitious rootlets to nourish the fresh shoots which soon ascend to the surface. The underground portion of *Honkenzja peploides* is often developed out of all proportion to the parts above, exceeding them occasionally as much as 24 times. Other species that grow in a similar manner, or have extensively spreading roots are *Viola Curtisii, Convolvulus Soldanella, Silene maritima, Euphorbia portlandica, Ononis spinosa,* and *Equisetum variegatum.* Certain species attempt to preserve a space for themselves by producing an intricate network of runners just beneath the surface, which are held down at short intervals by rootlets, and from the upper side of which leaves and flowering shoots appear. Of this class are *Eleocharis uniglumis, Carex arenaria, Festuca rubra, Triticum junceum, T. pungens, T. repens, Elymus arenarius,* and *Ammophila arenaria.* These secure the shifting sands for a long enough period to enable other species to establish themselves, and, if this occurs in a damp hollow, there is gradually produced by their decay a thin layer of peaty soil. Mosses then grow luxuriantly, and all the conditions of a miniature peat bog result, in which some of our rarer and most interesting sand-dune plants are found.

B. *The Acid Humus,* or Peat-loving plants, next require consideration. It is principally to humic acid that peat owes its peculiar qualities. Peat is formed where vegetable matter decays without free access of oxygen, as on ground underlying stagnant water, and where alternate flooding and baking in a hot sun keeps down the number of worms and causes the surface to cake. A sandy soil overlaid with acid peat forms a heath, with its characteristic flora, consisting of many species which are strongly calciphobous. When such calciphobous plants are grown on a calcareous soil, according to Fliche and Grandeau, the absorption of iron is prejudiced and the formation of chlorophyll checked. According to Chatin and Sendtner, a mere sprinkling of water rich in lime kills *Drosera, Blechnum spicant, Allosorus crispus,* and *Sphagnum.* Characteristic calciphobous plants in West Lancashire, are shown below :—

Viola palustris	Brachyodus trichodes
Drosera rotundifolia	Dicranella cerviculata
,, anglica	Dicranodontium longirostre
Sarothamnus scoparius	Dicranum majus
Digitalis purpurea	Rhacomitrium lanuginosum
Vaccinium Myrtillus	Campylostelium saxicola
,, Vitis-idæa	Leptodontium flexifolium
Calluna vulgaris	Webera nutans
Pedicularis sylvatica	,, annotina
Rumex Acetosella	Bryum alpinum
Carex dioica	Fontinalis squamosa
Aira flexuosa	Hyocomium flagellare
Blechnum Spicant	Eurynchium rusciforme
Cryptogramme crispa	Plagiothecium elegans
Sphagnum Spp.	,, undulatum
Andreæa Spp.	Brachythecium plumosum
Dicranum fuscescens	Hypnum fluitans
Catharinea crispa	,, palustre
Oligotrichum hercynicum	,, ochraceum
Polytrichum commune	,, stramineum
,, strictum	,, sarmentosum
Tetraphis pellucida	Hylocomium loreum
,, Browniana	

A large proportion of our fell district consists of elevated heathlands, where very moist conditions usually prevail. Here the acid humus soil acts in a prejudicial manner on many plants, deterring others completely, and limiting the vegetation to a few tolerant species' which, in the absence of competition, become gregarious and cover large tracts. Many of them assume xerophytic characters, due, according to Schimper, to the humic acid acting in a similar way to salt in interfering with the free absorption of water. Indeed the Humic Acid Association has several characters in common with the Halophytes, depending largely on chemical agency acting in a similar manner. Both groups of plants exhibit more or less evidence of xerophytic structure ; both consist of many gregarious individuals and few species ; and both are characterised by the dominant colour at the height of the blooming season being a shade of purple. It is worth noting, however, in estimating the effect of physiological dryness due to chemical influence, and physical dryness due to the nature of the habitat, that our moorland plants, as also the sea plants, are greatly exposed to drying winds and bright sunshine at certain seasons, and these must have had their influence.

Plants with xerophytic structure would have the greatest chance of surviving the occasional short but severe dessications to which our moorlands are sometimes exposed in summer. Heather and *Juncus squarrosus* may partly derive their xerophytic characters from this cause. But the same can hardly apply to the Cranberry and Andromeda, which grow on very wet moors which remain moist in the driest seasons. It is probable that chemical influence may here be the active cause.

Most of our higher hills are thickly capped with peat, and the plant associations vary with the degree of moisture and the altitude. In the wettest places Sphagnum is usually the dominant plant, with or without Cotton-grass. In drier tracts Bilberry or Heather predominate, or a mixture of the two. On the lower slopes Bracken enters into the association, sometimes to the exclusion of the others ; but it generally thins out abruptly near the 1,200 feet contour. As this series occupies so large a proportion of the area of West Lancashire, we will briefly indicate the leading features of the chief plant associations which characterise it. These may be conveniently considered under the three groups recognised by Dr. W. G. Smith and Mr. C. E. Moss in their paper on Yorkshire plant associations.[*] These are the *Moss Moor,* the *Heath Moor,* and the *Grass Moor.* Generally speaking their relative altitudes coincide with the order in which they are placed above. But this is not invariably the case, as the nature of the association does not altogether depend upon altitude, but on the slope of the ground, the thickness of the peat-bed, and the degree of moisture. The peat is thicker, and the surface usually more level on the summit plateaux than elsewhere, and consequently in such situations we find the wet-loving group (charaterised by *Eriophorum*) most in evidence.

I. This is known as the MOSS MOOR GROUP, and the character of the vegetation varies considerably with the nature of the surface and altitude, and several distinct plant associations may be distinguished within its limits, of which the following are perhaps the most marked.

(a) *The Cloudberry Association.*—This occurs on the peaty summits of the fells, generally at altitudes exceeding 1,500 feet, and reaching to our highest limits. It is found in the bleakest and most exposed places, where the peat is damp and cold, and especially where the surface has a northerly or easterly aspect.

* " Geographical Distribution of Vegetation in Yorkshire." Geographical Journal. Vol. xxi. 1903, p. 375.

It is well seen on Botton Head Fell at 1,750 feet, and on the Leagram Fells along the Yorkshire boundary at 1,650 feet. In the latter locality, where the Cloudberry is very abundant, the following species were noted :—

Rubus Chamæmorus	} Abundant and dominant.
Scirpus cæspitosus	
Eriophorum vaginatum	} In considerable quantity and about equally
Calluna vulgaris	frequent.
Vaccinium Myrtillus	} In small quantity only.
Empetrum nigrum	

The two species found accompanying the Cloudberry almost invariably in West Lancashire are *Scirpus cæspitosus,* and *Eriophorum vaginatum.* The relative abundance of the other plants enumerated above varies in different situations, being dependent apparently on the elevation and wetness of the surface. Sometimes *Vaccinium Oxycoccos* and *Eriophorum polystachion* form subordinate members of the group, and occasionally *Vaccinium Vitis-idæa.* On a part of Botton Head Fell at about 1,700 feet, the vegetation is almost entirely Cloudberry.

(b) *The Sphagnum Association.*—This may be divided into two groups, depending chiefly on altitude for their peculiar features. The lowland type occurs on what is known in West Lancashire as a " Moss," and differs from the upland type in its richer and more varied flora, and in the presence of such shrubby vegetation as *Betula, Salix,* and *Myrica.* Such a moss is called by Schimper a " High Moor," from its tendency to become elevated in the centre. Its soil is exceedingly poor in nitrogenous matter, as owing to the limited supply of oxygen, nitrifying bacteria are very rare, as also are worms. Some of the plants have therefore had to solve the question of ways and means by becoming carnivorous. Of these we have in West Lancashire *Drosera rotundifolia, D. anglica, Pinguicula vulgaris,* and *Utricularia vulgaris.* A description of Cockerham Moss and its " High Moor " flora will be found on page 30. Here we are only concerned with the much simpler Upland Sphagnum Association, which has for its dominant species *Sphagnum recurvum.* On our wetter fells may be seen frequently recurring wide bright green and very boggy patches, consisting entirely of this species, but also occasionally accompanied by other Sphagna, such as *S. subnitens, S. palpillosum, S. rubellum,* and still more frequently by *Polytrichum commune.* Frequent members of this group, though usually marginal or very subordinate, are *Viola palustris, Vaccinium Oxycoccos, Juncus effusus, Eriophorum polystachion,*

Carex echinata, and *C. canescens.* On a portion of Tarnbrook Fell in which Sphagnum is predominant, *Rynchospora alba* occurs.

(c) *The Cotton-grass Association* also, like the Sphagnum one, occurs on the wetter parts of the fells, and the two frequently blend and become inseparable. Here the dominant species is *Eriophorum vaginatum.* Less green and inviting to the footsteps of the unwary traveller than the Sphagnum, this type of ground is sufficiently treacherous, but can be easily negotiated by stepping from tussock to tussock of the Cotton-grass. From a little distance, at the fruiting season, the latter seems to be the only vegetation where the group is most typically developed, and the frequently-recurring name " White Moss " indicates its most striking feature. Close search will usually, however, reveal some of the following subordinate species intermingled :—*Erica Tetralix, Vaccinium Oxycoccos, Drosera rotundifolia, Scirpus cæspitosus,* and *Sphagnum papillosum.*

All these were noted on such a moor at about 950ft. on Threaphaw Fell near the Trough of Bowland. In other localities *Viola palustris, Narthecium ossifragum, Andromeda polifolia,* and *Molinia cærulea* may often also be detected. The pools and runnels of water which intersect the Cotton-grass bog contain *Sphagnum cuspidatum* and *Hypnum fluitans,* the latter being exceedingly abundant and in various forms.

II. THE HEATH GROUP as a rule occupies a lower zone than the one we have already considered. But it is also dependent on the greater proximity of the rock to the surface, and on drainage secured by sloping ground. Therefore, the Cotton-grass Group may be considered as characteristic of the thickly peat-capped plateaux and the Heath Group of the drier and more frequently rocky slopes. An intermediate group between the Cotton-grass and Heath-groups is very prevalent on our fells, and may be designated the *Cotton-grass-with-Heather Association,* as these are the dominant species, one or the other slightly preponderating, or at times both becoming equally prevalent. *Sphagnum papillosum* is the most abundant moss and *Erica Tetralix, Drosera rotundifolia,* and *Aira flexuosa* not unfrequently join the group as subordinate members. More rarely *Listera cordata, Juncus squarrosus,* and *Festuca ovina* are found with them.

(d) *The Heather Association* covers very extensive tracts, and when pure, and in places where its growth is carefully supervised for grouse preserves, few of the higher plants except *Calluna vulgaris* are evident. There is usually an abundant undergrowth of mosses such as *Dicranum scoparium, Hypnum cupressiforme,* and *Hypnum Schreberi,* and in the damper places a few Sphagna. Where the heather is less well cared for, or has been temporarily checked by burning, the following often accompany it :—*Vaccinium Myrtillus, Erica cinerea, Melampyrum pratense, Aira flexuosa, Campylopus turfaceus, C. flexuosus, Webera nutans,* and *Plagiothecium undulatum ;* and in more broken or stony places *Vaccinium Vitis-idæa, Empetrum nigrum,* and *Hylocomium loreum* are common. If the ground be very rocky, *Cryptogramme crispa, Lastrea aristata,* and *Lycopodium Selago* occur in the crevices of the gritstone, and there is a luxuriant growth of such mosses as *Dicranum fuscescens* and *Plagiothecium elegans,* with the hepatics *Mylia Taylori, Bazzania trilobata, Lepidozia cupressina, L. Pearsoni, L. reptans,* and *Lophozia gracilis.* More rarely the latter are accompanied by *Sphenolobus minutus.* The lichens become prominent in this group, especially the *Cladoniæ,* and where the heather has been burnt off *Lecidea decolorans* appears in great abundance. On the lower slopes *Pteris aquilina* is a serious competitor with the heather, and frequently becomes the dominant species. Where grouse are a prime consideration the bracken is mowed periodically at considerable expense, and the heather is then enabled to overcome all rivals.

(e) *The Bilberry Association* is practically only a dependent of the preceding one. It is usually met with on the higher and drier slopes of the fells, especially where the underlying rock is very near the surface. The predominant plant is *Vaccinium Myrtillus,* which is usually accompanied by a little heather and *Aira flexuosa.* The subordinate members are *Empetrum nigrum* and *Vaccinium Vitis-idæa,* with sometimes a little *Galium saxatile* and *Nardus stricta.* On one of our fells where bracken and bilberry are the predominant species, and are much intermingled, *Trientalis europæa* is very abundant, its slender underground soboles spreading far between the layers of decayed bracken fronds. The Bilberry Association may be readily distinguished by its pleasant shade of light green, forming a line near the summit edges of the fells, and occasionally sending down tongues or forming extensive patches in the darker Heath Association which lies below it.

(f) *The Rush (Juncus squarrosus) and Grass Association* does not seem to have been separately distinguished by plant geographers, it being probably deemed to belong to the dry grass moor association. It usually occurs in scattered islands amongst the associations of this group described already, and within their range. These islands are frequently of small extent, but occasionally expand so as to cover considerable areas. They occur on most of our fells, and in the summer and early autumn are at once distinguishable from a considerable distance by their rich reddish-brown tint, which differs in shade from that assumed by Cotton-grass in winter, although approaching it. The following are the components, arranged in the order of their frequency, as noticed in a very extensive tract of such ground on Threaphaw Fell at an altitude of 1,000ft.

Juncus squarrosa Nardus stricta	} Dominant
Galium saxatile Potentilla silvestris Festuca ovina Rumex Acetosella	} Frequent but subordinate.
Vaccinium Myrtillus	Rare and scattered.

III. THE GRASS MOOR GROUP is very extensively represented over the lower slopes of the West Lancashire fells, following everywhere the edges of the moors. Attempts are occasionally made by the farmers to extend it as grazing ground into the adjacent heath moor. The quality of the grass is so poor, however, that it is hardly worth winning, and the frequency with which it relapses into its original heathery condition seems to indicate that it rarely pays to bring it under the influence of even this mild degree of cultivation. Many of the species already enumerated descend into the grass moor group, but the dominant species are the coarser grasses, such as *Nardus stricta* and *Aira flexuosa,* with *Festuca ovina, Agrostis vulgaris,* and a few sedges and rushes. *Rumex Acetosella, Pedicularis sylvatica,* and *Luzula multiflora* also occur, and there are very often scattered patches of *Ulex Gallii, Pteris aquilina, Lastrea Filix-mas* or *L. montana.* This association is the lowest of the moorland series, and brings us to the upland pasture and meadow lands already considered, in which the species increase very considerably in variety.

DISTRIBUTION OF THE MOSSES, HEPATICS, AND LICHENS.

Some of the mosses fall naturally into the groups in which the flowering plants have been divided, and were noticed in the preceding pages, but they require, as a class, a little more detailed consideration. The merest tyro at collecting soon finds that, like the Phanerogams, these plants, so much lower in the scale, have marked idiosyncracies as to the degree of moisture in the habitat, the kind of rock, or the nature of the soil on which they grow, or even as to the species of the tree upon whose bark they live. They seem, in addition, to be affected by more obscure and delicate influences than are the higher plants. Many curious problems connected with their distribution await solution. Who can explain why one species of *Splachnum* grows almost exclusively on cow's dung and another on the remains of sheep? Why do some *Orthotricha* occur on trees only and others on rocks only, while some are equally at home on either? How is it that in a wood a single tree may be thickly clothed with a moss, such as *Aulacomnium androgynum,* without a single specimen straying to other trees around, although their branches interlace with those of the favoured tree? Why does *Endocarpon miniatum* grow on the face of rocks and *E. rufescens* only in their angles or crevices? We frankly admit our inability to answer these and other whys and wherefores which constantly arise in connection with mosses and lichens.

We can see, however, that temperature, altitude, and moisture influence their distribution in exactly the same way as in the higher groups, and that, in addition, their range is remarkably affected by subtle atmospheric influences, amongst which the products of coal combustion have an especially deleterious action even when very largely diluted. We therefore find few mosses or lichens on trees in the vicinity of large towns, and that they increase in number and luxuriance in localities further away from them, or which are out of the track of prevailing winds which blow from smoke-vitiated districts. West Lancashire is somewhat unfortunately situated, for, although few of these plague-spots blemish its fair face, it receives smoke clouds from surrounding districts, which, attenuated though they are, have a markedly noxious effect on the cryptogamic vegetation, appreciably diminishing the number of species and retarding and stunting their growth. Some of the more delicate arboreal species, unable to withstand the concentrated solutions of smoke deposits which trickle down

the trunks of trees, are still able to survive on their horizontal branches, and many mosses will be found thriving on the latter much more luxuriantly than on perpendicular trunks.

The horizontal distribution of mosses and lichens has not been thoroughly worked out in Britain, and until the scattered work of our cryptogamic botanists is brought together in some such work as " Topographical Botany," it is impossible to refer our species to the types of distribution defined by Hewitt C. Watson. To obtain some idea of the general character of our Moss Flora we have adopted the following modification of his divisions :—

NORTHERN :—Species most frequent in Scotland and the North of England, and confined to hilly districts further south. This includes the Arctic, Highland, Scottish, and Intermediate types of Watson. Examples :—*Oligotrichum hercynicum, Hypnum sarmentosum, Kantia submersa.*

SOUTHERN :—A combination of the English, Germanic, and Atlantic types of Watson. Examples :—*Ditrichum tortile, Pleurochæte squarrosa, Lepidozia pinnata.*

BRITISH :—Equivalent to the same type of Watson. Examples:—*Tortula muralis, Bryum argenteum, Pellia epiphylla.*

The following table shows the number of West Lancashire species falling into each of these three classes, but the estimate will probably require revision when the range of some of the species throughout the country is better ascertained :—

TYPE.	Sphagna.	Mosses.	Hepatics.	Total.
Northern	6	59	20	85
Southern	6 (?)	22	2	30
British	19	219	72	310
Total ✳ ...	31	300	94	425

So little is known regarding the distribution of the lichens that no attempt can be made to show them in the above groups. We may remark, however, that some of the more boreal genera are well represented in West Lancashire, especially amongst the *Cladoniæ, Stereocaulei, Cetrariæ,* and *Gyrophoræ ;* but the arboreal species of lowland or sub-alpine woods and hedgerows are

✳　Additions whilst in the press have increased the totals for Cryptogams to the following :—Sphagna 32, Hepatics 100, Lichens 300.

neither plentiful nor well-developed. In the hilly districts of the county the mosses and lichens are more abundant and frequently more luxuriant than in the lowlands. Temperature has some influence in causing this, many of our species having a sub-arctic range of distribution, but the amount of moisture is, no doubt, the most important factor, the extremes of temperature not being very great within our area. The lowlands naturally possess a drier climate than the hills, and are rendered still drier by cultivation and drainage of the soil. Our North and East Divisions consist very largely of hilly tracts, with a relatively small area of cultivated ground : on the contrary, the West District nowhere exceeds an altitude of 200ft., and is very extensively cultivated. The cryptogamic flora is profoundly affected by these different conditions, as shown below, where we have tabulated the number of species in the two areas mentioned. The difference in the relative sizes of the districts included to some extent vitiates the comparison, and the more varied geological features in the larger area have also, of course, a considerable influence, and must be allowed for. The small extent of uncultivated ground and sparing exposure of rock in West also tend to reduce the number of species. In North and East also there is sufficient low country to support many of the lowland species.

	N. and E. Chiefly Hill-country.	W. Chiefly Lowland.
Sphagna	27	19
Mosses	241	64
Hepatics	91	29
Lichens	262	40
Total	629	152

The following table, perhaps, gives a more accurate reflex of the effect of altitude on the range of our Archegoniate plants, showing the number of species which find their upward limits at various elevations. The vertical range of the lichens is not worked out with sufficient completeness to allow of their inclusion.

Upward limit between—	Sphagna.	Mosses.	Hepatics.	Total.
Sea level and 100ft. ...	3	29	3	35
100 ,, 300ft. ...	1	35	9	45
300 ,, 500ft. ...	1	33	8	42
500 ,, 1000ft. ...	5	62	26	93
1000 ,, 1500ft. ...	11	56	21	88
1500 ,, 2080ft. ...	10	83	32	125

Although the numbers are tolerably uniform, as compared with the extreme difference shown in the first table, owing to disappearing lowland forms being replaced by more montane ones as we ascend, there is a distinct tendency for the number of species to increase with the altitude. Our hills do not rise high enough to exhibit the decrease which occurs after reaching the maximum at extreme altitudes on lofty alpine peaks. Although figures are not available to show it, the lichens appear to be distributed altitudinally in a similar manner to the mosses, the number of species increasing in the more hilly districts.

Both mosses and lichens were at one time regarded as entirely aerial feeders, and thought to derive all their nourishment by absorption into their leaves or fronds directly from the air, or solutions derived from it, and that the substratum contributed little or nothing towards their support, the radicular filaments being only supposed to fix the plants to their place of growth. Although they have no true root or descending axis, it has, however, become recognised that the adventitious rhizoids of mosses also function as roots and absorb nutrient material from the soil. These rhizoids spring from almost any part of the gametophyte, and are most important organs in the economy of the plants, serving to protect them from cold, to retain or absorb moisture, and even to reproduce the plant, as it has been proved that under the influence of light they are convertible into conferva-like threads containing chlorophyll and in no way differing from protonemal filaments capable of producing buds and new plants. Haberlandt found that the rhizoids of mosses were able to penetrate the substratum of bark much as the mycelium of a fungus would do.* In the case of lichens, if we

* " Beitrage zur Anatomie und Physiologie der Laubmoosen." 1886.

accept the " dual hypothesis " introduced by De Bary and Schwenderer, the fungoid portion of the " Consortium," as the compound thallus is termed, must be regarded as probably often saprophytic. Graphis scripta is described as beginning life as a filamentous fungus, ramifying through the outer cork of tree trunks ; then chlorophyll-bearing algæ get distributed through its tissues, and the whole structure becomes recognisable as a lichen. It is quite possible that many mosses also are saprophytic, rooting into the dead corky layer of bark, or dead vegetable and animal matter on the ground. None of our West Lancashire species appear to be truly parasitic, although one British moss genus, Buxbaumia, has been so regarded, and the almost total suppression of its green parts referred to this cause. Once we admit the probability of mosses and lichens deriving part of their food supply from the substratum upon which they grow, we are provided with some clue to the marked idiosyncrasies they show in their choice of habitat. It may be possible also to explain their absence from certain kinds of rock on the assumption that some of the chemical elements therein exert an injurious influence, similar to that of combustion products in the air. So little, however, is known of the more minute influences which affect these plants, that we can only at the present time attempt to show the more salient features of their distribution. They are therefore grouped in the following four admittedly crude classes, with a few illustrative examples of each, a perusal of which will no doubt give a good general idea of the nature of our cryptogamic flora.

A. Rupestral species.—The mosses and lichens of this group grow on rocks or in their crevices, and on boulders, walls, and stones. Some of them spread at times to gravelly earth which may be near, but even then they adhere to the larger particles, and require little or no pulverised weathered material or soil for their support. There is no reason to suppose that these plants are unable to assimilate minute proportions of the intractable material upon which they grow. Their rhizinæ have probably the same power of eating into hard rock as that exhibited by the root of the bean in Sach's experiment, which left the imprints of its roothairs on a piece of polished marble. The rhizinæ of mosses and lichens are no doubt able to secrete a similar acid solvent of calcareous rocks. Some of the lichens, such as Lecidea calcivora and Verrucaria terebrata, are indeed able to eat very deeply into the surface, so as to completely honeycomb it, and many other species erode the rock to a less

extent. Certain hard impervious rocks belonging to the older formations, and those of an igneous or metamorphic origin are absent from West Lancashire, and in consequence a few species partial to them do not occur, although our district lies well within their area of distribution both vertically and horizontally. Amongst them may be enumerated Grimmia leucophæa, G. ovata, G. torquata, Ulota Hutchinsiæ, Hedwigia ciliata, Dicranoweisia crispula, Glyphomitrium Daviesii and Hypnum eugyrium; others are rare and not luxuriant, such as Webera elongata, Rhacomitrium protensum, and Bryum filiforme. Three sub-sections of Rupestral species are, however, readily discernible within our area depending upon the kinds of rock they mostly affect, as shown briefly below.

(a) Calco-saxicolæ (Lime rock-loving species). These are most abundant on the limestone rocks of North, occurring more sparingly on the smaller area of similar strata in East, and still more rarely elsewhere on plaster, or the mortar of walls. The cryptogamic vegetation of rocks and walls in the limestone districts has a peculiar and characteristic appearance, entirely different from that of the sandstone areas. Compactly tufted or densely cushioned mosses abound, often hoary from the hyaline points of the leaves, and these may be regarded as xerophytic features. The more creeping species attain a luxuriance and verdure unequalled elsehere, covering the walls everywhere with beautiful festoons in every shade of green. May we not consider it probable that the lime, in addition to meeting any special metabolic requirements of the plants, combines with and renders innocuous the deleterious acid constituents of smoke impurities? The lichens also on calcareous rocks occur in lavish profusion. Everywhere the grey stone is blotched with black patches of Pannaria and Verrucaria, or dark gelatinous rosettes of the Collemacei. White and grey Lecanorei and Verrucariæ spread extensively, some of them deeply pitting the surface. These more sombre or colourless species are enlivened by an intermixture of orange and yellow Physcia and Placodium, or the ochrey films of Lecanora ochracea and lemon yellow of L. xantholyta. Amongst the greenish scaly crusts of Lecanora crassa may be seen the bluish cushions of Lecidea cœruleo-nigricans, the whole forming an exquisite blend of tints.

The following is an illustrative list of the more striking or characteristic species of this group, omitting those already named :— Fissidens decipiens, Grimmia pulvinata, Tortula *intermedia, Barbula sinuosa, Weisia crispata, W. verticillata, Trichostomum crispulum, T. nitidum, Zygodon Stirtoni, Orthotrichum rupestre, O. saxatile, Bartramia Œderi, Bryum murale, Neckera crispa, Orthothecium intricatum, Eurynchium tenellum, Metzgeria conjugata, Coleolejeunia calcarea, C. Rossettiana, Marchesinia Mackaii, Collema melænum, Collemodium Schraderi, Lecanora sympagea, L. irrubata, Urceolaria scruposa, Lecidea Metzleri, L. exanthematica, L. rimosa, Endocarpon miniatum, Verrucaria conoidea,* and many other *Verrucariæ.*

(b) *Silico-saxicolæ* (Siliceous rock-loving species). These attain their greatest development amongst the arenaceous hills of East, and many of them are strongly calciphobous. Thus *Lecanora sulphurea,* although frequent on silurian boulders built into limestone walls, never spreads from one to the other. Indeed, the kind of rock used in building our walls may be readily determined at a distance from the prevailing tone of colour imparted by the lichens growing upon them. Even the Millstone-grit and Silurian rocks (although both forms of sandstone) have their distinctive features, and in both the prevailing tones are duller and more uniform than those of the limestone. On the Millstone grits the mosses are chiefly dark tufts of *Rhacomitrium* or *Hypnum cupressiforme,* or at greater elevations black patches of *Andreæa* or dark-green cushions of *Dicranum fuscescens,* with lurid clumps of *Mylia* and *Bazzania.* There are many grey *Parmeliæ* and *Cladoniæ,* with coral-like *Sphærophorei,* on the rocks ; and on the walls smoky-looking patches of *Parmelia fuliginosa* and ragged fringes of *Platysma glaucum* and *Evernia furfuracea.* On the higher scars, flat-topped tabular blocks exhibit black scaly *Gyrophoræ,* dingy-green *Lecidea viridi-atra,* and mouse-coloured *L. rivulosa.* Sub-orbicular patches of *Pertusaria lactea* and *P. dealbata* enliven the general sadness of tone, and everywhere loose rocks and stones are covered with the greyish black-spotted thallus of *Lecidea contigua.*

On the Silurian series the fruticulose *Stereocaulons* invite attention, and greenish or yellowish shades are introduced by an abundance of *Lecanora sulphurea, L. polytropa, Lecidea geographica* and *Parmelia conspersa,* often beautifully commingled with grey species, such as *L. contigua* and *L. stellulata,* and reddish angular patches of *L. Dicksonii.* These various colours resemble a mosaic, or give the face of the rock the appearance of a coloured map, in which the black apothecia represent the towns, and the cracks of the thallus and black hypothalline lines resemble

the rivers and boundary lines. The Silurian rocks also display in abundance orbicular patches of *Lecanora ventosa* covered with reddish-brown apothecia.

The following, omitting some of those named above, are representatives of the group under consideration :—*Andreæa petrophila, A. Rothii, A. crassinervia, Tetrodontium Brownii, Seligeria recurvata, Brachyodus trichodes, Grimmia maritima, G. Doniana, Rhacomitrium aciculare, Ptychomitrium polyphyllum, Campylostelium saxicola, Fontinalis squamosa, Heterocladium heteropterum, Hyocomium flagellare, Hypnum palustre, Pellia Neesiana, Lophozia gracilis, Lepidozia pinneta, L. Pearsoni.*

Racodium rupestre, Ramalina scopulorum, R. cuspidata, Parmelia prolixa, P. lanata, P. tristis, Physcia aquila, Lecanora glaucoma, Pertusaria dealbata, Lecidea crustuleta, L. lucida, L. coarctata, Verrucaria halophila, V. margacea.

(c) *Saxicolæ* (General rock-loving species). These occur on both siliceous and calcareous rocks, not always with equal frequency perhaps, except in the case of a few cosmopolitan species, but sufficiently often on both kinds of rock to render it impossible to refer them to either of the more restricted classes already noticed. They are naturally as a rule more generally distributed than those of the other sub-sections, and a few examples will suffice.

Grimmia apocarpa, Tortula muralis, Barbula rigidula, Orthotrichum cupulatum, Bryum cæspiticium, Mnium serratum, Fontinalis antipyretica, Isothecium myurum, Eurynchium pumilum, Amblystegium irriguum, A. fluviatile.

Conocephalus conicus, Marchantia polymorpha, Lichina confinis, Peltidea aphthosa, Leproloma lanuginosa, Lecanora galactina, Lecidea plana, L. contigua, L. concentrica, Opegrapha saxicola.

B. TERRESTRIAL SPECIES.—As the plants of the last section often stray from their normal rock habitat to stiff soil banks or stony ground, so those of the present one, although ordinarily preferring earth, sometimes spread to neighbouring rocks or stones. But as a general rule they delight in a well-pulverised soil, and when growing, as they often do, on rock ledges, it will be found that there is a layer of earth beneath them overlying the rock. They form a large section of our mosses and hepatics, and

I

the lichens, although less numerous in species, are sometimes very gregarious, and, therefore, conspicuous where they occur. All seem to require some amount of humus in the soil, although those in the first sub-section require but a small amount. They are divisible into six groups, as shown below.

(d) *Arenophilæ* (Sand-loving mosses). — These are best represented on the coast sand-dunes of the West Division, although a considerable number occur by sandy roadsides and streamsides amongst the eastern hills. Their chief requirement is a large percentage of sand in the soil, and so long as this is available, a slight admixture of peat, clay, or lime does not seem to materially affect them. Many of the mosses possess in a marked degree the power of innovation, and if buried by shifting sand, soon throw up fresh branches, which bring them level with the surface, where some of those growing in more exposed places form smooth, compact cushions, which are of much greater depth than their superficial appearance indicates. The lichens are very few indeed, a few *Cladoniæ* and sometimes *Peltigera canina,* being the only species growing in such situations with us. The following are representative species of sand-loving mosses, several of them occurring only amongst the higher gritstone fells and others only on the sea coast :—*Catharinea crispa, Oligotrichum hercynicum, Pleuridium subulatum, Ditrichum homomallum, Rhacomitrium canescens, Tortula ruraliformis, T. vinealis, Barbula Hornschuchiana* (not yet found on clay with us), *Trichostomum flavo-virens, Webera annotina, W. erecta, Bryum pendulum, B. Warneum, B. lacustre, Brachythecium albicans.*

Eurynchium megapolitanum, Hypnum polygamum, H. lycopodioides, H. Sendtneri, H. Wilsoni, H. Patientiæ, Pellia endiviæfolia, Nardia scalaris, Kantia arguta, and *Peltigera canina* (sometimes).

(e) *Calciphilæ* (Calcareous soil-loving species).—These are most frequent on banks and field borders, in woods and marshes, and by roadsides in calcareous districts, or occasionally elsewhere in situations where a certain amount of lime is afforded them, as on parts of the sand hills, at the base of mortared walls, or by springs which well up through calcareous shales. Some of them are partial to mud-capped walls and earth-covered ledges on limestone rocks, in such situations looking like rock plants, but seldom or never actually growing on the bare surface. The following are typical illustrative species :—*Pottia recta, P. lanceolata, Tortula*

ambigua, Weisia microstoma, Funaria calcarea, Philonotis calcarea, Bryum roseum, Thuidium recognitum, Cylindrothecium concinnum, Camptothecium lutescens, Brachythecium glareosum, Hypnum elodes, H. chrysophyllum, H. molluscum, H. falcatum, and *Hylocomium rugosum.*

Riccia Lescureana, R. sorocarpa, Lophozia turbinata, Saccogyna viticulosa, Scapania aspera, Collema ceranoides, Leptogium lacerum, L. sinuatum, Solorina saccata, Peltigera rufescens, Lecidea cærulea-nigricans, L. lurida, and *Endocarpon rufescens.*

(f) *Pelophilæ* (Clay-loving species).—The remarkably impervious nature of clay and its great power of retaining absorbed water (only exceeded in this respect by peat), necessarily renders less frequent and of shorter duration the droughty periods to which mosses growing upon it are subject, and many mosses flourish on various soils in which it forms one of the component parts. Both on limestone and peat soils we find the character of the flora at once altered where this element is largely combined with either. The species which are markedly pelophilous, often growing on soils consisting almost entirely of clay, are not very numerous in West Lancashire, and are mentioned below. Were those also included which prefer clay and lime, or clay and peat, the list would be very much longer :—*Catharinea undulata, Dicranella rufescens, D. varia, D. Schreberi, Fissidens viridulus, F. incurvus, F. taxifolius, Tortula aloides, Barbula tophacea, B. cylindrica, Weisia squarrosa, Discelium nudum, Webera carnea, W. albicans, Bryum intermedium, Brachythecium Mildeanum, Pterigophyllum lucens, Hypnum aduncum, Riccia glauca, Aneura pinguis, Blasia pusilla, Kantia trichomanis, Scapania compacta, Aplozia sphærocarpa, Collema glaucescens.*

(g) *Geophilæ* (Soil-loving species).—There are many mosses which do not appear to thrive well on either sandy, clayey, or calcareous soils where the substances named are in great excess, but require in addition a considerable admixture of vegetable humus, or ordinary mould. Probably the sandy soils are too dry, and the clayey ones too wet and cold for many of them ; and it may be that some of the species are more actively saprophytic than those already considered, and require a more liberal allowance of organic matter. But whatever the reason, they delight in a rich and generous mould, such as is found in fields, pastures and deciduous woods, and the more aquatic forms frequent birch swamps, marshes and the earthy margins of ditches and ponds. They are widely distributed, and some of them very abundant, and the following examples are selected from a large number of species.

On bare ground in pastures, or tilled fields, grassy banks, or muddy ground by ponds : — *Pleuridium axillare, Fissidens bryoides, F. exilis, Ceratodon purpureus, Pottia Heimii, P. intermedia, P truncata, Acaulon muticum, Phascum cuspidatum, Weisia viridula, Ephemerum serratum, Physcomitrium pyriforme, Funaria hygrometrica, Bryum argenteum, Fossombronia pusilla, Lophocolea bidentata, L. heterophylla, Anthoceros lævis, Ricciella fluitans,* and *Cladonia pyxidata.*

In woods, thickets and swamps :—*Dicranum Bonjeanii, D. majus, Thuidium tamariscinum, T. delicatulum, Eurynchium piliferum, E. prælongum, Plagiothecium sylvaticum, Hypnum cuspidatum, H. cordifolium, H. purum, Aneura multifida,* and *Plagiochila asplenioides.*

(*h*) *Turfophilæ* (Peat-loving species). — The most extreme examples of this group are those which grow on pure peat, as *Dicranella cerviculata,* or on fallen pine needles in coniferous woods, as *Campylopus turfaceus ;* in fact these two mosses luxuriate where there is absolutely no admixture of soil with the peat. But many ericetal species may also be included, which delight in banks or ditches on or near moorlands, where they are subject to frequent flooding with water charged with the soluble constituents of peat. Many of them are markedly gregarious, and cover extensive areas on the fells and mosses. The following are examples :—*Sphagnum spp., Polytrichum spp., Dicranella cerviculata, Campylopus spp., Leucobryum glaucum, Leptodontium flexifolium, Aulacomnium palustre, Philonotis fontana, Breutelia arcuata, Webera nutans, Bryum pseudotriquetrum, Mnium subglobosum, Cinclidium stygium, Hypnum fluitans, H. exannulatum, H. revolvens, H. scorpioides, H. stramineum, H. sarmentosum, Aneura pinguis, Marsupella emarginata, Nardia compressa, N. hyalina, N. obovata, Aplozia cordifolia, Lophozia inflata, Cephalozia Lammersiana, C. connivens, C. lunulæfolia, C. fluitans, Odontoschisma spp., Kantia Sprengellii, K. submersa, Lepidozia Pearsoni, Ptilidium ciliare, Diplophyllum albicans, Scapania purpurascens, S. undulata, S. irrigua, Cladonia spp., Cladina spp., Cetraria Islandica, C. aculeata, Peltigera polydactyla, Lecidea decolorans,* and *L. uliginosa.*

(*i*) *Zoophilæ* (Species loving animal matter).—A small class growing on dung or animal remains. Possibly the number of mosses requiring some proportion of animal food may be greater than is usually thought, and should include some *Phasca* and *Pottiæ,* and other species of tilled fields, which are rare in situations where the soil is not periodically manured, or to which cattle or rabbits have not access. This is especially the case with *Ephemerum serratum* which grows in cultivated fields, or on anthills which are usually sprinkled with rabbits dung, on grassy declivities amongst our limestone scars. *Pottia truncata* ascends by moorland roadsides far above its usual habitats in the low country, no doubt carried on the feet of cattle or amongst mud on cart wheels. But it never appears to spread beyond the range of washings from the roadway, to situations where the supply of animal manure would be much more precarious. At present, however, the group must be regarded as being represented in West Lancashire by two species only, viz : — *Splachnum sphæricum* and *Tetraplodon mnioides,* both of which are very partial to the remains of sheep.

C. ARBOREO-TERRESTRIAL AND ARBOREO-RUPESTRAL SPECIES. — The occurrence of these plants, some of them equally luxuriant whether growing on tree trunks, stone walls, or earth, would almost lead to the conclusion that mosses are utterly indifferent as to the nature of their habitats, were the contrary not indicated by the peculiarities shown by the preceding groups. We have a large number of these arboreo-terrestrial species, and, as might be expected from their accommodating nature, many of them are very abundant. Examples may be seen in the following selection :—*Tetraphis pellucida, Dicranoweisia cirrata, Dicranum scoparium, Zygodon viridissimus, Orthotrichum diaphanum, Aulacomnium androgynum, Bryum capillare, Mnium hornum, Neckera complanata, Leucodon sciuroides, Brachythecium rutabulum, B. velutinum, Eurynchium crassinervium, E. myosuroides, E. confertum, E. myurum, Plagiothecium denticulatum, Amblystegium serpens, A. Juratzkæ, Hypnum riparium, H. cupressiforme, Metzgeria furcata, M. pubescens, Lophocolea heterophylla, Cephalozia bicuspidata, Lepidozia reptans, Madotheca lævigata, M. platyphylla, Frullania Tamarisci, Cladonia pyxidata, C. fimbriata, C. digitata, C. macilenta, Ramalina farinacea, Platysma glaucum, Evernia prunastri, E. furfuracea, Parmelia cetrarioides, P. lævigata, P. saxatilis, P. sulcata, P. Borreri, P. caperata, P. fuliginosa, P. physodes, Physcia parietina, P.*

pulverulenta, P. tenella, Lecanora subfusca, L. atra, L. tartarea, L. parella, Lecidea sanguinaria, L. parasema, L. canescens, L. myriocarpa, and *L. alboatra.*

D. ARBOREAL SPECIES.—The group of species which are exclusively lignicole, growing on tree trunks and palings, is not with us a very extensive one. They appear to be highly susceptible to the effect of smoke in the atmosphere, and the same applies to the lichens. The relative abundance of the members of this group affords a good index of the purity of the atmosphere. Near large manufacturing towns they are amongst the first plants to disappear, and as West Lancashire is within range of the smoke of more than one industrial district, nearly all the species are somewhat rare, and some require very careful searching for indeed, even in the most favourable parts of the vice-county. When found, they are frequently poorly developed and far from being as luxuriant as in districts more remote from manufacturing centres. *All* our species, so far as observed, are included in the following list :—*Tortula mutica, T. lævipila, T. papillosa, Zygodon conoideus, Ulota Drummondii, U. Bruchii, U. crispa, U. phyllantha* (never rupestral in West Lancashire), *Orthotrichum leiocarpum, O. Lyellii, O. stramineum, O. tenellum, O. pulchellum, Leskea polycarpa, Radula complanata, Microlejeunea ulicina, Frullania dilatata, Collema flaccidum, Calicium spp., Coniocybe pallida, Trachylia tympanella, Ramalina fraxinea, R. fastigiata, Usnea hirta, U. plicata, Alectoria jubata, Platysma ulophyllum, Parmelia scortea, P. exasperata, P. subaurifera, Parmeliopsis ambigua, Physcia ciliaris, P. venusta, P. pityrea, P. stellaris, P. aipolia, P. erosa, Coccocarpia plumbea, Lecanora allophana, L. rugosa, L. chlarona, L. varia, L. conizæa, L. conizæoides, L. expallens, L. pallescens, Pertusaria globulifera, P. amara, P. communis, P. velata, P. Wulfenii, P. leioplaca, Phlyctis agelæa, P. argena, Thelotrema lepadinum, Lecidea ostreata, L. flexuosa, L. quernea, L. disciformis, L. endoleuca, L. truncigena, L. parasitica, Opegrapha herpetica, O. atra, O. Turneri, O. varia, O. vulgata, Arthonia punctiformis, A. epipasta, A. Swartziana, A. astroidea, A. pruinosa, Graphis elegans, G. scripta, G. inusta, G. sophistica, Normandina pulchella, Verrucaria gemmata, V. epidermidis,* and *V. nitida.*

The groups named might readily be still further sub-divided according to their requirements as regards moisture, and the species enumerated will show how widely they vary in this

respect, some preferring the driest situations, others growing totally submerged ; but to enumerate is unnecessary, as the nature of the habitats given in the list will sufficiently indicate them. We have only one floating species, *Ricciella fluitans,* and that rarely fruits unless it is stranded on the mud at the margin of ponds, so that the floating state is perhaps purely accidental. The *Sphagna* sometimes float in large masses, through the decay of the lower parts of their stems. But when mosses are found floating unattached, they have always become so through accident, and although they will live thus for a considerable period, growing and putting out innovations, the condition is unnatural.

Mr. Dixon has recently shown that in Northamptonshire some of the species of mosses can only be regarded as colonists.* In West Lancashire the conditions are totally different, and such that we cannot avoid the conclusion that all our species are truly native. We formerly regarded *Leptobryum pyriforme* and *Lunularia cruciata* as open to suspicion of introduction, and there is no doubt that in some of their localities they have come with garden material. But the recent discovery of both in thoroughly natural habitats removes them from the "doubtful" category.

---

* " The Moss Flora of Northamptonshire." 1899.

LIST OF AUTHORITIES FOR DETAILS, BOOKS AND HERBARIA CONSULTED, AND MS. CONTRIBUTIONS.

ARMITAGE, Miss E.—Herbarium containing W. Lancs. mosses.

ASHFIELD, CHARLES JOSEPH.—"*The Flora of Preston and Neighbourhood.*" Part I., 1858; Part II., 1860; Part III., 1862; Part IV., 1864. In the *Transactions of the Nat. Hist. Soc. of Lancs. and Chesh.* Contains a good list of the rarer phanerogams of the district, which includes part of S. Lancs. Also a list of plants in *Botanist's Chronicle*, 1864.

BAILEY, CHARLES, F.L.S.—Numerous records of phanerogams in *Bot. Record and Brit. Exch. Club Reports*, and in *Top. Bot.* Also, "*On the Adventitious Vegetation of the Sand-hills of St. Anne's*," in *Proc. Manchester Lit. and Phil. Soc.*, 1902. Also MSS. notes and specimens.

BAKER, E. G., F.L.S.—"*Some British Violets.*" *Journ. Bot.*, 1901. Refers to the occurrence of *Viola carpatica* in W. Lancs.

BAKER, J. G., F.R.S.—"*Flora of the Lake District.*" Records several localities for plants in W. Lancs.

BEANLAND, J.—Specimens of W. Lancs. phanerogams.

BEESLEY, HENRY.—MSS. lists and specimens of phanerogams, mosses and hepatics.

BENNETT, ARTHUR, F.L.S.—"*Supplement to Topographical Botany,*" 1905 *(Journal of Botany)*.

BOSWELL, J. T.—See Syme.

BRAITHWAITE, ROBERT, M.D., F.L.S.—"*British Moss Flora.*" A few W. Lancs. mosses noticed.

BREAKELL, ARTHUR.—Notes and specimens of W. Lancs. plants.

BRITTEN, JAMES. F.L.S.—"*Lunaria rediviva DC. in W. Lancs.*" *Naturalist*, 1864, p. 203.

BRITTEN AND RENDLE.—"*List of British Seed Plants*," 1907.

CAMDEN.—"*Brittaniæ*," Ed. by E. B. Gibson (folio), 1695. List of plants compiled by John Ray, p. 803.

CARR, AMOS.—Records in *Bot. Rec. Club Reports*.

CASH, JAMES.—Specimens of W Lancs. mosses and MS. notes.

CURWEN, EDWIN.—Notes and specimens of W. Lancs. plants.

DALLMAN, A. A., F.C.S.—Notes and specimens of W. Lancs. plants. Also Bibliographical notes.

DERHAM, WILLIAM.—"*Ray's Correspondence*," octavo, 1718.

DIXON, H. N., F.L.S.—"*Weisia crispata in Britain*" and "*Three new varieties of Hypnum fluitans,*" *Journ. Bot.*, 1899. "*Student's Handbook of British Mosses,*" Ed. II., 1904.

DOBSON, WILLIAM.—"*Rambles by the Ribble,*" 2 series, 1877 and 1884. Incidental references to many of the plants mentioned by Ashfield, who sometimes accompanied Dobson on his rambles.

FISHER, HARRY.—"*Victoria History of Lancashire,*" 1906. Several new records for W. Lancs. plants.

FLORA OF STONYHURST, 1891.—List of the higher plants of the district, embracing contiguous parts of W. Yorks. and S. Lancs. Published anonymously, but the principal contributors were Rev. John Gerard, F.L.S., S.J., and Rev. C. A. Newdigate, S.J.

HAMILTON, W. P.—List of "*Mosses near Lancaster,*" *The Naturalist*, p. 28, 1898.

HANBURY, F. J., F.L.S.—"*Notes on British Hieracia,*" *Journ. Bot.*, 1894. Records some W. Lancs. species.

HIERN, WILLIAM.—Records in *Top. Bot.*

HORNBY, Rev. PHIPPS J.—MSS. lists and specimens of plants of the Fylde district, etc.

HORRELL, E. CHAS., F.L.S.—"*The European Sphagnaceæ,*" *Journ. Bot.*, 1900. Contains the earliest notices of some of the W. Lancs. sphagna.

HUDSON, WILLIAM.—"*Flora Anglica,*" 1762.

JACOB, R., Rev.—Records of several W. Lancs. lichens in Mudd's "*Manual of Brit. Lichens,*" 1861.

JENKINSON, JAMES.—"*Descriptions of British Plants,*" London and Kendal, 8vo. (dated Yealand, 10th April, 1775). Many of the plants recorded at this early date still survive and bear testimony to the accuracy of the Author.

KING, F. C.—Records in *Bot. Rec. Club's Reports* and specimens in Herb. Wheldon.

KIRBY, WILLIAM.—MS. notes on Silverdale plants.

LAWSON, THOMAS.—"*Letter to Ray*" in *Ray's Corresp.*, dated April 9th, 1688.

LEES, Dr. F. A.—Records in *Top. Bot.* and *Bot. Rec. Club Reports.* "*Florula of Bare,*" *Naturalist*, 1899. "*West Lancashire Indigenes,*" *Naturalist*, 1900.

LEIGH, Dr. CHAS.—"*Nat. Hist. of Lancs. Chesh., and the Peak,*" 1700.

LEWIS, JOHN HARBORD, F.L.S.—Records in *Top. Bot.*

LINTON, Rev. E. F., M.A.—Records in *Top. Bot.* Note on "*W. Lancs. plants,*" *Journ Bot.*, p. 87, 1900, and various comments on W. Lancs. plants in letters to the Authors.

MARSHALL, Rev. E. S.—"*Additions to the Flora of Lancashire,*" *Journ. Bot.*, p. 136, 1896.

MARTINDALE, J. A.—Incidental references to W. Lancs. lichens in papers on "*Westmorland Lichens,*" *Naturalist*, 1897.

MARTYN, T.—"*Plantæ Cantabrigiensis,*" 1763.

MASON, Rev. W. W.—Notes and specimens of W. Lancs. plants. Sometimes accompanied the Authors on field excursions.

MELVILL, J. COSMO.—Records in *Top. Bot.*

MOSS, JOHN.—Notes and specimens from the Fylde districts.

MUDD, WILLIAM.—"*Manual of British Lichens,*" 1861.

NEWMAN, EDWARD.—"*History of British Ferns,*" 1844.

PARKINSON, JOHN.—"*Theatrum Botanicum,*" 1645.

PEARSON, — .—Records of plants in Ashfield's "*Flora of Preston and Neighbourhood.*"

PEARSON, W. H.—"*British Hepaticæ.*" W. Lancs. localities for a few species recorded.

PETTY, LISTER. — "*Plants of Leck and Neighbourhood,*" *Naturalist*, 1893. "*Some Plants of Silverdale,*" *Naturalist*, 1902. Also notes and specimens from the Silverdale district.

PHYTOLOGIST, THE.—Short notes on W. Lancs. plants.

PICKARD, JOSEPH F.—Various records and specimens of W. Lancs. plants.

PRESTON SCIENTIFIC SOCIETY.—"*Flora of Preston and Neighbourhood,*" 1903. List in tabular form of wild and cultivated phanerogams, ferns, and mosses. The district covered embraces parts of S. Lancs. also.

RAY, JOHN.—"*Historia plantarum,*" 1686-1704. "*Synopsis method. stirp. Brittanic,*" 1690-1724.

RIDDELSDELL, Rev. H. J., M.A.—"*The Motley Herbarium,*" *Naturalist*, 1902. The herbarium referred to contains some old W. Lancs. specimens.

ROWSE, JOSEPH S.—Specimens of phanerogams in Herb. Wheldon.

SALMON, C. E., and H. S. THOMPSON.—"*West Lancashire Notes,*" *Journ. Bot.*, p. 203, 1902. Contains several new W. Lancs. records. Mr. Salmon also kindly supplied a list of W. Lancs. records for the species of Limonium.

SEARLE, HENRY.—Specimens of W. Lancs. plants.

SIMPSON, S.—Notes in "*Phytologist,*" 1843, et. seq., and in *Top. Bot.*

SMYTHE, Rev. JOHN.—First recorded *Lastrea rigida*, in "*The Phytologist,*" 1843.

STABLER, GEORGE.—"*Hepaticæ and Musci of Westmorland,*" *The Naturalist*, 1896-8. Contains references to a few W. Lancs. mosses and hepatics.

SYME, J. T. BOSWELL.—Records in *Top. Bot.*

WATSON, HEWITT COTTRELL.—"*Topographical Botany,*" Ed. II., 1883. Contains several personal records, in addition to those of other botanists.

WHELDON, JAMES ALFRED, F.L.S.—Records in *Bot. Exch.* and *Moss Exch. Club Reports.* "*West Lancashire Flora Notes,*" *Naturalist*, p. 1, 1900. "*Lancashire and Cheshire Rubi,*" *Journ. Bot.*, p. 401, 1898. "*Mosses of the Mersey Province,*" *Naturalist*, p. 69, 1900. "*The North of England Hartidia,*" *Naturalist*, p. 65, 1902. With Mr. A. Wilson in the papers marked with an asterisk below.

WILSON, ALBERT, F.L.S., F.R. Met. Soc.—Records in *Bot. Rec., Bot. Exch.,* and *Moss Exch. Club Reports.* "*The great*

Smoke Cloud of the North of England and its influence on Plants," *Halifax Naturalist* (as read before Brit. Assoc. meeting, 1900). *"Additions to the Flora of West Lancashire,"* *Journ. Bot.,* p. 40, 1900 ; p. 22, 1901 ; p. 346, 1902 ; p. 94, 1905 ; p. 99, 1906. *"The Mosses of West Lancashire,"* *Journ. Bot.,* p. 465, 1899 ; p. 294, 1901 ; p. 412, 1902. *"Notes on the Flora of Over Wyresdale,"* *Naturalist,* p. 357, 1901. *"New West Lancashire Mosses,"* *Naturalist,* p. 35, 1900. *"West Lancashire Lichens,"* *Journ. Bot.,* p. 255, 1904.

WILSON, Mrs. A.—Several records of rarer plants.

WILSON, Mrs. C.—Note on *Myosurus* and information regarding other old plant records and localities.

WILSON, SYDNEY.—Notes and specimens of plants and information as to localities and local plant names. Sometimes accompanied the Authors on field excursions.

WILSON, JOHN.—*"A Synopsis of British Plants in Mr. Ray's Method,"* Newcastle-on-Tyne, 1744. Records apparently quoted from Ray.

WOOD, J. B.—Record for *Pyrola rotundifolia* in Dickinson's *"Flora of Liverpool"* (Supplement), 1855, p. 15.

WOOD, Rev. W.—Records of phanerogams in Turner and Dillwyn's *"Botanists' Guide,"* 1805.

WORSDELL, W. C.—MS. notes of phanerogams.

ARRANGEMENT AND ABBREVIATIONS ADOPTED IN THE FLORA.

In the arrangement of the species we have not thought it necessary to depart from that in common use in Britain, *i.e.,* for the higher plants, *"The London Catalogue"*; for the Mosses, Dixon's *"Students' Handbook"* ; for the Liverworts, Macvicar's *"Exchange Club Catalogue"* ; and for the Lichens the works of Leighton and Crombie.

NAMES.—The scientific names have been revised so as to accord with the rules of the recent International Botanical Congress. We are indebted to Mr. Britten for a list of the necessary alterations, which will appear in the *"British Museum Catalogue of British Seed Plants"* by Messrs. Britten and Rendle. The scientific name is followed by the so-called English name, after which the local or vernacular name is given wherever possible. Where the latter differs from the ordinary "English name" in general use, it is enclosed in inverted commas. The names of aliens, casuals, and doubtful plants are printed in lighter type than those of the higher plants of citizenship. A few species which it is deemed desirable to exclude until further evidence of their occurrence is obtained, are enclosed in brackets.

CITIZENSHIP.—The status of the plant in West Lancashire only is intended. It must not be inferred, however, that a plant said to be native is so in every station quoted, but only that it occurs in some part of our district where there is no reason to suspect that it has been introduced.

AREA.—The figures refer to the eight districts defined on page 5. They indicate, with few exceptions, that the authors have themselves seen the plant in the district named. In a few instances only, second-hand evidence has been accepted, when it seemed thoroughly reliable or was accompanied by a voucher specimen.

HABITAT.—The characteristic situations in which species grow within our area have been indicated without regard to book statements as to the class of habitat they usually occupy elsewhere.

FREQUENCY.—In the case of plants shown by the census numbers to be generally distributed, the terms used are to be

applied generally. Where the census numbers indicate that a plant is restricted to certain districts, the terms of frequency have reference to those districts only in which the species is shown to occur.

FIRST RECORDS.—These are submitted as being the earliest *published* records with which we are acquainted. It is possible that a more exhaustive search of periodical literature than we have been able to make would result in the discovery of earlier records for some of the species. The same would apply to a few rare botanical works to which we have not had access. Whenever possible, the locality on which a first record is based is quoted in its sequence with the later records, and it is then usually distinguished by having the date appended. Unfortunately, we have been unable to supply this information with many of the records from "Topographical Botany."

LOCALITIES.—Where no authority for these is quoted, the Authors jointly are responsible for them. When recorded by other botanists, the following abbreviations are used.

ABBREVIATIONS.

A. A. D. ...	A. A. Dallman, of Liverpool.
A. B. ...	Arthur Breakell, of Garstang.
Ashfield ...	Charles Joseph Ashfield, of Preston.
Carr ...	Amos Carr, of Sheffield.
Cash ...	James Cash, of Manchester.
C. B. ...	Charles Bailey, of St. Annes-on-Sea.
E. C. ...	Edwin Curwen, of Garstang.
E. F. L. ...	E. F. Linton.
E. S. M. ...	E. S. Marshall, of Taunton.
F. A. L. ...	F. A. Lees, of Leeds.
H. B. ...	Henry Beesley, of Preston.
Holt ...	G. A. Holt, of Manchester.
J. A. M. ...	J. A. Martindale, of Stavely.
Jenkinson ...	James Jenkinson, of Yealand.
J. F. P. ...	Joseph F. Pickard, of Leeds.
King ...	F. C. King, formerly of Preston.
Lawson ...	Thomas Lawson, a correspondent of Ray.
Lewis ...	John Harbord Lewis, late of Liverpool.
L. P. ...	Lister Petty, of Ulverston.
Moss ...	John Moss, of St. Michael's-on-Wyre.
P. J. H. ...	Phipps J. Hornby, of St. Michael's-on-Wyre.
Ray ...	John Ray, "The Father of English Botany."
Searle ...	Henry Searle, of Ashton-under-Lyne.

Stabler ...	George Stabler, of Levens.
S. & T. ...	C. E. Salmon and H. S. Thompson.
S. Wi. ...	Sydney Wilson, of Perth, late of Garstang.
W. C. W....	W. C. Worsdell.
Wh. ...	J. A. Wheldon, of Liverpool.
Wi. ...	Albert Wilson, of Ilkley.
W. K. ...	William Kirkby, of Leeds.
W. P. H....	William P. Hamilton, of Shrewsbury.
W. W. M....	W. Wright Mason, of Bootle, Lancs.

OTHER ABBREVIATIONS.

A.	Annual.
B.	Biennial.
B.E.C. Rep. ...	Botanical Exchange Club Report.
B.R.C. Rep. ...	Botanical Record Club Report.
Bot. Guide ...	Turner & Dillwyn's "Botanists Guide."
Brit. Pl. ...	Jenkinson's "Descriptions of British Plants."
Dist.	District.
E.	East Division.
Fl. Prest. ...	Ashfield's "Flora of Preston."
Fl. Prest. N. ...	Flora of Preston Neighbourhood (Preston Scientific Society).
Fl. St.... ...	Flora of Stonyhurst.
Herb.	Herbarium.
Journ. Bot. ...	Journal of Botany.
M.E.C. Rep. ...	Moss Exchange Club Reports.
N.	North Division.
P.	Perennial.
Sh.	Shrub.
Sp.	Specimen.
T.	Tree.
Top. Bot. ...	Topographical Botany, 3rd Ed.
W.	West Division.
! after locality	Seen there by the Authors.
! after sp. ...	Specimen seen by the Authors.

CLASS I.—DICOTYLEDONES.

I. RANUNCULACEÆ.

Clematis Vitalba, L.—Traveller's Joy. Alien. First record, *Lees, Naturalist*, 1899, p. 299.

N. 3. Sown or planted amongst Yews near Bare, 1899. *F.A.L.* Under the wall dividing Lancashire from Westmorland in the Silverdale district, *W.K.*

Thalictrum dunense, Dum.—Lesser Meadow Rue. Scottish type. Native. Area, 5. Range, only at 25 feet. P. July.

Sandy sea shores. Very rare. First record: *Ashfield, Fl. Prest.*, 1860.

W. 5. On the beach at Lytham, *Ashfield*, 1860. Not seen recently.

Thalictrum montanum, Wallr.—Meadow Rue. Scottish type. Native. Area, 2. Range, only seen at 700 feet. P. July.

Scars and crevices of Mountain Limestone. Very rare. First record: *S. Wilson, Journ. Bot.*, 1900, p. 41, as *T. collinum*.

N. 2. Dalton Crag, where it was detected by *S. Wilson*, in company with the Authors, June, 1899.

Thalictrum collinum, Wallr.—Meadow Rue. Scottish type. Native. Area, 1. Range, 160-175 feet.

Damp bushy places and wooded river banks. Rare. First record: *Lees, Fl. of W. Yorks*, p. 112.

N. 1. "Wenning banks from Clapdale down to Wennington," *F.A.L.*

Thalictrum flavum, L.—Meadow Rue. English type. Native. Area, 3, 4, 5. Range, 0-50 feet. P. June–July.

Watery places and ditch sides in the lowlands. Rare. First record: *Ashfield, Fl. Prest.*, 1864.

N. 3. Ditch near Silverdale Station, *Wi.* Silverdale Moss, *L.P.*
W. 4. Plentiful in ditches near Pilling, and in several places near St. Michael's, *Ashfield*, 1864. Still at Pilling, *Wi.* Wardless, *H.B.* Rawcliffe Moss, *P.J.H.* Near Winmarleigh.
5. Bank of the Wyre opposite Whitehall, *A.B.*

Our plant seems to agree best with *T. sphærocarpum* (Lej.), having a close panicle and ovoid carpels.

Anemone nemorosa, L.—Wood Anemone, "Billy Button." British type. Native. Area, 1, 2, 3, 4, 5, 6, 7, 8. Range, 0-1,050 feet. P. March–May.

Woods, hedgebanks, and occasionally in pastures. Common, except in parts of West. First record: *Linton, B.R.C. Rep.* for 1874, p. 80.

It ascends to 1,050 feet on Leck Fell in North, and to 760 feet on Greenside, Wyresdale in East.

[*Anemone Pulsatilla*, L.—Pasque Flower. Incognit.

In the *Phytologist* for Jan., 1862, the following passage occurs in an article on the "*Catalogue of plants cultivated by Coolinson*":—"Mr. Knowlton found millions of *Pulsatillas* growing everywhere in the grass, from a mile south of Lancaster, on the way to Little Purton." Ashfield suggests that this locality was near Londesborough in Yorkshire. There is no locality at all likely for this plant near Lancaster.]

Anemone ranunculoides, L.—Alien. First record, *P. J. Hornby, Journ. Bot.*, 1900, p. 40.

E. 8. Redscar, near Preston, *P.J.H.*

Adonis annua, L.—Pheasant's Eye. Alien. First record: *Lees, Naturalist*, 1900, p. 299.

N. 3. Casually on waste ground near Bare, *F.A.L.*

Myosurus minimus, L.—Alien. First record: The present one.

N. 3. Reported from near Slyne, where it may have been of accidental origin, about 50 years ago, by *Mrs O. Wilson*.

Ranunculus divaricatus, Schrank. (*R. circinatus*, Sibth.).— English type. Native. Area, 2, 3, 4, 5, 6, 7. Range, 0-150 feet. P. June-Aug.

Canals and slow streams, especially frequent in the Preston and Kendal canal. First record: *Wheldon, Journ. Bot.*, 1900, p. 41.

N. 2. In the canal near Borwick.
3. Near Slyne, *Wi.*
W. 4. Canal near Cock Hall, *Wi.* Canal between Winmarleigh and Garstang, and near Cabus; canal near Glasson, *Wh.*
5. Canal near Barton, *Wi.* Ditches near Marton Mere, *Wh.* St. Michael's, *F.P.N.*
E. 6. Canal near Lancaster, June, 1899, *Wh.*
7. Canal near Garstang, *Wi.*

Ranunculus fluitans, Lam.—English type. Native. Area, 8. Range, only at about 75 feet. P. June–July.

Rivers. Very rare. First record: *Fl. St.*, 1891, p. 5.

E. 8. In the Ribble at Ree Deep, not flowering, *Fl. St.*

Ranunculus Drouettii, F. Schultz.—British type. Native. Area, 1, 3, 4, 5, 8. Range, 0-400 feet. P. May–July.

J

Ponds and ditches. Rare. First record: *Bailey, B.C.R. Rep.*, 1883, p. 183.

N. 1. Mouth of the Greeta near Tunstall, *Wi.*
3. Ditch near Arnside Tower, 1882, *C.B.* Ditch near Yealand, *Wi.* Heysham Moss; ditch between Carnforth and Bolton-le-Sands, *Wh.*
W. 4. Ditch near Winmarleigh. Near Wardleys, *Wh.*
5. Pond near Catforth, *Wi.* Near Catterall, *Wi.* Near Preston, *Wh.*
E. 8. Stonyhurst Mill pond, *Fl. St.*

Ranunculus heterophyllus, Weber.—Water Crowfoot. British type. Native. Area, 2, 3, 4, 5, 6. Range, 0-300 feet. P. May–July.

Ditches, ponds and slow streams. Frequent in the low country; rare in East. First record: *Bailey, B.R.C. Rep.*, 1883, p. 203.

N. 2. Between Caton and Aughton; near Dunnald Mill Hole; near Morecambe, *Wh.*
3. Pool near Silverdale, *F.A.L.* Silverdale Moss, *Wi.*
W. 4. Ditches near Pilling, *Wh.*
5. Pond near Little Marton, *Wh.*
E. 6. Canal near Lancaster, *Wi.* In the Lune near Claughton, *Wi.*

Var. *submersus*, Hiern.

N. 3. Ditch by the railway between Silverdale and Arnside, 1881, *Bailey*. Canal between Skerton and Hest Bank, *Wi.*
W. 4. Ditch at Cockerham, *Wi.*
5. Ditch near Marton Mere, *Wh.*

Ranunculus peltatus, Schrank.—Water Crowfoot. British type. Native. Area, 1, 3, 4, 5, 8. Range, 0-250 feet. P. May–July.

First record: *Hiern. Top. Bot.*, 1883.

N. 1. Marshy broads near Tunstall, *Wi.*
3. In the stream from Burton Well, *L.P.* Near Thrang End, *Wi.* Ditches on Heysham Moss, in plenty.
W. 4. Ditches near Pilling, *Wi.*
5. Wood Plumpton, *P.J.H.* Sowerby Marshes, *Wi.*
E. 8. In the reservoir at Grimsargh, *Wh.* Quarry near Thornley, *Wi.*

Ranunculus Baudotii, Godr.—Water Crowfoot. English type. Native. Area, 3, 4, 5. Range, 0-25 feet. P. May–Aug.

Ponds and ditches near the sea. Rare. First record: *Hiern., Top. Bot.*, 1883.

N. 3. Muddy pools on Overton Marsh, *Wi.* Bolton-le-Sands, *Wi.*
W. 4. On the mud of a drying-up ditch between Cockerham and Cockersand, *Wi.*
5. Pond on the sandhills near St. Annes, 1899; very fine in the old river bed near Preston and on the salt marsh near Fleetwood, *Wh.*

Ranunculus Lenormandi, F. Schultz.—English type. Native. Area, 2, 4, 6, 7, 8. Range, 0-1,400 feet. P. June-Aug.

Ponds and rill sides. Frequent in the hill country; rare in the lowlands. First record: *Linton, B.R.C. Rep.*, 1874, p. 80.

N. 2. Whittington Moor and near Arkholme, *Wi.* Near Lord's Lot Wood, *Wh.*
W. 4. Moss-side, St. Michael's, *P.J.H.*
E. 6. Frequent in this district, ascending to 1,400 feet on Botton Head Fell.
7. Marshaw, Tarnbrook and Lower Heasdale, *Wi.* Between Abbeystead and Dolphinholme, *Wh.* Roadside above Catshaw. Ascends to 1,100 feet in Gavell's Clough, Over Wyresdale.
8. Below Composition Hill and near Kemple End! *Fl. St.*, 1891. Jeffrey Hill, *A.A.D.*

Ranunculus hederaceus, L.— Ivy-leaved Crowfoot. British type. Native. Area, 1, 2, 3, 4, 5, 6, 7, 8. Range, 0-600 feet. P. May–Sept.

Muddy ditches and pools. Frequent. First record: *Ashfield, Fl. Prest.*, 1858. "Ribbleton Moor."

Ranunculus sceleratus, L.—Celery-leaved Buttercup. British type. Native. Area, 3, 4, 5, 8. Range, 0-150 feet. P. June–Sept.

Ditches and pond-sides in the low country, principally near the sea. Rather uncommon. First record: *Lawson, Ray's Corresp.*, 1718.

N. 3. By Middleton near Lancaster, *Lawson*, 1688. Still frequent in the Heysham Peninsula; Carnforth, *Wi.* Near the Pool, Silverdale, *L.P.*
W. 4. Upper Thurnham; about Cockerham, and in other places near the sea, *Wi.* Salt marsh south of Glasson, *Wh.*
5. Blackpool, Myerscough, and in a pit near Barton, *Wi.* Greave's Town and Ashton Marsh, *Ashfield* About Freckleton and Kirkham, and very fine in Fleetwood Marsh, *Wh.*
E. 8. Near Barton Mill, *Wi.*

Ranunculus Flammula, L.—Lesser Spearwort. British type. Native. Area, 1, 2, 3, 4, 5, 6, 7, 8. Range, 0-1,950 feet. P. May–Sept.

Ditches and watery places. Common. First record: *Linton, B.R.C. Rep.*, 1874, p. 80.

Var. *pseudo-reptans*, Syme, is a small form occurring in spongy bogs amongst the hills—on Greygarth Fell, in Hindburn, near Catshaw, on Tarnbrook Fell, and below Black Clough, Marshaw.

[*Ranunculus Lingua*, L.—Great Spearwort. Incognit.

"I was informed that it grew in ditches between Kirkham and Blackpool." *Ashfield*, 1858. We have not been able to obtain further confirmation of this, and, although the locality is a very probable one, we do not feel able to include it on the above evidence, which, it will be noted, is not that of Ashfield himself. It occurs in S. Lancashire in similar ditches near the above.]

Ranunculus auricomus, L.—Goldilocks. British type. Native. Area, 1, 2, 3, 4, 5, 6, 7, 8. Range, 0-500 feet. P. May.

Hedgebanks and woods. Frequent in the limestone districts of North; much rarer elsewhere. Sub-xerophilous in W. Lancs. First record: *Ashfield, Fl. Prest.*, 1858.

N. 1. Frequent throughout the district, *Wi.*
   2. Whittington and Over Kellet, *Wi.*
   3. Frequent about Silverdale and Yealand.
W. 4. Between Stodday and Galgate, *Wi.* Kirkland, *P.J.H.*
   5. Meadow between Cottam Mill towards Ashton, *Ashfield*, 1858. Plantation near Ashton, *A.A.D.*
E. 6. Wennington, *Wi.* Near Cowkins and Wray.
   7. Near Garstang, *E.C.*
   8. Tunbrook Wood and Fishwick, *Ashfield*. Riddings Farm Meadow, *Fl. St.* Near Greystoneley, *Wi.*

Ranunculus repens, L.—Creeping Buttercup. "Cat-clawks." British type. Native. Area, 1, 2, 3, 4, 5, 6, 7, 8. Range, 0-2,040 feet. P. June-Aug.

Fields, banks and waste places, generally in damp situations. Very common. First record: *Linton, B.R.C. Rep.*, 1874, p. 80.

Ascends in North to 2,040 feet on Greygarth Fell; in East to 1,100 feet in Gavell's Clough, Wyresdale.

Ranunculus bulbosus, L.—Buttercup. British type. Native. Area, 1, 2, 3, 4, 5, 6, 7, 8. Range, 0-870 feet. P. May-June.

Meadows and pastures. Common, except in the hill districts of East. First record: *Linton, B.R.C. Rep.* for 1874, p. 80.

In North it ascends to 870 feet on Dalton Crag.

Ranunculus acris, L.—Buttercup. British type. Native. Area, 1, 2, 3, 4, 5, 6, 7, 8. Range, 0-1,930 feet. P. June-Sept.

Pastures and grassy places. Very common. First record: *Top. Bot.*, 1883.

Of the segregates classed under *R. acris* Linn., the commonest W. Lancs. form is *R. Boreanus* Jord., which appears to be scattered throughout the county. *R. tomophyllus* Jord. has been noted at Abbeystead and Silverdale, but it is connected with *R. Boreanus* by many intermediate forms. *R. rectus* Bor. appears to be rare. It occurs at Caton and Dolphinholme. *R. vulgatus* Jord. is uncertain as a W. Lancashire species, although a very downy plant from Easegill, with the elongated beak of the carpels strongly hooked, may perhaps be referred to it.

Ranunculus Sardous, Crantz.—English type. Native probably. Area, 5.

Sandy ground near the sea. Very rare. First record: *Boswell* in *Top. Bot.*, 1883. We have been unable to ascertain where Boswell's station for the plant was.

W. 5. Pasture near the canal, Ashton, *Fl. Prest. N.*

Ranunculus Ficaria, L.—Pilewort. Lesser Celandine. British type. Native. Area, 1, 2, 3, 4, 5, 6, 7, 8. Range, 0-1,200 feet. P. March-May.

Hedgebanks and damp shady places. Very common. First record: *Linton, B.R.C. Rep.*, 1874, p. 80.

Only the var. *divergens* Bab. occurs in W. Lancashire. It varies a good deal in the degree of proximity of the lower leaf lobes, but we have never seen forms in which they distinctly overlap as in the var. *incumbens* Schultz. Both *Uromyces ficariæ* (Schum.) and *U. poæ* (Rabh.) occur on its leaves and petioles.

Caltha palustris, L.—Marsh Marigold. "Mayflower" or "May Buttercup." British type. Native. Area, 1, 2, 3, 4, 5, 6, 7, 8. Range, 0-1,150 feet. P. March-May.

Damp meadows and marshy places. Common. First record: *Linton, B.R.C. Rep.* for 1874, p. 80.

Var. *minor*, Syme.—Spongy bogs about springs on the fells. Rare. 7. Springs near the foot of Gavell's Clough, on the White side of Tarnbrook Fell, alt. 1,150 feet; June, 1902.

Var. *Guerangerii*, Boreau.—Near Dolphinholme, June, 1900, *Wh.*

Trollius europæus, L.—Globe Flower, "May Buttercup." Scottish type. Native. Area, 1, 3, 6, 7, 8. Range, 650-50 feet. P. May-June.

Damp upland meadows and wooded banks of stony river beds. Rare. First record: *Ashfield, Fl. Prest.*, 1866.

N. 1. Gorge of the river Greeta.
   3. By the Leighton Beck, *Wi.*

E. 6. Bank of Lune between Caton and Halton, *Wi.*
   7. Meadows at Oakenclough, *Wi.* Higher Brock, *H.B.*
   8. Wood between Sale Wheel and Dinckley Hall, 1864, *Ashfield*. Riddings Farm Meadow and Harmer Ridding, *Fl. St.*

Helleborus viridis, L.— Green Hellebore. Germanic type. Native. Area, 3, 4, 5, 7. Range, 50-300 feet. P. March-April.

Woods and hedge-banks. Rare. First record: *Jenkinson, Brit. Plants*, 1775.

N. 3. About Leighton Hall, Yealand, plentifully, *Jenkinson*, 1775. Near the Cove, Silverdale, abundant, *R. Parkinson* (1870). Still there, *Wi.*
W. 4. Humblescough near Cabus, *A.B.*
   5. Between Woodplumpton Road and Cottam Hall, *Ashfield*. Still there in 1901, *H.B.* (sp.).
E. 7. Bank of the river Brock, *Wi.*

We regard it as truly native on the limestone of North; most likely only a denizen in the other divisions.

Helleborus fœtidus, L.—Stinking Hellebore. Bear's Foot. English type. Denizen or native? Area, 3, 5. Range, 100-150 feet. P. Feb.-April.

Limestone woods and thickets. Very rare. First record: *H. Beesley* and *Wilson, Journ. Bot.*, 1902, p. 346.

N. 3. Gatebarrow Woods, near Silverdale, 1901, *Wi.*
W. 5. Barton, near Preston, *H.B.* (Sp.)

Aquilegia vulgaris, L.—Columbine. English type. Native. Area, 2, 3, 4, 6, 7, 8. Range, 0-400 feet. P. June.

Woods and thickets, mostly on the limestone, and only native there. Rather rare. First record: *Ashfield, Bot. Chron.*, 1864, p. 73.

N. 2. Over Kellet, *Wi.*
   3. Abundant, Silverdale, 1863, *Ashfield*. Still plentiful there. Bolton-le-Sands, *Wi.*
W. 4. Near Winmarleigh, *Wi.* Cabus near Garstang.
E. 6. Near Hornby, *Wi.*
   7. Woods in Brock Valley, *Ashfield*. Hedgerow between Bay Horse and Dolphinholme, *Mason* and *Wh.*
   8. Between Sale Wheel and Dinckley Hall, *Ashfield*. Naturalised at Stonyhurst, *Fl. St.* Inglewhite Lane, *A.A.D.*

Aconitum Napellus, L.—Aconite. Monkshood. Alien. First record: *Fl. St.*, 1891.

N. 2. Subspontaneous on the banks of the Wash Dub Brook.
E. 8. Garden escape, Bailey Hall Wood, 1891, *Fl. St.*

Actæa spicata, L.—Bane Berry. Herb Christopher. Intermediate type. Native. Area, 1. Range, 1,050-850 feet. P. May-June.

Rock crevices and woods on limestone. Very rare. First record: *Wilson, Journ. Bot.*, 1900, p. 41.

N. 1. In a pot-hole on Leck Fell, amongst bushes on an inaccessible rock ledge, 1888. Limestone crevices in Ease Gill, *Wi.*

BERBERIDACEÆ.

Berberis vulgaris, L.—Barberry. English type. Denizen. Area, 3, 5. Range, 0-230 feet. Sh. May-June.

Hedges and thickets. Rare. First record: *Wilson, B.R.C. Rep.*, 1883.

N. 3. Whin's Brow, Silverdale, several plants, 1882, *Wi.*
W. 5. One tree in a hedge, Lea, near Preston, *A.A.D.*
E. 8. "I am told it grows in a hedge near the Preston Cemetery," *Ashfield*, 1858.

NYMPHÆACEÆ.

Nymphæa lutea, L.— Yellow Water-lily. "Tarn Tulip." "Brandy Bottle." English type. Native. Area, 3, 4, 5, 6, 7, 8. Range, 0-120 feet. P. June-July.

Sluggish streams and ponds. Abundant in some parts of the low country. First record: *Jenkinson, Brit. Plants*, 1775, p. 113.

N. 3. "About Hairs Water, within the liberties of Yeland Redman," *Jenkinson*, 1775. Still there, and in a pond near Yealand, *Wi.*
W. 4, 5. Common in both districts.
E. 6. Marsh by the Lune, near Claughton, *Wi.*
   7. Ponds near Garstang, *Wi.*
   8. Stonyhurst Millpond, introduced, *Fl. St.*

Castalia alba, Greene.—White Water-lily. British type. Native. Area, 3, 4, 5, 7. Range, 0-80 feet. P. June-July.

Ponds and slow waters in the low country. Rare. First record: *Jenkinson, Brit. Plants*, 1775, p. 113.

N. 3. "About Hairs Water," *Jenkinson*, 1775. Still there.
W. 4. Canal near Garstang, *Wi.* Cabus.
   5. Pond between Woodplumpton Church and Carr's Green, *Ashfield*. Ponds near Poulton-le-Fylde, *S. Wi.*
E. 7. Canal and ponds near Garstang, *Wi.*

PAPAVERACEÆ.

Papaver Rhœas, L.—Corn Poppy. British type. Colonist. Area, 2, 3, 4, 5, 7, 8. Range, 0-300 feet. A. June–August.

Cultivated fields and waste ground. Rather frequent in North and West, especially in the tracts nearest the sea; rare elsewhere. First record: *Linton, B.R.C. Rep.*, 1874, p. 80.

N. 2. Ascends to 200 feet near Over Kellet, *Wi.*
   3. Near Silverdale Station, *Wi.* Oatfield, near Cray Green, *L.P.* Plentiful in fields near Morecambe and Torrisholme, *Wh.*
W. 4. Cornfield, near Pilling, *Jones* and *Wh.*
   5. Salwick, Kirkham, and Lytham, sparingly, *Wh.*
E. 7. Near Garstang, *Wi.*
   8. Cross Gill's Farm, *Fl. St.*

Var. *strigosum*, Boeun.—Alien. Occurred casually by the side of a newly-made road near Morecambe, *Wh.*, 1899.

Papaver dubium, L.—Corn Poppy. British type. Colonist. Area, 2, 3, 4, 5, 6, 7, 8. Range, 0-400 feet. A. June–August.

Cultivated fields, gardens and waste ground. Common, except in East. First record: *Linton, B.R.C. Rep.*, 1874, p. 80.

N. and W. Common, except in District 1.
E. 6. Caton, *Wi.*
   7. Near Garstang, *Wi.*
   8. Fulwood, *Wh.* Cornfields near Ashton, *Fl. Prest. N.*

Papaver Argemone, L.—British type. Colonist? Area, 5. Range, 0-50 feet. A. June–August.

Cornfields. Rare. First record: *Wheldon, Journ. Bot.*, 1900, p. 41.

W. 5. Cornfield near St. Annes, 1898, *Wh.* Lytham, *Fl. Prest. N.*

Meconopsis cambrica, Vig.—Welsh Poppy. Alien.

N. 3. Near Bare, "about poultry runs," *F.A.L.*
E. 8. By the Hodder, near Whitewell, but introduced, *Ashfield*, 1863. Garden escape near Leagram, *Fl. St.*

Glaucium flavum, Crantz.— Horned Poppy. English type. Native. Area, 3, 4, 5. Range, 0-25 feet. B. June–Sept.

Shingly places on the sea coast. Rare, and diminishing. First record: *Jenkinson, Brit. Plants*, 1775, p. 111.

N. 3. "I found it near Warton, Lancashire," *Jenkinson,* 1775. Shore north of Keer, *W.C.W.* Near Bolton-le-Sands, *Wi.* Very abundant between Hest Bank and Morecambe; still occurs there. Sea shore near Silverdale, *Ashfield.* Not seen there recently, but, of course, quite likely to have occurred.
W. 4. Occasionally between Wardless and Fleetwood, *Ashfield.* Shingle between Glasson Dock and Cockersand Abbey, in two localities, *Wi.*
   5. Near Lytham, *H.B.*

Chelidonium majus, L.—Greater Celandine. English type. Denizen. Area, 1, 2, 3, 4, 5, 6, 7, 8. Range, 0-300 feet. P. May–August.

Hedgebanks and roadsides; generally near houses. Frequent. First record: *Ashfield, Fl. Prest.* 1858. "Greave's Town, Ashton."

FUMARIACEÆ.

Corydalis lutea, D.C.—Alien. First record: *Ashfield, Fl. Prest.*, 1860.

Garden weed at Garstang, 1860. *Mr. Pearson (Ashfield).* Little Singleton, *P.J.H.* Barton, *Fl. Prest.*, N.

Corydalis claviculata, D.C.—British type. Native. Area, 4, 5, 6, 7, 8. Range, 0-500 feet. A. June–August.

Rocky and sandy woods and banks, on grit. Rare. First record: *Ashfield, Fl. Prest.*, 1858.

W. 4. Moss Side near St. Michael's *P.J.H.*
   5. Between St. Michael's and Inskip, *Wi.*
E. 6. Near Tatham, Hindburn, *Wi.*
   7. Roadside between Beacon Fell and the Brock, *Wi.* Damas Gill, and between Damas Gill and Hay Horse, *Wi.*
   8. Kemple End and Leagram, *Fl. St.* Brockholes, *Ashfield* and *Fl. Prest. N.*

Fumaria capreolata, L. (*F. pallidiflora, Jord.*).—Ramping Fumitory. British type. Colonist. Area, 3, 4, 5, 6, 7. Range, 0-400 feet or higher. A. June–October.

Cultivated land or waste ground in its vicinity. Not unfrequent in the low country. First record: *Amos Carr, B.R.C. Rep.*, 1879, p. 52.

Fumaria purpurea, Pugsley. — Purple-flowered Fumitory. British type. Colonist. Area, 3, 4. Range, 0-200 feet or higher. A. June–October.

Cultivated fields and waste ground. Not common. First record: The present one.

N. 3. Near Silverdale, *Wh. & Wi.* Turnip field at the foot of Warton Crag, *Wh.*
W. 4. Plentiful on a newly-made embankment near Preesall, *Wh.*, 1900.

This is the most handsome of our fumitories, notwithstanding the fact that its flowers are a little smaller than those of the preceding. All our examples were named by Mr. Pugsley.

Fumaria Boræi, Jord.—British type. Colonist. Area, 1, 3, 4, 5, 8. Range, 0-400 feet? A. June–October.

Cultivated ground and hedgerows. Frequent? First record: *Fl. St.*, 1891. "Garden weed at Stonyhurst."

Var. *serotina*, Clavaud. (= *F. muralis*, Lond. auct. angl. ex-parte). With *F. purpurea* in the Warton Crag locality, Sept., 1906, *Wh.* (teste *Pugsley*). Near Carnforth, *Wi.*

Owing to recent changes in the nomenclature in this genus, it is probable that the distribution of this species shown above is only approximately correct. Some of the older records may prove to belong to *F. purpurea, Pugs.*

Fumaria Bastardi, Bor. (*F. confusa*, Jord.)— British type. Colonist. Area, 5. Range, only seen at about 50 feet. A. June–October.

Cultivated fields and waste places. Rare or overlooked. First record: *E. S. Marshall* in *Journ. Bot.*, 1896, p. 136.

W. 5. Near Little Eccleston, 1895, *E.S.M.* (teste *Pugsley*).

Fumaria officinalis, L.—Common Fumitory. British type. Colonist. Area, 1, 2, 3, 4, 5, 7, 8. Range, 0-400 feet or higher. A. June–September.

Cultivated fields and gardens. Common. First record: *Linton, B.R.C. Rep.*, 1874, p. 81.

CRUCIFERÆ.

Cheiranthus Cheiri, L.—Wallflower. British type. Denizen. Area, 3. Range, 0-50 feet. B. May.

Limestone cliffs by the sea. Rare. First record: *Ashfield, Bot. Chron.*, 1864, p. 73.

N. 3. Cliff between Silverdale and Arnside Point, *Ashfield*, 1864. Still there, *Petty*, 1898; and later !

Radicula Nasturtium-aquaticum, Britt and Rendle (*Nasturtium officinale, R. Br.*).— Water Cress. "Crash." British

type. Native. Area, 1, 2, 3, 4, 5, 6, 7, 8. Range, 0-600 feet. P. May–October.

Watery places and springs, and by slow flowing streams. Common. First record: *Linton, B.R.C. Rep.*, 1874, p. 81.

Radicula sylvestris, Druce.—English type. Native. Area, 5, 8. Range, 0-200 feet. P. June–September.

Stream sides and damp places. Rather rare. First record: *Ashfield, Fl. Prest.*, 1858.

W. 5. Seashore between Freckleton and Lytham, *Wi.* Wall above stone delph, Preston, *Ashfield.*
E. 8. Abundant by the river near Preston, *Wh.*

Radicula palustris, Moench (*N. terrestre, Sm.*)—English type. Native. Area, 1, 2, 3, 4, 5, 6, 7, 8. Range, 0-400 feet or higher. P. June–September.

Wet places, ditches, and stream sides. Much more frequent than the preceding. First record: *Ashfield, Fl. Prest.*, 1858.

N. 1. By the Lune near Tunstall, *Wi.*
   2. Near Dunnald Mill Hole.
   3. Canal near Bolton-le-Sands, *Wi.* Ponds near Overton.
W. 4. Near Pilling, *Wi.* Winmarleigh.
   5. Between Greaves Town and Ashton Marsh, *Ashfield.* Woodplumpton Brook, *Wi.* Fleetwood, *Wi.* Between Lytham and the Guide House Inn, *S.* and *T.* Marton Mere, *Wh.*
E. 6. Near Caton, *Wi.*
   7. Near Garstang, *Wi.* Near Abbeystead at 400 feet.
   8. Inglewhite, near Garstang, *Wi.*

Barbarea lyrata, Aschers (*B. vulgaris R. Br.*).—Winter Cress. British type. Native. Area, 1, 2, 3, 4, 5, 6, 7, 8. Range, 0-500 feet. B. May–July.

Hedgebanks, streamsides, and waste places. Frequent in the low country, rare amongst the hills. First record: *Ashfield, Fl. Prest.*, 1858.

Barbarea præcox, Aschers.—Alien. First record: *Journ. Bot.*, 1902, p. 346.

W. 5. Garden ground near St. Michael's, *P.J.H.*

Arabis hirsuta, Scop. — Hairy Rock Cress. British type. Native. Area, 1, 2, 3, 8. Range, 0-1,120 feet. P. May–July.

Dry rocks, walls, and stony banks. Common in the limestone districts; rare in others. First record: *Ashfield, Bot. Chron.*, 1864, p. 73.

N. 1. Ease Gill, *L.P.* Leck Fell, ascending to 1,120 feet, and near Wrayton, *Wi.*
2. Near Over Kellet, *Wi.* Dalton Crag.
3. Not uncommon about Silverdale, *Ashfield*, 1864. Still frequent there. Between Carnforth and Bolton-le-Sands, *Wi.*
E. 8. Leagram, *Fl. St.* Gorge of the Hodder below Whitewell. Near Chipping, *Wh.*

Cardamine amara, L.—Bitter Cress. British type. Native. Area, 1, 2, 3, 4, 5, 6, 7, 8. Range, 0-500 feet. P. May–June.

Shaded marshy places and wet woods. Not uncommon. First record: *Ashfield, Fl. Prest.*, 1858.

N. 1. Leck and Wennington, *Wi.* Banks of the Lune, *Wi.* Ease Gill.
2. Wood by the Lune, north of Caton, *Wi.*
3. By the canal near Hest Bank, *Wi.* Near Middleton, *Wi.*
W. 4. Catterall's Farm, *P.J.H.* Canal between Stodday and Galgate, *Wi.*
5. Near Sowerby, *P.J.H.* Ashton, *Fl. Prest. N.*
E. 6. Near Caton, *Wi.* Roeburndale.
7. Brock Valley, frequent, *Wi.* Near Garstang, *Wi.*
8. Inglewhite, *P.J.H.* Near Whitewell, *Wh.* Tunbrook Wood and Longridge Fell, *A.A.D.* Thornley and Greystoneley, *Wi.*

Cardamine pratensis, L.—Lady's Smock. Cuckoo Flower. "Bird's Eye." "Mayflower." British type. Native. Area, 1, 2, 3, 4, 5, 6, 7, 8. Range, 0-1,120 feet. P. April–June.

Damp meadows and grassy places. Common. First record: *Linton, B.R.C. Rep.*, 1874, p. 81.

In North it ascends to 1,120 feet on Leck Fell, and in East to 800 feet on the White side of Tarnbrook Fell.

Cardamine hirsuta, L.—Hairy Bitter-cress. British type. Native. Area, 1, 2, 3, 4, 5, 6, 7, 8. Range, 0-2,030 feet. A. April–July.

Walls, rocks, and banks. Common. First record: *Ashfield, Fl. Prest.*, 1858. "Abundant on banks by the side of Pope Lane."

Cardamine flexuosa, With.—Wood Bitter-cress. British type. Native. Area, 1, 2, 4, 6, 7, 8. Range, 100-1,200 feet. A. April–July.

Moist banks by streams and in woods. Frequent in the fell districts. First record: *Lewis, B.R.C. Rep.*, 1884, p. 7. "Thornley."

It ascends to 1,200 feet on Mallowdale Fell.

Cardamine impatiens, L.—Atlantic type. Native. Area, 3. Range, 200-300 feet. A. June–July.

Rocky woods on limestone. Rare. First record: *Melvill, Top. Bot.*, 1883, p. 44.

N. 3. Middlebarrow Wood near Silverdale, *J. C. Melvill*, 1868.

Alyssum incanum, L.—Alien.

W. 5. Sandhills near Lytham, *Fl. Prest. N.*, 1903, p. 8.

Alyssum maritimum, L.—Alien.

W. 5. Sea shore near Ansdell, *Fl. Prest. N.*, 1903, p. 8.

Hesperis matronalis, L.—Dame's Violet. Alien. First record: *Melvill, Naturalist*, 1902, p. 36.

N. 3. Hawes Tarn,, *J. C. Melvill*, 1868.
W. 5. Catforth, *A.A.D., Ashfield, Fl. Prest. N.*
E. 8. Whitewell, on the banks of the Hodder, *Wh.* Fulwood, *Fl. Prest. N.*

Isatis tinctoria, L.—Woad. Alien. First record: *Jenkinson, Brit. Pl.*, 1775.

N. 3. Roadside in Yealand Storrs, and in several other places in and about the Yealands, *Jenkinson*, 1775.

Draba muralis, L.—Speedwell-leaved Whitlow-grass. Local type. Native. Area, 2, 8. Range, 200-350 feet. A. April–May.

Shady limestone rocks. Very rare. First record: *Wilson, Journ. Bot.*, 1902, p. 347.

N. 2. On limestone between Kirkby Lonsdale and Whittington, May, 1901, *Wi.*
E. 8. Limestone rocks in the Hodder Valley.

Draba incana, L.—Twisted-podded Whitlow-grass. Highland type. Native. Area, 1. Range, 1,250 or higher, to 800 feet. P. June–July.

Rocky scars and rough mountain pastures on limestone. Very rare. First record: *Wilson, B.R.C. Rep.*, 1887, p. 86.

N. 1. On the Scar limestone of Leck Fell and Ease Gill, and with it occurs also the var. *confusa* Ehrh., *Wi.*

Draba verna, L.—Whitlow-grass. British type. Native. Area, 1, 3, 4, 5, 8. Range, 0-2,050 feet. A. March–May.

Rocks, walls, and dry banks. Frequent in North and some parts of West, rare in East. First record: *Ashfield, Fl. Prest.*, 1860. "On the beach at Lytham, between the windmill and the custom-house."

Draba præcox, Stev.—Whitlow-grass. British type. Native. Area, 3, 4, 5, 8. Range, 0-800 feet or higher. A. March–May.

Limestone rocks and dry sand-dunes by the sea. Rare. First record: *Bailey, B.R.C. Rep.*, 1883, p. 184.

N. 3. Silverdale, *Wi.*
W. 4. Between Pilling and Knott End, *Wh.*
5. Lytham! and Blackpool, 1883, *Bailey.* St. Annes, *H.B.* (Sp.)
E. 8. Near Whitewell.

Cochlearia officinalis, L.—Common Scurvy-grass. British type. Native. Area, 3, 4, 5, 6 (8). Range, 0-25 feet. B. May–July.

Sea shores, salt marshes, and on the banks of tidal rivers. Frequent. First record: *Ashfield, Fl. Prest.*, 1858.

N. 3. Silverdale coast, *Wi.* Carnforth and Ovangle, *Wh.*
W. 4. Frequent on the Cockerham coast and about the Wyre estuary, *Wi.*
5. Ashton Marsh, *Ashfield*, 1858.
E. 6. Lune banks near the Lancaster Carriage Works, *Wi.*
8. Ribble banks near Stonyhurst, *Fl. St.* Red Scar, *Fl. Prest. N.* (These were probably *C. alpina* brought down from the Yorkshire hills.)

Cochlearia alpina, H. C. Wats.—Highland type. Native. Area, 1. Range, 2,040-100 feet. B. May–August.

Mountain rills and river banks. Rare. First record: *Wilson, B.R.C. Rep.*, 1884, p. 6.

N. 1. Greygarth Fell, Leck Fell, and Ease Gill; also descending to the Lune banks near Kirkby Lonsdale, and the left bank of the Greeta near Wrayton, *Wi.* Gorge of the Greeta.

Plentiful on the S. Lancs. side of the Ribble about Chatburn and Clitheroe. *A.A.D.* and *Wh.*

[*Cochlearia grœnlandica,* L.—Incognit.

Mr. Riddelsdell reports in the *Naturalist*, 1902, p. 344, that a specimen exists in the Motley Herbarium labelled "Poulton-le-Sands, 1840." There is no later record, and we bracket it, therefore, as being probably extinct now.]

Cochlearia danica, L.—English type. Native. Area, 3, 4, 5. Range, 0-30 feet. A. and B. April–June.

Sandy places and dry hedgebanks near the sea. Rare. First record: *Ashfield, Fl. Prest.*, 1858.

N. 3. Hedgebanks near Middleton, *Wi.*
W. 4. Hedgebank near Pilling, *Wi.*
5. Blackpool, abundantly, also near Lytham, *Ashfield*, 1858. Still at Blackpool, towards Southshore, 1899, and near St. Annes, 1898, *Wh.* Lytham, *Fl. Prest. N.*, 1903.

The flowers of this plant have a sweet honey-like odour, which is, however, so faint as to be hardly perceptible unless they are massed together.

Cochlearia anglica, L.—English Scurvy-grass. English type. Native. Area, 3, 4, 5. Range, 0-25 feet. A. and B. May–June.

Muddy salt marshes. Frequent. First record: *Watson, Top. Bot.*, 1883.

N. 3. Salt Marsh, Ovangle, *Wh.* Near Middleton, *Wi.* Basil Point, *Wh.*
W. 4. Cocker estuary near Cockerham and estuary of the Lune, *Wi.* Wardleys-on-Wyre, *Wh.*
5. Ribble banks near Preston Docks and on Freckleton Marshes, *Wh.* Lea, *Fl. Prest. N.*

Cochlearia Armoracia, L.—Alien. First record: *Fl. St.*, 1891.

W. 5. Waste ground by Preston Docks, *Wh.* Rail bank, Ashton, *Fl. Prest. N.*
E. 8. Garden escape at Stonyhurst, 1891, *Fl. St.*

Sisymbrium Thalianum, J. Gay.—Thale-cress. British type. Native. Area, 1, 2, 3, 5, 6. Range, 0-500 feet. A. April–May.

Dry hedgebanks and walls, mostly on limestone. Not uncommon in North, rare elsewhere. First record: *Ashfield, Fl. Prest.*, 1858.

N. 1. Near Burrow, *Wi.* Near Ireby and Cowan Bridge.
2. Near Arkholme, Over Kellet, and Nether Kellet, *Wi.*
3. Yealand; and near the Keer estuary, Carnforth.
W. 5. Lytham, towards Blackpool, *Ashfield*, 1858. Wall on the sandhills North of Lytham, 1896, *Wh.* Ashton, *Fl. Prest. N.*
E. 6. Between Hornby and Wennington, *Wi.* Near Lowgill, Hindburn.

Sisymbrium officinale, Scop.—Hedge Mustard. British type. Native. Area, 1, 2, 3, 4, 5, 6, 7, 8. Range, 0-620 feet or higher. A. June–August.

Roadsides and waste places. Common. First record: *Linton, B.R.C. Rep.*, 1874, p. 81.

The var. *leiocarpum* D.C. is much less common, and perhaps merely an alien. It is abundant between Fulwood and Ribbleton, *Wh.*; and on the shore near Guide House Inn and near Freckleton, 1901, *S.* and *T.* Waste ground near Bay Horse Railway Station.

Sisymbrium Sophia, L.—Flix Weed. English type. Native. Area, 3, 4, 5. Range, 0-200 feet. A. June–August.

Waste ground near the sea. Rare. First record: *Jenkinson, Brit. Pl.*, 1775.

N. 3. About Leighton Hall, *Jenkinson*, 1775.

W. 4. Knott End, *S.* and *T.*

5. Near Lytham, *Wh.*

Sisymbrium Pannonicum, Jacq.—Alien. First record : *Wheldon, Journ. Bot.*, 1901, p. 23.

4. Glasson Dock, one or two plants, *Wh.*, 1906.

W. 5. Plentiful about Fleetwood Docks, June, 1900, and one or two plants near Preston Docks, *Wh.*, July, 1900. Railbanks, near St. Annes, *C.B.*, Aug., 1901.

Alliaria officinalis, Andrz. — Hedge Garlic. British type. Native. Area, 1, 2, 3, 4, 5, 6, 7, 8. Range, 0-610 feet or higher. A. May–June.

Hedge banks and roadsides. Common. First record : *Wilson, B.R.C. Rep.*, 1887, p. 87. " Garstang."

In North it ascends to 610 feet near Leck.

Erysimum perfoliatum, Crantz.—Alien. First record : *Wheldon, Journ. Bot.*, 1902, p. 347.

W. 5. Fleetwood and Preston Docks, *Wh.*, 1900.

Erysimum cheiranthoides.—Alien. First record : *E. F. Linton, B.R.C. Rep.*, 1874, p. 81. No locality quoted.

Camelina sativa, Crantz. — Gold of Pleasure. Alien. First record : *Ashfield, Fl. Prest.*, 1860.

W. 5. Marsh End, Preston, on a piece of uncultivated ground, no doubt introduced, *Ashfield*, 1860.

Brassica oleracea, L.—Alien. " By Bolton Sandsides." *Jenkinson's Brit. Pl.*, 1775.

Occurs occasionally as a garden outcast on rubbish heaps and the borders of cultivated fields. Jenkinson's plant was probably not the true wild cabbage.

Brassica campestris, L.—Rape. English type. Colonist. Area, 4, 5, 6, 7, 8. Range, 0-400 feet or higher. A. or B. July–August.

Borders of cultivated fields. Frequent. First record : *Linton, B.R.C. Rep.*, 1874, p. 81.

The three forms, rape, turnip, and swede, all occur, but only in a semi-wild state.

Brassica monensis, Huds.—Isle of Man Cabbage. Atlantic type. Native. Area, 4, 5. Range, 0-30 feet. B. or P. July–September.

Sandy sea shores. Very rare. First record : *Ashfield, Fl. Prest.*, 1860.

W. 4. Preesall, *Ashfield*, 1860.

5. Beach at Lytham, *Ashfield*. Seen yearly at Lytham, 1896-1905, *Wh.*

Sinapis nigra, L.—Black Mustard. English type. Colonist. Area, 1, 3, 4, 5. Range, 0-50 feet. A. June–August.

Banks of rivers, roadsides, and waste ground. Not uncommon in the low-lying Fylde district. First record : *Ashfield, Fl. Prest.*, 1860.

N. 1. Bank of the Lune, near Melling, *Wi.*

3. Fields between Middleton and Sunderland in the Heysham Peninsula.

W. 4. Wrampool, near Pilling, *Wi.*

5. Cornfield between Poulton and Bispham, *Ashfield*. Woodplumpton Brook, *Wi.* Preston Docks ; by the roadside between Wrea Green and Lytham and near Wyre Docks, Fleetwood, *Wh.*

Sinapis arvensis, L.—Charlock. Locally " Ketlock " in North, and " Runch " in S. Lancashire. British type. Colonist. Area, 1, 2, 3, 4, 5, 6, 7, 8. Range, 0-700 feet. A. June-September.

Cultivated fields. Very common, except in the moorland districts of East. First record : *Linton, B.R.C. Rep.*, 1874, p. 81.

It occurs with both smooth and bristly pods.

Sinapis alba, L.—White Mustard. English type. Colonist. Area, 4, 7. Range, 0-100 feet. A. July–August.

Waste ground and cornfields. Rare. First record : *Linton, B.R.C. Rep.*, 1874, p. 81.

W. 4. Stalmine, *Wh.*

E. 7. Near Garstang, *Wi.*

[*Diplotaxis tenuifolia*, DC.—Probably an error.

Mr. Linton cannot trace the specimen on which the record in Top. Bot. was founded, nor recollect the locality.]

Diplotaxis muralis, DC. — Wall Mustard. English type. Native. Area, 5. Range, 0-50 feet. A. June–September.

Sandy ground near the sea. First record : *Wheldon, Journ. Bot.*, 1900, p. 42.

W. 5. Between St. Annes and Lytham, *Wh.*, 1899. Also in the field near Lytham Vicarage, *Wh.* Near All Saints' Road, St. Annes, *C.B.*, 1906. Near Preston Dock, *Fl. Prest. N.*

The var. *Babingtonii*, *Syme*, occurs more commonly than the type.

K

Capsella Bursa-pastoris, L.—Shepherd's Purse. British type. Native. Area, 1, 2, 3, 4, 5, 6, 7, 8. Range, 0-600 feet or higher. A. March–November.

Waste places, cultivated fields, and roadsides. Very common. First record : *Linton, B.R.C. Rep.*, 1874, p. 81.

The greatest altitute of which we have any note is 600 feet near Lower Emmetts, Over Wyresdale. It is occasionally attacked by the white mould *Cystopus candidus, Lév.*

Coronopus procumbens, Gilib.—Wart Cress. English type. Native. Area, 5, 7. Range, 0-100 feet. A. July–September.

Waste ground and roadsides in the low country. Rare. First record : *Top. Bot.*, 1883 (no authority quoted).

W. 5. Between Blackpool and Cleveleys, *Wi.*, 1895. Little Bispham, *Wh.*

E. 7. Near Garstang, *Wi.*

Lunaria rediviva, DC.—Honesty. Alien. First record : *Britten, Naturalist*, 1864, p. 203.

N. 3. Sands of Morecambe Bay, *James Britten*, June, 1864.

A native of Germany. Not seen since.

Lepidium sativum, L.—Cress. Alien. First record : *Wheldon, Journ. Bot.*, 1900, p. 42.

N. 3. Plentiful on rubbish heaps by a newly-made road at Morecambe, *Wh.*, July, 1899.

W. 5. Near Preston Docks, *Fl. Prest. N.* Near St. Annes, *C.B.*

Lepidium campestre, R. Br. — Field Pepperwort. British type. Native. Area, 4, 5, 6, 7. Range, 0-200 feet. B. July–August.

Roadsides and waste places. Not common. First record : *Linton, B.R.C. Rep.*, 1874, p. 81.

W. 4. Tarnacre, *P.J.H.* Near Winmarleigh.

5. Catterall, *Wi.* Fleetwood, *Wh.*, Ashton, *Fl. Prest. N.*

E. 6. Near Galgate, *Wi.*

7. Sandholme Mill, Garstang, *Wi.*

Lepidium ruderale, L.—Alien. First record : *Bailey, Journ. Bot.*, 1902, p. 347.

W. 5. Waste ground off St. David's Road, St. Annes, *C.B.* Aug., 1901. Near Preston Docks, *Fl. Prest. N.*

Lepidium Draba, L.—Alien. First record : *Lees, Naturalist*, 1900, p. 246.

N. 3. Near Morecambe, *F.A.L.*, 1900.

W. 5. Fine and plentiful in waste ground near Wyre Docks, Fleetwood, *Wh.*, June, 1900. Old river bed at Preston ! *H.B.* Waste heap on the sandhills near Ormerod House, St. Annes, *C.B.*, May, 1905.

Lepidium heterophyllum, Benth. (*L. Smithii*, Hook).—Hairy Pepperwort. British type. Native. Area, 3, 5. Range, 0-100 feet. P. June–August.

Dry banks and roadsides. Rare. First record : *Ashfield, Fl. Prest.*, 1858.

N. 3. Near Old Church, Silverdale, *L.P.*

W. 5. Between Greave's Town and Ashton Marsh, *Ashfield*, 1858. Near Preston Docks, *A.A.D.*

Thlaspi arvense, L.—Penny Cress. British type. Colonist. Area, 5. Range, 0-50 feet. A. May–July.

Cultivated fields. Very rare. First record : *Syme, Top. Bot.*, 1883.

W. 5. Ashton, *Fl. Prest. N.*, 1903. Waste ground (formerly sandhills) near St. Annes, with other aliens, *C.B.*, 1906.

It is very doubtfully given the rank of colonist, and is a very rare plant not only with us, but in South Lancashire also.

Cakile maritima, Scop.—Sea Rocket. British type. Native. Area, 3, 4, 5. Range, 0-30 feet. A. July–September.

Sandy sea shores. Not common. First record : *Ashfield, Fl. Prest.*, 1858.

N. 3. Far Naze, *Wi.* Silverdale Cove, *L.P.*

W. 4. Coast near Pilling and Knott End, *Wi.*

5. Between Blackpool and Fleetwood, *Wi.* Lytham, *Ashfield*, 1858. Very fine at St. Annes, where the erection of palings serves to protect it, *Wh.*

[*Crambe maritima*, L.—Sea-kale Incognit.

Reported in *Flora of Preston and Neighbourhood*, p. 8, from Lytham. Confirmation desirable.]

Raphanus Raphanistrum, L—Wild Radish. British type. Colonist. Area, 3, 4, 5. Range, 0-100 feet. A. June–July.

Rather rare. First record : *Wheldon, Journ. Bot.*, 1901, p. 23.

W. 3. Overton and Heysham, *Wi.*

4. Cornfield near Preesall, June, 1900, and Knott End, *Wh.*

5. St. Michael's, *P.J.H.* Lea and Salwick, *Fl. Prest. N.* We have only observed the yellow-flowered form.

Raphanus maritimus, Sm.— Sea Radish. Altantic type. Native. Area, 5. Range, 0-30 feet. B. July–August.

Recorded by *Syme, Top. Bot.,* 1883.

W. 5. Lytham, *Fl. Prest. N.*
We have sought this in vain, and fear it is now extinct.

Rapistrum orientale, L., and *Rapistrum rugosum,* DC.—Were found with other aliens on waste ground, formerly sandhills, near St. Annes, 1906, *C.B.*

RESEDACEÆ.

Reseda lutea, L.—Wild Mignonette. Alien. First record : *Wheldon, Journ. Bot.,* 1901, p. 23.

N. 3. Rail bank near Warton, *Wi.*
W. 4. Casually by the Wyre, near Churchtown, *Wi.*
    5. Waste ground near Preston Docks, *Wh.,* July, 1900. St. Annes, *C.B.*

Reseda Luteola, L.—Weld. British type. Native. Area, 1, 2, 3, 4, 5, 6, 7, 8. Range, 0-400 feet. B. June–August.

Dry banks and waste places, mostly on calcareous soil. Not uncommon in North, more rare elsewhere. First record : *Ashfield, Fl. Prest.,* 1858. "Between Naze Point and Lytham."

In most of its stations in West and East it has probably been introduced, occurring about docks and on rail banks, but it seems to be native on the sea coast and in North.

CISTACEÆ.

Helianthemum Chamæcistus, Mill.—Rock Rose. British type. Native. Area, 2, 3. Range, 0-500 feet. P. June–September.

Limestone cliffs and banks. Common in parts of North, being one of the more frequent species of the scar limestone. First record : *Ashfield, Bot. Chron.,* 1864, p. 73.

N. 2. Over Kellet, *Wi.*
    3. Very abundant in the Silverdale district, *Ashfield,* 1864.

Still occurs in numerous localities, ascending to 500 feet on Warton Crag. We have not been able to ascertain what species is referred to in the following note from Jenkinson's British Plants :—"*Helianthemum.* Dwarf cistus, or little sun-flower, which is procumbent and shrubby, the stipulæ are lanceolated, the leaves oblong, turned back and hairy. The first variety has larger leaves and a white flower ; the second variety has a broader leaf and a rosy flower. On Yealand Common, plentifully."

    3. Lane near Hawes Tarn and Magstone Wood, *L.P.* Frequent about Silverdale, *Wi.* Warton and Yealand, and on scars by the road below Warton Crag leading to Silverdale, infested with *Puccinia violæ* (Schum.), *Wh.*
E. 8. Sale Wheel, *Fl. St.*

Viola sylvestris, Lamarck.—Dog Violet. British type. Native. Area, 1, 2, 3, 8. Range, 0-800 feet. P. March–May.

Woods and banks. Frequent in the calcareous districts ; and probably does not occur elsewhere. First record : *Lees, B.R.C. Rep.,* 1882, p. 132.

N. 1. Near Wrayton and in Ease Gill, *Wi.* Gorge of the Greeta.
    2. Near Arkholme, *F.A.L.*
    3. Silverdale and Leighton Beck, *Wi.* Carnforth.
E. 8. Gorge of the Hodder below Whitewell, and in Fence Wood.

Viola Riviniana, Reich.—Dog Violet. British type. Native. Area, 1, 2, 3, 4, 5, 6, 7, 8. Range, 0-1,600 feet. P. March–June.

Hedgebanks and woods. Common. First record : *Lees, B.R.C. Rep.,* 1882, p. 132.

It ascends in North to 1,350 feet on rocks above the waterfalls in Upper Ease Gill, and in East to 1,600 feet on Botton Head Fell.

Viola canina, L. (*V. ericetorum,* Schrad).—Dog Violet. British type. Native. Area, 3, 5. Range, 0-30 feet. P.

Damp sandy places. Rare. First record : *Wheldon, Journ. Bot.,* 1900, p. 42.

N. 3. Sandy peat by Hawes Water, *Wi.*
W. 5. Sandhills at Lytham, *Wh.,* 1896.

The small form with cordate leaves *V. flavicornis* (Sm.) is reported from the Lytham sandhills in *Fl. Prest. N.,* 1903, p. 10.

Viola tricolor, L.—Wild Pansy. British type. Native. Area, 1, 2, 3, 4, 5, 6, 7, 8. Range, 0-350 feet. A. May–September.

Cultivated fields and gardens. Common in the low country. First record : *Top. Bot.,* 1883.

Of the segregates included under this species, Mr. E. G. Baker and Prof. Borbas referred a large blue flowered form abundant in fields reclaimed from Cockerham Moss to *Viola carpatica, Borbas.** It occurs also in several other moss land districts, as at Thrang Moss, and near Garstang, and Thornley. *Viola Lloydii, Jord.,* is abundant with the last about Cockerham and in cornfields on the Carrs near Bruna Hill, Garstang.

Viola tricolor occurs with the next species on shallow soil overlying limestone rocks on Warton Crag, with every appearance of being truly native, in the company of *Arenaria verna, Cerastium semidecandrum,* and *Carex præcox.*

* Vide E. G. Baker, "Some British Violets," *Journ. Bot.,* 1901, p. 10, for notes on both, *V. Pesneanii* and *V. carpatica.*

VIOLACEÆ.

Viola palustris, L.—Marsh Violet. British type. Native. Area, 1, 2, 3, 5, 6, 7, 8. Range, 0-1,870 feet. P. May–June.

Swampy places and wet fields. Frequent, especially amongst the hills. First record : *Ashfield, Fl. Prest.,* 1858.

N. 1. Greygarth Fell, up to 1,870 feet, *Wi.* Ease Gill and Ireby Fell.
    2. Docker ; Gressingham ; Whittington and Arkholme moors ; and near Henridden, *Wi.*
    3. Thrang Moss, *Wi.* Hawes Tarn, *L.P.* Heysham Moss.
W. 5. Carrs Green Common, near Catforth, *Wi.*
E. 6. Harter Beck, *Wi.* Greenbank Fell.
    7. Nickey Nook, *Ashfield.* Blaze Moss, Marshaw ; Abbeystead ; Bleasdale, *Wi.* Gavell's Clough, at 1,200 feet.
    8. Knowl Green, *H.B.* Longridge Fell and Chipping, *Fl. Prest. N.*

Viola odorata, L.—Sweet Violet. English type. Native. Area, 1, 2, 3, 4, 5, 6, 7, 8. Range, 0-300 feet. P. March–April.

Hedgebanks and roadsides. Frequent in North ; rather rare in East and West, and probably not always native there. First record : *Ashfield, Fl. Prest.,* 1858.

N. 1. Near Cantsfield.
    2. Near Skerton, *Wi.*
    3. Frequent in this district.
W. 4. Winmarleigh, *S. Wi.* Nateby.
    5. Ashton, *Fl. Prest. N.* Myerscough, *Wi.* Trunnah, *S. Wi.*
E. 6. Near Wennington and Wray, *Wi.*
    7. Claughton, near Garstang, and near Scorton, *Wi.*
    8. Ribble bank below Hacking Boat, *Fl. St.* About Fishwick, 1858-60, *Ashfield.*

f. alba (Lange).—White Violet.

Between Warton and Yealand, and near Silverdale Cove, *Wh.* Banks of the Hodder and Ribble to Sale Wheel ; Saddle Hill, *Fl. St.*

Viola hirta, L.—Hairy Violet. English type. Native. Area, 1, 2, 3, 8. Range, 0-700 feet. P. April–May.

Dry banks and woods. Principally on the xerophilous rocks of North, where it is abundant. First record : *Wilson, B.R.C. Rep.,* 1881-2, p. 248.

N. 1. Leck, and near Wrayton, *Wi.*
    2. Near Kirkby Lonsdale ; Over Kellet ; Dalton Houses, *Wi.* Ascends to 700 feet on Dalton Crag.

Viola arvensis, Murr.—British type. Native. Area, 2, 3, 4, 5, 6, 7, 8. Range, 0-100 feet or higher. A. May–August.

Cultivated ground, and by roadsides. Common.

A form nearly allied to *V. mentita, Jord.* (teste E. G. Baker) occurs in fields near Cockerham Moss, *D. A. Jones* and *Wh. V. Déséglisei, Jord.* Near Lytham, *Wh.,* 1903. Plants which apparently belong to this sub-species are not unfrequent in the low country.

Viola Curtisii, Forster.—Wild Pansy. Altantic type. Native. Area, 5. Range, 0-30 feet. P. May–August.

Coast sandhills. Rare and decreasing. First record : *Melvill, Exch. Club Rep.,* 1880.

W. 5. Frequent about St. Annes and Lytham, and with the flowers variable in colour.* Mr. E. G. Baker refers our yellow-flowered plant to var. *Forsteri, H. C. Wats. forma.* A form with large wholly blue flowers, which seems to be the most frequent variety, according to the same authority, only differs from *V. Pesneanii, Lloyd and Foucaud,* in its elongated spur. Messrs. Salmon and Thompson appear to have found a form approaching this, but diverging in being less pubescent and having parti-coloured flowers. Mr. Baker says that *V. subulosa, Boreau,* which differs in having a much longer, narrower leaf lamina, has been recorded from Lytham. Blackpool, *G. Webster.* [Cultivated land near Ashton, *Fl. Prest. N.* Probably one of the preceding ?]

[*Viola lutea,* Huds.—Incognit.

The record in *Top. Bot.,* 1883, was an error. The plant was *Viola Curtisii.*]

POLYGALACEÆ.

Polygala vulgaris, L.—Milkwort. British type. Native. Area, 1, 2, 3, 4, 5, 6, 7, 8. Range, 0-1,490 feet. P. May–August.

Dry grassy banks. Not common, except in the limestone districts of North. First record : *Ashfield,* 1858, although it is doubtful whether he discriminated between this and the next, as the latter is not mentioned in his lists.

N. 1. Frequent in the Leck district.
    2. Whittington Moor.
    3. Thrang End and Carnforth, *Wi.* Heysham.
W. 5. Near Myerscough, *Wi.*
E. 6. Near Wray, *Wi.*
    7. Abbeystead Reservoir banks, and Tarnbrook Fell, where it ascends to 1,490 feet.
    8. Longridge, *Ashfield,* 1858.

* C. E. Salmon and H. S. Thompson, "West Lancashire Notes," *Journ. Bot.,* 1902, p. 295.

Polygala oxyptera, Reichb.—English type. Native. Area, 5. Range, 0-40 feet. P. May–August.

Dry sandy ground amongst the coast sand dunes. Rare. First record: *Marshall, Journ. Bot.,* 1896, p. 136.

> W. 5. Sandhills west of Lytham, *E.S.M.,* 1895. Still occurs on the sandhills between Lytham and St. Annes, *Wh.*

Polygala serpyllacea, Weihe.—Milkwort. British type. Native. Area, 1, 2, 3, 5, 6, 7, 8. Range, 0-1,450 feet or higher. P. May–September.

Heaths and upland pastures. Abundant in North and East, where it is the most common species ; less common in West. First record: *Linton, B.R.C. Rep.,* 1874, p. 81.

It ascends in North to 1,450 feet on Greygarth Fell.

CARYOPHYLLACEÆ.

Saponaria officinalis, L.—Soapwort. English type. Denizen. Area, 1, 2, 3, 4, 5, 6, 8. Range, 0-200 feet. P. August–September.

Hedges and river banks. Rather rare. First record: *Lawson,* in *Ray's Corresp.,* 1718.

> N. 1. Lune banks near Melling and Nether Burrow, *Wi.*
> 2. By the Lune near Arkholme.
> 3. "Carnforth in Lancashire," *Ray* 1718. Still there in 1903, *Wi.* In Warton Town and in Yealand Redmayne, plentifully, *Jenkinson.* Coast between Keer estuary and Bolton-le-Sands.
> W. 4. Near Pilling, *Wi.*
> 5. Near Blackpool, *Rev. W. Wood,* 1805 (in *Bot. Guide*). Near Wharles, *Wi.*
> E. 6. By the Lune, near Halton and Caton, and on the large island in the river near Lancaster, *Wh.*
> 8. Near Preston, and Ribchester Bridge, *Ashfield.* Still there, *Wh.* Alston, *E.F.L.* Bank of Ribble from Ree Deep downwards, *Fl. St.*

The var. *puberula, Wiersb.,* occurs more rarely ; it has been observed at Caton and Halton (*Wh.*) and near Nether Burrow (*Wi.*). The anthers of this variety are very frequently attacked by the fungus *Ustilago violacea, Pers.*

Silene conica, L.—Alien. First record : The present one.

> N. 1. Field at the north end of Little Hawes Water, *W.C.W.*

Silene latifolia, Britt. and Rendle (*S. inflata Sm.*).—Bladder Campion. British type. Native. Area, 1, 2, 3, 4, 5, 6, 7, 8. Range, 0-250 feet. P. June–August.

Dry banks, roadsides, and waste places. Frequent in North and West, rare in East. First record : *Linton, B.R.C. Rep.,* 1874, p. 81.

> N.   Widely distributed, *Wi.*
> W. 4. Moss Side, St. Michael's, *F.J.H.* Cabus.
> 5. St. Annes, *Wi.* Ashton Marsh, *Fl. Prest. N.*
> E. 6. By the Lune near Caton, *Wh.*
> 7. Near Garstang, *Wi.*
> 8. Ribble bank below the Aqueduct ; below Trough's, etc., *Fl. St.*

The var. *puberula, Syme,* is reported from below Trough's in the Flora of Stonyhurst.

Silene maritima, With.—Sea Campion. British type. Native. Area, 3, 4, 5. Range, 0-50 feet. P. June–August.

Sea shores, generally in rocky or shingly places. Frequent, especially in North. First record : *Ashfield, Fl. Prest.,* 1858.

> N. 3. Frequent on the coast from Morecambe to Silverdale, *Wi.* Bare and Hest Bank, *F.A.L.* Thusshouse sands, *Wh.*
> W. 4. Knott End and Pilling, *Wi.* Wyre bank, near Coatwalls, *Wh.*
> 5. Lytham, *Ashfield,* 1858. Still there and at Fleetwood, *Wh.*

Silene noctiflora, L.—Catchfly. Alien. First record : The present one.

> A few plants on the Lytham sandhills, with other aliens, *Wh.,* 1903.

Lychnis alba, Mill.—White Campion. British type. Native. Area, 1, 2, 3, 4, 5, 6, 8. Range, 0-100 feet. B. June–September.

Fields and roadsides. Frequent in parts of West and North, rare in East. First record : *Linton, B.R.C. Rep.,* 1874, p. 81.

A very hairy form with pink flowers occurs rarely, which is probably the hybrid *L. alba × dioica.* Its anthers are sometimes attacked by *Ustilago violacea, Pers.*

Lychnis dioica, L. — Red Campion. " Sweet William," " Soldier's Buttons." " Lousy Betty," on both sides of the Wyre. British type. Native. Area, 1, 2, 3, 4, 5, 6, 7 8. Range, 0-900 feet or higher. P. May–July.

Woods, hedgebanks, and shady places. Very common. First record : *Linton, B.R.C. Rep.,* 1874, p. 81.

Lychnis Flos-cuculi, L.—Ragged Robin. British type. Native. Area, 1, 2, 3, 4, 5, 6, 7, 8. Range, 0-890 feet or higher. P. June–July.

Marshy meadows, stream sides, and swampy places. Frequent. First record : *Linton, B.R.C. Rep.,* 1874, p. 81.

It ascends in East to 890 feet on Whitmoor

Lychnis Githago, Scop.—Corn Cockle. British type. Colonist. Area, 3, 4, 5, 8. Range, 0-100 feet. A. June–August.

Cornfields and waste places. Not common. First record : *Jenkinson, Brit. Pl.,* 1775, p. 97.

> N. 3. "In the cornfields about Yealand Redmayne I found it once or twice," *Jenkinson,* 1775.
> W. 4. Rawcliffe, Stalmine, and Garstang ! *Ashfield.* Abundant in 1902 at Cogie Hill, *A.B.*
> 5. Bispham, Poulton, and St. Michael's, *Ashfield.* Fleetwood ; near Preston Docks ; and in a cornfield near Lytham, *Wh.*
> E. 8. Casually at Greenfield, near Stonyhurst, *Fl. St.*

Cerastium tetrandrum, Curt. — Mouse-ear. British type. Native. Area, 3, 5. Range, 0-50 feet. A. May–June.

Sandy sea shores. Abundant on the sandhill tract. First record : *Top. Bot.,* 1883 (but the locality was Southport, beyond our borders, Mr. Linton informs us).

> N. 3. Near Bolton-le-Sands and near Middleton, *Wi.*
> W. 5. Frequent on the sandhills at Lytham and St. Annes, *Wh.* Abundant near the old river bed, Ashton, *A.A.D.*

Cerastium semidecandrum, L.—British type. Native. Area, 3, 5. Range, 0-100 feet. A. April–May.

Sandy ground, dry banks and walls. Rare. First record : *Wheldon, Journ. Bot.,* 1900, p. 42.

> N. 3. Silverdale, *W. W. Mason.* North side of Warton Crag.
> W. 5. Between Lytham and S. Annes, 1899. *Wh.* Near Stockenbridge, *P.J.H.*

Cerastium viscosum, L.—British type. Native. Area, 1, 2, 3, 4, 5, 6, 7, 8. Range, 0-900 feet or higher. A. May–August.

Roadsides, banks and fields. Frequent in North and East, but less so than the next species. First record : *Linton, B.R.C. Rep.,* 1874, p. 81.

Cerastium vulgatum, L.—Mouse-ear Chickweed. British type. Native. Area, 1, 2, 3, 4, 5, 6, 7, 8. Range, 0-2,050 feet. B. or P. May–September.

Fields, roadsides, waste places and moorlands. Very common. First record : *Linton, B.R.C. Rep.,* 1874, p. 81.

It ascends to 2,050 feet on Greygarth Fell.

Cerastium arvense, L.—Field Chickweed. British type. Native. Area, 4. Range, ? P. May–July.

Dry, sandy or calcareous banks and fields. Very rare. First record : *Syme, Top. Bot.,* 1883.

> W. 4. Cockerham, *Fl. Prest. N.*

Stellaria nemorum, L.—Wood Stitchwort. Scottish type. Native. Area, 1, 2, 5, 6, 7, 8. Range, 0-600 feet. P. May–June.

Woods and shady banks near streams. Frequent in the dales. First record : *Ashfield, Fl. Prest.,* 1858.

> N. 1. Near Tunstall and Kirkby Lonsdale, *Wi.* Ease Gill and Gorge of the Greeta.
> 2. Frequent by the Lune, between Halton and Aughton. On the bank or the river Keer, *Wi.*
> W. 5. South Meadow Lane, near Preston, *Ashfield,* 1858 (locality now built upon, *Wi.*). Bank of Barton Brook, *Wi.*
> E. 6. Below Botton Mill, and near Cowcins, Hindburn ; near Mill Houses ; frequent in Roeburndale ; south bank of the Lune both east and west of Caton, *Wi.* Lower Salter.
> 7. Wooded bank of the Wyre, one mile above Garstang, and on the banks of the Brock, *Wi.* Near Abbeystead Reservoir and other localities in Wyresdale.
> 8. Abundant in woods near Stonyhurst, *Fl. St.* Hodder Valley above Higher Bridge and Whitewell, *Wh.* Near Chipping Camp, *Wi.*

Stellaria media, Vill.— Chickweed. British type. Native. Area, 1, 2, 3, 4, 5, 6, 7, 8. Range, 0-2,050 feet. A. February–November.

Cultivated ground, roadsides and waste places. Very common. First record : *Linton, B.R.C. Rep.,* 1874, p. 81.

It ascends to 2,050 feet on Greygarth Fell, *Wi.*

Var. *neglecta (Weihe).*

> N. 2. Near Docker, *Wi.*
> E. 7. Bleasdale, *Wi.*

Stellaria Holostea, L.—Starwort. Stitchwort. " Bread and Cheese" and " Pretty Betty " in parts of the Fylde. British type. Native. Area, 1, 2, 3, 4, 5, 6, 7, 8. Range, 0-1,420 feet. P. April–June.

Woods and hedgebanks. Common. First record : *Linton, B.R.C. Rep.,* 1874, p. 81.

It ascends in North to 1,420 feet in Upper Ease Gill, and in East to 910 feet near Haylot.

Stellaria graminea, L.—Lesser Starwort. British type. Native. Area, 1, 2, 3, 4, 5, 6, 7, 8. Range, 0-990 feet or higher. P. May–July.

Hedgerows, heathy pastures and bushy places. Frequent. First record: *Linton, B.R.C. Rep.*, 1874, p. 81.

Stellaria uliginosa, Murr. — Bog Starwort. British type. Native. Area, 1, 2, 3, 4, 5, 6, 7, 8. Range, 0-2,050 feet. A. June–July.

Damp grassy places and bogs. Frequent. First record: *Ray's Corresp.*, 1718. "Stellaria aquatica, *Parkinson*. In the ditches of Middleton Moss, Lancaster, where I saw it in flower."

It ascends to 2,050 feet on Greygarth Fell in North and to 1,100 feet on the slopes above Gavell's Clough in East.

Minuartia verna, Hiern.—Vernal Sandwort. Intermediate type. Native. Area, 2, 3. Range, 0-500 feet. P. May–August.

Limestone hills of North. Frequent there; not found elsewhere. First record: *Ashfield, Bot. Chron.*, 1864, p. 73, or *Grindon, Naturalist,* 1864.

N. 2. Kellet seeds near Carnforth, *Hartley, Wh.*, and *Wi.*
3. Hilly pastures, frequent, *Ashfield*, 1864. Near Silverdale, *L. H. Grindon,* 1864. Limestone crags north of Warton! *Lewis.* Hills south of Leighton Beck, and in several other localities, *Wi.* West side of Warton Crag, *Wh.*

Arenaria trinervia, L.—British type. Native. Area, 1, 2, 3, 4, 5, 6, 7, 8. Range, 0-1,400 feet. A. May–July.

Hedge banks and damp shady places. Frequent. First record: *Ashfield, Fl. Prest.*, 1858. "Nickey Nook."

It ascends to 1,400 feet on Mallowdale Fell.

Arenaria serpyllifolia, L.—Thyme-leaved Sandwort. British type. Native. Area, 1, 2, 3, 4, 5, 6, 7, 8. Range, 0-—?. A. June–September.

Dry banks, sandy places, walls, etc. Not uncommon. First record: *Ashfield, Fl. Prest.*, 1858. "Ashton Marsh."

Var. *macrocarpa,* Lloyd (A. Lloydii, Jord.).

N. 3. Sea shore near Bare, *Wh.* Coast between Carnforth and Bolton-le-Sands, *Wi.* Hest Bank, on shingle, *Wh.*
W. 5. St. Annes, July, 1898, and near Lytham, *Wh.* Near Preston Docks, *Fl. Prest. N.*

Arenaria leptoclados, Guss.—British type. Native. Area, 1, 2, 3, 4, 5, 6. Range, 0-280 feet or higher. A. June–August.

Dry, sandy ground. Rare. First record: *Wheldon, Journ. Bot.,* 1900, p. 42.

Sagina maritima, Don. — Sea Pearlwort. British type. Native. Area, 3. Range, 0-30 feet. A. June–September.

Damp sandy or muddy places on the sea coast. Rare. First record: The Authors, *B.E.C., Rep.,* 1906, p. 215.

N. 3. Salt marsh near Middleton in the Heysham Peninsula, July, 1906. Sparingly on shingle at Basil Point, *Wh.*

Some of our specimens are much drawn out and are referred to var. *stricta* (*Gren.* and *Godr.*) by Mr. G. C. Druce.

Sagina nodosa, Fenzl.—Knotted Spurrey. British type. Native. Area, 1, 2, 3, 4, 5, 6, 7, 8. Range, 0-1,150 feet. P. July–August.

Damp sandy or peaty places. Not uncommon, especially amongst the hills. First record: *Ashfield, Fl. Prest.,* 1858. "Near Lytham and on Ashton Marsh."

It ascends in North to 1,150 feet on Ireby Fell. The glandular form (*S. glandulosa, Bess*) occurs at Longridge and near Preston, and is probably as frequent as the type, but we have not kept separate records of its distribution.

Spergula arvensis, L.—Field Spurrey. "Yorr." British type. Colonist. Area, 1, 2, 3, 4, 5, 6, 7, 8. Range, 0-700 feet. A. June–August.

Cultivated fields. Common. First record: *Linton, B.R.C. Rep.,* 1874, p. 81.

Both *S. vulgaris* (*Boenn*) and *S. sativa* (*Boenn*) occur, but we have not ascertained their separate areas of distribution.

Alsine rubra, Crantz.—Red Sandwort. British type. Native. Area, 4, 6. Range, 0-350 feet. A. June–September.

Sandy and rocky places. Rare. First record: *Linton, B.R.C. Rep.,* 1874, p. 81. No locality given.

W. 4. Between stones on the canal bank, Glasson, *Wh.*
E. 6. Near Lancaster, *Wi.*

Alsine media, Crantz.—Sea Sand Spurrey. English type. Native. Area, 3, 4, 5. Range, 0-25 feet. P. June–September.

Muddy salt marshes. Not common. First record: *Wilson, Journ. Bot.,* 1900, p. 42.

N. 3. Silverdale shore! *L.P.* Overton, *Wi.*
W. 4. Pilling salt marsh, *Wi.*, 1895. Bank of Wyre, opposite Preesall, *Wh.* Knott End, *P.J.H.*
5. Freckleton salt marsh and by the river at Preston Docks, *Wh.* Shore east of Lytham, *S.* and *T.*

N. 1. Near Cowan Bridge (alt. 280 feet).
2. By the Canal near Borwick.
3. Sea shore near Bare, *Wh.*, July, 1899. Side of Bottoms Lane, near Silverdale, *L.P.* Near Carnforth.
W. 4. Glasson, *Wh.*
5. Near Preston Docks and on the river embankment on Ashton Marsh, *Wh.*
E. 6. By the Lune, near Caton, *Wh.*

This and the preceding species grow together on the sandy shore near Bare, and with them a plant the Rev. E. S. Marshall thought to be var. *scabra, Rouy.* and *Fouc.*, which is believed to be a hybrid.

Honkeneja peploides, *Ehrh.*—Sea Sandwort. Sea Purslane. British type. Native. Area, 3, 4, 5. Range, 0-25 feet. P. June–September.

Sea shores. Frequent. First record: *Linton, B.R.C. Rep.,* 1874, p. 81.

N. 3. Silverdale shore, *Wi.* Between Bare and Hest Bank, *F.A.L.*
W. 4. Pilling and Knott End, *Wi.*
5. The Naze, Freckleton, 1871, *E.F.L.* Little Bispham, *Wh.* Rossall, *P.J.H.* Preston Dock, *Fl. Prest. N.*

Sagina Reuteri, Boiss.—Alien. First record: The present one.

Plentiful in Preston Dockyard on waste ground near the River Ribble, 1906, *Wh.*

Sagina apetala, Ard.—English type. Native. Area, 1, 2, 3, 4, 5, 6, 7. Range, 0-500 feet or higher. A. May–September.

Walls, footpaths and dry gravelly places. Not very common. First record: *Linton, B.R.C. Rep.,* 1874, p. 81.

N. 1. Near Wennington, *Wi.*
2. Near Carnforth, *Wi.*
3. Carnforth, 1880, *F.A.L.* In Morecambe Harbour and on the sea embankment at Basil Point, *Wh.* Near the quarry and on the summit of Warton Crag, *Wh.*
W. 4. Near Lancaster, *Wh.* Wall near Galgate, *Wi.*
5. Blackpool, *H. Searle.* St. Annes, *Wh.*
E. 6. Near Lancaster Asylum, *Wh.*
7. Near Garsiang and Catterall Station, *Wi.*

Some of our specimens have very glandular peduncles; in others they are nearly glabrous.

Sagina procumbens, L.—Pearlwort. British type. Native. Area, 1, 2, 3, 4, 5, 6, 7, 8. Range, 0-2,050. A. May–September.

Damp sandy places, garden paths, damp rocks, and walls. Very common. First record: *Linton, B.R.C. Rep.,* 1874, p. 81.

Ascends to 2,050 feet on both grit and limestone rocks on Greygarth Fell, *Wi.*

Var. *neglecta* (Kindb.).

N. 3. Lune estuary, *Wi.*
W. 4. Salt marsh at Pilling, *Wi.* Wyre bank near Coatwalls, *Wh.*
5. Preston Docks, *Fl. Prest. N.*

Alsine marginata, Reich.—Sea Sand Spurrey. British type. Native. Area, 3, 4, 5. Range, 0-25 feet. A. June–September.

Sea shores and salt marshes. Common. First record: *Ashfield, Fl. Prest.,* 1858.

N. 3. On the shore at Silverdale! *L.P.* Overton and Sunderland.
W. 4. Knott End! *P.J.H.*
5. Ashton Marsh! *Ashfield,* 1858; Freckleton Marsh, *Wh.*

PORTULACEÆ.

Claytonia sibirica, L.—First record: *P. J. Hornby, Journ. Bot.,* 1902, p. 347.

W. 5. Well-established near Eccleston Springs, Great Eccleston, *P.J.H.*

The climate of Lancashire evidently agrees with this native of N.W. America, and it thrives well in gardens.

Montia fontana, L.—Water Blinks. British type. Native. Area, 1, 5, 6, 7, 8. Range, 100-2,050 feet. A. May–July.

Wet peaty or sandy places, and in and about springs on the hills. Common in East, less plentiful elsewhere. First record: *Ashfield, Fl. Prest.,* 1858. "Near Cottam Hall."

It ascends Greygarth Fell to 2,050 feet, *Wi.* The two states known as var. *minor* (*All.*) and var. *major* (*All.*) both occur, and seem to be dependent on the place of growth.

HYPERICACEÆ.

Hypericum Androsæmum, L.—Tutsan. English type. Native. Area, 3, 5, 6, 7, 8. Range, 0-400 feet. P. June–July.

Damp woods and thickets. Rare. First record: *Ashfield, Fl. Prest.,* 1858.

N. 3. Gatebarrow Wood, Silverdale, *Wi.*
W. 5. Wood near Freckleton, *Ashfield,* 1858.
E. 6. Near Caton, *W.C.W.*
7. Nickey Nook and ravine above Greenhalgh Castle, *Mr. Pearson* (in *Ashfield, Fl. Prest.*). Grizedale and Stricklands Lane, near Garstang, *Wi.* Lower Brock Valley, *H.B.*
8. In a wood near Higher Brockholes, *Ashfield,* 1858. Dean Brook and near the cascades, *Fl. St.*

Hypericum perforatum, L.—St. John's Wort. British type. Area, 1, 2, 3, 4, 5, 6, 7, 8. Range, 0-600 feet or higher. P. July–August.

Banks and bushy places. Frequent. First record : *Linton, B.R.C. Rep.,* 1874, p. 81.

Hypericum maculatum, Crantz (*H. dubium, Leers*).—British type. Native. Area, 1, 2, 3, 6, 8. Range, 50-150 feet. P. July–August.

River banks, hedges and thickets. Rather rare. First record : *Wilson, Journ. Bot.,* 1901, p. 23.

N. 1. Near Nether Burrow and Tunstall, *Wi.*
   2. Bank of the Lune near Arkholme, 1900, *Wi.* Between Caton and Aughton and at Crook of Lune.
   3. Silverdale, *W. W. Mason* (" Vict. Histy. of Lancs.").
E. 6. By the Lune near Caton, *J.F.P.* Crook of Lune, *Wi.*
   8. By the Ribble near Alston, *Wh.*

Hypericum quadrangulum, L.—Square-stemmed St. John's Wort. British type. Native. Area, 1, 2, 3, 4, 5, 6, 7, 8. Range, 0—? P. July–August.

Ditch sides and marshy places. Common. First record : *Linton, B.R.C. Rep.,* 1874, p. 81.

Hypericum humifusum, L.—Trailing St. John's Wort. British type. Native. Area, 1, 2, 3, 4, 5, 6, 7, 8. Range, 0-900 feet. P. July–August.

Dry banks and sandy ground. Frequent. First record : *Ashfield, Fl. Prest.,* 1858. " Between Sion Hill and Fulwood Barracks."

Hypericum pulchrum, L.—Small St. John's Wort. British type. Native. Area, 1, 2, 3, 4, 5, 6, 7, 8. Range, 0-1,390 feet. P. July–August.

Heathy and rocky places, especially amongst the hills. Frequent. First record : *Ashfield, Fl. Prest.,* 1858. " Between Scorton and Nickey Nook."

It ascends in North to 1,390 feet, on rocks by the waterfalls in Upper Ease Gill ; and in East to 970 feet on Greenside, above Tarnbrook.

Hypericum hirsutum, L.—Hairy St. John's Wort. British type. Native. Area, 1, 2, 3, (5,) 6, 8. Range, 0-600 feet or higher. P. July–August.

Woods and banks. Common in the limestone districts ; rare elsewhere. First record : *Ashfield, Fl. Prest.,* 1858.

N. 1. Near Wennington and Melling, *Wi.*
   2. Borwick Quarry and near Kirkby Lonsdale, *Wi.*
   3. Eaves Wood, *L.P.* Stony ground near Woodwell, *Wh.* Abundant on the limestone hills.
W. 5. Casually as a garden weed at St. Michael's, *P.J.H.*
E. 6. Crook of Lune, *Wi.*
   8. Redscar, *Ashfield,* 1858 ; Gorge of the Hodder below Whitewell.

Hypericum montanum, L.—Mountain St. John's Wort. British type. Native. Area, 2, 3, (8). Range, 100-500 feet. P. July–August.

Dry rocky banks and woods on limestone. Rare. First record : *Linton, B.R.C. Rep.,* 874, p. 81.

N. 2. Kellet seeds, *Wi.*
   3. Rough pastures between Barton Well and Bank Well, Silverdale, *L.P.* Warton Crag, Thrang End, and Gatebarrow Wood, *Wi.* Near Yealand.
E. 8. Fulwood and Longridge, *Fl. Prest.* N. We have not seen a specimen. In our experience it only occurs in the limestone districts.

Hypericum elodes, L.—Marsh St. John's Wort. English type. Native. Area, 8. Range, ? P. July–August.

Wet places on peaty ground. First record : *Ashfield, Fl. Prest.,* 1858.

E. 8. Margins of shallow pools on Ribbleton Moor, *Ashfield.* Perhaps now extinct.

MALVACEÆ.

Malva moschata, L.—Musk Mallow. British type. Native. Area, 1, 2, 3, 4, 5, 6, 7. Range, 0-400 feet. P. June–August.

Dry banks in the low country. Frequent. First record : *Ashfield, Fl. Prest.,* 1860. " Coast between Sunderland and Heysham and at Overton."

Malva sylvestris, L.—Common Mallow. British type. Native. Area, 2, 3, 4, 5, 6, 7, 8. Range, 0-300 feet. P. June–September.

Roadsides and waste places. Frequent. First record : *Linton, B.R.C. Rep.,* 1874, p. 81.

L

N. 2. Carnforth and Nether Kellet, *Wi.*
   3. Outside Bottoms, Silverdale, *L.P.* Ings Point ; Overton ; Heysham ; between Carnforth and Bolton-le-Sands, and ascending to 200 feet on Warton Crag, *Wi.*
W. 4. Bonehill, Winmarleigh, *S. Wi.* Cockersand, *Wi.*
   5. Blackpool, *Wi.* Ashton, *Fl. Prest. N.*
E. 6. Near Wray, *Wi.*
   7. Garstang, *Wi.*
   8. Stydd, *Fl. St.,* 1891. Fulwood, *Wh.*

Malva rotundifolia, L.—Dwarf Mallow. English type. Native. Area, 4, 5, 8. Range, 0-250 feet. P. July–September.

Hedgebanks and roadsides, often near houses, and perhaps not native in all its stations. Rare. First record : *Ashfield, Fl. Prest.,* 1860.

W. 4. Bank of the Wyre near Rawcliffe, *Pearson.*
   5. Near St. Michael's, *Ashfield,* 1860. Between Blackpool and Marton Mere, *Wi.* St. Annes sandhills, *C.B.* Ashton, *Fl. Prest. N.*
E. 8. Alston, *Wh.*

Malva pusilla, Smith (*M. borealis Wallm.*).—Alien. First record : *F. C. King, B.R.C. Rep.,* 1884.

N. 2. Near Halton, *Wi.*
   3. Morecambe, *Wh.* Silverdale, *L.P.*
W. 4. Wardless, *Wh.*
   5. Building land (formerly sandhills) between Park Road and Orchard Road, St. Annes, with the preceding species, *C.B.*
E. 8. Waste land near Ribchester, Aug., 1883, *King.*

TILIACEÆ.

Tilia cordata, Mill.—Lime. English type. Native. Area, 2, 3. Range, 0-200 feet. T. July.

Rocky woods on limestone. Rare. First record : *Wilson, Journ. Bot.,* 1900, p. 42.

N. 2. Near Aughton.
   3. Thrang Wood and Warton Crag, 1888, *Wi.* Limestone hills south of Leighton Beck, *Wi.* Woodwell, *W.K.* Heald Brow Woods, *Wi.*

Tilia europæa, L.—Alien. Occasionally planted. There is a very fine tree on the Tarnbrook Wyre at 500 feet.

LINACEÆ.

Radiola linoides, Roth.—All seed. British type. Native. Area, 2. Range, ?-300 feet. A. July–August.

Damp sandy ground. Very rare. First record : *Linton, B.R.C. Rep.,* 1874, p. 81.

N. 2. Arkholme Moor at 300 feet, 900, *Wi.*

Linum catharticum, L.—Purging or Mountain Flax. British type. Native. Area, 1, 2, 3, 4, 5, 6, 7, 8. Range, 0-1,250 feet. A. June–September.

Dry pastures, heathy banks, and grassy places. Common, especially upon limestone. First record : *Ashfield, Fl. Prest.,* 1858. " Naze Point."

It ascends to 1,250 feet on Salter Fell
On Warton Crag it was observed to be attacked by the fungus *Melampsora lini* (*Pers.*).

Linum usitatissimum, L.—Flax. Alien. First record : *Pearson,* in *Ashfield, Fl. Prest.,* 1858.

N. 1. Hipping Hall, Leck, *L.P.*
W. 5. Field at Lingert, near Garstang, *Pearson.* Waste land near Preston Docks, *Wh.* Ashton, *Fl. Prest. N.*
E. 8. Casually near Leagram, 1867, and near Lower Bridge, 1888, *Fl. St.*

Linum perenne, L.—Alien. First record : *Pearson,* in *Ashfield, Fl. Prest.,* 1860.

W. 4. " Stubble fields near Stalmine 1851, *Mr. Pearson,*" *Ashfield.*

GERANIACEÆ.

Geranium sanguineum, L.—Red Cranesbill. British type. Native. Area, 3, 5. Range, 0-400 feet. P. July.

Dry limestone scars and rocky thickets, and on the sand dunes. Frequent. First record : *Jenkinson, Brit. Plants,* 1775, p. 163.

N. 3. In Cringlebarrow Wood ! *Jenkinson.* Dry woods and rocky places near Silverdale, *Ashfield.* Gatebarrow and south of Cold Well, Leighton Beck, *Wi.* Eaves Wood and Waterslack Wood, *L.P.*
W. 5. Fleetwood, 1842, *Hailstone.*
Between Naze Point and the Guide's House and near Fleetwood, *Ashfield.* Sand-hills between St. Annes and Fairhaven, *A. .D.* Lytham, *Fl. Prest. N.*

Geranium phæum, L.—Dusky Cranesbill. Alien. First record:
Ashfield, Fl. Prest., 1858.

> N. 1. Roadsides near Burrow, *Miss Maudsley*.
> W. 5. Near St. Michael's, *Pearson*. Still there, *P.J.H.* Lane from Greaves-
> town to Ashton Marsh, *Ashfield*.
> E. 7. Old lane near Barton, *Ashfield*.
> 8. Near Higher Bridge, Shire Lane, and Longridge, *Fl. St.*

Geranium sylvaticum, L.—Wood Cranesbill. Scottish type.
Area, 1, 2, 6. Range, 50-1,120 feet. P. May–July.

Wooded river banks, bushy places, and meadows amongst the
hills. Rare. First record: *Wilson, B.R.C. Rep.*, 1887, p. 88.

> N. 1. Ease Gill and Leck Fell at 1,120 feet, *Wi*. Meadow near Nether
> Burrow, *Wi*. Gorge of the Greeta.
> 2. Bank of the Lune below Kirkby Lonsdale, and on an island in the Lune
> near Arkholme.
> E. 6. Wooded bank of the Lune near Halton, 1886, *Wi*. Meadows near
> Wennington, *Wi*.

Geranium pratense, L.—Meadow Cranesbill. British type.
Native. Area, 1, 2, 3, 4, 5, 6, 7, 8. Range, 0-400 feet. P.
June–August.

Moist meadows, banks, and roadsides. Common in North ;
much less frequent elsewhere. First record: *Linton, B.R.C.
Rep.*, 1874, p. 81.

> N. 1. Ireby, Melling, Nether Burrow, Wennington, etc. ; very frequent, *Wi*.
> 2. Over Kellet, Carnforth, and other places, *Wi*.
> 3. Middlebarrow, *Wi*. Silverdale Moss Road, *L.P.* Fields at the foot of
> Warton Crag, *Wh*.
> W. 4. Canal side between Galgate and Glasson, *Wi*. Cock Hall and Galgate,
> *Wi*.
> 5. Woodplumpton and Catforth, *Wi*.
> E. 6. By the Lune near Caton, *Wh*.
> 7. Claughton, near Garstang, *E.C.*
> 8. Higher Bridge Island, *Fl. St.* By the Ribble near Preston, *Wh*. Barton
> Mill and near Whitewell, *Wi*.

Geranium pyrenaicum, Burm. fil. — Mountain Cranesbill.
English type. Denizen. Area, 5, 7. Range, 0-80 feet. P.
June–July.

Banks and meadows. Rare. First record: *Wilson, B.R.C.
Rep.*, 1883, p. 248.

> W. 5. Lytham. *Motley Herbarium (Naturalist, 1902).*
> E. 7. Near Garstang, 1882, *Wi*.

Geranium molle, L.—Dove's-foot Cranesbill. British type.
Native. Area, 1, 2, 3, 4, 5, 6, 7, 8. Range, 0-400 feet. A.
April–September.

Roadsides, fields and waste places. Frequent. First record:
Ashfield, Fl. Prest., 1858. "Ashton Marsh."

A form with white flowers is not infrequent.

Geranium pusillum, L.—Small-flowered Cranesbill. English
type. Native? Area, 2, 3, 5. Range, 0-100 feet. A. June–
September.

Roadsides and cultivated ground. Rare. First record:
Wheldon, Journ. Bot., 1901, p. 23.

> N. 2. Roadside between Caton and Halton, *Wh.*, 1900.
> 3. Near Bolton-le-Sands, *L.P.*
> W. 5. Lytham, *Fl. Prest. N.*

[*Geranium rotundifolium*, L.—Incognit.

"I am informed that this grows in Kirkham Churchyard," *Ashfield, Fl. Prest.*]

Geranium dissectum, L.—Jagged Cranesbill. British type.
Native. Area, 1, 2, 3, 4, 5, 6, 7, 8. Range, 0-500 feet. A. June–
September.

Dry banks and waysides. Frequent, especially in North.
First record: *Ashfield, Fl. Prest.*, 1858. "Between Greaves-
town and Ashton Marsh."

Geranium columbinum, L.—Long-stalked Cranesbill. English
type. Native. Area, 2, 3. Range, 0-400 feet. A. June–July.

Dry calcareous banks and woods. Not uncommon on the
limestone in North. First record: *Petty, Silv. Plants, Naturalist,*
1902, p. 38.

> N. 2. On Kellet seeds at 400 feet, *Wi*.
> 3. Frequent in the Silverdale district, *Wi*. Near the Cemetery, Cove, and
> foot of Heald Brow, Silverdale, *L.P.*, 1901. Warton Wood, *Wh*.

Geranium lucidum, L.—Shining Cranesbill. British type.
Native. Area, 1, 2, 3, 5, 7, 8. Range, 0-1,200 feet. A. May–
August.

Rocks, walls and stony places, mostly on limestone.
Frequent in North ; rare elsewhere. First record: *Ashfield,
Bot. Chron.*, 1864.

> N. 1. About Wennington and the Ease Gill district, ascending to 1,200 feet on
> Leck Fell, *Wi*.
> 2. Near Arkholme, Whittington, and Kirkby Lonsdale, *Wi*.
> 3. Old walls, Silverdale, very abundant ! *Ashfield*.
> W. 5. Near the railway, Ashton, *Fl. Prest. N.*
> E. 7. Hedgerow near Bay Horse, *W.W.M.* and *Wh*.
> 8. Near Stonyhurst Church and by the Lower Bridge Road, *Fl. St.*
> Gorge of the Hodder below Whitewell.

Geranium Robertianum, L.—Herb Robert. British type.
Native. Area, 1, 2, 3, 4, 5, 6, 7, 8. Range, 0-1,420 feet. B.
May–September.

Woods, hedgebanks and shady places. Common. First
record: *Linton, B.R.C. Rep.*, 1874, p. 81.

It ascends in North to 1,420 feet on rocks by the waterfalls in Upper Ease Gill,
and in East to 1,280 feet on Mallowdale Fell.

Var. *purpureum* (Forster).

> N. 3. Shore North of Bare, *F.A.L.*

A native of Carnforth informs us that it is called locally by the children
"Mother-will-die," and rejected from nosegays from a superstition that if taken into
a house the mother will die within the year. For a list of other plants to which
this curious bit of folk-lore attaches in the North of England, see *Naturalist*, 1902,
p. 4.

Erodium cicutarium, L'Herit.—Stork's Bill. British type.
Native. Area, 3, 4, 5. Range, 0-50 feet. A. June–September.

Sandy ground. Not common, except on the sandhill tract of
West. First record: *Ashfield, Fl. Prest.*, 1858.

> N. 3. A few plants near Jenny Brown's Point, *L.P.* Near Overton, *Wi*.
> W. 4. Near Preesall, *Wh*. Cockerham coast, *Wi*.
> 5. Lytham, *Ashfield*, 1858. Fleetwood and St. Annes, *Wh*. Near the
> Railway, Ashton, *Fl. Prest. N.*
> A white flowered form occurs at Lytham occasionally.

Erodium moschatum, L'Herit.—Musky Stork's Bill. Alien.
First record: *Jenkinson, Brit. Pl.*, 1775.

> N. 3. "I found it, too, in Yealand, but not much of it," *Jenkinson*.

Oxalis Acetosella, L.—Wood Sorrel. "Bread and Cheese" at
Dolphinholme. "Shamrock" in the Fylde. British type.
Native. Area, 1, 2, 3, 4, 5, 6, 7, 8. Range, 0-2,040 feet. P.
April–May.

Woods and shady places. Common. First record : *Linton,
B.R.C. Rep.*, 1874, p. 81.

It ascends to 2,040 feet on Greygarth Fell, *Wi*.

Impatiens Noli-tangere, L.—Touch-me-not or Yellow Balsam.
Local type. Denizen. Area, 2. Range, only found at about
300 feet. A. July–September.

Moist woods and shady glens. Rare. First record: *Baker,
Fl. Lake Dist.*, 1885, p. 64.

> N. 2. Ghyll near Whittington Hall (*Hindson*). This locality is quoted as in
> Westmorland in *Fl. Lake Dist.*, but it is a mile or more over the
> Lancashire border.

AQUIFOLIACEÆ.

Ilex aquifolium, L.—Holly. British type. Native. Area,
1, 2, 3, 4, 5, 6, 7, 8. Range, 0-1,400 feet. T. May–June.

Woods and hedges, and cloughs on the fells. Common,
especially in East. First record: *Ashfield, Fl. Prest.*, 1858.
"Longridge, apparently naturalized."

It ascends to 1,000 feet in Catshaw Greave and to 1,400 feet on Hell Crag above
Tarnbrook Fell. The fungus *Trochila ilicis, Cronan*, is extremely common on fallen
holly leaves.

CELASTRACEÆ.

Euonymus europæus, L.—Spindle Tree. Local name at Kellet
"Needle Tree." English type. Native. Area, 2, 3. Range,
0-500 feet. Small T. May–June.

Woods and bushy places on the limestone, where it is clearly
indigenous. Frequent. First record: *Ashfield, Bot. Chron.*,
1864, p. 73.

> N. 2. Over Kellet, ascending to 400 feet on Kellet seeds, *Wi*.
> 3. Silverdale woods and hedges, frequent, *Ashfield*, 1864. Gatebarrow,
> Middlebarrow, Thrang Wood, etc., *Wi*.

RHAMNACEÆ.

Rhamnus catharticus, L.—Buckthorn. English type. Native.
Area, 2, 3. Range, 0-500 feet. Sh. May–June.

Woods, bushy places, and hedgerows on the limestone.
Frequent. First record: *Ashfield, Bot. Chron.*, 1864, p. 73.

> N. 2. Near Over Kellet, *Wi*.
> 3. Frequent at Silverdale ! *Ashfield*. Also on hills south of Leighton
> Beck ; about Warton Crag ; and Bolton-le-Sands, *Wi*.

Rhamnus Frangula, L.—Alder Buckthorn. English type. Native. Area, 3 (8). Range, 0-400 feet. Small T. May–June.

Woods and peaty thickets on the limestone. Not unfrequent in parts of North ; very rare and not indigenous elsewhere. First record : *Ashfield, Fl. Prest.*, 1860.

> N. 3. Limestone hills about Silverdale and to the south of Leighton Beck ; Warton Crag, *Wi.*
> E. 8. Near Beacon Fell, by the road from Derby Arms, one shrub, 1859, *Ashfield.*

SAPINDACEÆ.

Acer Pseudo-platanus, L.—Sycamore. British type. Denizen. Area, 1, 2, 3, 4, 5, 6, 7, 8. Range, 0-840 feet or higher. T. May.

Woods, parks, and hedgerows. Very common. First record : *Linton, B.R.C. Rep.*, 1874, p. 81.?

> It occurs on Dalton Crag at 840 feet.
> Its leaves are rarely free from the black spots of the fungus *Rhytisma acerinum* Pers.).

Acer campestre, L. — Maple. Alien. First record : *Linton, B.R.C. Rep.*, 1874, p. 81. (Without locality).

> E. 8. A single shrub in a hedge near Grimsargh, 1903, *Wh.*

LEGUMINOSÆ.

Lupinus nootkatensis, Down ex-Sims, Bot. Mag., 1810.—Alien. First record : *Wheldon, Journ. Bot.*, 1902, p. 347.

> W. 5. Plentiful on rail banks between Salwick and Kirkham, and well-established. First seen in 1899, but probably planted, *Wh.*

Genista anglica, L.—Needle Furze or Petty Whin. British type. Native. Area, 2, 8. Range, 0-200 feet. Sh. June.

On heathy ground at low elevations. Rare. First record : *Ashfield, Fl. Prest.*, 1858.

> N. 2. Bog near Docker, May, 1901, *Wi.* The only existing locality known.
> E. 8. Ribbleton Moor, near Preston, *Ashfield*, 1858. "Grew in small quantities in a small portion of the moor nearest to Preston, on the left-hand side of the road from that place." Locality now drained.

Genista tinctoria, L.—Greenweed. English type. Native. Area, 2, 3, 4, 5, 6, 7, 8. Range, 0-750 feet. P. July–August.

Fields, heaths and banks, generally on a clay soil. Not common. First record : *Ashfield, Fl. Prest.*, 1858.

> N. 2. On Gressingham Moor and in the bog near Docker, *Wi.*
> 3. Silverdale, *Wi.* Morecambbe, *P.J.H.*
> W. 4. Preesall Marsh, *Wh.*
> 5. Near Naze Point, *Ashfield.* Near Ashton, *Fl. Prest. N.* Between Freckleton and Lytham abundantly, and near Garstang, *Wi.*
> E. 6. Mill Houses near Wray, *Wi.*
> 7. Near Garstang, *Wi.* Various localities in Wyresdale, ascending to 750 feet on Greenside.
> 8. Between Grimsargh and Goosnargh, *Ashfield*, 1858. Cross Gills Farm, near St. John's Well, Ribchester, and Longridge Fell Quarries ! *F. St.* Buckley Delph, *Dobson.*

Ulex europæus, L.—Furze, Gorse or Whin. British type. Native. Area, 1, 2, 3, 4, 5, 6, 7, 8. Range, 0-950 feet. Sh. January–May.

Banks, commons and heathy fields. Common. First record : *Linton, B.R.C. Rep.*, 1874, p. 81.

> It ascends to 950 feet on Burnslack Fell.

Ulex Gallii, Planch.—Autumnal Gorse. English type. Native. Area, 2, 3, 5, 6, 7, 8. Range, 0-1,100 feet. Sh. August–October.

In similar situations to the preceding species. Not unfrequent in East. First record : *Wilson, B.R.C. Rep.*, 1884, p. 8.

> N. 2. Hills east of Kellet, abundant, *Wi.* Gressingham Moor.
> 3. Between Heysham and Far Naze, *Wi.*
> W. 5. Near Cleveleys and Wrea Green, *Wi.* Gravel pit near Marton, *Wh.*
> E. 6. Lower Hindburn, *Wi.* Between Lancaster and Quernmore. Claughton Moor.
> 7. Common near Bay Horse, 1883 ; near Garstang ; Parlick ; Bleasdale ; and Nether Wyresdale, *Wi.*
> 8. White Stone Clough at 1,100 feet, and near Crowshaw Reservoir.

Sarothamnus vulgaris, Wimm.—Broom. British type. Native. Area, 1, 2, 3, 4, 5, 6, 7, 8. Range, 0-890 feet. Sh. May–June.

Sandy or gravelly banks, and bushy places. Frequent. First record : *Ashfield, Fl. Prest.*, 1858. "Melling's Wood."

> Ascends to 890 feet in Foxdale.

Ononis spinosa, L.—Rest Harrow. British type. Native. Area, 3, 4, 5. Range, 0-50 feet. P. June–September.

Dry banks and fields near the coast. Rare. First record : *Wilson, B.R.C. Rep.*, 1887, p. 89.

> N. 3. Between Carnforth and Bolton-le-Sands, *Wi.*
> W. 4. Fields near the sea coast, Pilling, *Wi.*, 1884. Abundant in a field near Hale's Hall, Rawcliffe, *S. Wi.* Wyre banks near Shovels Inn, *Wh.* ; Cartford Bridge, *Wi.*
> 5. Between Lytham and Freckleton, *Wi.* Near the Ribble below Preston, *Wh.* Ashton, *Fl. Prest. N.*

Ononis repens, L.—Rest Harrow. British type. Native. Area, 1, 2, 3, 4, 5, 6, 7, 8. Range, 0-300 feet. P. June–September.

Dry banks and sandy places. Frequent, especially on the sea coast of West and North. First record : *Linton, B.R.C. Rep.*, 1874, p. 81.

> N. 1. Leck, *L.P.* Melling, *Wi.* Gorge of the Greeta.
> 2. Whittington Moor, and Nether Kellet, *Wi.*
> 3. Near Silverdale Station, *L.P.* Carnforth, *Wi.* Bare ! *F.A.L.*
> W. 4. Wyre banks near Preesall, *Wh.* Cabus.
> 5. Lytham and St. Annes, *Wi.* Ashton, *Fl. Prest. N.* Fleetwood, *Wh.*
> E. 6. Near Caton, *Wh.*
> 7. Near Forton, *E.C.*
> 8. Hacking Boat, near Stonyhurst, *Fl. St.*
> All the specimens we have seen are referable to the var. *inermis, Lange.*

Medicago sativa, L.—Lucerne. Alien. First record : *W. Kirkby, Naturalist*, 1901, p. 316.

> N. 3. Cornfield near Cove House, Silverdale, *W.K.*
> W. 5. Waste ground about Wyre Dock, Fleetwood, *Wh.*, 1900. Ashton Marsh, *H.B.*

Medicago lupulina, L.—Medick. British type. Native. Area, 1, 2, 3, 4, 5, 6, 7, 8. Range, 0-500 feet or higher. A. or B. May–August.

Fields, waste ground and roadsides. Very common. First record : *Linton, B.R.C. Rep.*, 1874, p. 81.

Medicago denticulata, Willd.—Alien. First record : *Fl. St.*, 1891.

> E. 8. Seminary garden, Stonyhurst, *Fl. St.*, 1886.

Melilotus alba, Desr.—Alien. First record : *Fl. Prest., N.*, 1903, p. 14.

> W. 5. Near Preston Docks, *Fl. Prest. N.*
> On building land (formerly sandhills), St. Annes, *C.B.* (Sp. !).

Melilotus indica, All.—Alien. First record : *Wheldon, Journ. Bot.*, 1901, p. 24.

> W. 5. Waste ground by the railway in Fleetwood Docks, *Wh.* Near Preston Docks ! *Fl. Prest. N.*

Melilotus Petitpierreana, Hayne (*M. arvensis, Wallr.*).—Meliot. Alien. First record : The present one.

> W. 5. Sandhills off the South Drive, St. Annes, *C.B.*, 1904 (Sp.!).
> This and the three preceding species were probably introduced with grain.

Melilotus officinalis, Lam.—Melilot. English type. Native. Area, 1, 2, 3, 4, 5, 6. Range, 0-250 feet. A. June–August.

Fields, banks and waste places. Rather rare. First record : *Ashfield, Fl. Prest.*, 1858.

> N. 1. By the Lune near Melling and on the left bank of the Greeta near Wrayton, *Wi.*
> 2. Near Arkholme and Skerton, *Wi.*
> 3. Near Carnforth and Silverdale, *Wi.* Between Morecambe and Snatchems, *Wh.*
> W. 4. Rail banks near Lancaster, *Wi.*
> 5. Naze Point and Lytham, *Ashfield*, 1858. Still there ! Ashton, *Fl. Prest. N.*
> E. 6. Bank of the Lune near Caton, *Wh.*

Trifolium pratense, L.—Red Clover. British type. Native. Area, 1, 2, 3, 4, 5, 6, 7, 8. Range, 0-900 feet or higher. P. May–August.

Meadows and banks. Very common, both as a wild plant and as an escape from cultivation. First record : *Linton, B.R.C. Rep.*, 1874, p. 81.

> Ascends to 800 feet in Wyresdale, on Greenside above Tarnbrook, and to 900 feet by the Tatham Beck in Hindburndale.

Trifolium medium, L.—Zig-zag Clover. British type. Native. Area, 1, 2, 3, 4, 5, 6, 7, 8. Range, 0-500 feet. P. July–August.

Hedgebanks and bushy places. Very frequent, but much less common than the preceding species. First record : *Ashfield, Fl. Prest.*, 1860. "Between Poulton and Bispham."

Trifolium arvense, L.—Hare's-foot Clover. British type. Native. Area, 3, 4, 5. Range 0-50 feet. A. July–September.

Dry sandy banks and fields near the sea. Rather rare, but locally plentiful. First record : *Ashfield, Fl. Prest.*, 1858.

N. 3. Poulton-le-Sands, 1840. *Motley Herbarium (Nat.,* 1902).

W. 4. Pilling and Knott End, *Wi.*

    5. Lytham ! *Ashfield,* 1858.

Trifolium striatum, L.—English type. Native. Area, 5. Range, 0–30 feet. A. June–July.

Sandy bank near the sea. Very rare. First record: *Ashfield, Fl. Prest.,* 1866.

W. 5. "I believe this plant is occasionally to be found on the beach at Lytham. In 1862 I found a *Trifolium* there in considerable quantities, which I believe to be this, but it was so far past maturity that I could not with certainty determine the species. I searched in the same place several times in 1864–5, and did not find one specimen." *Ashfield.* Still occurs at Lytham, *A.A.D.* Banks of the Ribble, *Fl. Prest. N.*

[*Trifolium suffocatum,* L.—Incognit.

"This plant is stated in Part II. (of *Fl. of Preston*) to grow on the beach at Lytham, but I believe incorrectly ; at any rate, I have not been able to meet with it there for many years." *Ashfield, Fl. Prest. N.,* 1866.]

Trifolium hybridum, L.—Alsike Clover. Alien. First record: *Fl. St.,* 1891.

N. 1. Near Cantsfield, *Wi.*

    3. Bolton-le-Sands, *L.P.* Waste ground near Heysham Harbour.

W. 4. Near Cockerham, *Wi.*

    5. About Preston and Fleetwood Docks, *Wh.*

    7. Near Garstang, *Wi.*

E. 8. Field near Hacking Boat, *Fl. St.,* 1887. Near Grimsargh, *Wh.*

Trifolium incarnatum, L.—Alien. First record: The present one.

W. 5. Near St. Michael's, *P.J.H.*

Trifolium repens, L.—White Clover. British type. Native. Area, 1, 2, 3, 4, 5, 6, 7, 8. Range, 0–2,040 feet. P. June–September.

Fields, banks and grassy places. Very common. First record: *Linton, B.R.C. Rep.,* 1874, p. 81.

Trifolium fragiferum, L.—Strawberry-headed Clover. English type. Native. Area, 3, 4, 5. Range, 0–50 feet. P. June–August.

Damp sandy places, especially near the sea coast. Frequent. First record: *Ashfield, Fl. Prest.,* 1858.

N. 3. Near Overton, *Wi.* Basil Point, very fine, *Wh.*

W. 4. Coast near Pilling, Glasson, and Cockersand, *Wi.* Bank of the Wyre, near Preesall, *Wh.*

    5. Ashton Marsh, *Ashfield,* 1858. Between Naze Point and Lytham, *Ashfield.* Still there, *Wi.* Near Ansdell, *Wh.*

Trifolium procumbens, L.—Hop Trefoil. British type. Native. Area, 1, 2, 3, 4, 5, 6, 7. Range, 0–400 feet. A. June–August.

Sandy fields and dry grassy places. Frequent. First record: *Ashfield, Fl. Prest.,* 1858. "Ashton Marsh."

Trifolium dubium, Sibth.—Lesser Trefoil. English type. Native. Area, 1, 2, 3, 4, 5, 6, 7, 8. Range, 0–1,100 feet. A. June–August.

Fields, roadsides, and dry grassy places. Very common. First record: *Ashfield, Fl. Prest.,* 1858. "Between Naze Point and Lytham."

Ascends to 1,100 feet on Leck Fell.

Anthyllis vulneraria, L. — Lady's Fingers. British type. Native. Area, 2, 3, 4, 5. Range, ?. P. June–August.

Hills and dry banks on calcareous soil. Common in parts of North, rarer elsewhere. First record: *Top. Bot.,* 1883.

Lotus corniculatus, L.—Bird's Foot Trefoil. "Lambsfoot." British type. Native. Area, 1, 2, 3, 4, 5, 6, 7, 8. Range, 0–910 feet or higher. P. May–September.

Dry banks and fields. Very common. First record: *Linton, B.R.C. Rep.,* 1874, p. 81.

In North it ascends to 880 feet on Dalton Crag and in East to 910 feet in Foxdale.

Var. *crassifolius,* Pers. Frequent on the coast.

N. 3. Bare, *F.A.L.* Silverdale, *L.P.* Near Middleton, *Wi.*

W. 5. Lytham and St. Annes, *Wh.*

Lotus uliginosus, Schkuhr.—Hairy Bird's-foot Trefoil. British type. Native. Area, 1, 2, 3, 4, 5, 6, 7, 8. Range, 0–990 feet or higher. P. July–August.

Moist thickets and banks of ditches. Common. First record: *Linton, B.R.C. Rep.,* 1874, p. 81.

It ascends to 910 feet on Haylot Fell and in Foxdale.

Ornithopus perpusilius, L. — Bird's Foot. British type. Native. Area, 4. Range, 0–30 feet. A. June–July.

Heathy ground. Rare. First record: *P. J. Hornby, Journ. Bot.,* 1902, p. 347.

W. 4. Moss-side near St. Michael's, *P.J.H.,* 1901.

Hippocrepis comosa, L.—Horse-shoe Vetch. English type. Native. Area, 2, 3. Range, 0–500 feet. P. June–July.

Limestone hills, particularly on exposed cliffs. Rare. First record: *Wilson, Journ. Bot.,* 1900.

N. 2. Over Kellet, *Wi.*

    3. Warton Crag, *Wi.,* June, 1899. Silverdale Cove, *Wh.* Rocks near Jenny Brown's Point, *L.P.*

Vicia hirsuta, S. F. Gray.—Hairy Tare. British type. Colonist. Area, 3, 4, 5, 8. Range, 0–400 feet. A. July–August.

Cornfields and sandy banks. Rather frequent in West, rarer elsewhere. First record: *Ashfield, Fl. Prest.,* 1858.

N. 3. Trowbarrow and other places near Silverdale ! *L.P.*

W. 4. Pilling and near Preesall, *Wi.* Turnover, *P.J.H.*

    5. Ashton, *H.B.* Fleetwood, *Wh.* Old river bed, Preston, *Fl. Prest. N.*

E. 8. Frenchwood, Preston, *Ashfield,* 1858. Stonyhurst and Bolton Roughs, *Fl. St.*

Vicia pseudo-cracca, L.—Alien. First record: *Wheldon, Journ. Bot.,* 1902, p. 347.

W. 5. Waste ground near Fleetwood Docks, *Wh.,* 1900.

Vicia cracca, L.—Tufted Vetch. British type. Native. Area, 1, 2, 3, 4, 5, 6, 7, 8. Range, ?. P. July–August.

Hedges and bushy places. Very common. First record: *Linton, B.R.C. Rep.,* 1874, p. 82.

Vicia villosa, Roth.—Alien. First record: *Wheldon, Journ. Bot.,* 1901, p. 24.

W. 5. Fleetwood Docks *(the type),* June, 1900, *Wh.* Sandhills near St. Annes (the var. *glabrescens, Koch.*), *C.B.,* 1902.

Vicia sepium, L.—Bush Vetch. British type. Native. Area, 1, 2, 3, 4, 5, 6, 7, 8. Range, 0–1,420 feet. P. May–July.

Hedgebanks and thickets. Very common. First record: *Linton, B.R.C. Rep.,* 1874, p. 82.

It ascends to 1,420 feet in North, on rocks by the falls of Upper Ease Gill, and to 900 feet in East by the Tatham Beck.

Vicia sativa, L.—Common Vetch. British type. Colonist. Area, 1, 2, 4, 7, 8. Range, 0–500 feet or higher. A. May–July.

Waste ground and fields. Frequent. First record: *Linton, B.R.C. Rep.,* 1874, p. 82.

This and other species of *Vicia* are called in the vernacular "Fitches," a corruption of Vetches.

Vicia augustifolia, L.—Wild Vetch. British type. Native. Area, 2, 3, 4, 6, 7. Range, 0–400 feet or higher. A. May–July.

Sandy banks and fields. Frequent. First record: The present one.

N. 2. Near Dalton Houses.

    3. Near Heysham, *Wi.* Near Yealand.

W. 4. Knott End and Pilling, *Wi.* Preesall, *Wh.* Coast bank near Cockerham, *Mrs. A. Wilson.* By the canal, Glasson, *Wh.*

E. 6. Near Wray, *Wh.* and *Wi.* Near Galgate, *Wi.*

    7. Bruna Hill, near Garstang, *Wi.*

Vicia lathyroides, L.—Spring Vetch. British type. Native. Area, 5. Range, 0–30 feet. A. April–May.

Dry sandy places near the sea. Very rare. First record: *Wheldon, Journ. Bot.,* 1900, p. 43.

W. 5. Lytham and St. Annes, *Wh.,* May, 1899. Sea embankment near Fairhaven, *C.B.*

Lathyrus pratensis, L.—Meadow Vetchling. British type. Native. Area, 1, 2, 3, 4, 5, 6, 7, 8. Range, ?. P. July–August.

Moist grassy places, fields and banks. Very common. First record: *Linton, B.R.C. Rep.,* 1874, p. 82.

Lathyrus montanus, Bernh.—Mountain Vetchling. British type. Native. Area, 1, 2, 3, 5, 6, 7, 8. Range, 0–1,000 feet. P. May–June.

Woods, thickets, and grassy banks. Frequent. First record: *Ashfield, Fl. Prest.,* 1858.

N. 1. Wennington, Wrayton, and Ease Gill, *Wi.*

    2. Arkholme, *Wi.*

W. 3. Heysham, *Wi.*

    5. Road to Lancaster beyond Plungington, *Ashfield,* 1858.

E. 6. Near Wray and on Whitmoor, *Wi.* Cowkins, Hindburn.

    7. Deer Clough at 1,000 feet, and near Lower Emmetts, etc.

    8. Sion Hill, *Ashfield,* 1858.

Faba vulgaris is cultivated up to a height of 550 feet in Lower Hindburn, and the fungus *Uromyces appendiculata, Lév.,* occurs at that altitude on its leaves.

ROSACEÆ.

Prunus spinosa, L.—Blackthorn. Sloe. "Slaythorn." British type. Native. Area, 1, 2, 3, 4, 5, 6, 7, 8. Range, 0-1,120 feet. Sh. April–May.

Hedges and thickets, apparently most at home on the limestone tracts, but by no means restricted to them. Common, especially in North, where it fruits freely. First record: *Linton, B.R.C. Rep.,* 1874, p. 82.

Ascends to 1,120 feet in a pot-hole on Leck Fell.

Var. *macrocarpa,* Wallr.

W. 3. One or two bushes in a thicket near the canal south of Hest Bank, *F.A.L.,* Oct., 1899.

Prunus Padus, L.—Bird Cherry. Egg Berry. Scottish type. Native. Area, 1, 2, 3, 4, 6, 7, 8. Range, 0-1,070 feet. T. May–June.

Woods, hedges, and streamsides. Frequent amongst the hills; very rare in West. First record: *Ashfield, Fl. Prest.,* 1858. "Between Longridge and Knowl Green."

Ascends in North to 1,070 feet on Leck Fell.

Prunus Avium, L.—Wild Cherry. "Merry." English type. Native. Area, 1, 2, 3, 4, 5, 6, 7, 8. Range, 0-790 feet. T. May. Woods and thickets. Frequent, especially in some parts of East and North; rarer in West. First record: *Wilson, B.R.C. Rep.,* 1884, p. 9. "Woods near Garstang."

Spiræa Ulmaria, L.—Meadow-sweet. British type. Native. Area, 1, 2, 3, 4, 5, 6, 7, 8. Range, 0-1,420 feet. P. June–August.

Damp meadows, swampy thickets and streamsides. Very common, ascending to 1,420 feet in Ease Gill. First record: *Linton, B.R.C. Rep.,* 1874, p. 82.

The depauperate form known as var. *denudata, Bœnn.,* occurs near Bare (*F.A.L.*) and near Abbeystead.

Spiræa Filipendula, L. — Dropwort. English type. Native. Area, 3. Range, 0-300 feet or higher. P. June–July.

Dry pastures, banks and grassy places. Not unfrequent on the limestone tracts of North, to which it appears to be restricted. Likely to occur in Dist. 2. First record: *Simpson, Top. Bot.,* 1883.

N. 3. Pasture near Silverdale Station, *Wi.,* 1881. Limestone hills north of Carnforth and near Leighton Beck, *Wi.* Rough ground between Bank Well and Burton Well, Silverdale, *L.P.*

Rubus Idæus, L.—Raspberry. British type. Native. Area, 1, 2, 3, 4, 5, 6, 7, 8. Range, 0-1,160 feet. P. May–June.

Damp woods and bushy places. Common in North and East. First record: *Ashfield, Fl. Prest.,* 1858. "By the Brock and between Loud Bridge and Beacon Fell."

Occurs in North up to 1,160 feet in a pot hole on Leck Fell, and in East to 800 feet on Greenside, Over Wyresdale.

Rubus fissus, Lindl.—Scottish type. Native. Area, 2, 3, 5, 6, 7, 8. Range, 0-690 feet. Sh.

Damp thickets, woods and by ditches. Not very frequent, but locally abundant. First record: *Wilson, Journ. Bot.,* 1898, p. 401.

N. 2. Arkholme Moor, *Wi.*
    3. Thrang Moss, *Wi.*
W. 5. Wood between St. Michael's and Catforth, *Wi.*
E. 6. Heights, *Wi.* Near Ivah.
    7. Near Garstang, 1898, and Bailey Hey, *Wi.* Frequent about Abbeystead, on the roadside, by the reservoir, and by the Wyre lower down. On the south slope of Wardstone at 690 feet.
    8. By the Hodder, near Whitewell.

Rubus suberectus, Anders.—English type. Native. Area, 2, 6, 8. Range, 100-600 feet. Sh.

Boggy woods and thickets. Rare. First record: *Fl. St.,* 1891, p. 15.

N. 2. Kellet Park Wood, near Capernwray.
E. 6. Boggy thicket above Botton Mill, Hindburn, *Wi.*
    8. Sale Wheel, *Fl. St.,* 1891.

Rubus plicatus, W. & N.—British type. Native. Area, 1, 3, 4, 5. Range, 0-220 feet. Sh. June–July.

On damp heaths and ditch sides. Rather rare. First record: The Authors, *Journ. Bot.,* 1901, p. 24.

N. 1. Roman road near Cowan Bridge.
    3. Heysham Moss. Storrs Moss, *Wi.*
W. 4. Rawcliff Moss, *Wi.* Plentiful on Cockerham Moss, June, 1900.
    5. Between Ansdell and Wrea Green, *Wi.*

M

Rubus incurvatus, Bab.—English type? Native. Area, 5. Range, ? Sh.

Thickets and heaths. Very rare. First record: *Marshall, Journ. Bot.,* 1896. p. 136.

W. 5. Near Inskip, *E.S.M.,* 1895.
    The Rev. W. M. Rogers, in a letter to the authors in 1899, says of Mr. Marshall's plant:—"Not what I call typical *R. incurvatus,* but that form which prevails in Derbyshire, Salop, and Lancashire, hardly differing from the type except in its laxer panicle, crowded strongly falcate prickles, and somewhat thinner and less lobate leaves—characters suggesting a divergence from type towards *R. Colemanni.*"

Rubus Lindleianus, Lees. — British type. Native. Area, 1, 2, 3, 4, 5, 6, 7, 8. Range, 0-500 feet or higher. Sh.

Hedges and thickets. Very common. First record: *Wilson, Journ. Bot.,* 1898, p. 401. "Scorton and Garstang."

Rubus argenteus, W. & N. (*R. erythrinus,* Genev.)—Atlantic type? Native. Area, 4, 5, 7. Range, 0-200 feet or higher. Sh.

Hedges and thickets. Rare. First record: *Marshall, Journ. Bot.,* 1896, p. 136.

W. 4. Frequent about Little Eccleston and Elswick, *E.S.M.,* 1895. Bushy bank of the Wyre near Knott End and Coatwalls, *Wh.*
    5. Bank of the Wyre near Fleetwood, *Wh.*
E. 7. Lower Grizedale, near Garstang.

Rubus rhamnifolius, W. and N. — British type. Native. Area, 3, 6, 7. Range, 0-500 feet. Sh.

Hedges and wood borders. Rare. First record: *Wilson, Journ. Bot.,* 1898, p. 401.

N. 3. Bare, *F.A.L.*
E. 7. Near Garstang, *Wi.,* 1898.

Sub-sp. *Bakeri,* F. A. Lees.

E. 6. Claughton Moor, near Caton, *Wi.,* 1905.

Rubus Silurum, A. Ley.—Atlantic type? Native. Area, 5, 6, 7, 8. Range, 0-900 feet. Sh.

Hilly slopes and riversides. Frequent. First record: *Marshall, Journ. Bot.,* 1896, p. 136.

W. 5. Between Eccleston and Inskip, *E.S.M.,* 1895.
E. 6. Lancaster Moor, *Wh.* Roeburndale and Hindburn, *Wi.* Between Caton and Caton Moor, and near Quernmore.
    7. By the Wyre near Scorton, Barnacre, and Over Wyresdale. Near Bailey Hey, Bleasdale, *Wi.* Upper Claughton.
    8. Near Hurst Green.

Rubus Scheutzii, Lindeb.—Scottish type? Native. Area 7. Range, 200-600 feet. Sh.

Roadsides and stony places in upland districts. Rare. First record: *Wilson, Journ. Bot.,* 1898, p. 401.

E. 7. Roadside near Catshaw, Wyresdale (alt. 600 feet), *Wi.,* 1898. Between Five Lanes and Dolphinholme, *Wh.* Near Lower Emmetts and on Barnacre Moor.

Rubus pulcherrimus, Neum.—British type. Native. Area, 1, 2, 3, 4, 5, 6, 7, 8. Range, 0-500 feet. Sh.

Hedges, thickets, and borders of woods. Frequent. First record: *Wheldon, Journ. Bot.,* 1898, p. 401. "Silverdale and Yealand Conyers."

Rubus Lindebergii, P. J. Muell.—English type. Native. Area, 3, 5. Range, 0-200 feet. Sh.

Banks and thickets. Rare or overlooked? First record: *Marshall, Journ. Bot.,* 1896, p. 136.

N. 3. Gatebarrow Wood.
W. 5. Between Inskip and Elswick, *E.S.M.,* 1895.

Rubus bracteatus (Bagn), Rogers.— English type. Native. Area, 1, 2, 3, 4, 5, 6, 7, 8. Range, 0-500 feet or higher. Sh. July–August.

Bushy banks and hedgerows. Rather frequent. First record: *Wheldon, Journ. Bot.,* 1898, p. 401.

N. 1. Roman road near Cowan Bridge.
    2. River side near Skerton, *Wh.*
    3. Near Heysham Harbour.
W. 4. Between Scorton and Winmareigh. Staining, *Wh.*
    5. Near Kirkham, *Wh.*
    6. Near Wray.
E. 7. Barnacre, near Garstang; Upper Claughton.
    8. Longridge Fell, *Wh.,* 1898. Near Hurst Green.

Rubus Selmeri, Lindeb.—British type. Native. Area, 1, 2, 3, 4, 5, 6, 7, 8. Range, 0-930 feet. Sh.

Hedgerows and thickets, especially on a sandy or gravelly soil. Common. First record: *Wilson, Journ. Bot.,* 1898, p. 401. "Pilling."

It ascends to 930 feet near the Tatham Beck in Hindburndale.

Rubus gratus, Focke.—British type. Native. Area, 8. Range, ? Sh.

Hedges and thickets. Rare. First record: *Wheldon, Journ. Bot.*, 1898, p. 401.

> E. 8.   Hodder Valley, near Kemple End, 1898, and near Preston Wives, *Wh.*

Rubus rusticanus, Merc.—British type. Native. Area, 1, 2, 3, 4, 5, 6, 7, 8. Range, 0-480 feet. Sh.

Hedges and thickets. Common at low altitudes, and especially so near the coast on calcareous or clayey soils. First record: *Wilson, Journ. Bot.*, 1898, p. 401. "Near Scotforth."

> It ascends to 480 feet on Warton Crag, in North, and to 330 feet near Ellel, in East.
> A curious form occurs near Lea Toll Gate (District 5), with narrower flat leaflets and broader, more pyramidal panicles, of which Mr. Rogers writes:—"A well-marked form I have occasionally met with, but have no separate name for." The fungus *Phragmidium rubi* (*Pers.*) is frequent on this species in North, and *P. violaceum* occurs on its leaves near Cowan Bridge.

Rubus macrophyllus, W. & N. — British type. Native. Area, 2, 7, 8. Range, 100-300 feet. Sh.

Shady woods and thickets. Rare. First record: *Lewis, B.R.C. Rep.*, 1884, p. 10.

> N. 2.   Wood near Over Kellet.
> E. 7.   Near Upper Claughton—somewhat doubtful.
>   8.   Between the Alston Arms and Thornley, *Lewis.* Elston Wood, *Wh.*

[*Rubus Salteri,* Bab.—Near Preston, *Linton.*

> As Mr. Rogers does not admit this in his census of British Rubi for V.C. 60, we think it desirable to obtain modern confirmation before inserting it.]

Rubus Sprengellii, Weihe.—English type. Native. Area, 2, 3, 5, 6, 7, 8. Range, 0-600 feet. Sh.

Woods, thickets and hedgerows. Frequent in East; rarer elsewhere. First record: *Wheldon, Journ. Bot.*, 1898, p. 401.

> N. 2.   Wood near Over Kellet.
>   3.   Near Morecambe and Silverdale, *Wh.*
> W. 5.   Near Preston, *Wh.*
> E. 6.   Lancaster Moor, *Wh.* Plentiful near Ellel, *Wi.*
>   7.   Bleasdale, *Wi.* Barnacre Moor and Upper Claughton.
>   8.   Longridge, 1898, *Wh.* Near Goosnargh, Grimsargh, Alston, Dutton Lea, and Knowl Green, *Wh.*

Rubus pyramidalis, Kalt.—British type. Native. Area, 3. Range, only seen at about 30 feet. Sh. July–August.

Hedges. Apparently rare. First record: *Wheldon, B.E.C. Rep.* for 1899 (pub. 1901), p. 602.

> N. 3.   Between Morecambe and Snatchems, July, 1899, *Wh. (teste Liuton).*

Rubus leucostachys, Schleich.—English type. Native. Area, 1, 2, 3, 4, 6, 7, 8. Range, 0-400 feet. Sh. July–August.

Woods, thickets and bushy banks. Frequent. First record: *Wheldon, B.E.C. Rep.*, 1898 (pub. 1900), p. 401.

> N. 1.   Cowan Bridge, *Wi.*
>   2.   Wood near Halton, *Wh.* Between Caton and Aughton; Kellet Seeds.
>   3.   Silverdale, frequent, 1898, and near Morecambe, *Wh.* Near Carnforth.
> W. 4.   Wardleys, *Wh.*
> E. 6.   Near Lancaster, *Wh.* Claughton, near Caton, *Wi.*
>   7.   Dolphinholme.
>   8.   Frequent between Fulwood and Ribbleton and in the Longridge and Ribchester districts, *Wh.*

Rubus cinerosus, Rogers.—English type? Native. Area, 4, 5, 7. Range, 0-500 feet. Sh. August.

Thickets and gravelly banks. Rare. First record: *Wheldon, Journ. Bot.*, 1901, p. 24.

> W. 4.   Banks of the Wyre about Preesall and Knott End, 1900, *Wh.*
>   5.   By the Wyre above Fleetwood, *Wh.*
> E. 7.   Barnacre, near Garstang.

Rubus criniger, Linton.—English type? Native. Area, 6, 7. Range, 350-500 feet.

Bushy places. Rare. First record: The Authors, *Journ. Bot.*, 1905, p. 94.

> E. 6.   Claughton Moor, near Caton, *Wi.*
>   7.   Roadside at Claughton, near Garstang, Sept., 1904.

Rubus mucronatus, Blox.—British type. Native. Area, 1, 2, 4, 5, 7. Range, 1-400 feet. Sh.

Hedges and bushy places. Rare. First record: *Wheldon, Journ. Bot.*, 1902, p. 347.

> N. 1.   Near Nether Burrow, "a form with exceptionally few stalked glands and acicles on the stem," teste *W. M. Rogers.*
>   2.   Gressingham Moor.
> W. 4.   Knott End, July, 1901, and by the canal between Bay Horse and Glasson, *Wh.*
>   5.   Near Little Marton, *Wh.*
> E. 7.   Near Abbeystead Reservoir, and Upper Claughton, near Garstang.
> Forms occur which seem somewhat intermediate between this and *R. cinerosus* in the Knott End district, where both parents grow.

Rubus infestus, Weihe.—English type? Native. Area, 2, 6, 7, 8. Range, 0-350 feet or higher. Sh.

Hedges and bushy places. Rare, and always in small quantity. First record: *Wheldon, Journ. Bot.*, 1898, p. 401.

> N. 2.   Borders of Kellet Park Wood.
> E. 6.   Roadside on Caton Moor.
>   7.   Dolphinholme, *Wh.* Upper Claughton, near Garstang.
>   8.   Near Preston, *Wh.*, 1898.

Rubus Drejeri, Jens.—English type. Native. Area, 8. Range, only seen at about 600 feet. Sh.

Stony ground by moorland roadsides. Rare. First record: *Wheldon, Journ. Bot.*, 1901, p. 24.

> E. 8.   Roadside near the reservoir above Longridge, *Wh.*, July, 1900.
> The Rev. W. M. Rogers says, "You need not hesitate to put it to this, as a form going off from type towards my var. *Leyanus*, but still under the type."

Rubus Radula, W.—British type. Native. Area, 2, 3, 4. Range, 0-400 feet. Sh.

Sandy ground in the low country. Very rare. First record: *Lees, Naturalist*, 1901.

> N. 2.   Very abundant in the wood on Kellet Seeds. "The type," W. M. Rogers, *Wh.*
>   3.   Near Bare, *F.A.L.*, Oct., 1899.

Var. *echinatoides*, Rogers.

> W. 4.   Near Cockerham, *Wi.*, 1904.
> Somewhat off type, but cannot be kept apart from the var. *echinatoides, Rogers* (teste *W.M.R.*).

Rubus podophyllus, P. J. Muell.—English type. Native. Area, 5. Range, only seen at about 50 feet. Sh.

Bushy stony places. Very rare. First record: *Marshall, Journ. Bot.*, 1896, p. 136.

> W. 5.   About Inskip and Elswick, *E.S.M.*, 1895.

[*Rubus pallidus,* W. and N.—Incognit.

> Recorded from Fulwood and Redscar near Preston, in B.R.C. Rep., 1883, *W. R. & E. F. Linton.* As this was hardly distinguished from *R. dasyphyllus* at that date, we bracket it for further confirmation. We found a plant near Abbeystead which Mr. Rogers suggested might be referred to *R. pallidus*, but the specimens were collected too late for satisfactory determination.]

Rubus rosaceus (Aggreg).—British type. Native. Area, 1, 3, 4, 6, 7, 8. Range, ? Sh.

Woods and shady places. Not common. First record: *Wheldon, Naturalist*, 1902, p. 39.

> N. 1.   Near Nether Burrow.
>   3.   Near Silverdale, *Wh.*, 1900.
> W. 4.   Cabus, near Garstang.
> E. 7.   Woods by the Wyre, near Dolphinholme.

Var. *hystrix*, W. & N.

> E. 6.   Near Quernmore Park, *Wh.*
>   7.   Upper Claughton, near Garstang.

Var. *infecundus*, Rogers.

> E. 7.   Foot of Grizedale, near Garstang.
>   8.   Longridge Fell, 1898, *Wh.*

Rubus dasyphyllus, Rogers.—British type. Native. Area, 1, 2, 3, 4, 5, 6, 7, 8. Range, 0-880 feet. Sh.

Woods, bushy places, hedgerows and at the foot of walls on the fells. Common, especially in the more hilly districts. First record: *Wilson, Journ. Bot.*, 1898, p. 401.

> Ascends to 880 feet on Dalton Crag.

Rubus dumetorum, W. & N.—English type? Native. Area, 1, 2, 3, 4, 5, 6, 7, 8. Range, 0-400 feet or higher. Sh.

Hedges and waste ground. Common. First record: *Linton, B.R.C. Rep.*, 1884, p. 10.

Var. *ferox*, Weihe.

> N. 2.   Between Caton and Aughton and between Docker and Borwick.

Var. *diversifolius* (Lindl.)—

> W. 5.   Near Preston, *E.F.L.* Freckleton, *Wh.*
> Occurs probably in other districts, but intermediate forms which are difficult to refer to any of the varieties are numerous.

Var. *concinnus*, Warren.

> W. 5.   Between Preston and Freckleton, 1899, and between St. Annes and Kirkham, *Wh.*
> E. 6.   Near Quernmore Park, *Wh.*

Rubus corylifolius, Sm.—British type. Native. Area, 1, 2, 3, 4, 5, 6, 7, 8. Range, 0-400 feet or higher. Sh.

Hedges and thickets. Abundant. First record: *Wheldon, Journ. Bot.*, 1898, p. 401. "Silverdale." The above distribution refers to our common form, the var. *sublustris*, Lees.

Var. *cyclophyllus* (Lindeb.).

N. 3.　By the canal near Yealand.

W. 4.　Knott End, *Wh.*, 1902.

Hybrids between *R. corylifolius* and other *Rubi* are not infrequent; the most common appear to be between it and *R. Lindleianus* or *R. cæsius*.

Rubus cæsius, L.—Dewberry. British type. Native. Area, 1, 2, 3, 4, 5, 6, 7, 8. Range, 0-500 feet. Sh.

Hedges and waste places. Common, especially on the sand-hills and in the calcareous districts of North. First record: *Linton, B.R.C. Rep.*, 1874, p. 82.

Mr. Rogers refers our sandhill plant to var. *aquaticus, W.* and *N. R. cæsius* × *Idæus* occurs near Hurst Green, *Wh.*, 1899.

Rubus Chamæmorus, L.—Cloudberry. Knot-berry or Knout-berry. Highland type. Native. Area, 1, 6, 7, 8. Range, 2,060-1,250 feet. P. June.

High peat moors, rarely descending below 1,400 feet. Not unfrequent on the loftier fells of North and East. Fruit rather rare. First record: *Gerarde's "Herball*," 1636.

N. 1.　"It groweth naturally in a black moist earth or mosse, whereof the countrie maketh a fewel we call turfe and that upon the tops of wet fells and mountains among the heath, mosse and brake as. . . . . on Greygarth a high fell on the edge of Lancashire and Stainmoor a like place in Westmerland and other such like high places," *John Gerarde.* Abundant still on the peat over grit rocks of Greygarth Fell between 1,400 and 2,060 feet, *Wi.*

E. 6.　Summit ridge moors to west and north-west of Slaidburn, *F.A.L.* Summit and north slopes of Botton Head Fell, very abundant ; near the head of Middle Gill, *Wi.* ; and descending in Dale Gill, Hindburn, to 1,250 feet ; also in Upper Foxdale and on the east slopes of Wardstone.

7.　Hazelhurst Fell, *Mrs. C. Wilson* and *A. Wilson*, 1875 ; still there. White Moss at the head of Grizedale, *Wi.*

8.　Leagram Fells, *Fl. St.* Very abundant for a mile or more on the east side of Fairsnape Fell at about 1,650 feet, *Wi.*

Rubus saxatilis, L.—Stone Bramble. Scottish type. Native. Area, 1, 2, 3, 6, 7, 8. Range, 1,000-100 feet. P. June–July.

Stony woods and gills, generally in calcareous districts. Not uncommon. First record: *Wilson, B.R.C. Rep.*, 1887, p. 90.

N. 1.　Several places in Ease Gill and in the Gorge of the Greeta, *Wi.*

2.　Between Kirkby Lonsdale and Whittington and on Kellet Seeds, *Wi.*

3.　Rocky wooded ground about Silverdale, 1884, and in Gatebarrow Wood, *Wi.*

E. 6.　In Hindburn by the falls above Mill Houses, *Wi.* Deep gulley on Mallowdale Fell (alt. 700 feet) and by the Harter Beck, Roeburndale.

7.　Gorge of the Wyre below Abbeystead, *Wi.*

8.　Bolton and Hodder Roughs, *Fl. St.*

Geum urbanum, L.—Common Avens. "Black Bobs." British type. Area, 1, 2, 3, 4, 5, 6, 7, 8. Range, 0-1,420 feet. P. June–August.

Woods and hedgebanks. Very common. First record: *Linton, B.R.C. Rep.*, 1874, p. 82

1.　Ascends to 1,420 feet in Upper Ease Gill.

G. rivale × urbanum (G. intermedium, Ehrh.)—First record : *Fl. Stonyhurst*, 1891.

N. 1.　Between Whoop Hall and Kirkby Lonsdale Station, *L.P.*

E. 8.　Seminary Wood, 1886, and near Chaigley, *Fl. St.* Seen in this vicinity in 1897, *Wh.* Longridge, *Fl. Prest. N.*

Geum rivale, L.—Water Avens. British type. Native. Area, 1, 2, 3, 4, 5, 6, 7, 8. Range, 0-1,350 feet. P. May–June.

Damp woods, meadows and streamsides. Frequent amongst the hills. First record : *Ashfield, Bot. Chron.*, 1864, p. 74. "Marshes (near Silverdale) frequent."

Ascends in North to 1,350 feet in Ease Gill.

Fragaria moschata, Duchesne.—Strawberry.

Occurred casually on a waste heap near the shore, St. Annes, 1906, *C.B.* (Sp. !). Not unfrequently thrown out of gardens, but never establishes itself.

Fragaria vesca, L.—Wild Strawberry. British type. Native. Area, 1, 2, 3, 4, 5, 6, 7, 8. Range, 0-1,400 feet. P. May–June.

Dry hedgebanks, woods and rocky places. Frequent, especially in North and East. First record: *Ashfield, Fl. Prest.*, 1858. "Redscar."

It ascends to 1,400 feet in Ease Gill and also to the same height on Mallowdale Fell.

Potentilla sterilis, Garcke.—Barren Strawberry. British type. Native. Area, 1, 2, 3, 4, 5, 6, 7, 8. Range, 0-1,280 feet. P. March–May.

Hedge banks and woods. Common. First record: *Linton, B.R.C. Rep.*, 1874, p. 82.

In North it ascends to 1,280 feet on Ireby Fell.

Potentilla verna, L.—Intermediate type. Native. Area, 3, 8. Range, 100-400 feet ? P. April–May.

Dry upland pastures. Very rare. First record: *Ashfield, Fl. Prest.*, 1858.

N. 3.　Silverdale, *Miss Susan Beaver (Baker's Fl. Lake Dist.).*

W. 8.　"Said to grow in dry pastures near Preston, but I have not yet met with it," *Ashfield*, 1858. Pastures near Longridge, *H.B.*

Specimens of Mr. Beesley's Longridge plant have been submitted to Mr. A. Bennett, who agrees that they are referable to this species, and not to *P. alpestris (Hall.)*

Potentilla silvestris, Neck.—Tormentil. British type. Native. Area, 1, 2, 3, 4, 5, 6, 7, 8. Range, 0-2,080 feet. P. June–August.

Dry banks and heathy places. Common. First record: *Ashfield, Fl. Prest.*, 1858. "Longridge."

Ascends to 2,080 feet on Greygarth Fell and to 1,730 feet on Wardstone.

Potentilla procumbens (Sibth.). — English type. Native. Area, 2, 4, 5, 6, 7, 8. Range, 0-400 feet or higher. P. June–August.

Hedgebanks, grassy roadsides and heaths. Frequent. First record: *Fl. St.*, 1891, p. 16.

N. 2.　Arkholme, *Wi.*

W. 4.　On the moss near Cuckoo Farm, St. Michael's, *P.J.H.*

5.　Ashton, *Fl. Prest. N.*

E. 6.　Between Caton and Lancaster Moor, *Wh.*

7.　Between Dolphinholme and Abbeystead, *Wh.* Between Bay Horse and Dolphinholme, *W.W.M.* and *Wh.*

8.　Quarry Road and Kemple End, *Fl. St.*

Potentilla reptans, L.—Creeping Cinquefoil. English type. Native. Area, 1, 2, 3, 4, 5, 6, 7, 8. Range, 0-400 feet or higher. P. June–September.

Banks, waste places and roadsides. Common. First record: *Ashfield, Fl. Prest.*, 1858. "Cadeley."

Potentilla mixta Nolte, a hybrid between *P. reptans* and *P. procumbens*, was found near Inskip, 1895, *E.S.M.*

Potentilla Anserina, L.—Silver Weed. British type. Native. Area, 1, 2, 3, 4, 5, 6, 7, 8. Range, 0-780 feet. P. June–July.

Roadsides and waste places. Common. First record: *Linton, B.R.C. Rep.*, 1874, p. 82.

[*Potentilla fruticosa*, L.—Shrubby Cinquefoil. Incognit.

N. 2.　"Sparingly on limestone rocks 100 yards due east of Nether Kellet Church, eight miles beyond Lancaster," *Thomas Williams, Science Gossip,* 1870, p. 19.

We have searched for this in vain. The locality is not really an improbable one, but further confirmation is required.]

Comarum palustre, L.— Marsh Cinquefoil. British type. Native. Area, 1, 2, 3, 4, 5, 6, 7, 8. Range, 0-550 feet. P. July.

Bogs, marshes and margins of slow streams. Rather rare. First record: *Ashfield, Fl. Prest.*, 1858.

N. 1.　Marshes near Tunstall, *Wi.*

2.　Whittington Moor and bog near Docker, abundant. Bogs near Over Kellet.

3.　Little Hawes Water and near Bolton-le-Sands, *Wi.*

W. 4.　Tarnacre, *P.J.H.*

5.　Carrs Green Common, *Wi.* Ashton, *Fl. Prest. N.*

E. 6.　Mill Dam, Quernmore, *Wi.*

7.　Near Garstang, *Wi.*

8.　Ribbleton Moor, *Ashfield.* Composition Hill, *Fl. St.*

Alchemilla arvensis, Scop.— Parsley Piert. British type. Native. Area, 1, 2, 3, 4, 5, 7, 8. Range, 0-600 feet or higher. A. May–August.

Stony or gravelly fields and banks. Rather rare. First record: *Jenkinson, Brit. Pl.*, 1775. p. 19. "*Aphanes*—On Yealand common, plentifully."

Alchemilla vulgaris, L.—Lady's Mantle. British type. Native. Area, 1, 2, 3, 4, 5, 6, 7, 8. Range, 0-1,930 feet. P. June–August.

Fields, banks and rocky or grassy places. Frequent. First record: *Linton, B.R.C. Rep.*, 1874, p. 82.

Var. *alpestris* (Schmidt.).

N. 1.　Ease Gill at 1,390 feet, June, 1901, and near Ireby.

7.　Near Abbeystead Reservoir, *W.W.M.* and *Wh.*

Var. *filicaulis*, Buser.

N. 2.　Kellet Seeds.

3.　Near Silverdale, *Wh.*

E. 7.　Near Abbeystead.

8.　Fairsnape Fell.

Agrimonia Eupatoria, L.—Agrimony. British type. Native. Area, 1, 2, 3, 4, 5, 6, 7, 8. Range, 0-550 feet. P. June–August.

Dry banks, roadsides and bushy places. Not uncommon. First record: *Linton, B.R.C. Rep.*, 1874, p. 82.

Agrimonia odorata, Mill.—Agrimony. English type. Native. Area, 1, 2, 6. Range, 0-200 feet. P. June–July.

Rare. First record: *Wilson, Journ. Bot.*, 1901, p. 24.

N. 1. Near Melling, *Wi.*
　　2. By the Lune near Arkholme, *Wi.*
E. 6. By the Hindburn between Wray and Mill Houses, *Wh.*

Poterium Sanguisorba, L.—Salad Burnet. English type. Native. Area, 1, 2, 3, 8. Range, 0-900 feet or higher. P. June–July.

Dry banks and rocks. Abundant on the scar limestone hills of North. First record: *Ashfield, Bot. Chron.*, 1864, p. 74.

N. 1. Leck, *L.P.* Ease Gill and Gorge of the Greeta.
　　2. Near Whittington, and frequent about Over and Nether Kellet, *Wi.* Between Caton and Aughton.
　　3. Abundant on the hills north of Carnforth, ascending to 490 feet on Warton Crag.
E. 8. Near Hacking Boat and Trough's Ferry, *Fl. St.*

Sanguisorba officinalis, L.—Great Burnet. Intermediate type. Native. Area, 1, 2, 3, 4, 5, 6, 7, 8. Range, 0-1,000 feet. P. June–August.

Damp meadows and banks of ditches and streams. Frequent. First record: *Linton, B.R.C. Rep.*, 1874, p. 82, but there is a specimen in Herb., C. E. Salmon, dated 1843, collected near Lancaster by R. Pryor.

Rosa spinosissima, L.—Burnet Rose. British type. Native. Area, 2, 3, 5. Range, 0-500 feet. Small Sh. June.

Limestone hills and sand-dunes. Locally abundant. First record: *Linton, B.R.C. Rep.*, 1874, p. 82.

N. 2. Over Kellet, *Wi.*
　　3. Common in the Silverdale district and in the Heysham Peninsula about Far Naze, *Wi.*
W. 5. St. Annes, *Wh.*

Rosa involuta, Sm., is reported from near Kirkham in the Flora of the Preston Scientific Society, 1903, p. 18.

Rosa villosa, L.—Wild Rose. British type. Native. Area, 1, 2, 3, 5, 6, 7, 8. Range, 50-1,100 feet. Sh. June.

Hedges and thickets. Frequent amongst the hills; rare elsewhere. First record: *Lewis, B.R.C. Rep.*, 1884, p. 11.

N. 1, 2, 3. Frequent, ascending in Middle Ease Gill to 1,100 feet.
　　5. Kirkham, Blackpool and Marton, *Wh.*
　　6. Frequent in Hindburn and Roeburndale, ascending to 930 feet by the Tatham Beck.
　　7. Calder Valley near Garstang, near Claughton, and in Bleasdale, *Wi.* Frequent in both branches of Upper Wyresdale and about Catshaw and Barnacre.
　　8. Between Thornley and Alston! *Lewis*, 1883. Kemple End! *Fl. St.* Near Grimsargh, *Wh.* Dinkling Green, *Wi.*

The distribution shown is that of aggregate *Rosa mollis*. Of the numerous forms recently defined by the Rev. A. Ley in Journ. Bot. (1907, p. 200) the following have been ascertained to occur, but of course we have had no opportunity to determine their range and distribution. Mr. Ley has very kindly examined a large number of our specimens. *R. mollis*, Sm., is abundant amongst the hills of North and East, and its variety *recondita* (Puget) occurs at Longridge, Preston, etc., and var. *cærulea*, Woods, on the banks of the Tarnbrook Wyre. *R. submollis*, Ley, which is probably frequent, is recorded by Mr. Ley in Journ. Bot. (loc. cit.), and we have collected it near Abbeystead. *R. omissa*, Déséglé, always as the var. *resinosoides*, Crép, is frequent. *R. pseudo-rubiginosa*, Lejeune, at Wrea Green (*Wh.*), Barnacre (*Wi.* & *Wh.*) and near Garstang (*Wi.*). *R. suberecta*, Ley, is probably frequent; we have specimens from Hest Bank, near Wrea Green, and Abbeystead. *R. Sherardi*, Davies, grows near Garstang, and probably elsewhere (*Wi.*). *R. uncinata*, Ley, occurs not unfrequently in lanes in the vicinity of Blackpool and Marton (*Wh.*). *R. Andrzeiovii*, Steven. Between Tarnbrook and Abbeystead.

Rosa tomentosa, Sm.—Downy-leaved Rose. British type. Native. Area, 1, 2, 3, 4, 5, 6, 7, 8. Range, 0-900 feet. Sh. June.

Hedges, woods, and thickets. More frequent in the low country than the preceding, but like it, attaining its greatest frequency amongst the hills. First record: *Linton, B.R.C. Rep.*, 1874, p. 82.

N. 1, 2, 3. Generally distributed throughout the districts.
W. 4. Hedges near Stalmine, *Wh.* Rawcliffe Moss, *S. Wi.*
　　5. Hedges near Wrea Green and Marton, *Wh.*
E. 6. Common in Hindburn and Roeburndale, ascending to 900 feet, *Wi.*
　　7. Wyresdale, frequent, and near Garstang, *Wi.* Near Bay Horse, *Wh.* About Catshaw and Barnacre.
　　8. Frequent about Longridge and Alston, *Wh.*

R. scabriuscula, Sm., occurs near Garstang, and a plant with very straight prickles, but the general characters of *R. tomentosa*, is probably a hybrid with *R. mollis*.

Rosa eglanteria, L.—Sweet Briar. English type. Probably not native, although occurring away from gardens, and looking thoroughly wild in the only locality in which it has been noted. Perhaps best regarded as a denizen.

Bushy limestone banks. Very rare. First record: The present one.

N. 3. Warton Crag, on the slope facing the sea, Oct., 1906, *Wh.* (teste *W. M. Rogers* and *A. Ley*).

Rosa micrantha, Sm.—Small Scentless Rose. English type. Native. Area, 3. Range, ? Sh.

Hedges and thickets. Rare. First record: The present one.

N. 3. Frequent at Silverdale, 1877-81, *E. S. Pickard*.

Rosa canina, L.—Dog Rose, Wild Rose. British type. Native. Area, 1, 2, 3, 4, 5, 6, 7, 8. Range, 0-1,190 feet. Sh. June.

Hedges and thickets. Common. First record: *Linton, B.R.C. Rep.*, 1874, p. 82. The distribution of the segregates has been only partially worked out.

In the district south of Lancaster the fruits are called "Heps"; further north "Shoups," which in Westmorland becomes "Choups."

Var. *lutetiana* (Leman).—Area, 1, 2, 3, 5, 8.

N. 1. Roadside above Leck, at 500 feet.
　　2. Near Over Kellet.
　　3. Silverdale.
W. 5. Near Staining and Freckleton, *Wh.*
E. 7. Over Wyresdale at 450 feet.
　　8. Preston, 1883, *E.F.L.* Ribchester, *Wh.*

Forms between vars. *dumalis* and *lutetiana* are common, one such occurring in Middle Ease Gill at 1,000 feet.

Var. *sphærica* (Gren.).—Area, 1, 3, 4, 5.

N. 1. Middle Ease Gill at 1,000 feet.
　　3. Warton, *Wh.*
W. 4. Wardless, 1884, *E.F.L.*
　　5. Hedges near Freckleton, *Wh.*

Var. *dumalis* (Bechst.).—Area, 1, 2, 3, 4, 5, 6, 7, 8. General and common.

The *F. verticillacantha* (Merat) is reported from near Grimsargh by the Rev. E. F. Linton.

Var. *vinacea* (Baker).—Area, 3, 5.

N. 3. Warton, 1903, and near Silverdale.
W. 5. Freckleton, *Wh.*

Var. *urbica* (Leman).—Area, 2, 3, 4, 5, 7, 8.

N. 2. Between Caton and Aughton and near Carnforth.
　　3. Silverdale. Near Basil Point and near Overton, and with it *R. platyphylla*, Rau., *Wh.*
W. 4. Wyre banks near Preesall, a glandular form, *Wh.* Near Garstang, *Wi.*
　　5. Freckleton, Wrea Green and Marton, *Wh.*
E. 7. Near Claughton, *Wi.* Barnacre and Sandholme, near Garstang.
　　8. Grimsargh, *E.F.L.*, 1883. Fulwood, Alston, and Ribbleton, *Wh.*

Var. *arvatica* (Baker).—Area, 2, 3, 5, 8.

N. 2. Near Carnforth.
　　3. Near Overton, *Wh.*
W. 5. Near Wrea Green, *Wh.* Near Garstang, *Wi.*
E. 8. Between Alston and Thornley, *Lewis*, 1883.

Var. *dumetorum* (Thuill).—Area, 7.

E. 7. Barnacre, near Garstang, Sept., 1902.

Rosa tomentella, Leman.—Dog Rose. English type. Native. Area, 8. Range, only seen at about 250 feet. Sh.

Hedges. Rare. First record: *Linton, B.R.C. Rep.*, 1884, p. 12.

E. 8. Grimsargh, near Preston, *E.F.L.*, 1883.

Rosa glauca (Vill.).—Dog Rose. Intermediate type. Area, 1, 2, 3, 4, 5, 6, 7, 8. Range, 0-920 feet. Sh. June.

Hedges and thickets, principally in upland districts. Frequent. First record: The authors, *Journ. Bot.*, 1901.

N. 1. Frequent, ascending in Ease Gill to 920 feet.
　　2. Between Caton and Aughton and near Over Kellet.
　　3. By the canal near Yealand (var. *Reuteri*, *Godet*) and on Warton Crag, *Wh.* Near Basil Point (var. *subcristata*), *Wh.*
W. 4. Near Garstang, *Wi.* Preesall and Staining (var. *Reuteri*), *Wh.*
　　5. Sparingly, near Kirkham, *Wh.*
E. 6. Near Ivah, Hindburn.
　　7. Frequent, ascending in Wyresdale to 800 feet. The var. *subcanina*, Christ., occurs near Lower Emmetts.
　　8. North side of Longridge Fell, *Wi.*

The segregates grouped under this species have not been separately worked out, but our forms appear mostly to belong to var. *subcristata*, Baker, and var. *Reuteri*, *Godet*.

Rosa arvensis, Huds.—Trailing Rose. English type. Native. Area, 1, 2, 3, 4, 5, 6, 7, 8. Range, 0-630 feet. Sh. June–July.

Hedgebanks and thickets. Very common in the low country ; rarer amongst the hills. First record : *Ashfield, Fl. Prest.,* 1858. " Salwick Brook, near Lea Road."

Var. *bibracteata* (Bast.).

> E. 8. Redscar, near Preston, *E.F.L.,* 1883.

Var. *gallicoides,* Baker.

> E. 8. Between Longridge and Grimsargh and near Alston a handsome, strong form occurs which, according to Mr. Rogers, approaches this variety, *Wh.*

[*Rosa systyla,* Woods, is reported from near Kirkham in the *Fl. Prest. N.,* 1903, p. 18. Further confirmation would be 'esirable.]

Rosa pomifera, Herrm.—Alien. First record : *Fl. St.,* 1891.

> E. 8. Alston Rake quarry, near Chaigley, introduced, *Fl. St.,* 1891.

Pyrus rupicola, Syme.—White Beam-tree. English type. Native. Area, 3. Range, 0-500 feet. T. May–June.

Limestone scars and rocky woods of North. First record : *Lawson, Ray. Fascic.,* 1688, p. 26, as *Aria Theophrasti,* Ger. *B.E.C. Rep.,* 1872, p. 14. *Bailey,* as *P. rupicola,* Syme.

> N. 3. Silverdale, *T. Lawson,* 1688. Still there, *Bailey,* 1871. From shore rocks to high up Castle Barrow, say, 250 feet, *L.P.* Warton Crag, ascending to 500 feet.
>
> Not yet seen in Dist. 2, but likely to occur on Dalton Crag or Kellet Seeds.

Pyrus Aucuparia, Ehrh. — Mountain Ash. Rowan Tree. " Wickens." British type. Native. Area, 1, 2, 3, 6, 7, 8. Range, 0-1,520 feet. T. May.

Woods, thickets, and rocky banks of streams. Common in the dales, ascending to 1,440 feet in Ease Gill, 1,500 feet in Dale Gill, Hindburndale, and 1,520 feet near the head of Gavells Clough, Wyresdale. It is a great ornament of the sheltered gullies on the gritstone moors. First record : *Ashfield, Fl. Prest.,* 1858. " Longridge."

Pyrus communis, L.—Alien. First record : *Ashfield, Fl. Prest.,* 1860.

> N. 3. A large tree away from houses, between the Keer bank and Warton, *Wi.*
> W. 4. Stainall, in a hedge on the road to Stalmine, *Ashfield,* 1860.
>   5. Catforth, *H.B.*

Pyrus Malus, L.—Crab-tree. English type. Native. Area, 1, 2, 3, 4, 5, 6, 7, 8. Range, 0-380 feet. Small T. May.

Woods and hedges. Not uncommon. First record : *Linton, B.R.C. Rep.,* 1874, p. 82.

> The two forms, var. *acerba, DC.,* and var. *mitis, Wallr.,* both occur. Of the two, the former or glabrous form is the more frequent, and ascends to 880 feet on Whitmoor.

Cotoneaster microphylla, Wall.—Alien.

> Has been seen as a garden escape near Silverdale ! It has been erroneously reported as *C. integerrimus, Medic.,* from this district.

Cratægus Oxyacantha, L.—Hawthorn. May or May-blossom. British type. Native. Area, 1, 2, 3, 4, 5, 6, 7, 8. Range, 0-1,160 feet. Sh. or small T. May–June.

Commonly planted in hedgerows, but also native in woods and thickets, especially amongst the hills, ascending in North to 1,160 feet on Leck Fell and in East to 1,000 feet on Mallowdale Fell, and to 1,120 feet on Wolf Fell. Our common plant is *C. monogynia, Jacq.*

Var. *oxyacanthoides* (Thuill).

> N. 1. Near Leck, *L.P.*
> W. 5. Ashton, in hedges, *Fl. Prest. N.*
> E. 6. Ascends to 1,100 feet on Mallowdale Fell, where its large oval fruits, tipped with the multiple styles, are quite ripe when those of *C. monogynia* at a similar elevation are still hard and just commencing to change colour.
>
> The fruits of this bush are known as " haws " in some parts of the county and as " haigs " in others. The latter is possibly derived from " Hay-eggs," which Mr. Moss informs us is the name given on both sides of the Wyre near St. Michael's.

SAXIFRAGACEÆ.

Saxifraga umbrosa, L.—London Pride. " Prince of Wales' Feathers." Atlantic type. Area, 6 (7). Denizen or Native possibly. Range, only at about 500 feet. P. June.

By waterfalls and damp shady banks of streams. Very rare. First record : The Authors, *Journ. Bot.,* 1900, p. 43.

> E. 6. By a waterfall near Botton, Hindburn, Oct., 1899.
>   7. [" Near Lower Lea, Wyresdale, by a brook running into the Wyre," *Ashfield, Fl. Prest.,* Part I. " I believe this should be *S. hirsute,*" ditto, Part II.]

Saxifraga tridactylites, L.—Rue-leaved Saxifrage. British type. Native. Area, 1, 2, 3, 5, 6, 8. Range, 0-1,050 feet or higher. A. April–June.

N

Walls, limestone rocks, and scars. Frequent in North and on the sandhills of West. Rather rare elsewhere, and so far unrecorded for Districts 4 and 7. First record : *Ashfield, Fl. Prest.,* 1860. " Near Ashton Church and Overton."

> It ascends in North to 1,025 feet on Greygarth Fell.

Saxifraga granulata, L.—White Meadow Saxifrage. British type. Native. Area, 3, 8. Range, 100-200 feet. P. May.

Gravelly banks. Very rare. First record : *Jenkinson, Brit. Pl.,* 1775, p. 88.

> N. 3. " In Mr. Townley's wood called Grisedale nigh the side of the coppice wood, but not plentiful," *Jenkinson.*
> E. 8. Dean Brook and Sale Wheel, *Fl. St.* Near Fulwood, 1901, *A.A.D.*

Saxifraga hypnoides, L.—Mossy Saxifrage. Scottish type. Native. Area, 1. Range, 2,030-1,000 feet. P. May–June.

Wet mountain rocks and banks. Rare. First record : *John Wilson, Synops. Brit. Plants,* 1744, p. 238 (as *"Saxifraga Muscosa trifido folio R."*).

> N. 1. " At Ease Gill Kirk, about three miles from Kirkby Lonsdale," *John Wilson,* 1744. Still there and in other places in Ease Gill, and on Greygarth, Ireby, and Leck Fells, *Wi.*

Chrysosplenium oppositifolium, L.—Golden Saxifrage. British type. Native. Area, 1, 2, 3, 5, 6, 7, 8. Range, 0-1,850 feet. P. March–May.

Wet shady banks, rocks, and brooksides. Frequent, except in West. First record : *Jenkinson, Brit. Pl.,* 1775, p. 86. " About Leighton Hall."

Chrysosplenium alternifolium, L.—British type. Native. Area, 1, 5, 7, 8. Range, 0-1,150 feet. P. April–May.

Damp mossy banks and rocks. Rare. First record : *Ashfield, Fl. Prest.,* 1858.

> N. 1. Pot-holes of Leck Fell and in Ease Gill, *Wi.* Springs Wood and Gorge of the Greeta.
> W. 5. Ashton, *Fl. Prest. N.*
> E. 7. Brock Valley, *Wh.*
>   8. Several places on Ribbleton Moor and between Parlick and Chipping, *Ashfield,* 1858. By the Hodder below Whitewell, *Wi.* and *Wh.* Redscar, *Fl. Prest. N.*

Parnassia palustris, L.—Grass of Parnassus. Scottish type. Native. Area, 1, 2, 3, 5, 6, 7, 8. Range, 0-1,250 feet. P. July– September.

Upland swamps and rill sides, and damp sandy places near the sea. Not infrequent in suitable situations, and locally plentiful. First record : *Jenkinson, Brit. Fl.,* 1775, p. 61. " In several of the meadows belonging to Geo. Townley, Esq., of Leighton Hall."

> It ascends to 1,250 feet on Salter Fell. The plant of the sand-dune tract has a distinct facies, owing to its numerous short stems and larger flowers.

Ribes Grossularia, L.—Gooseberry. Intermediate type. Denizen. Area, 1, 2, 3, 6, 7, 8. Range, 0-500 feet. Sh. April– May.

Hedges and bushy places, generally bird-sown. Frequent in North, and occurs sporadically elsewhere. First record : *Linton, B.R.C. Rep.,* 1874, p. 82.

Ribes alpinum, L.—Tasteless Mountain Currant. Intermediate type. Denizen. Area, 5, 8. Range, 0-500 feet. Sh. April– May.

Hedges and thickets. Rare. First record : *Ashfield, Fl. Prest.,* 1858.

> W. 5. Between Penwortham and Woodplumpton and hedge at Hollinhead Fold, near Cottam, *Ashfield,* 1858.
> E. 8. Between Longridge and Chipping, *Ashfield.* Still there in 1901, *H.E.* Near Gibbon Bridge and Hurst Green, *Fl. St.*

Ribes rubrum, L.—Red Currant. Intermediate type. Denizen. Area, 1, 5, 6, 8. Range, 0-500 feet. Sh. April–May.

Woods and thickets by streams. Rare. First record : *Ashfield, Fl. Prest.,* 1858.

> N. 1. By Leck Beck ! *L.P.* Near Cantsfield, *Wi.*
> W. 5. Salwick Brook, Cottam Mill and several places near, *Ashfield,* 1858.
> E. 6. Wennington (garden escape) and near Halton Station, *Wi.*
>   8. Gorge of the Hodder, below Whitewell.

Var. *petræum* (Sm.).

> N. 1. Between Over Burrow and Kirkby Lonsdale Bridge.

Ribes nigrum, L.—Black Currant. Intermediate type. Denizen. Area, 1, 3, 5, 7, 8. Range, 0-500 feet or higher. Sh. May– June.

Swampy thickets and river sides. Rare. First record : *Ashfield, Fl. Prest.,* 1858.

N. 1.   Gorge of the Greeta.
    3.   Hedges near Burton Well, Silverdale, *L.P.*
W. 5.   Salwick Brook, *Ashfield*, 1858.
E. 7.   Brock Bottom, Barnacre and Lower Bleasdale, *Wi.*
    8.   Ribbleton Moor, *Ashfield.* Ree Deep, near Stonyhurst, *Fl. St.*

CRASSULACEÆ.

Sedum Telephium, L.—Orpine. Live Long. Local name
" Farewell-to-Summer." English type. Native ? Area, 1, 2, 3, 7.
Range, 0-300 feet. P. August-September.

Sandy banks and hedges. Rare. First record : *Wilson,
B.R.C. Rep.*, 1883, p. 248.

N. 1.   Near Nether Burrow, *Wi.*
    2.   Gressingham, *Wi.*
    3.   Sea cliffs near Heysham, *Wi.*
E. 7.   Sandholm Mill Bridge and Dandy Birks, near Garstang, *Wi.*, 1881.

Sedum purpureum, Tausch. (*S. Fabaria, Koch.*) — Orpine.
English type. Native. Area, 1, 2, 3. Range, 0-1,000 feet. P.
August.

Limestone rocks and scars. Rare. First record : *Wilson,
Journ. Bot.*, 1900, p. 43.

N. 1.   Near Leck and Ireby, *Wi.* Ease Gill.
    2.   Rocks near Over Kellet, *Wi.*
    3.   Near Leighton Beck, Gatebarrow Wood, and Thrang Wood, *Wi.*, 1899.
         Heald Brow, Silverdale.

Sedum album, L.—White Stone-crop. Alien. First record :
Fl. St., 1891.

N. 1.   Walls about Leck, *L.P*
E. 8.   The Barracks, Chaigley ; cottage wall near Chipping, introduced,
         Fl. St. 1891.

Sedum anglicum, Huds.—White Stone-crop. Atlantic type.
Native. Area, 3 (8). Range, 0-50 feet. P. June-August.

Rocky places and sandy banks on the seashore. Rare. First
record : *Ashfield, Fl. Prest.*, 1860.

N. 3.   Bank at Sunderland, between the Ferry-house and sea, *Ashfield*, 1860.
         Cliffs by the sea, Far Naze, 1885, now perhaps destroyed by making
         Heysham Harbour, *Wi.*
E. 8.   Higher Bridge, introduced, *Fl. St.*

Sedum acre, L.—Stone-crop. " Golden Moss." British type.
Native. Area, 1, 2, 3, 4, 5, 6, 7, 8. Range, 0-880 feet or higher.
P. June–July.

Old walls, limestone rocks, and dry banks. Common in North ;
more rare in East and West. First record : *Ashfield, Fl. Prest.*,
1858.

N. 1, 2, 3.   Very common, especially on limestone, ascending to 880 feet on
               Dalton Crag.
W. 4.   Cockersand Abbey, coast, and walls, *Wi.* Knott End, *Wh.*
    5.   Ashton Marsh, *Ashfield*, 1858. Still there, and common on the sand-
         hills, Lytham, St. Annes, etc., *Wh.*
E. 6.   Old walls at Wray, *Wi.*
    7.   Walls near Lower Core, perhaps introduced, *Wi.* Wall near Lea.
    8.   Elston, *Ashfield*, 1858.

Sedum sexangulare, L. — Yellow Stone-crop. Alien. First
record : *Petty, Naturalist*, 1902, p. 41.

N. 3.   Garden escape on a wall near Cray Green and railway bridge, Ford
         Lane, Silverdale, *L.P.*

Sempervivum tectorum, L.—House-leek. Alien. First record :
Naturalist, 1893.

N. 1.   On roof of outhouse at Collin Holme, *L.P.*
W. 4.   Cottage roofs at Preesall, *Wh.*
    5.   St. Michael's (both sides of Wyre), *P.J.H.*
E. 7.   Tarnbrook.

DROSERACEÆ.

Drosera rotundifolia, L. — Round-leaved Sundew. British
type. Native. Area, 1, 2, 3, 4, 6, 7, 8. Range, 0-1,550 feet.
P. July–September.

Peat bogs and mossy rill sides. Rare in North ; frequent on
the moors of East ; now rare in West, but before the bogs were
reclaimed it was common there. First record : *Jenkinson, Brit.
Pl.*, 1775, p. 63.

N. 1.   Ease Gill, *Wi.*
    2.   Arkholme, Whittington, and Gressingham Moors and in the bog near
         Docker, *Wi.*
    3.   White Moss, Yealand, *Jenkinson*, 1775. Thrang Moss, *Wi.* Heysham
         Moss.
W. 4.   Pilling Moss, *Ashfield.* Cockerham Moss.
E. 6, 7, 8.   Frequent on the fells, ascending to 1,550 feet on Tarnbrook Fell.

[*Drosera longifolia,* L.—Incognit. Pilling Moss, *Ashfield, Fl.
Prest.*, 1858.

This is quite likely to have occurred in the locality indicated by Ashfield.
As we have searched for it in vain on many different occasions, we fear it has become
extinct. It is, therefore, removed from our list until re-discovered in Ashfield's
locality or elsewhere.]

Drosera anglica, Huds. — Great Sundew. Scottish type.
Native. Area, (3), 4. Range, only known at 25 feet. P.
July–August.

Deep bogs, amongst sphagnum. Rare, and rapidly becoming
more so. First record : *Jenkinson, Brit. Pl.*, 1775, p. 64.

N. 3.   " Long-leaved sundew with radicated stems and oblong leaves. Both
         species are found plentifully on the White Moss, Yealand, Lancashire,"
         Jenkinson, 1775. Now extinct there.
W. 4.   Pilling Moss, *Ashfield.* Cockerham Moss, *Wi.*

HALORAGACEÆ.

Hippuris vulgaris, L.—Mare's Tail. British type. Native.
Area, 1, 2, 3, 4, 5, 6. Range, 0-100 feet. P. July.

Ponds, ditches, and slow streams. Rare. First record :
Ashfield, Fl. Prest., 1860.

N. 1.   Marshy broads near Tunstall, *Wi.*
    2.   In the Keer near Carnforth, *Wi.*
    3.   Ditch by the railway near the Westmorland border, Silverdale, *Wi.*
         Water Slack, Silverdale, *W.K.* In the river Keer near Carnforth,
         Wi. Ditches about Bare, *F.A.L.*
W. 4.   Canal at Galgate and Ellel Grange, *Wi.* And near Glasson, *Wh.*
    5.   Ditch near Lytham Lighthouse, *Ashfield*, 1860. Near Ansdell, *Wi.*
E. 6.   Canal near Lancaster, *Wi.*

Myriophyllum verticillatum, L. — Water Milfoil. English
type. Native. Area, 5. Range, only known at about 25 feet.
P. July–August.

Deep ditches. Very rare. First record : *Ashfield, Fl.
Prest.*, 1858.

W. 5.   Ashton Marsh, *Ashfield*, 1858.

Myriophyllum spicatum, L.—Water Milfoil. British type.
Native. Area, 1, 2, 3, 4, 5, 6, 7, 8. Range, 0-200 feet. P. June–
July.

Ponds, ditches, and canals. Frequent. First record : *Ashfield,
Fl. Prest.*, 1860. " Mill pond near Grimsargh."

Myriophyllum alterniflorum, DC.—Water Milfoil. British
type. Native. Area, 1, 2, 3, 6. Range, 30-170 feet. P. June–
August.

Slow streams and ditches. Rare. First record : *Wilson,
Journ. Bot.*, 1902, p. 348.

N. 1.   River Lune between Kirkby Lonsdale and Tunstall, *Wi.*, 1901.
    2.   River Lune near Arkholme.
    3.   Ditch near Silverdale Station, *Wi.*
E. 6.   Crook of Lune.

Callitriche palustris, L. — Water Starwort. British type.
Native. Area, 1, 2, 3, 4, 5, 6, 7, 8. Range, 0-1,100 feet. A. or
P. April–September.

Ponds, ditches, and slow streams. Very common. First
record : *Linton, B.R.C. Rep.*, 1874, p. 82.

The distribution given above is that of the aggregate *C.
palustris*, L. The following segregates occur.

Callitriche vernalis, Koch.—Water Starwort. British type.
Native. Area, 7. Range, uncertain. A. or P. April–September.

Ponds and ditches. First record : For the segregate, the
present one.

E. 7.   Near Bruna Hill, Garstang, *Wi.*

Rarely fruiting, and the above is the only fruiting example we have found,
although it is probably common in the low country in a sterile condition.

Callitriche polymorpha (Lönnr.)—Water Starwort. British type.
Native. Area, 3. Range, only seen at about 20 feet.

Ponds near the sea-coast. Rare. First record : The present
one.

N. 3.   In a pond near Middleton in the Heysham Peninsula.

Callitriche stagnalis (Scop.)—Water Starwort. British type.
Native. Area, 1, 2, 3, 4, 5, 6, 7. 8. Range, 0-1,100 feet.

Muddy ponds, ditches, and slow streams. Very common.
First record : *Petty, Plants of Silverdale. Nat.*, 1902, p. 41.

It ascends from sea-level near Overton to 1,100 feet in Gavell's Clough, Over
Wyresdale. It apparently fruits much more freely in our district than *C. vernalis,
Koch.*

Callitriche intermedia (Hoffm.) *C. hamulata* (Kütz.).—
British type. Native. Area, 2. Range, only seen at about
400 feet.

In deep, still waters. Very rare. First record : The present one.

N. 2.   In the stream from the large pond near Henridden, 1905.

Callitriche autumnalis, L.—British type. Native. Area, 2, 3, 4, 5, 6, 7. Range, 20-100 feet. P. August.

Canals and slow streams. Frequent in West, especially in the Preston and Kendal Canal. First record : *King, B.R.C. Rep.*, 1884, p. 14.

N. 2 and 3.   Canal near Borwick, etc., *Wi.*
W. 4.   Canal between Garstang and Cabus, *Wi.* Very fine in the Canal between Glasson and Galgate, *Wh.*
5.   Canal, Ashton-on-Ribble, *F.C.K.*, 1883. Catterall Mill Dam, *Wi.*
6.   Canal near Lancaster, *Wi.*
7.   Canal near Garstang.

LYTHRACEÆ.

Peplis Portula, L.—Water Purslane. British type. Native. Area, 3, 5, (8). Range, 0-50 feet. P. July–August.

Damp ground and watery places on heaths, etc. Rare. First record : *Ashfield, Fl. Prest.*, 1858.

N. 3.   Muddy pools near Overton, 1900, *Wi.*
W. 5.   Carrs Green Common, *Wi.*
E. 8.   Ditch on south side of Ribbleton Moor, *Ashfield*, 1858.   Now extinct, and for many years missing from our Flora until its re-discovery at Overton.

Lythrum Salicaria, L.—Purple Loosestrife. " Swaggering Sally." English type. Native. Area, 1, 2, 3, 4, 5, 6, 7, 8. Range, 0-600 feet. P. August.

Ditches, pool sides, and bogs. Common in the low country ; rarer elsewhere. First record : *Linton, B.R.C. Rep.*, 1874, p. 82.

ONAGRACEÆ.

Epilobium angustifolium, L.—Rose-bay. British type. Native. Area, 1, 3, 4, 5, 6, 7, 8. Range, 0-1,200 feet. P. July.

Woods, heaths, and mountain rocks. Rather rare. Not always native in the low country. First record : *Ashfield, Fl. Prest.*, 1858.

N. 1.   Indigenous about the pot-holes of Leck Fell and by the Leck Beck, *Wi.*
3.   Thrang Moss, *Wi.*   Warton Crag, *Wh.*
W. 4.   Pilling Moss, near Rawcliffe, *Ashfield*, 1858.   Moss Side, St. Michael's, *P.J.H.* Cockerham Moss, *Wh.*
5.   Near Catterall, *Wi.* Fleetwood Marsh, *Wh.*   Waste ground by Preston Docks, not native, *Wi.*
E. 6.   By the Lune near Caton, *Wh.*   Several localities in Roeburndale, *Wi.*
7.   Marshaw, S. *Wi.*   Abbeystead, *Wh.*
8.   Saddle Hill, *Fl. St.*   Near Chipping, S. *Wi.*   Greystoneley Glen, *Wi.*   Longridge, *Fl. Prest. N.*

Epilobium hirsutum, L. — Codlins-and-cream. Apple Pie. Hairy Willow Herb. English type. Native. Area, 1, 2, 3, 4, 5, 6, 7, 8. Range, 0-500 feet. P. July–August.

Ditches, damp thickets, and watery places. Very common in the low country, but rare amongst the hills. First record : *Linton, B.R.C. Rep.*, 1874, p. 82

Its rustic name near Garstang is " Flowering Sally," no doubt derived from " sallow," owing to its willow-like leaves. Another local name derived from its fruity odour is " Plum-pie Syrup," used in the vicinity of St. Michael's-on-Wyre. It is also sometimes called " Sweet William."

Epilobium parviflorum, Schreb. — Small-flowered Willow Herb. British type. Native. Area, 1, 2, 3, 4, 5, 6, 7, 8. Range, 0-500 feet or higher. P. July–September.

Ditches and wet places. Common. First record : *Top. Bot.*, 1883.

Epilobium montanum, L.—Common Willow Herb. British type. Native. Area, 1, 2, 3, 4, 5, 6, 7, 8. Range, 0-1,340 feet. P. June-September.

Damp ground, woods, and shady places. Very common. First record : *Linton, B.R.C. Rep.*, 1874, p. 82.

Ascends in North to 1,340 feet near the waterfalls in Upper Ease Gill, and to 870 feet on Dalton Crag.

Epilobium roseum, Schreb.—English type. Native ? Area ? Range, uncertain. P. July–August.

Damp places. Rare. First record : *Watson, Top. Bot.*, 1883.

We have never met with it, nor have we been able to ascertain where Watson found it.

Epilobium obscurum, Schreb.—British type. Native. Area, 2, 4, 5, 6, 7, 8. Range, 0-500 feet or higher. P. July–August.

Wet places, peaty bogs, and by brooks. Rather rare. First record : *Watson, Top. Bot.*, 1883.

N. 2.   By the Keer near Carnforth.
W. 4.   Cockerham Moss, and between Glasson and Galgate, *Wh.*
5.   Bank of the Wyre below Garstang, *Wi.*   Old river bed, Ashton, *Fl. Prest. N.*
6.   Near Galgate, *E.C.*
E. 7.   Ditches on the Carrs near Bruna Hill, Garstang, *Wi.*
8.   Dean Brook, and fell side near Chaigley Farm, *Fl. St.*

Epilobium palustre, L.—Marsh Willow Herb. British type. Native. Area, 1, 2, 3, 4, 5, 6, 7, 8. Range, 0-1,850 feet. P. July–August.

Peaty ditches, swamps, and marshy places, especially amongst the hills. Common. First record : *Linton, B.R.C. Rep.*, 1874, p. 82.

It ascends in North to 1,850 feet near the summit ridge of Greygarth Fell. The fungus *Puccinia epilobii DC.* is common upon its leaves.

Œnothera biennis, L.—Alien. Sandy seashore and shingly river banks. First record : *Bailey, Proc. Manch. Lit. and Phil. Soc.*, 1902.

N. 1.   Lune bank below Kirkby Lonsdale.
2.   By the Lune at Arkholme, *Wi.*
W. 5.   Sandhills at St. Annes ! *C.B.* Ansdell, *Wh.*

Œnothera Lamarkiana (Ser.), DC.—Alien. Sandhills and waste ground on the seashore. First record : *Bailey, B.E.C. Rep.*, 1905.

W. 5.   Abundant in several localities near St. Annes ! Ansdell ! and South Shore, *C.B.* (sp.).

Mr. Bailey has found several of the mutations of De Vries on the coast, including forms resembling *Œ. brevistyla* and *Œ. rubrinervis*, and also other forms referred to by him in a paper read before the Manchester Field Club. It is probable that some of the stations recorded for *Œ. biennis* refer to this species, if the two are specifically distinct, which seems doubtful. Is it not possible that one of these " species " is only a mutation of the other ?

Œnothera sinuata, L.—Alien. First record : The present one.

W. 5.   On waste ground with other casuals near St. Annes, *C.B.* (sp. !).

Circæa lutetiana, L.—Enchanter's Nightshade. British type. Native. Area, 1, 2, 3, 4, 5, 6, 7, 8. Range, 0-1,000 feet. P. July–August.

Woods and shady places. Frequent. First record : *Watson, Top. Bot.*, 1883.

It ascends in North to 870 feet on Dalton Crag and to 1,000 feet on Leck Fell.

Circæa alpina, L., var. *intermedia* (Ehrh).— Scottish type. Native. Area, 1, 6. Range, 50-700 feet. P. July–August.

Woods and shady banks, usually near rivers. Rare. First record : *Wilson, Journ. Bot.*, 1902, p. 348.

N. 1.   Bank of the Lune near Tunstall and between Kirkby Lonsdale and Nether Burrow, *Wi.* Springs Wood, Lower Ease Gill, and Gorge of the Greeta.
E. 6.   Wood below White Moss, Hindburn, Sept., 1901, and by the Lune near Caton, *Wi.*

CUCURBITACEÆ.

Bryonia dioica, Jacq.—White Briony. Alien. First record : *Pearson in Ashfield, Fl. Prest.*, 1860.

W. 4.   Hedges at White Boar, St. Michael's, *Mr. Pearson.*

UMBELLIFERÆ.

Hydrocotyle vulgaris, L.— Marsh Pennywort. White-rot. British type. Native. Area, 1, 2, 3, 4, 5, 6, 7, 8. Range, 0-800 feet. P. June–August.

Boggy places and swampy ground. Rather common, especially amongst the hills. First record : *Ashfield, Fl. Prest.*, 1858. " Marshy field near the river Loud between Longridge and Chipping."

Eryngium maritimum, L.—Sea Holly. British type. Native. Area, 3, 4, 5. Range, 0-20 feet. P. July–August.

Sandy and shingly sea-shores. Frequent. First record : *Jenkinson, Brit. Pl.*, 1775, p. 43

N. 3.   Sea shore, Poulton near Lancaster, *Jenkinson*, 1775.   This was Poulton-le-sands, now Morecambe. Between Sunderland Point and Far Naze, *Wi.*
W. 4.   Very sparingly at Knott End, *Wh.*   On several parts of the coast near Pilling and near Cockersand Abbey, *Wh.*
5.   Lytham, *Ashfield*, 1858.   Still grows between Lytham and Blackpool, *Wi.* And on to Fleetwood sparingly, *Wh.*

Sanicula europæa, L.—Wood Sanicle. British type. Native. Area, 1, 2, 3, 4, 6, 7, 8. Range, 0-500 feet or higher. P. June.

Woods and thickets. Frequent. First record : *Jenkinson, Brit. Pl.*, 1775, p. 44. " In all the woods about Yealand in Lancaster."

[*Smyrnium Olusatrum*, L.—Incognit.

W.   Reported to grow between Marton and Lytham, *Ashfield*, 1858.]

Conium maculatum, L.—Hemlock. British type. Native. Area, 1, 3, 4, 5, 8. Range, 0-500 feet. B. June–July.

Damp roadsides, waste places and shady banks. Not common. First record: *Linton, B.R.C. Rep.,* 1874, p. 83.

N. 1.   Near Cantsfield, *Wi.*
   3.   Carnforth, and between Thrang End and Silverdale Station, *Wi.* Near Bolton-le-Sands. Sea embankment below the quarry on Warton Crag, *Wh.*
W. 4.   Near Rawcliffe Church, *S. Wi.* Tarnacre, *P.J.H.* Wyre banks near Cartford Bridge, *Wi.*
   5.   Near Cleveleys and Marton Mere, *Wi.* Very fine on Fleetwood Marsh, *Wh.* Between Ansdell and Wrea Green, *Wi.*
E. 8.   Leagram, and near Stonyhurst Mill Pond, 1887, *Fl. St.*

[*Bupleurum tenuissimum*, L.—Incognit. Recorded by Syme in *Top. Bot.,* ed. 3.

It has not been seen recently, and nothing is known of its habitat or status, and therefore we bracket it as requiring further and recent confirmation.]

Bupleurum rotundifolium, L.—Thorough Wax. Alien.

W. 5.   A few plants near the Fleetwood Grain Elevator 1902, *Wh.* (*Journ. Bot.*, 1902, p. 348).

Apium graveolens, L.—Smallage. Wild Celery. English type. Native. Area, 3, 4, 5, 6, 8. Range, 0-25 feet. P. June–August.

Salt marshes and waste ground near the sea. Frequent. First record: *Ashfield, Fl. Prest.,* 1858.

N. 3.   Salt ditches near Overton, *Wi.* Near Thurshouse Sands, *Wh.* Between Oxcliffe and Heaton, *Wi.*
W. 4.   Between Skippool and Rawcliffe Marsh to the sea, *Ashfield.* Salt marshes about Pilling and Cockerham, *Wi.*
   5.   Between Greaves Town and Ashton Marsh, *Ashfield.* Lytham, *Wi.*
E. 6.   Near the Lune to the east of Lancaster, *Wi.*
   8.   Near Preston Cemetery, *Ashfield*, 1858.

Apium nodiflorum, Reichb. fil.—Water Parsnip. British type. Native. Area, 1, 2, 3, 4, 5, 6, 7, 8. Range, 0-300 feet. P. July–August.

Ditches and watery places. Common, especially in the low country. First record: *Linton, B.R.C. Rep.,* 1874, p. 82.

Apium inundatum, Reichb. fil.—British type. Native. Area, 1, 3, 4, 5, 8. Range, 0-200 feet. P. June–July.

Ponds and ditches. Rare. First record: *Pearson,* in *Ashfield's Fl. Prest.,* 1860.

N. 1.   Marshes near Tunstall, *Wi.*
   3.   Marsh between Yealand Storrs and Borwick, and near Silverdale, *Wi.*
W. 4.   Between Stainall and Stalmine, *Pearson,* 1860.
   5.   In a pit near Barton ; near Woodplumpton ; and on Carrs Green Common, near Catforth, *Wi.*
E. 8.   Pit on the east side of Ribbleton Moor, *Ashfield.*

Sium erectum, Huds.—Narrow-leaved Water Parsnip. English type. Native. Area, 2, 3, 4, 5, 7, 8. Range, 0-250 feet. P. July–September.

Ditches and watery places. Frequent. First record: *Ashfield, Fl. Prest.,* 1858.

N. 2.   Over Kellet, and near Halton, *Wi.*
   3.   Bare Lane, *F.A.L.*
W. 4.   Canal between Glasson and Galgate, *Wh.*
   5.   Ditch near Lytham, and between Kirkham and Little Marton, *Ashfield,* 1858.
E. 7.   Common near Garstang, *Wi.*
   8.   Pit near Whittingham, *Wh.*

Ægopodium Podagraria, L.—Gout Weed. British type. Native. Area, 1, 2, 3, 4, 5, 6, 7, 8. Range, 0-500 feet. P. July.

Hedges, banks of streams, and damp shady places, especially near houses. Frequent. First record: *Ashfield, Fl. Prest.,* 1858. "Near the Larches, Ashton."

Pimpinella Saxifraga, L.—Burnet Saxifrage. British type. Native. Area, 1, 2, 3, 4, 5, 6, 7, 8. Range, 0-620 feet or higher. P. July–September.

Dry fields, banks and waste ground. Common. First record: *Wilson, B.R.C. Rep.,* 1883, p. 248. "Near Garstang."

Ascends in North to 620 feet in Ease Gill.

Pimpinella Major, Huds.—Great Burnet Saxifrage. English type. Native. Area, 5, 7, 8. Range, 0-450 feet. P. July–September.

Rare generally, but locally abundant. First record: *Linton, B.R.C. Rep.,* 1874, p. 82.

W. 5.   Near Marton Mere, *Wi.*
E. 7.   Hedgebanks about Cleveley Mill ; also lower down the Wyre near Scorton, *Wi.*
   8.   Abundant in the Longridge district, 1884, *Wi.* Near Alston and Ribchester, *Wh.* Common in the Loud Valley, *Wi.*

This plant has a very peculiar distribution. It is not recorded for Cumberland, Westmorland or Lake Lancashire, nor for the limestone district of Silverdale ; but it is abundant over the country, from Grimsargh (east of Preston) to Bolton Abbey in West Yorkshire, by way of Longridge, Clitheroe, Thornton-in-Craven, and Skipton. It occurs at Ilkley, but very sparingly, near the river, and then across N.E. Yorkshire by way of York.

Carum majus, Britt. & Rendle. (*Conopodium denudatum,* Koch).—Pig Nut. Earth Nut. British type. Native. Area, 1, 2, 3, 4, 5, 6, 7, 8. Range, 0-1,030 feet. P. May–June.

Meadows and grassy places on sandy or gravelly soil. Abundant. First record: *Linton, B.R.C. Rep.,* 1874, p. 82.

In North it ascends to 1,000 feet on Leck Fell and in East to 1,030 feet on Haylot Fell.

Myrrhis odorata, Scop.—Sweet Cicely. Intermediate type. Native. Area, 1, 2, 3, 4, 5, 6, 7, 8. Range, 0-900 feet. P. May–June.

Banks of streams, ditches, and moist thickets. Rather frequent amongst the hills, rarer elsewhere. First record: *Ashfield, Fl. Prest.,* 1858. "Near Chipping, towards Whitewell."

Scandix Pecten, L.—Shepherd's Needle. British type. Colonist. Area, 5. Range, 0-100 feet. A. June–September.

Cornfields. Very rare. First record: *Ashfield, Fl. Prest.,* 1860.

W. 5.   Cornfields near Preston, *Ashfield,* 1860. Near St. Michael's, *Wi.*

Chærophyllum temulum, L.—Rough Chervil. British type. Native. Area, 1, 2, 3, 4, 5, 6, 7, 8. Range, 0-500 feet or higher. P. June–August.

Hedges, roadsides, and borders of fields. Very common. First record: *Linton, B.R.C. Rep.,* 1874, p. 82.

Chærophyllum sylvestre, L. — Cow Parsley. "Humlock." British type. Native. Area, 1, 2, 3, 4, 5, 6, 7, 8. Range, 0-500 feet or higher. P. May–June.

Hedgebanks, meadows, and borders of fields. Very common. First record: *Linton, B.R.C. Rep.,* 1874, p. 82.

Chærophyllum Anthriscus, Lam.—Beaked Parsley. British type. Native. Area, 4, 5, 8. Range, 0-150 feet. A. June.

Sandy ground, especially near the sea. Rare. First record: *Ashfield, Fl. Prest.,* 1858.

W. 4.   Knott End, *Wi.* Near Preesall, *Wh.*
   5.   St. Annes, *P.J.H.*
E. 8.   Near Fulwood Barracks, *Ashfield,* 1858.

Crithmum maritimum, L.—Samphire. English type. Native. Area, 3. Range, only seen at about 15 feet. P. August.

Sea cliffs. Very rare. First record: *Ashfield, Bot. Chron.,* 1864.

N. 3.   Between Silverdale and Arnside Point, *Ashfield,* 1864. This was probably just over the Westmorland boundary. Seen by the Authors further south, within the W. Lancashire area, in 1904.

Fœniculum vulgare, Mill.—Alien.

Has been seen subspontaneous here and there near the coast of District 5, but never becomes established.

Œnanthe fistulosa, L. — Water Dropwort. English type. Native. Area, 3, 4, 5. Range, 0-75 feet. P. July–September.

Ditches and marshes. Rare. First record: *Lawson,* in *Ray's Corresp.,* 1718.

N. 3.   In the ditches between Warton and Carnforth, *Lawson,* 1688. Damp ground on the coast near Middleton.
W. 4.   Ditches about Cockerham and Glasson, *Wi.*
   5.   Near Singleton, *P.J.H.* Salwick, *Fl. Prest. N.*

[*Œnanthe pimpinelloides*, L.—Incognit.

Reported in Ashfield's *Fl. Prest.,* 1858, from near Lytham. In Part III., however, it is stated to have been " probably " *Œ. Lachenalii.* The latter surmise was, no doubt, correct, as it still grows there.]

Œnanthe Lachenalii, Gmel.—English type. Native. Area, 3, 4, 5. Range, 0-25 feet. P. July–September.

Salt marshes and ditches near the sea. Frequent. First record: *Linton, B.R.C. Rep.,* 1874, p. 82.

N. 3.   Keer estuary, salt marsh east of Jenny Brown's Point, Overton, and Middleton, *Wi.*
W. 4.   Salt marshes about Pilling, Cockerham, and Glasson, *Wi.* Wyre banks, near Preesall, *Wh.*
   5.   Ribble bank, near Freckleton, and between Naze Point and Lytham, *Wi.* Ansdell, *S. & T.*

Œnanthe crocata, L.— Hemlock Dropwort. British type. Native. Area, 1, 2, 3, 4, 5, 6, 7, 8. Range, 0-700 feet. P. July–August.

Ditches, streamsides and marshes. Common, especially in East and West. First record: *Ashfield, Fl. Prest.*, 1860. " Near Grimsargh."

It ascends in East to 700 feet near Salter in Roeburndale.

Œnanthe aquatica, Poir. (*Œ. Phellandrium*, Lam.—Horsebane. English type. Native. Area, 5. Range, 0-25 feet. B. July–September.

Ditches and pits. Rare. First record: *Hornby, Journ. Bot.*, 1902, p. 348 (recorded in *Top. Bot.* without personal authority).

W. 5. In a pit near Stockenbridge, *P.J.H.*, 1901, and in another place about 500 yards away, *J. Moss.*

Æthusa Cynapium, L.—Fool's Parsley. British type. Colonist. Area, 1, 2, 3, 4, 5, 6, 7, 8. Range, 0-500 feet. A. June–August.

Cultivated ground and waste places. Common. First record: *Linton, B.R.C. Rep.*, 1874, p. 82.

Silaus flavescens, Bernh.—Sulphur-wort. English type. Native. Area, 1. Range, 200-300 feet. P. August–September.

Moist meadows and pastures. Rare. First record: *Wilson, Journ. Bot.*, 1901, p. 24.

N. 1. Frequent in pastures about Cantsfield, *Wi.*
This is likely to occur in District 8 as it is not uncommon on similar ground in the adjoining districts of West Yorkshire.

Angelica sylvestris, L.—Wild Angelica. " Kewse." British type. Native. Area, 1, 2, 3, 4, 5, 6, 7, 8. Range, 0-1,080 feet. P. July–September.

Ditchsides and swampy woods. Common. First record: *Jenkinson, Brit. Pl.*, 1775, p. 51. " In the woods belonging to Dalton Hall."

It ascends in North to 1,080 feet in Navel Pot, Leck Fell; and in East to 900 feet by the Tatham Beck.

Peucedanum Ostruthium, Koch.—Masterwort. Intermediate type. Denizen. Area, 5, 7. Range, 0-600 feet. P. June.

Roadsides and moist meadows, generally near farm buildings. Rare. First record: *Sydney Wilson, Journ. Bot.*, 1900, p. 44.

W. 5. Near St. Michael's, *P.J.H.*
E. 7. Roadside near an old barn, Oakenclough, 1888, *S. Wi.* Bank of brook near Low Moor Head, Wyresdale, *Wi.*

Pastinaca sativa, L.—Wild Parsnip. English type. Native. Area, 3, 4, 5, 6. Range, 0-100 feet. B. July–August.

Sandy banks and roadsides, principally near the sea coast. Rare. First record: *Ashfield, Fl. Prest.*, 1860.

N. 3. Morecambe, *Wi.* Near Thurshouse Sands, *Wh.*
W. 4. Cockerham, *Wi.*
5. Copse at Bispham, *Ashfield*, 1860. Still there, *Wh.* Here and there along the coast from Freckleton to Fleetwood.
E. 6. Near Lancaster, *Wh.*

Heracleum Sphondylium, L.— Cow Parsnip. Hog Weed. Kex. " Kewse." British type. Native. Area, 1, 2, 3, 4, 5, 6, 7, 8. Range, 0-1,050 feet. B July–August.

Woods, meadows and banks. Very common. First record: *Linton, B.R.C. Rep.*, 1874, p. 82.

It ascends to 1,050 feet on Leck Fell, and to 880 feet in Roeburndale.

Daucus Carota, L.—Wild Carrot. British type. Native. Area, 1, 2, 3, 4, 5, 6, 7, 8. Range, 0-400 feet. B. June–August.

Dry banks and pastures, especially in the limestone districts. Frequent. First record: *Ashfield, Fl. Prest.*, 1858. " Brookside at Fulwood."

Torilis Anthriscus, Bernh.—Hedge Parsley. British type. Native. Area, 1, 2, 3, 4, 5, 6, 7, 8. Range, 0-500 feet or higher. A. July–August.

Hedgebanks and roadsides. Common. First record: *Linton, B.R.C. Rep.*, 1874, p. 82.

Torilis nodosa, Gaert.—Knotted Hedge Parsley. English type. Native, 2, 4, 5. Range, 0-100 feet. A. May–June.

Dry banks, roadsides, and borders of fields. Rare. First record: *Wilson, Journ. Bot.*, 1901, p. 24.

N. 2. Near Borwick and Carnforth, 1900, *Wi.*
W. 4. Near Cockersand Abbey, *Wi.*
5. Near Lytham, *A.A.D.*

Caucalis latifolia, L.—Great Bur Parsley. Alien. First record: *Ashfield, Fl. Prest.*, 1858.

W. 5. Casually at Preston, 1858, *Ashfield.*

O

ARALIACEÆ.

Hedera Helix, L.—Ivy. British type. Native. Area, 1, 2, 3, 4, 5, 6, 7, 8. Range, 0-1,190 feet. Sh. June.

Climbing over trees, banks, rocks and walls. Very common. First record: *Linton, B.R.C. Rep.*, 1874, p. 82.

It ascends to 1,190 feet on Ireby Fell.

CORNACEÆ.

Cornus sanguinea, L.—Cornel. Dog-wood. English type. Native. Area, 1, 2, 3, 5, 7, 8. Range, 0-300 feet. Sh. June.

Woods, hedges and thickets, mostly on limestone. Not common, and not always indigenous in localities away from the calcareous woods of North, being sometimes planted in woods as a shelter for game. First record: *Ashfield, Fl. Prest.*, 1858.

N. 1. Hedges near Cantsfield, *Wi.*
2. Carnforth and Nether Kellet, *Wi.*
3. Frequent about Silverdale, *Wi.* Hest Bank, *F.A.L.*
W. 5. Lea, near Preston, *H.B.*
E. 7. Near Sandholme Mill, Garstang, *Wi.*
8. Hedges about Elston and Fishwick, *Ashfield*, 1858.

CAPRIFOLIACEÆ.

Adoxa moschatellina, L.—Moschatel. British type. Native. Area, 1, 2, 3, 4, 5, 6, 7, 8. Range, 0-1,050 feet. P. May.

Hedgebanks, woods and shady places. Frequent. First record: *Simpson, Top. Bot.*, 1883.

It ascends to 1,050 feet on Leck Fell.

Sambucus nigra, L.—Elder. " Burtree" or " Boatery." Called locally " Dog Tree" by farmers. British type. Native. Area, 1, 2, 3, 4, 5, 6, 7, 8. Range, 0-880 feet. T. June.

Woods and hedges. Common. First record: *Linton, B.R.C. Rep.*, 1874, p. 82.

It ascends to the summit of Dalton Crag.

Viburnum Opulus, L.—Guelder Rose. " Hag Berries " (fruit). British type. Native. Area, 1, 2, 3, 4, 5, 6, 7, 8. Range, 0-580 feet. Sh. June–July.

Woods, thickets and hedges. Frequent. First record: *Linton, B.R.C. Rep.*, 1874, p. 82.

It ascends in East to 580 feet on Greenside, near Tarnbrook.

Lonicera Periclymenum, L. — Honeysuckle. Woodbine. British type. Native. Area, 1, 2, 3, 4, 5, 6, 7, 8. Range, 0-1,100 feet. Sh. June–September.

Woods, thickets and hedges. Common. First record: *Linton, B.R.C. Rep.*, 1874, p. 82.

It ascends to 1,100 feet in Ease Gill, *Wi.*

Lonicera Xylosteum, L.—Alien. First record: *Petty, Naturalist*, 1900, p. 42.

N. 3. Eaves Wood, near Silverdale ; of garden origin, *L.P.*

RUBIACEÆ.

Galium boreale, L.—Cross-leaved Bedstraw. Highland type. Native. Area, 6. Range, 40-50 feet. P. July–August.

Rocky places by rivers. Very rare. First record: *Wilson, Journ. Bot.*, 1905, p. 95.

E. 6. By the Lune near Halton, 1903, *Wi.* Plentifully on rocky islands at the Crook of Lune near Caton.

The altitude (50 feet) is a very low one for this plant. It is well established, but was possibly originally brought down by floods from near Sedbergh (27 miles higher up the river), where it occurs in Westmorland and Yorkshire. Its associates in the localities near Caton are *Trollius europæus, Circæa alpina* var. *intermedia*, and *Geranium sylvaticum.*

Galium cruciatum, With.—Cross-wort. British type. Native. Area, 1, 2, 3, 4, 5, 6, 7, 8. Range, 0-500 feet or higher. P. April–June.

Hedgebanks, roadsides and bushy places. Common. First record: *Linton, B.R.C. Rep.*, 1874, p. 82.

Galium verum, L.—Yellow or Lady's Bedstraw. British type. Native. Area, 1, 2, 3, 4, 5, 6, 7, 8. Range, 0-860 feet. P. July–September.

Dry banks and grassy places. Frequent in calcareous districts and on sandy ground near the sea. Rare elsewhere and apparently absent from some of the dale districts. First record: *Ashfield, Fl. Prest.*, 1858.

N. 1, 2, 3. Frequent, especially in 3.
W. 4. Knott End, *Wi.* Stodday.
    5. Between Naze Point and Lytham, *Ashfield*, 1858. Still there. Fleet-wood, *Wh.*
E. 6. Between Halton and Caton.
    7. Canal bank near Garstang, *E.C.*
    8. Near Fulwood, *Ashfield.* Hacking Boat, Sale Wheel, Chipping, etc., *Fl. St.* Leagram, *Wh.*

This species seems to be absent over large areas. We have never seen it in Hindburn, Roeburndale, Littledale, Wyresdale, or Bleasdale. It ascends to 860 feet on Dalton Crag.

Galium Mollugo, L.—Hedge Bedstraw. English type. Native. Area, 1, 2, 3, 4, 5, 6, 7, 8. Range, 0–450 feet. P. July–August.

Dry hedgebanks and thickets, showing a decided predilection for calcareous soil. Like the last, quite absent from large tracts, but locally abundant in some parts of North. First record: *Ashfield, Fl. Prest.*, 1858.

N. 1, 2, 3. Frequent.
W. 4. Near Stodday, *Wi.* Near the "Old Holly," *E.C.*
    5. Catterall, *Wi.*
E. 6. Near Cator, *Wi.*
    7. In hedges near Beacon Fell, *Ashfield*, 1858. Scorton, *Wi.*
    8. Stonyhurst Mill Pond, *Fl. St.*, 1887.

Galium saxatile, L.—Heath Bedstraw. British type. Native. Area, 1, 2, 3, 4, 5, 6, 7, 8. Range, 0–2,030 feet. P. June–August.

Heaths, rocky and grassy places. Common, especially amongst the gritstone fells. First record: *Ashfield, Fl. Prest.*, 1858. " Ribbleton Moor."

Galium umbellatum, Lam. (*G. sylvestre, Poll.*). — Intermediate type. Native. Area, 1, 2, 3. Range, 1,930–200 feet. P. July.

Confined to the limestone hills and scars, where it is frequent. First record: *Wilson, Journ. Bot.,* 1900, p. 44.

N. 1. Easegill and Greygarth Fell, ascending to 1,930 feet, *Wi.*
    2. Over Kellet, *Wi.* Dalton Crag.
    3. Warton Crag, 1902, and near Jenny Brown's Point, *Wi.* Castlebarrow, *L.P.*

Galium palustre, L.—Water Bedstraw. British type. Native. Area, 1, 2, 3, 4, 5, 6, 7, 8. Range, 0–1,300 feet. P. June–July.

Marshes, ditches and bogs. Frequent, and in some parts abundant. First record: *Linton, B.R.C. Rep.*, 1874, p. 82.

Ascends to 1,300 feet on Salter Fell.

Var. *Witheringii* (Sm.).

N. 3. Silverdale, 1884, *Wi.*
W. 4. A large diffuse form of *G. palustre* occurs near Knott End, associated with *Rumex Hydrolapathum*, which Mr. A. Bennett thinks may be the var. *maximum*, Marsson, *Wh.*
E. 8. Near Greystoneley, *Wi.*

Galium Aparine, L.—Cleavers. Goose-grass. "Sticky Dick." "Robin run in th' hedge." British type. Native. Area, 1, 2, 3, 4, 5, 6, 7, 8. Range, 0–810 feet or higher. P. June–August.

Hedgebanks and thickets. Very common. First record: *Linton, B.R.C. Rep.*, 1874, p. 82.

It ascends to over 800 feet on Dalton Crag.

Galium uliginosum, L.—Marsh Bedstraw. British type. Native. Area, 1, 2, 3, 6. Range, 0–800 feet. P. July–August.

Marshes and ditch sides. Rare. First record: *Wilson, Journ. Bot.*, 1900, p. 44.

N. 1. Wrayton and Lower Ease Gill, *Wi.*
    2. Bog near Docker, *Wi.* Between Carnforth and Nether Kellet.
    3. Boggy ground near Borwick, *Wi.*
E. 6. Between Caton and Hornby, *Wi.* Clougha, *Wi.*

Asperula odorata, L. — Woodruff. British type. Native. Area, 1, 2, 3, 6, 7, 8. Range, 0–1,070 feet. P. May–June.

Woods, thickets and shady places amongst the hills. Frequent. First record: *Ashfield, Bot. Chron.*, 1864. "Silverdale, abundant."

It ascends to 1,070 feet in a pot-hole on Leck Fell.

Asperula cynanchica, L. — Quinancy-wort. English type. Native. Area, 2, 3. Range, 0–870 feet. P. July–August.

Dry, rocky, limestone banks and hills of North. Rare. First record: *Ashfield, Bot. Chron.*, 1864.

N. 2. Dalton Crag, ascending to 870 feet.
    3. Plentiful in the Silverdale district ; abundant below Trowbarrow, *Ashfield*, 1864. Still plentiful at Silverdale. Occurs also on Warton Crag, *Wi.*

Sherardia arvensis, L.—Field Madder. British type. Colonist. Area, 1, 2, 3, 4, 5, 7, 8. Range, 0–320 feet or higher. A. June–October.

Cultivated fields and waste places. Frequent. First record: *Ashfield, Fl. Prest.*, 1858. " Fields near Stydd Church."

It ascends in North to 320 feet on Kellet Seeds.

VALERIANACEÆ.

Valeriana dioica, L.—Marsh Valerian. English type. Native. Area, 1, 2, 3, 4, 5, 6, 7, 8. Range, 0–700 feet or higher. P. May–June.

Marshy and boggy places. Rather common amongst the hills. First record: *Ashfield, Fl. Prest.*, 1858. "Meadows near Cottam Hall ; near Nickey Nook."

Valeriana sambucifolia, Willd.—Wild Valerian. British type. Native. Area, 1, 2, 3, 4, 5, 6, 7, 8. Range, 0–1,420 feet. P. July–August.

Swampy places, ditches and stream sides. Common. First record: *Linton, B.R.C. Rep.*, p. 83.

Ascends in North to 1,420 feet in Upper Ease Gill, and in East to 890 feet in Foxdale.

Valerianella olitoria, Poll.—Lamb's Lettuce. British type. Native. Area, 2, 3, 4, 5. Range, 0–500 feet. A. May–June.

Dry banks, fields and rocks. Frequent amongst the limestone hills of North ; rare elsewhere. First record: *Ashfield, Fl. Prest.*, 1860.

N. 2. Over Kellet.
    3. Frequent on hilly ground about Silverdale, *Wi.* Heysham and Middleton.
W. 4. Roadside at Stainall, *Ashfield*, 1860. Canal bank, Stodday, *Wi.*
    5. Sandhills at St. Annes, *Wh.*

This appears to be truly native on the limestone of North, and on the sand-dune tract of West. In some of its other localities it is probably only an accidental introduction.

Valerianella dentata, Poll. — Corn Salad. English type. Colonist. Area, 3, 4, 5. Range, 0–200 feet. A. June–August.

Cultivated fields and waste ground. Rare. First record: *Ashfield, Fl. Prest.*, 1860.

N. 3. Near Leighton Beck, Warton, and Yealand, *Wi.*
W. 4. Cornfield between Cockerham and Glasson Dock, *Ashfield*, 1860. Near Turnover Farm, St. Michael's, *P.J.H.* Cabus near Garstang.
    5. Near Poulton Railway Station, *Ashfield.*

DIPSACEÆ.

Scabiosa Succisa, L.— Devil's-bit. British type. Native. Area, 1, 2, 3, 4, 5, 6, 7, 8. Range, 0–1,390 feet. P. July–September.

Damp pastures and banks. Common, especially amongst the hills. First record: *Jenkinson, Brit. Pl.*, 1775, p. 13. "On Mr. Townley's, of Leighton Hall, meadows adjoining the mosses."

Ascends in North to 1,390 feet in Upper Ease Gill, and in East to 900 feet in Roeburndale.

Scabiosa columbaria, L.— Small Scabious. English type. Native. Area, 1, 2, 3, 8. Range, 0–1,000 feet or higher. P. July–September.

Dry limestone hills and banks. Plentiful in North. First record: *Jenkinson, Brit. Pl.*, 1775, p. 13.

N. 1. Ease Gill.
    2. Over Kellet, *Wi.*
    3. In several dry pastures about Yealand, *Jenkinson.* Still there. Common about Silverdale, and on Warton Crag, *Wi.* Near Carnforth station, *Wh.* Poulton-le-Sands, *Motley Herbarium.*
E. 8. Limestone rocks near Dinkling Green.

Knautia arvensis, Coult. — Field Scabious. British type. Native. Area, 1, 2, 3, 4, 5, 6, 7, 8. Range, 0–400 feet. P. July–August.

Dry gravelly banks and fields. Most plentiful in North, occurring only occasionally in the other divisions. First record: *Linton, B.R.C. Rep.*, 1874, p. 83.

Dipsacus sylvestris, Huds.—Wild Teasel. Alien.

W. 5. Between Greaves Town and Ashton, *Ashfield*, 1858. River bank, Ashton, *Fl. Prest. N.*

It is possibly native here. It is very abundant on boulder clay on the banks of the Mersey near Speke in S. Lancashire, looking quite native.

COMPOSITÆ.

Eupatorium cannabinum, L.—Hemp Agrimony. British type. Native. Area, 1, 2, 3, 4, 5, 6, 7, 8. Range, 0–500 feet. P. August.

Ditch sides, marshy thickets and watery places. Common, especially in West and North. First record: *Ashfield, Fl. Prest.*, 1858. " By the Brock east of the railway."

Cyclachæna xanthifolia, Fresen (*Iva xanthifolia*, Nutt.).—Alien. First record : The present one.

W. 5. Sandhills near St. Annes with other aliens, 1906, *C.B.*

Solidago Virgaurea, L.—Golden Rod. British type. Native. Area, 1, 2, 3, 6, 7, 8. Range, 0–1,500 feet. P. July–September.

Woods and rocky banks. Frequent in North and East ; not seen in West. First record : *Wilson, B.R.C. Rep.*, 1883, p. 248. "Grizedale."

Ascends in North to 1,440 feet in Upper Ease Gill ; and in East to 1,500 feet on Botton Head Fell.

Var. *cambrica* (Huds.).

N. 1. Upper Ease Gill, 1899, *Wi.*
E. 6. By a waterfall in Whiteray Gill, at 980 feet, *Wi.*

Bellis perennis, L.—Daisy. British type. Native. Area, 1, 2, 3, 4, 5 6, 7, 8. Range, 0–2,080 feet. P. March–October.

Pastures and grassy places. Abundant. First record : *Linton, B.R.C. Rep.*, 1874, p. 83.

It ascends to 2,080 feet on the limestone of Greygarth Fell.

Aster Tripolium, L.—Sea Aster. Sea Starwort. British type. Native. Area, 3, 4, 5. Range, 0–25 feet. P. August–September.

Salt marshes and muddy estuaries. Common. First record : *Ashfield, Fl Prest.*, 1858. "Ashton Marsh."

A discoid form was found by Mr. Beesley near Knott End.

Erigeron acre, L.—Blue Flea-bane. English type. Native. Area, 3, 4, 5. Range, 0–240 feet. B. July–September.

Dry sandy banks and limestone hills. Rare. First record : *Ashfield, Fl. Prest.*, 1858.

N. 3. Silverdale, *Motley Herbarium.* Still occurs on limestone hills there, and to the south of Leighton Beck, *Wi.*
W. 4. Banks near Cockerham and Pilling, *Wi.*
5. Lytham, *Ashfield*, 1858. St. Annes, *Wi.*

Erigeron caucasicum, L.—Alien. Reported from waste ground near St. Peter's Churchyard, near Preston Docks. *Fl. Prest. N.*, 1903.

Filago germanica, L.—Cudweed. British type. Colonist. Area, 3, 8. Range, 0–400 feet. A. July–September.

Dry fields and sandy ground. Very rare. First record : *Fl. St.*, 1891, p. 22.

N. 3. A form approaching " *spathulata* " in a hen run near Bare, *F.A.L.*
E. 8. Leagram, *Fl. St.*, 1891.

Antennaria dioica, R. Br.—Cats-foot. Scottish type. Native. Area, 1, 3 (7). Range, 1,250–100 feet. P. June.

Rocky mountain pastures, mainly on limestone. Rare, and only in North. First record : *Jenkinson, Brit. Pl.*, 1775, p. 199.

N. 1. Leck Fell, and on Silurian (Coniston grit) rock, Lower Ease Gill, *Wi.*
3. Yealand Common, *Jenkinson*, 1775. Limestone hills near Silvercale, *Wi.*
[E. 7. Nickey Nook, *Ashfield.* Error?. A rather unlikely station, and not since confirmed.]

Gnaphalium sylvaticum, L. — Highland Cudweed. British type. Native. Area, 2, 3, 4, 5, 6, 7, 8. Range, 0–600 feet. P. July–August.

Dry heathy pastures, old quarries, and banks. Not common. First record : *Linton, B.R.C. Rep.*, 1874, p. 83.

N. 2. Wash Dub Wood.
3. Gatebarrow Wood.
W. 4. Fields near Cockerham Moss, *Wi.*
5. Knowles Pad near St. Michae 's, *P.J.H.*
E. 6. Quarries above Wray, *J.F.P.* Crook of Lune and near Ellel, *Wi.*
7. Sandholme near Garstang, *Wi.* Brock, *Fl. Prest. N.* Barnacre Moor. Near Abbeystead.
8. Bank of Hodder, *Fl. St.*

Gnaphalium uliginosum, L.—Marsh Cudweed. British type. Native. Area, 1, 2, 3, 4, 5, 6, 7, 8. Range, 0–780 feet. A. July–September.

Cultivated fields, roadsides, and waste ground, especially where water has stood. Common, except in parts of East. First record : *Linton, B.R.C. Rep.*, 1874, p. 83.

It ascends to nearly 800 feet in Wyresdale.

Inula vulgaris, Theirson.—Ploughman's Spikenard. English type. Native. Area, 2, 3, 5, 8. Range, 0–840 feet. B. July–August.

Rocky banks and thickets. Frequent on the limestone hills of districts 2 and 3 ; rare elsewhere. First record : *Jenkinson, Brit. Pl.*, 1775, p. 201.

N. 2. Between Poulton and Carnforth, *Ashfield.* Dalton Crag at 840 feet.
3. On Yealand Common, *Jenkinson*, 1775. Still occurs about Yealand, *Wi.* Frequent about Silverdale and the Leighton Beck district, *Wi.* Carnforth, and on the north side of Warton Crag, *Wh.*
W. 5. On the sandhills near Lytham, *Ashfield.* Still there and at St. Annes, *Wh.* Between Ansdell and St. Annes, *S. & T.*
E. 8. By the Hodder below Whitewell.

Inula Helenium, L.—Elecampane. English type. Denizen. Area, 3. Range, only known at about 250 feet. P. July–August.

Roadsides and waste places. Very rare. First record : *Robson, Turner,* and *Dillwyn, Botanists' Guide*, 1805.

N. 3. At Yealand, *Robson.* Still near Yealand, *Wi.*

Pulicaria dysenterica, Gray. — Fleabane. English type. Native. Area, 2, 3, 4, 5, 6, 7, 8. Range, 0–500 feet. P. August–September.

Damp roadsides, waste places, and by streams. Frequent. First record : *Linton, B.R.C. Rep.*, 1874, p. 83.

Bidens tripartita, L.—Bur Marigold. English type. Native. Area, 3, 4, 5, 7. Range, 0–80 feet. A. July–September.

Ditches, pondsides, and wet places in the low country. Not unfrequent in West, rare in North and East. First record : *Ashfield, Fl. Prest.*, 1860.

N. 3. Near Middleton, *Wi.*
W. 4. Churchtown, near Garstang, *Wi.* Cabus.
5. Canal side, Preston, *Ashfield*, 1860. Catterall and Shard Bridge, *Wi.* Stockenbridge, *P.J.H.*
E. 7. Near Garstang, *Wi.*

Bidens cernua, L.—Bur Marigold. English type. Native. Area, 3, 4, 5, 7, 8. Range, 0–400 feet. A. July–September.

In similar localities to the preceding. Frequent in West, rare in North and East. First record : *Ashfield, Fl. Prest.*, 1858.

N. 3. Canal between Hest Bank and Bolton-le-Sands, *Wh.*
W. 4. Pilling Moss, *Ashfield*, 1858. Cabus. Near Cockersand Abbey, *Wi.* Hudsons, near St. Michael's, *P.J.H.* By the canal near Glasson, *Wh.*
5. Ponds between Freckleton and Preston, near Kirkham, and about Marton and Staining, *Wh.* By the Wyre near Garstang.
E. 7. Near Garstang, *Wi.*
8. Leagram, *Fl. St.*

Achillea Millefolium, L. — Yarrow Milfoil. British type. Native. Area, 1, 2, 3, 4, 5, 6, 7, 8. Range, 0–1,930 feet. P. June September.

Pastures, grassy places and waste ground. Very common. First record : *Linton, B.R.C. Rep.*, 1874, p. 83.

Achillea Ptarmica, L.—Sneezewort. British type. Native. Area, 1, 2, 3, 4, 5, 6, 7, 8. Range, 0–1,000 feet. P. July–October.

Heathy places, waste ground and damp pastures. Frequent. First record : *Linton, B.R.C. Rep.*, 1874, p. 83.

Anthemis Cotula, L.—Stinking Chamomile. English type. Colonist. Area, 3, 4, 5, 8. Range, 0–400 feet. A. July–September.

Cultivated fields and waste places in the low country. Frequent. First record : *Syme, Top. Bot.*, 1883.

Chrysanthemum segetum, L.—Corn Marigold. English type. Colonist. Area, 3, 4, 5, 6, 7, 8. Range, 0–400 feet. A. June–September.

Occurs as a weed in cultivated fields. Frequent in West ; rarer in East and North. First record : *Linton, B.R.C. Rep.*, 1874, p. 83.

Chrysanthemum Leucanthemum, L.—Ox Eye Daisy. Dog Daisy. Marguerite. "Caton Heroes." British type. Native. Area, 1, 2, 3, 4, 5, 6, 7, 8. Range, 0–930 feet. P. June–August.

Fields, railway banks and grassy wastes. Very abundant in North ; less so in East and West. First record : *Linton, B.R.C. Rep.*, 1874, p. 83.

It ascends to 930 feet in Hindburndale.

Matricaria Parthenium, L.—Feverfew. Alien. Reported from districts 2, 3, 4, 5, 7, 8.

A garden outcast in most, if not all of its stations.

Matricaria inodora, L.—Mayweed. British type. Native. Area, 1, 2, 3, 4, 5, 6, 7, 8. Range, 0–600 feet. A. July–October.

Cultivated and waste ground. Common. First record : *Linton, B.R.C. Rep.*, 1874, p. 83.

Var. *salina*, Bab.—Frequent on the sea coast and on the shores of tidal river estuaries.

Matricaria discoidea, DC.—Alien. First record : *Salmon* and *Thompson*, *Journ. Bot.*, 1902, p. 294.

> N. 3. Footpath between Carnforth Station and the sea, *Wh.*
> W. 4. Very abundant about Glasson Dock, *Wh.*
> 5. Near Freckleton, 1901, *S. & T.* Barton, *Wi.*

Matricaria Chamomilla, L.—Corn Chamomile. English type. Colonist. Area, 1, 2, 3, 4, 5, 6, 7, 8. Range, 0–200 feet. A. July–September.

Cultivated fields and waste places in the low country. Frequent. First record : *Linton, B.R.C. Rep.*, 1874, p. 83.

Tanacetum vulgare, D.— Tansy. British type. Native. Area, 1, 2, 3, 4, 5, 6, 7, 8. Range, 0–500 feet. P. August.

Banks of rivers and streams ; damp roadsides and waste ground near villages. Not very common, and not always native ; but abundant locally. First record : *Ashfield, Fl. Prest.*, 1858.

> N. 1. Abundant about the beck near Ireby, *L.P.*
> 2. By the roadside near Aughton.
> 3. Near Warton and Bolton-le-Sands, *Wi.*
> W. 4. Cabus and St. Michael's, *Wi.* Preesall Salt Marsh, *Wh.* Near Cockerham, *Wi.*
> 5. Blackpool, St. Michael's, and Catterall, *Wi.* By the Wyre at Fleetwood, *Wh.*
> E. 6. Near Claughton, *Wi.*
> 7. Near Claughton and Garstang, *Wi.*
> 8. Ribble near Redscar, *Ashfield*, 1858. Higher Bridge Island ; Jumbles ; and Leagram, *Fl. St.*

Artemisia vulgaris, L.—Mugwort. British type. Native. Area, 1, 2, 3, 4, 5, 6, 8. Range, 0–400 feet. P. July–September.

Roadsides, thickets and waste places. Not very common. First record : *Linton, B.R.C. Rep.*, 1874, p. 83.

Artemisia maritima, L.—Sea Wormwood. English type. Native. Area, 3, 4, 5. Range, 0–20 feet. P. August–September.

Muddy salt marshes. Rather rare. First record : *Ashfield, Fl. Prest.*, 1858.

> N. 3. Near Bolton-le-Sands and at Sunderland Point, *Wi.*
> W. 4. From Stainall Marsh to the sea, *Pearson.* Still there, and at Knott End, Cockerham, and Glasson, *Wi.* Preesall Salt Marsh, *Wh.*
> 5. Skippool, near Poulton, *Ashfield*, 1858. Bank of the Wyre near Fleetwood, *Wh.*

Tussilago Farfara, L.—Coltsfoot. "Flapperdock." "Clatterdock." "Clay-leaf." British type. Native. Area, 1, 2, 3, 4, 5, 6, 7, 8. Range, 0–1,420 feet. P. February–April.

Roadsides, banks, and waste places, especially on clay. Abundant. First record : *Linton, B.R.C. Rep.*, 1874, p. 83.

It ascends to 1,420 feet in Ease Gill in North, and to 1,280 feet on Mallowdale Fell in East. *Puccinia poarum*, Niels, and *Coleosporium sonchi* (Pers.), are frequent on its leaves.

Petasites ovatus, Hill.—Butterbur. "Wild Rhubarb." British type. Native. Area, 1, 2, 3, 4, 5, 6, 7, 8. Range 0–400 feet or higher. P. March–April.

Swampy bushy places, river banks and about ditches. Common. First record : *Top. Bot.*, 1883.

Petasites fragrans, Presl.— Sweet Butterbur. Alien. First record : The Authors, *Journ. Bot.*, 1900, p. 44.

> N. 3. Roadside near Yealand.

Senecio vulgaris, L.—Groundsel. British type. Native. Area, 1, 2, 3, 4, 5, 6, 7, 8. Range, 0–700 feet or higher. A. January–December.

Cultivated fields, gardens and waste places. Abundant. First record : *Linton, B.R.C. Rep.*, 1874, p. 83.

Senecio sylvaticus, L.—Wood Groundsel. British type. Native. Area, 1, 2, 3, 4, 5, 6, 7, 8. Range, 0–700 feet. A. July–August.

Sandy banks and heaths. Frequent. First record : *Ashfield, Fl. Prest.*, 1858. "Pilling Moss."

Senecio viscosus, L.—Stinking Groundsel. Germanic type. Native ? Area, 3, 4. Range, 0–30 feet. A. July–August.

Waste ground, especially near the sea. Very rare. First record : *Lawson in Ray's Corresp.*, 1718.

> N. 3. "About Sunderland, nigh Lancaster," *Lawson*, 1688.
> W. 4. Pilling Moss, *Ashfield.*

Senecio erucifolius, L.—English type. Native. Area, 2, 3, 4, 5, 7, 8. Range, 0–400 feet. P. July–August.

Hedgebanks and bushy places. Not common. First record : *Ashfield, Fl. Prest.*, 1858.

> N. 2. Arkholme, *Wi.*
> 3. On the coast near Middleton.
> W. 4. Pilling, and Churchtown near Garstang, *Wi.* Preesall and Stalmine, *Wh.*
> 5. Between Naze Point and Lytham, *Ashfield*, 1858. Cleveleys and Catterall, *Wi.* Fleetwood Marsh, *Wh.*
> E. 7. Near Garstang, *Wi.*
> 8. Broughton, *Ashfield.* Inglewhite, *Wi.* Sale Wheel and Thornley Hall, *Fl. St.*

Senecio Jacobœa, L.—Ragwort. "Groundsward." British type. Native. Area, 1, 2, 3, 4, 5, 6, 7, 8. Range, 0–1,060 feet. P. July–August.

Roadsides, fields, sandhills and waste places. Very common. First record : *Linton, B.R.C. Rep.*, 1874, p. 83.

It ascends in North to 1,060 feet in Navel Pot, Leck Fell ; and in East to 850 feet in Hindburndale.

Senecio aquaticus, Huds.—Marsh Ragwort. British type. Native. Area, 1, 2, 3, 4, 5, 6, 7, 8. Range, 0–1,000 feet or higher. P. July–August.

Swampy fields and roadside ditches. Common. First record : *Linton, B.R.C. Rep.*, 1874, p. 83.

It ascends to 1,000 feet in Tatham, Hindburndale.

Senecio saracenicus, L.—Broad-leaved Ragwort. Intermediate type. Denizen. Area, 1, 6, 8. Range, 50–130 feet. P. July–August.

Moist meadows and river banks. Rare. First record : *Ashfield, Fl. Prest.*, 1858.

> N. 1. Banks of the Lune near Melling, *Wi.* Right bank of the Wenning near Hornby, *E.C.*
> E. 6. Bank of the Lune near Halton Station, *Wi.*
> 8. Ribble bank near Redscar, *Ashfield*, 1858. Seen there in 1875, *E.F.L.* Near Hacking Boat, *Fl. St.*

[*Senecio palustris*, DC.—Incognit.

Reported to grow abundantly on Pilling Moss in Ray's list in *Camden's* "*Britannica*," p. 803, 1695. Also in *Ray's Syn.* " In the ditches about Pillin Moss, Lancashire." The record is repeated by John Wilson, Sir J. Smith, Ashfield, etc., but none of them give first-hand evidence of its occurrence.]

Carlina vulgaris, L.—Carline Thistle. English type. Native. Area, 1, 2, 3, 5, 6, 7. Range, 0–800 feet. B. July–October.

Dry pastures and sandy banks. Not uncommon, especially on the limestone and amongst the coast sand-hills. First record : *Ashfield, Fl. Prest.*, 1858.

> N. 1, 2, 3. Frequent, ascending to 500 feet on Warton Crag, and 750 feet on Leck Fell.
> W. 5. Lytham, *Wh.* St. Annes, *Wi.*
> E. 6. Clougha at 800 feet.
> 7. Nickey Nook, *Ashfield*, 1858. Still there, and near Claughton, *Wi.*

Arctium minus, Bernh.—Burdock. British type. Native. Area, 1, 2, 3, 4, 5, 7, 8. Range, 0–840 feet or higher. P. July–August.

Woods, thickets and roadsides. Common. First record : The present one ?

Arctium nemorosum, Lej.—British type. Native. Area, 4. Range, uncertain. B. July.

In similar situations to the preceding. Very rare. First record : *Salmon & Thompson, Journ. Bot.*, 1902, p. 294.

> W. 4. Roadside near Pilling, *S. & T.*, August, 1901.

Arctium intermedium, Lange.—British type. Native. Area, 1, 2, 3, 7, 8. Range, 0–500 feet or higher. B. August.

Shady woods and thickets. Frequent in parts of East. First record : *Wheldon, Journ. Bot.*, 1902, p. 348.

> N. 1. Near Wennington.
> 2. Between Caton and Aughton. Kellet Seeds, *Wh.*
> 3. Woods on Warton Crag, ascending to the summit, *Wh.*
> E. 7. Abbeystead 1901, *Wh.* And in several other localities in Over Wyresdale.
> 8. Near Alston, *Wh.*

Carduus tenuiflorus, Curt.—Slender-flowered Thistle. English type. Native. Area, 3, 4, 5. Range, 0–450 feet. A. or B. July–August.

Damp sandy ground and waste places near the sea coast. Rare. First record : *Ashfield, Fl. Prest.*, 1860.

> N. 3. Near Brown's House, Silverdale, and between Far Naze and Sunderland Point, *Wi.* Near Carnforth, and on Warton Crag, *Wh.* Coast near Bare, *F.A.L.*
> W. 4. Near Cockersand Abbey, *Wi.*
> 5. Coast near Bispham, *Ashfield*, 1860. Near Preston Dock, *Fl. Prest. N.*

[*Carduus nutans*, L.—Incognit.

" In many places about Preston," *Ashfield.* No doubt an error.]

Carduus crispus, L.—Welted Thistle. British type. Native. Area, 1, 2, 3, 4, 5, 6, 8. Range, 0–400 feet. A. or B. June–August.

Fields, roadsides, and banks. Frequent on the limestone of North ; rare in East and West. First record : *Ashfield, Fl. Prest.*, 1860.

   N. 1. Near Wennington and Tunstall, *Wi.*
      2. Borwick Quarry, *Wi.*
      3. Frequent in the Silverdale and Leighton Beck districts, *Wi.* Sea embankment N.W. of Carnforth, *Wh.*
   W. 4. Glasson Dock and Cockersand Abbey, *Ashfield*, 1860. Stodday and Galgate, *Wi.* Between Winmarleigh and Garstang.
   E. 5. Waste ground near Preston Docks, *Wh.*
      6. Between Lancaster and Caton, *Wi.*
      8. Saddle Hill, *Fl. St.* Ribble banks above Preston, *Wh.* Near Whitewell, *Wi.*

Carduus lanceolatus, L. — Spear Thistle. British type. Native. Area, 1, 2, 3, 4, 5, 6, 7, 8. Range, 0–2,000 feet. B. July–September.

Fields, roadsides, and waste ground. Common. First record : *Linton, B.R.C. Rep.*, 1874, p. 83.

   It is occasionally attacked by *Puccinia hieracii* (Schum).

[*Carduus eriophorus*, L.—Incognit.

   Reported for V.C. 60 in Top. Bot. 1883, but the record is marked with a query. We think it is an error. It has not been confirmed, and so conspicuous a plant is hardly likely to have been overlooked.]

Carduus palustris, L.—Marsh Thistle. British type. Native. Area, 1, 2, 3, 4, 5, 6, 7, 8. Range, 0–1,520 feet. B. July–August.

Wet pastures and marshy places. Very common. First record : *Linton, B.R.C. Rep.*, 1874, p. 83.

   It ascends to 1,520 feet in North on Greygarth Fell ; and to 1,350 feet in East on Salter Fell.

Carduus heterophyllus, L. — Melancholy Thistle. Scottish type. Native. Area, 6, 7, 8. Range, 500–100 feet. P. June–July.

Damp mountain pastures, meadows, and bushy banks of streams. Rare. First record : *Bibby, Fl. Prest.*, 1860.

   E. 6. Banks of the Roeburndale River about Salter, *Wi.* Meadow at Quernmore, *E.C.*

    7. Near Lower Emmetts by the Tarnbrook Wyre, *W.W.M.* In a second locality near the above at 500 feet.
    8. Reported to Ashfield by Mr. J. Bibby from the Ribble side beyond Redscar. "I have this year (1859) searched the locality without success," *Ashfield.* A few plants on Higher Bridge Island, 1886. Not seen since, *Fl. St.*, 1891. Leagram, 1903, *Wi.*

Carduus arvensis, Robs.—Field Thistle. British type. Native. Area, 1, 2, 3, 4, 5, 6, 7, 8. Range, 0–800 feet or higher. P. July–September.

Pastures, roadsides, and waste places. Very common. First record : *Linton, B.R.C. Rep.*, 1874, p. 83. Ascends to 800 feet on Greenside, Wyresdale.

   This species often suffers severely from the ravages of the fungus *Puccinia suaveolens* (Pers.).

   The var. *setosum*, C. A. May, occurs in fields at Heaton near Lancaster (Dist. 3). The farmer has observed it there for many years, *E.C.* (Sp.).

Serratula tinctoria, L. — Saw-wort. English type. Native. Area, 2, 3, 5. Range, 0–300 feet. P. July–August.

Rocky woods and bushy places. Rare. First record : *Ashfield, Bot. Chron.*, 1864.

   N. 2. Near Halton Mill, *Wh.*
      3. Wood behind Smithy, Silverdale, *Ashfield*, 1864. Still there in 1901, *L.P.* Gatebarrow Wood, *Wi.*
   W. 5. Cottam, near Preston, *H.B.* (Sp.) Freckleton and Ashton, *A.A.D.*

Centaurea nigra, L. — Black Knapweed. "Hard Heads." British type. Native. Area, 1, 2, 3, 4, 5, 6, 7, 8. Range, 0–930 feet. P. June–September.

Meadows, pastures, and roadsides. Common. First record : *Linton, B.R.C. Rep.*, 1874, p. 83.

   The rayed form (*f. radiata*, Williams) occurs occasionally, as, for instance, in the woods near Abbeystead, but is much rarer than the type.

Centaurea solstitialis, L.—Alien. Reported in *Fl. Prest. N.* from cultivated land near Ashton.

Centaurea Scabiosa, L. — Great Knapweed. British type. Native. Area, 2, 3. Range, 0–250 feet. P. July–August.

Thickets and banks on the limestone. Rare, but locally abundant. First record : *Lewis, Top. Bot.*, 1883.

   N. 2. Near Carnforth, *Wi.*
      3. Silverdale, in numerous localities, and in the neighbourhood of Leighton Beck, *Wi.* Bank south of Warton, *Lewis.* Bolton-le-Sands, *L.P.* Near Carnforth Railway Station, *Wh.* Ascends to 250 feet on Warton Crag.

P

Centaurea Cyanus, L. — Corn Bluebottle. British type. Colonist. Area, 3, 5, 7. Range, 0–100 feet. A. June–August.

Cultivated fields. Rare. First record : *Petty, Naturalist*, 1902.

   N. 3. Silverdale ! *L.P.*
   W. 5. Near St. Michael's, *P.J.H.* Sandhills, Fairhaven, *A.A.D.* Lytham, *Fl. Prest. N.*
   E. 7. Near Garstang, *Wi.*

Cichorium Intybus, L. — Chicory. English type. Colonist. Area, 4, 5. Range, 0–100 feet. P. July–August.

Roadsides and waste ground. Rare, and perhaps hardly entitled to the rank of colonist. First record : *Pearson, Fl. Prest.*, 1860.

   W. 4. Near the Old Hall, Winmarleigh, *J. Pearson*, 1860. Near Bay Horse, *Wi.*
      5. Near St. Michael's Hall, *P.J.H.* Near Blackpool, *Wi.* On ballast, Preston Docks, and grassy banks round Fairhaven Lake, *A.A.D.*

Lapsana communis, L.—Nipplewort. British type. Native. Area, 1, 2, 3, 4, 5, 6, 7, 8. Range, 0–550 feet. A. July–September.

Hedgebanks and roadsides. Very common. First record : *Linton, B.R.C. Rep.*, 1874, p. 83.

   It ascends to 550 feet in Wyresdale.

Helminthia echioides, Gaertn.—Ox Tongue. English type. Native. Area, 5. Range, only seen at 25 feet. B. July–September.

Clayey banks. Rare. First record : *Wilson, Journ. Bot.*, 1900, p. 44.

   W. 5. Clay bank by the sea between Naze Point and Lytham, 1888, *Wi.*

[*Crepis taraxacifolia*, Thuill.—Reported from fields near St. Michael's, in *Fl. Prest. N.*, 1903, p. 24.

Crepis capillaris, Wallr. (*C. virens*, L.). — Hawk's Beard. British type. Native. Area, 1, 2, 3, 4, 5, 6, 7, 8. Range, 0–1,390 feet. A. June–September.

Dry banks, cultivated ground, and waste places. Common. First record : *Linton, B.R.C. Rep.*, 1874, p. 83.

   Ascends in North to 1,390 feet in Upper Ease Gill. A small slender state is plentiful in North on the sea coast, having the root crown much divided and leaves almost or quite entire. It is probably the *var. integrifolia*, Coss and Germ.

Crepis paludosa, Moench. — Succory-leaved Hawkweed. Scottish type. Native. Area, 1, 2, 4, 6, 7, 8. Range, 60–1,390 feet. P. July–August.

Damp shady places, swamps and banks of streams. Common amongst the hills of East ; rare elsewhere. First record : *Linton, B.R.C. Rep.*, 1874, p. 83.

   N. 1. Wrayton and Leck districts, and in Ease Gill, ascending to 1,390 feet, *Wi.*
   W. 4. Wood near Cock Hall, *S. Wi.*
   E. 6. Frequent in Foxdale, Roeburndale, etc., ascending to 1,280 feet.
      7. Nickey Nook, 1874, *E.F.L.* And many other localities.
      8. Near Chipping, *Wh.*

   Puccinia lapsanæ (Schultz) occasionally occurs on the leaves of this species.

Hieracium Pilosella, L.—Mouse-ear. British type. Native. Area, 1, 2, 3, 4, 5, 6, 7, 8. Range, 0–1,090 feet. P. June–August.

Dry fields and grassy banks. Common on the limestone ; rarer elsewhere. First record : *Linton, B.R.C. Rep.*, 1874, p. 83.

   In North it ascends to 1,090 feet on Leck Fell ; in East to 890 feet in Foxdale.

Hieracium aurantiacum, L.—Alien. First record : *Fl. Prest. N.*, 1903.

   W. 5. Copses near Ashton, planted, *Fl. Prest. N.* On the golf links east of the railway, St. Annes, July, 1906. *Miss Wood* (Sp. in Herb. C. Bailey). May have spread from the plants introduced at Ashton.

Hieracium sylvaticum, Gouan. (*H. murorum*, Auct. Angl.).—Golden Lungwort. British type. Native. Area, 1, 3, 7, 8. Range, 0–1,390 feet. B. July–August.

Dry woods, banks, and rocks, especially on limestone. Not common. First record : *Linton, B.R.C. Rep.*, 1874, p. 83.

   N. 1. Ease Gill and Leck Fell, ascending to 1,390 feet, *Wi.*
      3. Silverdale, *Wi.* Rocks near the Cove, "what I took to be this," *L.P.* Warton Crag, ascending to 480 feet, *Wi.*
   E. 7. Rocks by the Wyre, two miles above Dolphinholme.
      8. By the Hodder near Mitton Bridge, *Wh.*

Var. *micracladium*, F. N. Williams (subsp. *micracladium*, Dahlst.).

   N. 1. On Yoredale limestone rocks in Ease Gill, at 1,130 feet, July, 1906, *Wi.*

Hieracium Orarium, Lindeb., var. *ravusculum,* Williams.— British type. Native. Area, 5, 6. Range, 0–100 feet or higher. B. July–August.

Dry sandy banks near the sea and on shady walls. Rare. First record : *Marshall, Journ Bot.,* 1896, p. 136.

W. 5. Coast sandhills near St. Annes, 1895, *E.S.M.*
E. 6. Bank of the Lune near Halton.

Hieracium vulgatum, Fries. — Hawkweed. British type. Native. Area, 1, 2, 3, 4, 6, 7, 8. Range, 0–1,000 feet. P. June–September.

Walls, rocks, and banks. Frequent, especially in East and North. First record : *Linton, B.E.C. Rep.,* 1891.

Ascends in North to 820 feet on Dalton Crag, and in East to 990 feet in Foxdale.

Hieracium sciaphilum, Uechtritz. — British type. Native. Area, 1, 3, 7, 8. Range, 50–1,000 feet. P. June–September.

Walls, rocks, and stony places. Widely, but thinly distributed in North and East. First record : *Wilson, Journ. Bot.,* 1900, p. 44.

N. 1. Ease Gill.
3. Near Leighton Beck, *Wi.*
E. 7. Lower Bleasdale, near Garstang, 1898, *Wi.* Deer Clough at 1,000 feet.
8. Tootle Heights Quarry near Longridge, *Wh.*

Hieracium diaphanoides, Lindeb. — British type. Native. Area, 6. Range, 40–700 feet. P. July–August.

Shady rocks and walls by rivers. Rare. First record : The Authors, *Journ. Bot.,* 1905, p. 95.

E. 6. South bank of the Lune above Halton, 1904. Rocks by the Harter Beck, Roeburndale, ascending to 700 feet.

Hieracium duriceps, Hanb., var. *cravoniense,* Hanb.—Scottish type. Area, 1, 6. Range, 50–1,390 feet. P. June–August.

Shady rocks and stony places by rivers. Rare. First record : *Wilson, Journ. Bot.,* 1894, p. 230. "*Notes on British Hieracia,*" by F. J. Hanbury.

N. 1. Ease Gill near Leck, 1888, ascending to 1,390 feet near the upper falls, *Wi.* Wenning banks near Wennington. Rocks by the Greeta above Wrayton.
E. 6. Damp wall by the Lune, Halton, *Wi.* Rocks by the Foxdale Beck at 910 feet.

[*H. diaphanum,* Fr.—Incognit.

Reported from near Longridge in B.E.C. Report, 1892. "The specimen was no doubt a weather-worn *H. vulgatum* with the involucre denuded of hair and down by a wet and smoky climate." *E. F. Linton, Journ. Bot.,* 1900, p. 87.]

Hieracium gothicum (Backh.). — Highland type. Native. Area, 1. Range, only seen at about 1,100 feet. P. July–September.

Crevices of limestone rocks. Very rare. First record : *Wilson, Journ. Bot.,* 1905, p. 95.

N. 1. In a pot-hole on Leck Fell, 1904, *Wi.*

Hieracium tridentatum, Fries.—English type. Native. Area, 1, 6. Range, 50–1,300 feet. P. July–September.

Banks and borders of woods. Rare. First record : *Wilson, Journ. Bot.,* 1900, p. 44.

N. 1. Ease Gill 1899, and left bank of the Greeta near Wrayton, *Wi.*
E. 6. Rocks by the Lune near Caton, *Wi.*

Hieracium umbellatum, L.—British type. Native. Area, 2, 5, 6. Range, 0–200 feet. P. July–September.

Heathy places, thickets, and hedgebanks, on a sandy or gravelly soil. Very rare. First record : *Linton, B.R.C. Rep.,* 1874, p. 83.

N. 2. Hedgebanks near Docker.
W. 5. Lytham towards where St. Annes now is, 1873, *E.F.L.*
E. 6. Near Ellel, 1906, *Wi.*

Var. *coronopifolium* (Bernh.).

E. 5. Sandhills between Lytham and St. Annes, *S. & T.* Ansdell, *Wi.*

Hieracium boreale, Fries.—British type. Native. Area, 1, 2, 3, 4, 5, 6, 7, 8. Range, ? P. July–September.

Heathy and stony places, and on railway and hedgebanks. Common in East ; less frequent in North and West. First record : *Dobson, "Rambles by the Ribble,"* as *H. sabaudum,* 1877.

N. 1. Wrayton and Leck, *Wi.*
2. Lune banks near Halton, *Wh.* Arkholme Moor and Borwick, *Wi.*
3. Gatebarrow Wood.
W. 4. Near Ashton Hall, *Wh.* Glaciated rocks south of Galgate, *Wi.*
E. 6, 7, 8. Generally distributed. A form with green phyllaries occurs near Dutton Lea and Alston, *Wh.*

Hypochæris radicata, L.—Cat's-ear. British type. Native. Area, 1, 2, 3, 4, 5, 6, 7, 8. Range, 0–930 feet. P. July–August.

Banks, pastures, and grassy places. Very common. First record : *Linton, B.R.C. Rep.,* 1874, p. 83.

It ascends to 930 feet near the Tatham Beck.

Thrincia nudicaulis, Britten (*Leontodon hirtus,* L.)—Hairy Hawkbit. English type. Native. Area, 1, 2, 3, 4, 5, 6, 7, 8. Range, 0–600 feet. P. July–September.

Dry sandy banks and fields. Frequent, especially near the coast. First record : *Wilson, Journ. Bot.,* 1900, p. 44. "Bleasdale."

It ascends to 600 feet near Clougha Scar.

Leontodon hispidus, L. — Rough Hawkbit. English type. Native. Area, 1, 2, 3, 4, 5, 6, 7, 8. Range, 0–900 feet. P. July–August.

Fields, banks, and grassy places. Common. First record : *Linton, B.R.C. Rep.,* 1874, p. 83.

Leontodon autumnalis, L. — Autumnal Hawkbit. British type. Native. Area, 1, 2, 3, 4, 5, 6, 7, 8. Range, 0–1,850 feet. P. August–October.

Meadows, pastures, and grassy places. Common. First record : *Fl. St.,* 1891, p. 24.

It ascends to 1,850 feet on Greygarth Fell.

Taraxacum officinale, Weber. — Dandelion. The fruiting heads called "clocks," from a game children play with them. British type. Native. Area, 1, 2, 3, 4, 5, 6, 7, 8. Range, 0–2,040 feet. P. April–October.

Meadows, roadsides, and grassy places. Very common. First record : *Linton, B.R.C. Rep.,* 1874, p. 83.

The following sub-species occur :—

T. dens-leonis, Desf.—Area, 1, 2, 3, 4, 5, 6, 7, 8. Fields and roadsides. Very common.

T. erythrospermum, Andrz. — Area, 2, 3, 5. Sub-xerophilous. Dry gravelly places on limestone, or amongst blown sand on

the seashore. Local. First record : *Wheldon, Journ. Bot.,* 1900, p. 44.

N. 2. Over Kellet, *Wi.*
3. Warton Crag, *Wi.*
W. 5. Sand-hills between Lytham and St. Annes, 1898, *Wh.*

T. corniculatum, DC.—Area, 1, 3. Xerophilous ? Crevices of limestone rocks. Rare. First record : The Authors, *Journ. Bot.,* 1900, p. 44.

N. 1. Ease Gill Kirk, 1899.
3. Cringlebarrow. Warton Crag.

T. lævigatum, DC.—Area, 3, 5. Sandy or gravelly places and waste ground. Rare ? First record : *Wheldon, Journ. Bot.,* 1900, p. 44.

N. 3. Warton Crag.
W. 5. Roadside near Lytham, 1898, *Wh.*

T. palustre, DC.—Area, 1, 7. Wet grassy places and bogs. Frequent ? First record : The Authors, *Journ. Bot.,* 1900, p. 44.

N. 1. Ease Gill, 1899.
E. 7. Marshaw Fell.

Lactuca virosa, L.—Alien. First record : *Wheldon, Journ. Bot.,* 1901, p. 25.

W. 5. Abundant by the Wyre near Fleetwood Docks, 1900, *Wh.* St. Annes-on-Sea, *C.B.*

Lactuca muralis, Fresen. — Wall Lettuce. English type. Native. Area, 1, 2, 3, 5, 6, 7, 8. Range, 0–1,080 feet. P. June–August.

Shady walls, rocks, and banks. Frequent in North and East, especially on limestone ; very rare in West. First record : *Ashfield, Fl. Prest.,* 1858. "Claughton Hall."

In North it ascends to 1,080 feet at Navel Pot, Leck Fell, and to 880 feet on Dalton Crag ; in East to 850 feet near Haylot. The only record in West is in cultivated ground near St. Michael's on Wyre, where it is probably not truly native.

Sonchus oleraceus, L. — Sowthistle. British type. Native. Area, 1, 2, 3, 4, 5, 6, 7, 8. Range, 0–400 feet or higher. A. July–August.

Cultivated ground and waste places. Common. First record : *Linton, B.R.C. Rep.,* 1874, p. 83.

Sonchus asper, Hill. — Rough Sowthistle. British type. Native. Area, 1, 2, 3, 4, 5, 6, 7, 8. Range, 0–750 feet. A. July–September.

Roadsides and waste places. Common. First record: *Top. Bot.*, 1883.

Sonchus arvensis, L. — Field Sowthistle. British type. Colonist. Area, 2, 3, 4, 5, 6, 7, 8. Range 0–550 feet. P. August.

Cultivated fields. Frequent in the low country. First record: *Linton, B.R.C. Rep.*, 1874, p. 83.

Tragopogon minus, Mill. — Goat's-beard. British type. Native. Area, 1, 2, 3, 4, 5, 6, 7, 8. Range, 0–400 feet. B. June–July.

Dry banks and waste ground. Frequent. First record: *Ashfield, Fl. Prest.*, 1858. "Between Naze Point and Lytham."

Tragopogon porrifolium L.—Purple Goat's-beard. Alien.

W. 4. Near the Canal Bridge, Garstang, a few plants only, in 1851, *Mr. Pearson in Ashfield's Fl. Prest.*

Ambrosia artemisiæfolia, L. — Alien. First record: *Bailey, Memoirs Manch. Lit. & Phil. Soc.*, Vol. XLVII. (1902), No. 2.

W. 5. Several places on the sandhills near St. Annes ! C.B.

Mr. Bailey, who regards this as a thoroughly naturalized species at St. Annes, has procured evidence of its occurrence there for over eighteen years. It is a native of N. America, as also is *Ambrosia acanthicarpa*, Hook, a single specimen of which was found at St. Annes by Mr. Bailey.

Ambrosia trifida, L.—Alien. First record: *Wheldon, Journ. Bot.*, 1901, p. 25.

W. 5. Waste ground in Preston Dockyard, *Wh.* With the preceding near St. Annes, *C.B.* (Sp. !). A native of North America.

CAMPANULACEÆ.

Jasione montana, L. — Sheep's Scabious. British type. Native. Area, 1, 2, 3, 4, 5, 6, 7, 8. Range, 0–700 feet. B. July.

Sandy banks, heathy places, and old quarries. Frequent. First record: *Linton, B.R.C. Rep.*, 1874, p. 83.

N. 1. Lower Ease Gill, *Wi.*
  2. Arkholme Moor, and between Caton and Nether Kellet.
  3. Near Heysham, *Wi.*
W. 4. Rawcliffe Moss, *Wi.*
  5. Roadside near Sowerby, *P.J.H.*
E. 6. Lower Hindburn Bridge and near Ellel, *Wi.*
  7. Foot of Grizedale, near Garstang.
  8. Longridge Quarries, *Fl. St.*, 1891. Still plentiful there, *Wh.*

Wahlenbergia hederacea, Reichb. — Ivy-leaved Bell-flower. Atlantic type. Native. Area, (4), 7. Range, 0–200 feet. P. July–August.

Sheltered, damp mossy or marshy banks amongst the hills. Very rare. First record: *Pearson in Ashfield, Fl. Prest.*, 1860.

W. 4. Nateby Moss, 1854, *Mr. Pearson.* Now gone.
E. 7. Nickey Nook, *Ashfield*, 1860. Barnacre, near Garstang, seen in 1878 and 1880, but looked for without success in 1895 and since, *Wi.*

Campanula rapunculoides, L.—Alien. First record: *Miss K. Pickard, Naturalist*, April, 1897.

N. 3. Little Hawes Tarn, 1881, *Miss K. Pickard.*
W. 4. A garden escape on hedgebanks near Pilling, *Wi.*
  5. Between Lytham and Ansdell, *S. & T.*

[*Campanula Trachelium*, L.—Recorded by Mr. E. F. Linton in *B.R.C. Rep.*, 1874, p. 83, without locality. We know nothing of its status as a West Lancashire plant.]

Campanulata latifolia, L.—Giant Bell-flower. Scottish type. Native. Area, 1, 2, 3, 5, 6, 7, 8. Range, 0–530 feet or higher. P. July–August.

Hedgebanks and damp bushy places amongst the hills. Frequent in North and East and in some districts abundant, forming a conspicuous ornament of the lanes. Rare in West. First record: *Jenkinson, Brit. Pl.*, 1775, p. 32.

N. & E. About Yealand in abundance, *Jenkinson.* Generally distributed, ascending to 530 feet on Longridge Fell.
W. 5. Bank of the river below Brock Village and near Catterall, *Wi.*

Campanula rotundifolia, L. — Harebell. Bluebell. British type. Native. Area, 1, 2, 3, 4, 5, 6, 7, 8. Range, 0–920 feet. P. July–August.

Dry fields, banks, and grassy places. First record: *Linton, B.R.C. Rep.*, 1874, p. 83.

It ascends to 920 feet or higher in Foxdale and Hindburndale.

VACCINIACEÆ.

Vaccinium Vitis-idæa, L.—Cowberry. Whortleberry. "Flowering Box." Highland type. Native. Area, 1, 6, 7, 8. Range, 2,030–700 feet. P. June–July.

Gritstone rocks and peatbogs on the higher parts of the fells of North and East. First record: *Ashfield, Fl. Prest.*, 1860.

N. 1. Leck Fell, *L.P.* Greygarth Fell, from 2,030 feet downwards.
E. 6, 7, 8. Frequent on all the gritstone moors. Very fine on Wolfhole Crag and Clougha, and on the rocky parts of the Wyresdale Fells. Sometimes fruits freely in sheltered situations.

Vaccinium Oxycoccos, L.—Cranberry. British type. Native. Area, 1, 2, 3, 4, 6, 7, 8. Range, 1,820–20 feet. P. June–August.

Swampy moors and peatbogs amongst sphagnum. Common in some parts of East; formerly so on the *mosses* of West, but now becoming scarce owing to reclamation of the soil. First record: *Ashfield, Fl. Prest.*, 1858.

N. 1. Leck Fell, *Wi.*
  2. Bog near Docker, *Wh. & Wi.* Arkholme Moor and near Over Kellet, *Wi.*
  3. Heysham Moss.
W. 4. Pilling Moss, *Ashfield*, 1858. Cockerham Moss, abundant.
E. 6. Hindburn and Roeburndale, ascending to 1,820 feet on Wardstone and to 1,650 feet on Mallowdale Fell.
  7. Wyresdale and Bleasdale Fells, common.
  8. Longridge Fell ! *Ashfield.* Leagram Fells, *Wi.*

Vaccinium Myrtillus, L. — Bilberry. Bleaberry. Locally Whinberry and Wimberry, the latter no doubt a corruption of Whinberry. British type. Native. Area, 1, 2, 3, 6, 7, 8. Range, 200–2,030 feet. Sh. May.

Rocky woods, heaths, and upland banks. Abundant on all the fells of East, often covering large tracts. Scarce in North and very rare, if at all, in West. First record: *Ashfield, Fl. Prest.*, 1858. "Longridge Fell."

ERICACEÆ.

Andromeda polifolia, L. — Bog-bell. Intermediate type. Native. Area, 2, 3, 4, 6, 7, 8. Range, 25–1,730 feet. P. June–August.

Wet moors and peat bogs. Rather frequent on the fells of East; rare in North; rapidly becoming very rare in West. First record: *Lawson*, in *Ray's Corresp.*, 1718.

N. 2. Bog near Docker and Whittington Moor.
  3. "I have found it on the White Moss, Yealand." *Jenkinson*, 1775. 'On Middleton Moss by Lancaster," *Lawson*, 1688. Heysham Moss.
W. 4. Pilling Moss, *Ashfield*, 1858. Still grows on Cockerham Moss.
E. 6 and 7. Frequent, and in some parts abundant, as on Blaze Moss above Marshaw. It is also very plentiful at 1,500 feet on White Moss, Catshaw Greave Head, and ascends to 1,700 feet on Botton Head Fell and 1,730 feet on Wardstore.
  8. Longridge Fell, *Fl. St.*

Calluna vulgaris, Hull. — Ling. Heather. British type. Native. Area, 1, 2, 3, 4, 5, 6, 7, 8. Range, 20–2,030 feet. P. July–September.

Heaths, moors, and sandy or peaty banks. Exceedingly abundant on the fells of East, covering wide tracts of country. Formerly abundant in West on Cockerham Moss, but now becoming scarce, although it still grows there, and between Ansdell and Wrea Green. On the dry limestone hills of North it occurs much more sparingly than in East, although it is abundant on Leck Fell. First record: *Ashfield, Fl. Prest.*, 1858. "Longridge Fell."

Erica Tetralix, L. — Cross-leaved Heath. British type. Native. Area, 1, 2, 3, 4, 6, 7, 8. Range, 25–1,820 feet. P. July–August.

Boggy moors and wet places on heaths. Common on the fells of East. Rare in North, and rapidly becoming rare in West owing to the reclamation of the bogs. First record: *Ashfield, Fl. Prest.*, 1858. "Longridge Fell."

Erica cinerea, L. — Purple Heath. British type. Native. Area, 1, 2, 3, 4, 6, 7, 8. Range, 50–1,400 feet or higher. P. July–October.

Moors and dry stony and sandy banks. Very common in East; rare in most parts of North and West. First record: *Ashfield, Fl. Prest.*, 1858. "Longridge Fell."

Pyrola rotundifolia, L. — Winter-green. Germanic type. Native. Area, 5. Range, only at about 25 feet. P. July–August.

Marshy places amongst *Salix repens*, on the sandhills. Formerly plentiful locally, but now very rare, if not quite extinct. First record: *J. B. Wood*, in *Supplement to Dickenson's Flora of Liverpool*, 1855, p. 15.

> W. 5. Abundant at Lytham, *J.B.W.*, 1855. Between Lytham and the Lighthouse, *Ashfield*, 1858. Between Ansdell and St. Annes, 1901, *S. & T.*
>
> Our plant is the var. *maritima*, Kenyon.

Monotropa Hypopitys, L.—Yellow Bird's-nest. Germanic type. Native. Area, 2, 5. Range, 25–130 feet. P. July– August.

Very rare. First record: *Wilson, B.R.C. Rep.*, 1883, p. 248.

> N. 2. Holme area below Kirkby Lonsdale Bridge, two plants seen in 1836, *Hindson (in Baker's Fl. Lake District)*.
>
> W. 5. Sparingly amongst *Salix repens* on the sandhills near St. Annes, and likely to become extinct soon, *Wi.*

PLUMBAGINACEÆ.

Limonium vulgare, Mill (= *Statice Limonium*, L. p.p.).— Sea Lavender. English type. Native. Area, 3, 4, 5. Range, 0–25 feet. P. August–October.

Muddy sea shores and salt marshes. Not uncommon, and locally abundant. First record: *Ashfield, Fl. Prest.*, 1858.

> N. 3. Near Overton, and on the sea-coast between Far Naze and Sunderland Point, *Wi.*
>
> W. 4. Knott End, 1858, and between Wardless and Fleetwood, *Ashfield*. Still in these localities and on the coast near Cockerham, *Wi.* Wardless, 1874, *C. Bailey (Herb. Brit. Mus.)*. Saltmarsh near Glasson, *Wh.*
>
> 5. Fleetwood, 1848, *J.C. (Herb. S. H. Bickham)*. Between Skippool and Fleetwood, *Ashfield*. Still there, *Wh.*

Var. *pyramidale* (Syme).

> W. 4. Saltmarsh by the Wyre near Preesall, with the type and two next species, *Wh.*

Limonium humile, Mill. (= *Statice bahusiensis*, Fries.).— English type. Native. Area, 3, 4, 5. Range, 0–25 feet. P. August–October.

In similar situations to the preceding, but less common. First record: *Syme, Top. Bot.*, 1883.

> N. 3. Between Far Naze and Sunderland Point, and about the Lune Estuary, *Wi.*

> W. 4. Aldcliffe Marsh near Lancaster, 1846, *T. Westcombe,* (*Herb. Brit. Mus.*). Knott End, 1901, *S. & T.* Sea-coast near Cockerham, *Wi.* Saltmarsh by the Wyre, near Preesall, *Wh.*
>
> 5. Fleetwood-on-Wyre, 1841, *J. T. Syme (Herb. Watson)*. Wyre banks above Fleetwood, *Wh.*
>
> f. *nanum*. W. 4. Wardless, 1874, *C. Bailey (Herb. Edinburgh)*.

Limonium Neumani, C.E.S. (= *L. humile* × *vulgare*).

> W. 4. Near Preesall, *Wh.*
>
> 5. Blackpool, 1874, *W. R. Linton (in Herb., C. E. Salmon)*.

Limonium binervosum, C.E.S. (= *Statice binervosa*, G. E. Sm.).—Atlantic type. Native. Area, 3, 4, 5. Range, 0–25 feet. P. July–September.

Sea coast and river estuaries, in drier, more sandy and stony places than the two preceding, which prefer muddy situations. Not common. First record: *Syme, Top. Bot.*, 1883.

> N. 3. Between Far Naze and Sunderland Point, *Wi.* Basil Point, *E.C.*
>
> W. 4. Near Stainall-on-Wyre, 1843, *J. B. Syme (Herb. F. J. Hanbury)*. Knott End ! *F. C. King*, 1883 (*Herb. E. F. Linton*), and 1901, *S. & T.* Still there ! Between Pilling and Knott End, 1884, *Wi.* Saltmarsh near Preesall, *Wh.*
>
> 5. Fleetwood-on-Wyre, 1841 (*Herb. Watson*). Banks of Wyre, *C. J. Ashfield*, 1866 (*Herb. F. J. Hanbury*). Still there, *Wh.*

Statice Armeria, L.—Sea Pink. Thrift. Local name at Carnforth, "Sea Weed." British type. Native. Area, 3, 4, 5. Range, 0–25 feet. P. May–August.

Sea coast, mud flats, and salt marshes. Locally abundant. First record: *Ashfield, Fl. Prest.*, 1858. "Ashton Marsh."

PRIMULACEÆ.

Hottonia palustris, L.—Water Violet. English type. Native. Area, 4, 5, (7). Range, 0–50 feet. P. June.

Slow-flowing drains, ponds, and ditches in the low country. Rare, and only in West. First record: *Ashfield, Fl. Prest.*, 1858.

> W. 4. Ditches from Cockerham Church to the sea, *Ashfield*. Still there, *Wi.* Between Winmarleigh and Pilling, and between Cockerham and Glasson, *Wi.*
>
> 5. Ditches between Lytham and Marton Moss, *Ashfield*, 1858. Ditches and ponds near Staining and Marton, *Wi.*
>
> E. 7. Pond near Garstang (Planted), *Wi.*

Primula acaulis, L.—Primrose. British type. Native. Area, 1, 2, 3, 4, 5, 6, 7, 8. Range, 0–1,200 feet. P. March–May.

Hedgebanks, railway banks, woods and thickets. Common ; and particularly abundant in some parts of North as about Wennington, etc. First record: *Ashfield Fl. Prest.*?

It ascends to the pot-holes of Leck Fell.

The hybrid, *P. acaulis* × *veris* occurs occasionally and has been recorded as "*P. elatior*" by Ashfield in *Fl. Prest.* and *Bot. Chron.* from localities near Redscar and Silverdale.

Primula veris, L.—Cowslip. Often corrupted, "Keslop." British type. Native. Area, 1, 2, 3, 4, 5, 6, 7, 8. Range, 0–500 feet. P. April–June.

Meadows, pastures, grassy banks and open woods. Abundant on the limestone of North ; less common in East and West, and sometimes in these divisions imported by canals or with railway ballast. First record: *Linton, B.R.C. Rep.*, 1874, p. 84.

Primula farinosa, L.—Bird's-eye Primrose. Intermediate type. Native. Area, 1, 2, 3, 7, 8. Range, 25–800 feet. P. June.

Damp mountain pastures, and boggy ground, mostly on limestone. Rare, but abundant locally. First record: *Jenkinson, Brit. Pl.*, 1775, p. 27.

> N. 1. Lower Ease Gill, and boggy field near Old Wennington, *Wi.*
>
> 2. Between Carnforth and Nether Kellet.
>
> 3. " In the meadows about Leighton Hall, the seat of Geo. Townley, Esq., Lancashire, plentifully," *Jenkinson*, 1775. Hawes Tarn, *Ashfield* (*Bot. Chron.*, 1864). Still occurs in both these old stations, as well as on Thrang Moss and near Borwick, *Wi.*
>
> E. 7. Brock rills, Fairsnape Fell, *Wi.* Field below Harrisend Fell above Wyresdale Fishery, *A.B.*
>
> 8. Bolton Roughs, and near Lickhurst, Leagram, *Fl. St.*

Lysimachia vulgaris, L.—Yellow Loosestrife. English type. Native. Area, 1, 2, 3, 4, 5, 7, 8. Range, 0–200 feet. P. July– August.

Marshes, swampy thickets and ditches in the low country. Not uncommon. First record: *Pearson*, in *Ashfield Fl. Prest.*, 1860. "Winmarleigh Moss."

Lysimachia Nummularia, L.—Moneywort. English type. Native. Area, 1, 3, 5, 6, 8. Range, 0–150 feet. P. June–July.

Damp banks and marshes in the low country. Rare, but locally abundant. First record: *Ashfield, Fl. Prest.*, 1860.

> N. 1. Marshy broads near Tunstall, *Wi.*
>
> 3. Between Silverdale and Jenny Brown's Point, *Wi.*
>
> W. 5. Banks of Woodplumpton Brook, etc., *Wi.* South of St. Michael's, *P.J.H.* Ashton and Barton, *Fl. Prest. N.*
>
> E. 6. Lune bank near Lancaster, and marsh between Caton and Hornby, *Wi.*
>
> 8. Lower Brockholes, very abundant, *Ashfield*, 1860. Near Higher Bridge Cottages, naturalized, *Fl. St.* Redscar, *P.J.H.*

Lysimachia nemorum, L.—Wood Pimpernel. British type. Native. Area, 1, 2, 3, 4, 5, 6, 7, 8. Range, 0–1,390 feet. P. May–August.

Damp shady woods and banks. Common. First record : *Ashfield, Fl. Prest.*, 1860. "Redscar."

Ascends in North to 1,390 feet in Upper Ease Gill.

Trientalis europæa, L.—Chickweed Winter-green. Scottish type. Native. Area, 7. Range, 1,050–750 feet. P. June.

Boggy moorlands amongst bilberry, bracken, and heather. Very rare, but locally abundant. First record: *Wilson, Journ. Bot.*, 1901, p. 25.

> E. 7. Marshaw Fell, 1900, *Wi.* Extends over a considerable area on both sides of Black Clough, and on the adjacent moors.

Glaux maritima, L.—Sea Milkwort. British type. Native. Area, 3, 4, 5. Range, 0–25 feet. P. June–August.

Sea shores, muddy saltmarshes and tidal ditches. Common. First record : *Ashfield, Fl. Prest.*, 1858. "Lytham and Blackpool"!

Anagallis arvensis, L.—Field Pimpernel. British type. Colonist. Area, 2, 3, 4, 5, 6, 7, 8. Range, 0–400 feet. A. May–October.

Cultivated land and waste places. Frequent except in East. First record: *Linton, B.R.C. Rep.*, 1874, p. 84.

Anagallis cœrulea, Schreb.—Blue Pimpernel. English type. Colonist or alien? Area, 5. Range, 0–25 feet. A. July– October.

Waste places. First record : *Pearson*, in *Ashfield, Fl. Prest.*, 1860.

> W. 4. Glasson Dock, 1906, *Wh.*
>
> 5. " Mr. Pearson found a few plants on the bank of the Wyre opposite Preesall in 1851."

The Glasson plant has the long and distant cilia edging the petals which distinguish *A. cærulea* from blue flowered forms of *A. arvensis*, in which the petals and are ciliate but closely and irregularly denticulate.

Anagallis tenella, Lightf.—Bog Pimpernel. British type. Native. Area, 1, 2, 3, 5, 6, 7. 8. Range, 0–900 feet. P. July–August.

Marshes, boggy rill sides, and swampy cloughs amongst the fells; and on damp sands near the sea coast. Frequent, in some parts abundant. First record: *Jenkinson, Brit. Pl., 1775.*

N. 1. Ease Gill, *Wi.*
2. Arkholme Moor, *Wi.*
3. "Mossy ground on Yealand Common plentifully," *Jenkinson,* 1775. Hawes Tarn, *Ashfield.* Still there !
W. 5. Sandhills on Lytham Coast ! *Ashfield.* St. Annes.
E. 6, 7, 8. Abundant amongst the fells.

Samolus Valerandi, L.—Brook-weed. English type. Native. Area, 3, 4, 5. Range. 0–25 feet. P. July–September.

Tidal ditches, and damp sandy or muddy places near the sea. Not uncommon. First record: *Lawson, in Ray's Corresp., 1718.*

N. 3. "Between Bare and Fulton nigh Lancaster, on the seabank," *Lawson,* 1688. Ings Point and Overton, *Wi.*
W. 4. Cockerham, *Wi.*
5. St. Annes, *Wi.* Lytham, *Fl. Prest. N.*

Centunculus minimus, L.—Chaff-weed. English type. Native. Area, 2. Range, 300–400 feet. A. June–July.

Damp, sandy and gravelly places. Very rare. First record: *Wilson, Journ. Bot., 1901, p. 25.*

N. 2. Arkholme Moor, 1901, *Wi.* And since in several localities on different parts of the hill.

OLEACEÆ.

Fraxinus excelsior, L.—Ash. British type. Native. Area, 1, 2, 3, 4, 5, 6, 7, 8. Range, 0–1,070 feet. T. April–May.

Woods and hedges. Common. First record: *Linton, B.R.C. Rep., 1874, p. 83.*

It ascends to 1,070 feet on Leck Fell.

Ligustrum vulgare, L.—Privet. English type. Native. Area, 2, 3, 8. Range, 0–500 feet. Sh. June.

Limestone rocks and scars of North, where it is indigenous; occasionally planted in hedges elsewhere. Frequent. First record: *Ashfield, Fl. Prest.,* 1358, in a suspicious locality near Preston. In 1864, as a native, n *Bot. Chron.* Probably native in all the stations given below.

N. 2. Dalton Crag and near Over Eellet, *Wi.*
3. Apparently wild near Silverdale, *Ashfield,* 1864. Still occurs frequently in this district, ascending to 500 feet on Warton Crag.
E. 8. Gorge of the Hodder below Whitewell.

APOCYNACEÆ.

Vinca major, L.—Greater Periwinkle. Alien. First record: *W. Kirkby, Naturalist,* 1902.

N. 3. On the rocks at Jenny Browns Point, *W.K.*

Vinca minor, L.—Periwinkle. Alien. First record: *Ashfield, Fl. Prest.,* 1858.

N. 3. Near Silverdale, *Wi.*
W. 5. On Ashton Marsh, not truly wild, *Ashfield,* 1858. Still there, 1901, *H.B.* Thickets near Lea and Ashton, *Fl. Prest. N.*
E. 8. Near Hurst Green Church, *Fl. St.*

GENTIANACEÆ.

Centaurion umbellatum, Adans. (*Erythræa Centaurium,* Pers.—Centaury. British type. Native. Area, 1, 2, 3, 4, 5, 6, 7, 8. Range, 0–500 feet. A. July–September.

Dry pastures and banks. Frequent, but principally on the limestone and in sandy ground near the sea coast. First record: *Ashfield, Fl. Prest.,* 1858. "Near the Guide's House, below Freckleton."

Centaurion vulgare, Rafn. (*E. littoralis,* Fr.).—British type. Native. Area, 3, 5. Range, 0–50 feet. A. July–September.

Damp sandy shores. Frequent on the sandhills. First record: *Linton, B.R.C. Rep.,* 1874, p. 83.

N. 3. Coast near Heysham and Middleton, *Wi.*
W. 5. Lytham 1873, *E.F.L.* Still there and at St. Annes.

Centaurion pulchellum, Druce.—Dwarf Centaury. English type. Native. Area, 5. Range, 0–50 feet. A. July–September.

Q

In similar situations to the preceding, but less common. First record: *Ashfield, Fl. Prest.,* 1860.

W. 5. Bispham near Blackpool, sparingly, *Ashfield,* 1860. Sandhills near St. Annes, *Wi.*

Blackstonia perfoliata, Huds.—Yellow Centaury or Yellow-wort. English type. Native. Area, 4, 5. Range, 0–40 feet. B. July-August.

Clay fields and gravelly banks. Very rare. First record: *Pearson, in Ashfield's Fl. Prest.,* 1860.

W. 4. Between Skippool and Wardless, *Pearson,* 1860.
5. Between Ribblebank and Preston, *Ashfield.* Lytham, *Fl. Prest. N.*

Gentiana Pneumonanthe, L.—Marsh Gentian. English type. Native. Area, 7, 8. Range, 150–400 feet. P. August–September.

Damp moors and heathy ground. Very rare. First record: *Ashfield, Fl. Prest.,* 1858.

E. 7. Upper Claughton, *Wilson, B.R.C. Rep.,* 1887. Still there.
8. Abundant on Ribbleton Moor, *Ashfield,* 1858. The moor was drained in 1860 and the plant is now almost gone. A specimen was sent us from this vicinity in 1901 by *H. Beesley.*

Gentiana Amarella, L. — Autumnal Gentian. British type. Native. Area, 1, 2, 3, 5, 8. Range, 0–600 feet or higher. A. August–September.

Dry pastures and banks, on limestone especially, and on the coast sandhills. First record: *Jenkinson, Brit. Pl.,* 1775, p. 42.

N. 1. Sandy ground above the bank of Ireby Beck at 600 feet, *Wi.*
2. Kellet Seeds, *Wi.*
3. Found in Crowashes, plentifully, *Jenkinson,* 1775. Coast north of Bare, *F.A.L.* Middlebarrow and Gatebarrow, *Wi.* Castlebarrow, *L.P.* North side of Warton Crag, *Wh.*
W. 5. A white flowered form on Ansdell sandhills, *S. & T.* St. Annes, *Wh.*
E. 8. Formerly on Ribbleton Moor, *Mrs. C. Wilson.* Near Lickhurst and Leagram, *Fl. St.*

Gentiana campestris, L. — Field Gentian. British type. Native. Area, 3, 5, 8. Range, 0–500 feet or higher. A. September.

In similar localities to the preceding. Rare. First record: *Robson ; Turner & Dillwyn, Botanists' Guide,* 1805.

N. 3. On Yealand and Warton Commons, *Robson.* Near Leighton Beck, *Wi.* Silverdale, *W.K.,* and later by *J.F.P.* North side of Warton Crag, *Wh.*
W. 5. Lytham, *Ashfield,* 1858 (perhaps the next species was what Ashfield saw).
E. 8. Field below the Almshouses, Stonyhurst, *Fl. St.*

Gentiana baltica, Murb. — Scottish type. Native. Area, 5. Range, 0–25 feet. A. August–October.

Sandhills near the sea. Rare. First record: *Marshall, Bot. Exch. Club Rep.,* 1895.

W. 5. Sandhills between Lytham and St. Annes in quantity, *E.S.M.* Still there but not very plentiful, *Wh.*

Menyanthes trifoliata, L.—Bog Bean. British type. Native. Area, 1, 2, 3, 4, 5, 6, 7, 8. Range, 0–1,000 feet. P. May–June.

Boggy margins of pools and in swamps. Frequent, especially amongst the hills. First record: *Jenkinson, Brit. Pl.,* 1775, p. 27. "About New Dyke and Tewfit-field Farm, Yealand."

POLEMONIACEÆ.

Polemonium cæruleum, L.—Jacob's Ladder. Alien. A garden escape in all its localities. First record: *Fl. St.,* 1891.

W. 4. St. Michael's Hall farm, *P.J.E.*
E. 8. Stonyhurst Mill Pond ; Sale Wheel, 1887 ; Hodder bank opposite Bashall Lodge, 1890, *Fl. St.*

BORAGINACEÆ.

Cynoglossum officinale, L. — Hound's Tongue. " Stinking Roger." English type. Native. Area, 2, 3, 5. Range, 0–800 feet. B. June–August.

Coast sandhills and rough stony places on limestone. Not common. First record: *Ashfield, Fl. Prest.,* 1858.

N. 2. Dalton Crag, sparingly, ascending to 800 feet, *Wi.*
3. Morecambe Sands, 1874, *Roger Earnshaw.* Silverdale, *L.P.* Near Leighton Beck and Carnforth, *Wh.*
W. 5. Very abundant between Lytham and Blackpool, *Ashfield,* 1858. Still frequent, Lytham to South Shore, *Wh.*

Symphytum officinale, L.—Comfrey. English type. Denizen. Area, 1, 2, 3, 4, 5, 6, 8. Range, 0–300 feet. P. June–July.

River banks and damp thickets. Rare. First record: *Ashfield, Fl. Prest.,* 1858.

N. 1. Near Wennington and Kirkby Lonsdale, *Wi.* Near Tunstall.
    2. Lune side below Kirkby Lonsdale (the type).
    3. Near Bolton-le-Sands, *Wi.*
W. 4. Not unfrequent in ditches about Pilling where it is really wild, *Ashfield,* 1858. Near Winmarleigh (var. *patens*, Sibth).
    5. Vicarage sandbank, St. Michael's, *P.J.H.* Near Preston, *Ashfield,* 1858.
E. 6. Bank of the Lune near Caton (the type), *Wh.*
    8. Gas Wood, Stonyhurst, an escape, *Fl. St.*

Symphytum asperrimum, Willd.—Alien. First record: *Petty, The Naturalist,* 1902, p. 45.

N. 3. Eaves Wood and Townfield Pastures, Silverdale, *L.P.*

Borago officinalis, L.—Borage. Alien. First record: *Pearson, Ashfield's Fl. Prest.,* 1860.

Garden weed at Garstang and St. Michael's, *Pearson.* Near Preston Docks, *Fl. Prest. N.*

Anchusa sempervirens, L.—Alkanet. Alien. First record: *Ashfield, Fl. Prest.,* 1858.

N. 2. Well established on a hedgebank near Docker, but not native, 1903, *Wh.* & *Wi.* Nether Kellet, *Wi.*
    3. Near Thrang End, *Wi.*
W. 5. Ashton, *Fl. Prest. N.* Waste land, formerly sandhills, near St. Annes, *C.B.*
E. 8. At Ribbleton, but doubtfully wild, *Ashfield.*

Anchusa italica, Retz —Alien. First record: The present one.

With the preceding species near St. Annes, 1906, *C.B.*

Lycopsis arvensis, L. — Bugloss. British type. Colonist. Area, 3, 5. Range, 0–50 feet. A. June–August.

Dry fields and sandy places near the coast. Rare. First record: *Ashfield, Fl. Prest.,* 1858.

N. 3. About Sunderland Point and Far Naze, *Wi.* Coast near Bare, *F.A.L.*
W. 5. Beach at Lytham, *Ashfield,* 1858. Near Blackpool, *Wi.* Fleetwood, *Wh.*

Lycopsis orientalis, L.—Alien. First record: *Bailey, Memoirs Manch. Lit. & Phil. Soc.,* 1907.

W. 5. Waste ground, formerly sandhills, near St. Annes, 1906, *C.B.*

Pulmonaria officinalis, L.—Lungwort. Alien.

N. 2. Near Docker, *Wi.*
    3. Between Warton and Yealand, 1902, *Wh.*
W. 5. Near Preston, *H.B.* In each instance a garden outcast, although looking well established.

Echinospermum Lappula, Lehm.—Alien. First record: *Bailey, Memoirs Manch. Lit. & Phil. Soc.,* 1907.

W. 5. On building land, formerly sandhills, near St. Annes, September, 1906, *C.B.* A native of Italy.

Myosotis cæspitosa, F. Schultz. — Tufted Forget-me-not. British type. Native. Area, 1, 2, 3, 4, 5, 6, 7, 8. Range, 0–1,100 feet. P. May–September.

Roadsides, ditches, and wet places. Common. First record: *Linton, B.R.C. Rep.,* 1874, p. 84.

It ascends to about 1,100 feet on Tarnbrook Fell.

Myosotis scorpioides, L. — Forget-me-not. British type. Native. Area, 1, 2, 3, 4, 5, 6, 7, 8. Range, 0–580 feet. P. June–September.

Ditches, streamsides and watery places. Frequent. First record: *Watson, Top. Bot.,* 1883.

Ascends to 580 feet on Greenside, Wyresdale.

Myosotis repens, Don. — Creeping Forget-me-not. British type. Native. Area, 1, 6, 7, 8. Range, 600–1,300 feet or higher. P. June–September.

About springs, and in rivulets and watery places amongst the hills. Not common. First record: *Wilson, Journ. Bot.,* 1900, p. 45.

N. 1. Leck, *Wi.* Irehy Fell and Ease Gill at 1,150 feet.
E. 6. Dale Gill, Upper Hindburn, *Wi.* Mallowdale Fell at 1,300 feet ; Clougha.
    7. Fairsnape Fell Clough, 1899, *Wi.* Marshaw and Tarnbrook Fells and Calder Valley above Oakenclough.
    8. Clough on east side of Burnslack Fell, *Wi.* North side of Longridge Fell, *Wh.*

Myosotis sylvatica, Hoffm.—Wood Forget-me-not. English type. Native. Area, 1, 6, 8. Range, 50–800 feet. P. May–July.

Wooded river banks and shady glens amongst the hills. Rare. First record: *Ashfield, Fl. Prest.,* 1858.

N. 1. Ease Gill ; bank of Lune near Nether Burrow and lower down near Hornby, *Wi.*
E. 6. Lune banks between Halton and Caton, *Wh.*
    8. Woods about Redscar, *Ashfield,* 1858. Wooded banks of the Hodder near Whitewell, *Wi.*

Myosotis arvensis, Hill.—Field Forget-me-not. British type. Native. Area, 1, 2, 3, 4, 5, 6, 7, 8. Range, 0–700 feet. A. June–August.

Cultivated ground, banks and waste places. Frequent. First record: *Linton, B.R.C. Rep.,* 1874, p. 84.

Myosotis collina, Hoffm.—Early Forget-me-not. British type. Native. Area, 3, 5, 7. Range, 0–400 feet. A. April–June.

Hedgebanks and dry sandy places. Not common. First record: *Ashfield, Fl. Prest.,* 1858.

N. 3. Silverdale.
W. 5. "Reported to grow on Lytham sandhills," *Ashfield,* 1858. Lytham, St. Annes and South Shore, *Wi.*
E. 7. Grizedale, near Garstang, *Wi.*

Myosotis versicolor, Sm.—British type. Native. Area, 1, 2, 3, 4, 5, 6, 7, 8. Range, 0–500 feet. A. May–June.

Dry sandy fields, cultivated and waste ground. Frequent. First record: *Ashfield, Fl. Prest.,* 1858. "By the Dow Brook below the camp at Kirkham."

Lithospermum officinale, L. — Common Gromwell. "Nipbone." British type. Native. Area, 1, 3, 5. Range, 0–300 feet. P. June–August.

Rocky banks, and rough bushy places, chiefly on limestone. Rare. First record: *Jenkinson, Brit. Pl.,* 1775, p. 21.

N. 1. Wennington.
    3. "In the Tarns, a common field between Yealand and Burton," *Jenkinson,* 1775. Between Warton and Silverdale in several places, *Wi.* Between the Tarns, Silverdale, *Ashfield.* Still occurs here and there in the Silverdale district, *Wi.* Near Carnforth, *Wh.*
W. 5. Larbrick, *P.J.H.* Orchard in Skarve Green Lane, Fulwood, *A.A.D.*

Lithospermum arvense, L.—Corn Gromwell. British type. Colonist. Area, 3, 4. Range, 0–200 feet. A. June–July.

Cultivated fields. Rare. First record: *Ashfield, Fl. Prest.,* 1866.

N. 3. Near Carnforth, *Wh.*
W. 4. Cornfields between Knott End and Pilling, plentiful, *Ashfield,* 1866.

Echium vulgare, L.—Viper's Bugloss. British type. Native. Area, 5. Range, 0–50 feet. B. June–August.

Sandy ground, especially near the sea. Rare. First record: *Wheldon, Victoria History of Lancashire,* 1906, p. 55.

W. 5. South bank of the Wyre near Fleetwood, 1903, *Wh.* Near Lytham, *A.A.D.* St. Annes, *H.B., C.B.* (sp.)

CONVOLVULACEÆ.

Convolvulus sepium, L.—Great Bindweed. "Devil Filary." English type. Native. Area, 1, 2, 3, 4, 5, 6, 7, 8. Range, 0–400 feet. P. July–August.

Hedges and thickets. Rather frequent in the low country. First record: *Ashfield, Fl. Prest.,* 1860. "Near Preston."

"The rose or pink colored variety grows in a hedge between Freckleton and the Ribble in considerable quantities." The form referred to by Ashfield also occurs by the Canal near Glasson, *Wh.*

Convolvulus Soldanella, L.—Sea Bindweed. English type. Native. Area, 4, 5. Range, 0–30 feet. P. July–August.

Sandy seashores. Very rare—and rapidly decreasing. First record: *Ashfield, Fl. Prest.,* 1858.

W. 4. Between Fleetwood and Wardless, *Ashfield.*
    5. Between Lytham and Blackpool, and between Blackpool and Fleetwood, *Ashfield.* South Shore, *Wi.* Little Bispham and near Fleetwood, *Wh.* Rossall, *P.J.H.*

Convolvulus arvensis, L.—Small Bindweed. English type. Native. Area, 2, 3, 4, 5, 8. Range, 0–75 feet. P. July–September.

Roadsides, banks and waste ground. Rather rare, and in East almost unknown. First record: *Ashfield, Fl. Prest.,* 1858.

N. 2. Roadside between Carnforth and Kellet, and canal bank near Carnforth, *Wi.*
    3. Bare, *F.A.L.* Railbank near Trowbarrow, *L.P.* Carnforth, *Wi.*
W. 4. Hedgebank between Lancaster and Stodday, and Wyre banks near Churchtown, *Wi.*
    5. Sandhills, St. Annes ! *C.B.*
E. 8. Near Preston, *Ashfield,* 1858.

Cuscuta Epithymum, Murr.—Lesser Dodder. English type. Native. Area, 5. Range, only seen at 25 feet. A. July–September.

Coast sandhills. Very rare. First record: *Salmon and Thompson, Journ. Bot.,* 1902, p. 295.

W. 5. Near the lake, Ansdell, 1901, *S. & T.* Previously found here by *C. Bailey,* but not in a condition to determine with certainty, and therefore unrecorded.

SOLANACEÆ.

Solanum Dulcamara, L.—Bittersweet. Woody Nightshade. "Robin-run-i'-th'-hedge." English type. Native. Area, 1, 2, 3, 4, 5, 6, 7, 8. Range, 0–830 feet. P. June–August.

Damp or shady hedgebanks, and bushy places. Common in the low country. First record : *Linton, B.R.C. Rep.,* 1874, p. 83.

It ascends to 830 feet on Dalton Crag.

Solanum nigrum, L. — Black Nightshade. English type. Colonist (or perhaps only alien). Area, 1, 5, 7. Range, 0–100 feet. A. July–August.

Cultivated and waste ground. Rare. First record : *Syme, Top. Bot.,* 1883.

N. 1.   Bank of the Wenning below Hornby, *E.C.*
W. 5.   Ballast heaps near Preston Docks, 1898, *Wh.* Ashton, *Fl. Prest. N.*
E. 7.   Garden near Garstang, *E.C.*

Solanum rostratum, Dunal.—Alien. First record : The present one.

W. 5.   On building land, formerly sandhills, near St. Annes, October, 1906, *C.B.* (sp. !).

This handsome yellow-flowered plant belongs to the prickly section of the genus, and is a native of the United States. It has occurred several times as a casual in South Lancashire also.

Solanum triflorum, Nutt.—Alien. First record : The present one.

W. 5.   With the preceding at St. Annes, *C.B.* (sp. !).

Lycopersicum esculentum, Mill.—Occasionally flowers on waste heaps near towns, but never ripens its fruit.

Atropa Belladonna, L.—Deadly Nightshade. English type. Native. Area, 3, 5. Range, 0–200 feet. P. June–August.

Woods and thickets, only indigenous on the limestone of North. Rare. First record : *Jenkinson, Brit. Pl.,* 1775, p. 35.

N. 3.   About Warton Lane End and Throng Wood ; Yealand Common in several places, *Jenkinson,* 1775. Still grows near Warton, *Wi.,* and in the thicket on the left-hand side of the road to Hawes Tarn from Silverdale, *Wh.* Near Challan Hall, on White Scar and Trowbarrow, *W. Deason.* Amongst brushwood on the shore bank, *Silverdale, L.P.*
W. 5.   A large plant near Ribby Hall, Kirkham (probably planted), *A.A.D.*

Datura Stramonium, L.—Thornapple. Alien.

W. 5.   Occasionally on the beach at Lytham, *Ashfield,* 1860.

Hyoscyamus niger, L.—Henbane. English type. Native. Area, 3, 5. Range, 0–75 feet. A or B. July–August.

Waste ground near the sea. Rare. First record : *Wilson, B.R.C. Rep.,* 1883, p. 222.

N. 3.   On the sea coast at Hest Bank, 1881, *S. Wi.*
W. 5.   Between Lytham and the Guide House Inn, *S. & T.* Warton near Lytham, *H.B.* Kirkham, *Fl. Prest. N.* Near St. Annes, 1906, *C.B.* (sp.).

Lycium chinense, Mill.—Alien.

Occasionally planted near houses, and sometimes occurring in an apparently subspontaneous manner, especially near the sea coast.

SCROPHULARIACEÆ.

Verbascum Thapsus, L.—Common Mullein. English type. Native. Area, 1, 2, 3, 6, 7, 8. Range, 0–480 feet. B. June–August.

Dry rocky banks and hills. Frequent in the limestone district of North. Rare and not always native elsewhere. First record : *Ashfield, Bot. Chron.,* 1864.

N. 1.   Fell Lane near Leck, *Miss Maudsley.* Wennington, *Wi.* Lune bank below Kirkby Lonsdale.
2 and 3.   Very frequent. Silverdale, *Ashfield,* 1864.
E. 6.   Railbanks near Caton, *Wi.*
7.   Railbanks near Scorton Station, imported, *Wh.*
8.   Stonyhurst Mill, garden escape, *Fl. St.*

Verbascum Blattaria, L.—Moth Mullein. Alien.

Reported as a casual from near Bare by *F. A. Lees.*

Verbascum virgatum, Stokes.—Alien.

Between Wardless and Fleetwood, 1879, *Ashfield.*

Linaria Cymbalaria, Mill.—Ivy-leaved Toadflax. " Pedlar's Basket." " Creeping Jenny." British type. Denizen. Area, 1, 2, 3, 4, 5, 6, 7, 8. Range, 0–600 feet. P. June–October.

Old walls and bridges, especially near villages. Not unfrequent. First record : *Linton, B.R.C. Rep.,* 1874, p. 84.

Linaria repens, Mill. — Creeping Toadflax. English type. Denizen. Area, 3, 4. Range, 0–25 feet. P. July–September.

Rocky places and waste ground near the sea. Rare. First record : *J. H. Rossall in Baker's Fl. Lake Dist.,* 1885, p. 156.

N. 3.   At a cave called the Cow's Mouth, Morecambe Shore, near Arnside Knott, *Rossall.* Several places on the coast between the Westmorland boundary and Jenny Brown's Point, Silverdale, *Wi.*
W. 4.   Moss Side, garden escape, *P.J.H.*

Linaria vulgaris, Mill.—Toadflax. Butter-and-eggs. British type. Native. Area, 2, 3, 4, 5. Range, 0–100 feet. P. July–September.

Dry banks and waste ground. Frequent in the low country ; very rare in the hilly districts. First record : *Linton, B.R.C. Rep.,* 1874, p. 84.

Linaria minor, Desf.—English type. Colonist. Area, 1, 2, 3, 4, 5, 6, 7, 8. Range, 0–375 feet. A. June–September.

Railway lines, waste ground and shingles by rivers. Rather frequent. First record : *Petty, The Naturalist,* 1893, p. 98.

N. 1.   On rail line from Cowan Bridge to the Yorkshire boundary, *L.P.* On shingles by the Lune near Melling, *Wi.*
2.   By the railway at Arkholme, *Wi.*
3.   Railbanks near Silverdale, *L.P.*
W. 4.   Railway side near Winmarleigh. Glasson Dock, *Wi.*
5.   Railbanks near Fleetwood, *Wh.*
E. 6.   Wennington Station, *Wi.*
7.   Garstang and Catterall Station, *Wi.*
8.   Waste ground about Longridge station, *Wh.*

Antirrhinum majus, L.—Snap Dragon. Alien. First record :
Shore between Lytham and Fairhaven, 1901, *A.A.D.* (garden escape only).

Antirrhinum Orontium, L.—Alien. First record : *H. Beesley, Journ. Bot.,* 1902, p. 348.

W. 5.   Railway bank near Garstang, *H.B.*

Scrophularia aquatica, L.—Water Figwort. English type. Native. Area, 2, 3, 4, 5, 7, 8. Range, 0–100 feet. P. July–August.

Ditches, river banks and watery places. Not very common. First record : *Linton, B.R.C. Rep.,* 1874, p. 84.

N. 2.   By the Lune near Halton, *Wh.* On the banks of the Keer, *Wi.*
3.   Burton Well, Silverdale, *L.P.*
W. 4.   Cockerham, *Wi.* Near Ashton Hall, *Wh.*
5.   By the Wyre near Catterall, and near Inskip, *Wi.* Ashton, *Fl. Prest. N.*
E. 7.   Canal near Garstang, *Wi.*
8.   Banks of the Ribble above Preston, *Wh.*

Scrophularia umbrosa, Dum.—English type. Native. Area, 5, 8. Range, 0–400 feet. P. July–August.

Ditches and damp thickets. Rare. First record : *Top. Bot.,* 1883, but with a ? " Near Preston."

W. 5.   Between Lytham and the Guide House Inn, 1901, *S. & T.*
E. 8.   Near Knowl Green, 1899, *Wi.*

Scrophularia nodosa, L.—Knotted Figwort. British type. Native. Area, 1, 2, 3, 4, 5, 6, 7, 8. Range : 0–700 feet. P. July–September.

Damp hedgebanks, woods and streamsides. Common. First record : *Linton, B.R.C. Rep.,* 1874, p. 84.

It ascends to 700 feet on Dalton Crag.

Mimulus Langsdorffii, Donn.—Monkey Flower. British type. Denizen. Area, 1, 4, 5, 6, 7, 8. Range, 0–500 feet. P. June–August.

Streamsides, shingle beds in rivers, and wet stony places amongst the hills. Not unfrequent. First record : *Wilson, B.R.C. Rep.,* 1887, p. 102.

N. 1.   Leck Beck near Burrow, *L.P.* Very fine by the Cant Beck, Cantsfield, and on shingle by the Lune, Melling ; also in marshes near Turnstall, *Wi.*
W. 4 and 5.   Both banks of the Wyre near Catterall, *Wi.*
E. 6.   By the Lune near Halton Station and Caton.
7.   Abundant by the Wyre near Scorton, *H.B., Wh. & Wi.*
8.   Dinkling Green, Chipping, 1887, *Wi.* Frequent in the Hodder and Ribble districts, *Fl. St., Wh., H.B., &c.*

Limosella aquatica, L.—Mudwort. Germanic type. Native. Area, 3, 8. Range : 0–150 feet. A. July–September.

Muddy margins of pools and damp places where water has stood. Rare. First record : *Ashfield, Fl. Prest.,* 1858.

N. 3.   Brackish pools near Overton, and near Bolton-le-Sands, *Wi.*
E. 8.   Ribbleton Lane, *Ashfield.*

Digitalis purpurea, L. — Foxglove. British type. Native. Area, 1, 2, 3, 4, 5, 6, 7, 8. Range, 0–1,430 feet. B. June–August.

Woodland banks and hedgerows. Very common on gritstone, rare on limestone rocks. First record : *Ashfield, Fl. Prest.,* 1858. " Longridge."

Ascends in North to 1,430 feet in Upper Ease Gill ; and in East to 1,020 feet on Haylot Fell.

Veronica hederæfolia, L.—Ivy-leaved Speedwell. British type. Native. Area, 4, 5, 8. Range, 0–350 feet. P. April–July.

Hedgebanks and cultivated ground. Not frequent—almost rare. First record: *Ashfield, Fl. Prest.*, 1858.

W. 4. Near Cockerham, Galgate, Stodday and between Lancaster and Scotforth, *Wi.*

5. Ashton near Preston, *Ashfield*, 1858. Great Eccleston, *P.J.H.*

E. 8. Casual in Gas Wood, and Cross Gills Farm, Stonyhurst, *Fl. St.*

Veronica didyma, Ten. (*V. polita*, Fr.).—Grey Speedwell. British type. Colonist. Area, 2, 3, 5, 6, 7, 8. Range, 0–500 feet. A. April–October.

Cultivated and waste ground. Not uncommon, but less frequent than the next species. First record: *Ashfield, Fl. Prest.*, 1858. "Near Greave's Town, Ashton."

Veronica agrestis, L.—Field Speedwell. British type. Colonist. Area, 1, 2, 3, 4, 5, 6, 7, 8. Range, 0–600 feet or higher. A. April–October.

Cultivated and waste ground. Very common. First record: *Linton, B.R.C. Rep.*, 1874, p. 84.

Veronica Buxbaumii, Ten.—English type. Colonist. Area, 1, 3, 4, 5, 6, 7. Range, 0–150 feet. A. April–October. Cultivated ground. Not unfrequent. First record: The present one. The *Top. Bot.* record was an error, based on luxuriant *V. agrestis* (*fide E. F. L. in litt.*).

N. 1. Near Melling, *Wi.*

3. Heysham, *Wh.* Near Overton; Silverdale, *Wi.* Basil Point, *Wh.*

W. 4 and 5. Near Garstang, *Wi.*

E. 6. Caton, *Wh.*

7. Near Garstang, *Wi.*

Veronica arvensis, L.—Wall Speedwell. British type. Native. Area, 1, 2, 3, 4, 5, 6, 7, 8. Range, 0–1,930 feet. A. May–July.

Dry banks and fields. Common. First record: *Linton, B.R.C. Rep.*, 1874, p. 83.

Veronica serpyllifolia, L.—Thyme-leaved Speedwell. British type. Native. Area, 1, 2, 3, 4, 5, 6, 7, 8. Range, 0–2,030 feet. P. May–July.

Waste and cultivated ground, and by roadsides. Common. First record: *Linton, B.R.C. Rep.*, 1874, p. 83.

[**Veronica hybrida, L.**—Native (formerly). Area, (3). Range, ? P. July–September.

Limestone cliffs and dry banks. Very rare—now we fear quite extinct. First record: *Baker's Flora of the Lake District*, sub nom. *Veronica spicata*, L.

N. 3. "Gathered by Miss Beaver in Silverdale." *Baker's Fl. of the Lake Dist.*

We have carefully searched the Silverdale district for this, but without success. It still occurs at Humphrey Head in Lake Lancashire, four miles away across Morecambe Bay.]

Veronica officinalis, L.—Common Speedwell. British type. Native. Area, 1, 2, 3, 4, 5, 6, 7, 8. Range, 0–1,600 feet. P. May–August.

Dry banks and heathy pastures. Common. First record: *Ashfield, Fl. Prest.*, 1858. "Ribbleton Moor."

In North it ascends to 1,100 feet on Leck Fell; and in East to 1,600 feet in Antley Gill, Botton Head Fell.

Veronica Chamædrys, L. — Germander Speedwell. British type. Native. Area, 1, 2, 3, 4, 5, 6, 7, 8. Range, 0–1,920 feet. P. May–June.

Hedgebanks and grassy places. Common. First record: *Linton, B.R.C. Rep.*, 1874, p. 83.

Veronica montana, L.—British type. Native. Area, 1, 2, 3, 4, 5, 6, 7, 8. Range, 0–600 feet. P. May–July.

Damp shady woods. Common amongst the hills of East; rarer elsewhere. First record: *Ashfield, Fl. Prest.*, 1858. "Near Redscar."

Veronica scutellata, L.—British type. Native. Area, 1, 2, 3, 5, 6, 7, 8. Range, 0–1,150 feet. P. June–August.

Ditches, poolsides, and swampy places. Not frequent. First record: *Ashfield, Fl. Prest.*, 1858.

N. 1. Near Ireby and Middle Ease Gill, and in the marshes near Tunstall, *Wi.*

2. Near Docker and Henridden, *Wi.*

3. Near Overton.

W. 5. St. Michael's, *P.J.H.* Near Inskip, *Wi.*

E. 6. On Whitmoor at 890 feet, *Wi.* Hindburndale.

7. Barnacre; Lower Bleasdale; Abbeystead, *Wi.* Near Marshaw, and in Gavell's Clough at 1,150 feet.

8. Ribbleton Moor, *Ashfield*, 1858. Composition Hill, *Fl. St.* Near Whittingham, *Wh.*

Veronica Anagallis, L. — Water Speedwell. British type. Native. Area, 1, 2, 3, 4, 5. Range, 0–400 feet. P. June–August.

Ditches and watery places; not uncommon in the low country. First record: *Ashfield, Fl. Prest.*, 1858.

N. 1. Ireby, *Wi. & Wh.* Cantsfield and Melling, *Wi.*

2. Near Arkholme and Carnforth, *Wi.*

3. Bare, *F.A.L.* Silverdale, Bolton-le-Sands and Overton, *Wi.*

W. 4. Cockerham, *Wi.* Winmarleigh.

5. Ditches between Kirkham and Freckleton, *Ashfield*, 1858. Near Catforth, *Wi.*

Veronica Beccabunga, L.—Brooklime. British type. Native. Area, 1, 2, 3, 4, 5, 6, 7, 8. Range, 0–1,150 feet. P. June–August.

Ditches, brooks and watery places. Very common. First record: *Linton, B.R.C. Rep.*, 1874, p. 83.

In North it ascends to 1,150 feet on Ireby Fell; and in East to 1,100 feet by the Tatham Beck. The fungus, *Peronospora grisea*, De B., occurs occasionally upon its leaves.

Euphrasia officinalis, L.—Eyebright. British type. Native. Area, 1, 2, 3, 4, 5, 6, 7, 8. Range, 0–1,920 feet. A. June–September.

Roadsides, pastures, sand-dunes and heaths. Very common. First record: *Linton, B.R.C. Rep.*, 1874, p. 84.

The following segregates, regarded by many botanists as distinct species, have been ascertained to occur in West Lancashire.

E. nemorosa, H. Mart.—Area, 1, 2, 3, 4, 5, 6, 7, 8. Range, 0–800 feet or higher. Roadsides, pastures, and heaths. Very common, and much more plentiful than any of the succeeding forms. First record: *Wheldon, B.E.C. Rep.*, 1900, p. 641. "Lancaster Moor."

E. borealis, Towns.—Area, 1, 6, 7. Range, 600–1,920 feet. Grassy banks and heaths amongst the hills. Rare. First record: The Authors, *Journ. Bot.*, 1901, p. 25.

N. 1. Greygarth Fell at 1,920 feet, on limestone, *Wi.*

E. 6. Lower Salter, ascending to 1,250 feet on Salter Fell, September, 1900.

7. Catshaw Greave.

E. curta, Fries., including *var. glabrescens*, Wettst.—Area, 3, 4, 5. Range, 0–100 feet. Rare. First record: *Wheldon, Journ. Bot.*, 1901, p. 25.

N. 3. Heathy ground near Jenny Brown's Point, *Wi.* (var. *glabrescens*).

W. 4. Near Preesall, 1899 (the type), *Wi.*

5. Ansdell, *C. E. Salmon.* "The branching of this is very like *E. stricta*, Host, but the fruit and calyx are not those of stricta, and the whole plant is very setose"—*F. Townsend.* Plentiful on the sandbank near Little Bispham, 1900 (var. *glabrescens*, Wettst., teste Townsend), *Wh.* Carrs Green Common, *Wi.*

Odontites rubra, Gilib. (*Bartsia Odontites*, Huds.).—Red Eyebright. British type. Native. Area, 1, 2, 3, 4, 5, 6, 7, 8. Range, 0–500 feet or higher. A. July–September.

Damp roadsides, fields, and waste ground. Common. First record: *Linton, B.R.C. Rep.*, 1874, p. 84.

Both *O. verna*, Dum. and *O. serotina*, Dum. occur in West Lancashire, but their separate distribution has not been worked out. This species is sometimes almost smothered by the fungus *Coleosporium rhinanthacearum*, Lév., which deforms both stems and leaves.

Lasiopera viscosa, Hoffm.—Yellow Bartsia. Marsh Eyebright. Atlantic type. Native. Area, 5. Range, 0–25 feet. A. July–September.

Damp sandy ground, ditch banks, and margins of cultivated fields near the sea. Very rare. First record: *Ashfield, Fl. Prest.*, 1858.

W. 5. Ditch at the edge of the sandhills at Lytham next the cultivated ground, plentiful. Near Blackpool in marshy ground near the sea. It has also been found behind the sandhills near Lytham, *Ashfield*. St. Annes, August, 1880, *J. F. Rowse in Herb. Wheldon*. Near Ansdell, August, 1904, *Wi.*

Pedicularis palustris, L.—Marsh Lousewort. British type. Native. Area, 1, 2, 3, 5, 6, 7, 8. Range, 0–900 feet. P. June–August.

Spongy bogs and swampy places. Not very common. First record: *Petty, The Naturalist*, 1893, p. 99.

N. 1. Near Anneside, Leck Fell, *L.P.* Near Wrayton, *Wi.*

2. Bog near Docker and Whittington Moor.

3. Leighton Beck, Borwick and Carnforth, *Wi.*

W. 5. Near Lytham.

E. 6. Hindburndale; and on Whitmoor at 900 feet, *Wi.* Mallowdale Fell.

7. Fairsnape Fell, *Wh.* Hazelhurst, *Wi.*

8. Near Lickhurst, *Wi.*

Pedicularis sylvatica, L.—Lousewort. British type. Native. Area, 1, 2, 3, 6, 7, 8. Range, 0–1,050 feet. P. May–July.

Damp upland pastures and heaths. Common in East, frequent in North, rare or absent in West. First record: *Ray's Corresp.*, 1718. " Gunnerthwaite, Lancs."

It ascends to 1,050 feet on Salter Fell.

Rhinanthus Crista-galli, L.—Yellow Rattle. " Horse Pennies." " Penny Grass." British type. Native. Area, 1, 2, 3, 4, 5, 6, 7, 8. Range, 0–800 feet or higher. A. May–July.

Meadows and grassy places. Abundant. First record: *Linton, B.R.C. Rep.*, 1874, p. 84.

Melampyrum pratense, L.—Cow-wheat. British type. Native. Area, 1, 2, 3, 4, 6, 7, 8. Range, 0–1,470 feet. A. July–September.

Woodland banks and damp heaths. Frequent. First record: *Bailey, B.R.C. Rep.*, 1875.

> N. 1. Gorge of the Greeta, *Wi.*
>   2. Lords Lot Wood. Kellet seeds at 400 feet, and in a wood between Over Kellet and Capernwray, *Wi.*
>   3. The var. *latifolium*, Syme, is widely spread in the neighbourhood of Silverdale and Arnside, *C.B.*, 1874. Middlebarrow, and near Leighton Beck, *Wi.* Eaves Wood, Waterslack Wood, etc., *L.P.*
> W. 4. Cockerham Moss.
> E. 6. Near Botton, and at the falls above Mill Houses, *Wi.* At 1,350 feet on Mallowdale Fell, *Wh. & Wi.* Trough Brook Ghyll, *Wi.*
>   7. Nickey Nook, *P.J.H.* Black Clough at 1,100 feet, *Wi.* Tarnbrook Fell, at 1,470 feet.
>   8. On old Longridge road and near Hurst Green, *Fl. St.* Near Lickhurst, *Wi.*

OROBANCHACEÆ.

Orobanche minor, Sm.—Broomrape. Germanic type. Colonist. Area, 5, 7. Range, 0–75 feet. A. July–September.

Clover fields. Very rare. First record: *Wilson, B.R.C. Rep.*, 1887, p. 102.

> W. 5. Lytham, *Fl. Prest. N.*
> E. 7. Clover field at Sturzaker, near Garstang, 1886, *Wi.*

Lathræa Squamaria, L.—Toothwort. English type. Native. Area, 3, 8. Range, 0–500 feet. P. April–May.

Shady woods on river banks, etc. Rare. First record: *Jenkinson, Brit. Pl.*, 1775, p. 140–141.

> N. 3. " In Deepdale Wood in Grisedale, within the liberties of Yealand," *Jenkinson*, 1775. Seen there, on the roots of an ash tree, by the Authors in 1906.
> E. 8. Saddle Hill ; Hodder Bank above Beesley Wheel, *Fl. St.* Wooded bank of the Hodder below Whitewell.

Utricularia vulgaris, L.—Bladderwort. British type. Native. Area, 2, 3, 5, 6, 7, 8. Range, 0–350 feet. P. July–August.

Ponds and ditches in the low country. Rare. First record: *Ashfield, Fl. Prest.*, 1858.

> N. 2. Ditches near a swamp between Carnforth and Nether Kellet.
>   3. Storrs Moss, and ditch east of Silverdale Station, *Wi.*
> W. 5. " I have been informed that this plant has been found in the canal near Preston, and in pits near Lytham." *Ashfield.*
> E. 6. Near Quernmore, *S. Wi.*
>   7. Ponds, Stubbings Bridge, near Garstang, 1882, *Wi.*
>   8. Near Thornley Hall, *Fl. St.*

[*Utricularia minor,* L.—Incognt. " I found this also in the great meadow at Gunnerthwaite, belonging to Mr. Breakbane," *Jenkinson, Brit. Pl.*, 1775.

No recent confirmation, and perhaps now extinct.]

Pinguicula vulgaris, L.—Common Butterwort. Scottish type. Native. Area, 1, 2, 3, 4, 6, 7, 8. Range, 0–1,430 feet. P. June.

Sphagnum beds in bogs, and wet mossy banks ; occasionally on dripping rocks. First record: *Ashfield, Fl. Prest.*, 1860. " Beacon Fell."

It ascends in North to 1,430 feet in Upper Ease Gill.

VERBENACEÆ.

Verbena officinalis, L.—Vervain. English type. Native. Area, 1, 2, 3, 5. Range, 0–200 feet. P. July–August.

Roadsides, waste ground, and hedgebanks, generally on limestone. Not unfrequent in North, rare elsewhere. First record: *Ashfield, Fl. Prest.*, 1858.

> N. 1. Wennington, *Wi.*
>   2. Between Bolton-le-Sands and Carnforth, *Ashfield*, 1858.
>   3. Frequent about Silverdale and near Leighton Beck, *Wi.*
> W. 5. Woodplumpton and Ashton, *Ashfield.*

R

LABIATÆ.

Mentha spicata, L.—Spearmint. English type. Denizen. Area, 6. Range, only seen at about 40 feet. P. August–September.

Banks of streams and wet places. Rare. First record: *Wilson, Journ. Bot.*, 1900, p. 45.

> W. 5. Near St. Annes, *Fl. Prest. N.*
> E. 6. On the bank of the Lune near Halton, *Wi.*

Mentha piperita, L.—Peppermint. English type. Native. Area, 1, 4, 5, 7, 8. Range, 0–400 feet. P. August–September.

Watery places and sides of ditches. Rare, and not always indigenous. First record: *Wilson, B.R.C. Rep.*, 1887, p. 103.

> N. 1. Ireby, *Wi. & Wh.* Wennington, *Wi.*
> W. 4. Between Glasson and Galgate, *Wh.*
>   5. Cleveleys, and near Catterall, *Wi.*
> E. 7. In several localities near Garstang, 1886, *Wi.* Near Dolphinholme, *Wh.* Brock, *Fl. Prest. N.*
>   8. Near Hurst Green.

Mentha aquatica, L.—Common Mint. Wild Mint. British type. Native. Area, 1, 2, 3, 4, 5, 6, 7, 8. Range, 0–700 feet or higher. P. August–September.

Ditches, streamsides, and watery places. Common. First record: *Linton, B.R.C. Rep.*, 1874, p. 84.

Mentha sativa, L.—British type. Native. Area, 1, 2, 3, 4, 5, 6, 7, 8. Range, 0–500 feet. P. July–September.

Ditches, streamsides, and watery places. Very frequent. First record: *Fl. St.*, 1891. " Ribble banks at Sale Wheel and Hodder bank at Mab Wheel."

Mentha rubra, Huds.—Red-veined Mint. British type. Native. Area, 8. Range, 0–200 feet. P. August–September.

River banks. Very rare. First record: *Fl. St.*, 1891, p. 30.

> E. 8. Higher Bridge Island, *Fl. St.*

Mentha gentilis, L.—Bushy Red Mint. British type. Native. Area, 1, 2, 8. Range, 0–150 feet. P. August–September.

Backwaters and river banks where liable to inundation. Rare. First record: *Wheldon, B.E.C. Rep.*, 1899.

> N. 1. By the Lune below Kirkby Lonsdale.
>   2. Bank of the Lune between Caton and Aughton.
> E. 8. Banks of the Hodder near Mitton, 1899, *Wh.*

Mentha arvensis, L.—Field Mint. Corn Mint. British type. Native. Area, 2, 3, 4, 5, 6, 7, 8. Range, 0–400 feet or higher. P. August–September.

Cultivated fields and waste ground. Frequent in the low country ; rare in East. First record: *Linton, B.R.C. Rep.*, 1874, p. 84.

Badly infected with *Puccinia menthæ* (Pers.) near Alston.

Mentha Pulegium, L.—Penny-royal. English type. Native or Denizen? Area, 7? Range, uncertain. P. August–September.

Poolsides and marshy places. Very rare. First record: *Flora of Preston and Neighbourhood*, 1903.

> E. 7? Near Garstang, *H.B.*

Lycopus europæus, L.—Gipsy Wort. British type. Native. Area, 2, 3, 4, 5, 6, 7, 8. Range, 0–400 feet. P. July–September.

Ditches, pond sides, and watery places. Frequent in the low country. First record: *Top. Bot.*, 1883.

Origanum vulgare, L.—Marjoram. British type. Native. Area, 1, 2, 3, 6, 8. Range, 0–350 feet. P. July–September.

Dry calcareous banks and bushy places. Frequent on the limestone in North ; rare elsewhere. First record: *Jenkinson, Brit. Pl.*, 1775, p. 132.

> N. 1. Cantsfield, Nether Burrow, Melling, and Wennington, *Wi.* Near Kirkby Lonsdale.
>   2. About Borwick, *Jenkinson*, 1775. Frequent in this division.
>   3. Middlebarrow, *Wi.* ; and other localities about Silverdale and Warton, ascending to about 350 feet.
> E. 6. Lune bank above Caton, *Wi.*
>   8. Sale Wheel, *Fl. St.* Ribble banks near Ribchester, *Wh.*

Thymus Serpyllum, L.—Wild Thyme. British type. Native. Area, 1, 2, 3, 4, 5, 6, 7, 8. Range, 0–2,030 feet. P. July–August.

Dry, grassy banks, sandy commons, and limestone rocks. Common, especially in North. First record: *Linton, B.R.C. Rep.*, 1874, p. 84.

It ascends to 2,030 feet near the summit of Greygarth Fell.

Calamintha vulgare, L.—Wild Basil. British type. Native. Area, 1, 2, 3, 4, 5, 6, 8. Range, 0–500 feet. P. July–August.

Hedgebanks, woods, and thickets. Frequent in the limestone districts of North; less common in the other divisions. First record: *Ashfield, Fl. Prest.,* 1860.

> N. 1. Leck, Cantsfield, Tunstall and Wennington, and in Ease Gill, ascending to 500 feet, *Wi.*
> 2 and 3. Frequent in these districts.
> W. 4. Near Stodday, *Wi.*
> 5. Sandhills at Lytham, frequent, *Ashfield,* 1860.
> E. 6. Wray, *Wi.*
> 8. Saddle Hill and Beesley Wheel, *Fl. St.* Near Ribchester! *Fl. Prest. N.*

Calamintha Acinos, O. Kuntze.—Basil Thyme. British type. Native. Area, 3. Range, 0–300 feet. A. July–August.

Dry calcareous banks. Rare. First record: *Jenkinson, Brit Pl.,* 1775, p. 133.

> N. 3. On the top of Cringlebarrow, *Jenkinson,* 1775. Between Ings Point and Jenny Brown's Point, *Wi.*

Nepeta Cataria, L.—Cat Mint. English Type. Native. Area, 3, 4. Range, 0–100 feet. P. July–August.

Dry hedgebanks, native on the limestone of North; adventive elsewhere. Rare. First record: *Wilson, B.R.C. Rep.,* 1883, p. 248.

> N. 3. Hedgebank near Jenny Brown's Point, Silverdale, 1878, *Wi.* Ings Point, and shore near Carnforth, *Wi.*
> W. 4. Wardless-on-Wyre, *H.B.*

Nepeta hederacea, Trevis.—Ground Ivy. British type. Native. Area, 1, 2, 3, 4, 5, 6, 7, 8. Range, 0–460 feet or higher. P. April–May.

Dry hedgebanks and thickets. Common. First record: *Linton, B.R.C. Rep.,* 1874, p. 84.

Salvia verticillata, L.—Alien. First record: *Bailey, B.E.C. Rep.,* 1903, p. 21.

Sandhills north of St. Thomas's Church, St. Annes, 1903, *C.B.*

> With *Gypsophila,* &c., no doubt a garden outcast.

Salvia Verbenaca, L.—Wild Sage or English Clary. English type. Native probably. Area, 3. Range, 0–200 feet. P. June–July.

Dry pastures and banks on calcareous soil. Rare. First record: *Petty, Naturalist,* 1902, p. 47.

> N. 3. In two localities near Silverdale, 1901, *L.P.*

Scutellaria minor, Huds. — Lesser Skull-cap. English type. Native. Area, 2, 7. Range, 300–500 feet. P. June–September.

Wet heathy places. Very rare. First record: *Wilson, Journ. Bot.,* 1901, p. 25.

> N. 2. Whittington and Arkholme Moors, 1900, *S. Wi. & Wi.*
> W. 7. Found by Mr. John Moss near Nickey Nook, *P.J.H.*

Scutellaria galericulata, L.—Skull-cap. British type. Native. Area, 1, 2, 3, 4, 5, 6, 7, 8. Range, 0–580 feet. P. July–September.

Marshes and swamps, by ditches and canals, and damp sands on the sea-coast. Widely distributed but not very common. First record: *Jenkinson, Brit. Pl.,* 1775, p. 135. "About New-dyke, Thornbarrow and Yealand Conyers."

Prunella vulgaris, L.—Self-heal. British type. Native. Area, 1, 2, 3, 4, 5, 6, 7, 8. Range, 0–1,850 feet. P. July–September.

Pastures, roadsides and grassy places. Very common. First record: *Jenkinson, Brit. Pl.,* 1775, p. 136. He records a white flowered variety from near Dalton Hall.

> It ascends in North to 1,850 feet on Greygarth Fell, and in East to 1,250 feet on Salter Fell.

Marrubium vulgare, L.—Horehound. Alien. First record: *Jenkinson, Brit. Pl.,* 1775, p. 131.

> N. 3. About Warton plentifully, *Jenkinson.* Near Morecambe, casually, *Wh.*
> W. 5. St. Annes! *C.B.* Near Preston Docks, *Fl. Prest. N.*

Leonurus cardiaca, L.—Motherwort. Alien. First record: The present one.

> W. 5. Near Ansdell, September, 1901, *C.B.*

Stachys officinalis, Franch.—Betony. English type. Native. Area, 1, 2, 3, 4, 5, 6, 7, 8. Range, 0–860 feet. P. July–August.

Woodland banks and pastures. Common, especially in East and North. First record: *Linton, B.R.C. Rep.,* 1874, p. 84.

> It ascends to 860 feet in Hindburndale.

Stachys palustris, L.—Marsh Woundwort. British type. Native. Area, 1, 2, 3, 4, 5, 6, 7, 8. Range, 0–500 feet. P. July–September.

Ditches, streamsides, and marshy places; and occasionally in cultivated fields. Common. First record: *Linton, B.R.C. Rep.,* 1874, p. 84.

Stachys sylvatica, L. — Hedge Woundwort. British type. Native. Area, 1, 2, 3, 4, 5, 6, 7, 8. Range, 0–1,050 feet. P. July–August.

Hedgebanks, woods, and shady places. Very common. First record: *Linton, B.R.C. Rep.,* 1874, p. 84.

> It ascends to 1,050 feet in the shelter of the pot-holes of Leck Fell and to 820 feet in Roeburndale.

Stachys arvensis, L. — Corn Woundwort. Bristish type. Colonist. Area, 2, 3, 4, 5, 7. Range, 0–300 feet. A. July–September.

Cultivated fields and gardens. Not uncommon. First record: *Wilson, Journ. Bot.,* 1900, p. 45.

> N. 2. Near Nether Kellet, *Wi.*
> 3. Heald Brow, Silverdale, 1888, and near Middleton, *Wi.* Near Overton, *Wh.*
> W. 4. Fields near Preesall, *Wh.*
> 5. Freckleton and near St. Annes, *S. & T.* St. Michael's, *Fl. Prest. N.*
> E. 7. Sullam Side, *Wi.*

Galeopsis Tetrahit, L.—Hemp Nettle. British type. Native. Area, 1, 2, 3, 4, 5, 6, 7, 8. Range, 0–600 feet. A. July–August.

Cultivated fields, waste ground, and hedgebanks. Common. First record: *Linton, B.R.C. Rep.,* 1874, p. 84.

Galeopsis speciosa, Mill. — Large-flowered Hemp Nettle. English type. Colonist. Area, 4, 5, 7, 8. Range, 0–200 feet. A. July-August.

Cultivated fields. Not uncommon in the low country; very abundant in potato fields in the Fylde area. First record: *Ashfield, Fl. Prest.,* 1858.

> W. 4. Near Lancaster, 1843, *R. Pryor,* in *Herb. C. E. Salmon.* Pilling, *Wi.* Cockerham, abundant, *Wh. & Wi.* Near Stalmine, *Wh.*
> 5. Waste ground near Preston Docks, *Wh.* Between Lytham and Guide House Inn, *S. & T.*

> E. 7. Between Scorton and Nickey Nook, *Ashfield,* 1858. Near Garstang, *Wi.*
> 8. St. Mary's Pond, *Fl. St.*

Galeopsis Ladanum, L.—Red Hemp Nettle. English type. Colonist. Area, 3, 5. Range, 0–25 feet. A. July-October.

Sandy or shingly sea shores. Rare. First record: *Flora of Preston and Neighbourhood,* 1903, p. 32.

> N. 3. Sea-coast shingle at Bare, 1887, and again in 1895 (the segregate angustifolia, Ehrh.), *J. Reanland* (sp.)
> W. 5. Lytham, *Fl. Prest. N.*

Lamium amplexicaule, L.—Henbit. British type. Colonist. Area, 5. Range, 0–100 feet. A. May–August.

Cultivated fields and hedgebanks. Rare. First record: *Wheldon, Journ. Bot.,* 1901, p. 25.

> W. 5. Near Lytham, 1900, *Wh.* Barton, *Fl. Prest. N.*

Lamium mollucellifolium, Fr.—Dead Nettle. British type. Colonist. Area, 3, 5. Range, 0–100 feet. A. May–August.

Cultivated ground and waste places. Rare. First record: *Melvill, Top. Bot.,* 1883.

> N. 3. Silverdale, 1883, *J.C.M.*
> W. 5. Near St. Michael's, *P.J.H.* Barton, *Fl. Prest. N.*

Lamium hybridum, Vill. — Dead Nettle. British type. Colonist. Area, 1, 2, 3, 4, 5, 7, 8. Range, 0–430 feet. A. May–August.

Cultivated ground and waste places. Frequent in the low country. First record: *Ashfield, Fl. Prest.,* 1860.

> N. 1. Near Ireby at 430 feet, and near Wennington.
> 2. Near Arkholme and Carnforth, *Wi.*
> 3. Silverdale, *J.C.M.* Warton, *Wi.*
> W. 4. Preesall, Pilling, and Glasson, *Wh.* Cockerham Moss and Winmarleigh.
> 5. Lea, *Ashfield,* 1860. Fleetwood, *Wh.*
> E. 7. Near Garstang, *Wi.*
> 8. Garden weed at Stonyhurst, *Fl. St.* Near Ribchester, *Wh.*

Lamium purpureum, L.—Red Dead Nettle. British type. Native. Area, 1, 2, 3, 4, 5, 6, 7, 8. Range, 0–500 feet or higher. A. April–October.

Cultivated fields, roadsides, and waste ground. Frequent. First record: *Linton, B.R.C. Rep.,* 1874, p. 84.

Lamium maculatum, L.—Spotted Dead Nettle. Alien. First record: *Pearson,* in *Ashfield's Fl. Prest.,* 1858.

N. 1. Near Melling, *Wi.*
W. 4. Rawcliffe, *Pearson,* 1858. Knott End, well established, *Wh.*
E. 6. Near a farm between Botton and Wray, *Wi.*
    8. Near Ribchester Bridge, *Ashfield.* Alston, *Wh.* Thoroughly established on the road between Preston and Longridge, *Lord de Tabley* (Flora of Cheshire, 1899). Seen recently, *A.A.D.*

Lamium lævigatum, E.B.—Alien. First record: The present one.

W. 5. Along the foot of the wall on Lytham Sands, with other aliens. 1900 *Wh.*

Lamium album, L.—White Dead Nettle. British type. Native. Area, 1, 2, 3, 4, 5, 7. Range, 0–200 feet. P. April–October.

Hedgebanks and roadsides. Not very frequent; rare in East. First record: *Linton, B.R.C. Rep.,* 1874, p. 84.

N. 1. Cantsfield and Wennington, *Wi.*
    2. Borwick and Priest Hutton, *Wi.*
    3. Warton, *Wi.*
W. 4. Glasson, *Wh.* Galgate, *Wi.*
    5. Woodplumpton and St. Michael's, *Wi.* Kirkham, *Wh.* Ashton, *A.A.D.*
E. 7. Scorton and Garstang, *Wi.*

Lamium Galeobdolon, Crantz.—Weasel-snout. English type. Native. Area, 4, 8. Range, 0–250 feet. P. May–July.

Woods and shady places. Rare. First record: *Ashfield, Fl. Prest.,* 1858.

W. 4. Tarnacre. *P.J.H.*
E. 8. Mellings Wood and Tunbrook Wood, *Ashfield,* 1858, and in many of the woods between Preston and Ribchester, *Ashfield.* Still occurs! Sale Wheel, *Fl. St.* Tunbrook Wood, *A.A.D.*

Ballota nigra, L. — Black Horehound. Alien? First record: *Fl. Prest. N.,* 1903, p. 34-5.

N. 2. Hedgebank near a row of houses, Carnforth, *Wi.*
W. 5. Near Preston Docks, *Fl. Prest. N.*

Teucrium Scorodonia, L.—Wood Sage. British type. Native. Area, 1, 2, 3, 4, 5, 6, 7, 8. Range, 0–920 feet. P. July–September.

Dry woods, hedgebanks, and heathy ground. Common, except in West. First record: *Ashfield, Fl. Prest.,* 1858. "Sion Hill."

Plantago lanceolata, L.—Ribwort. "Ribbed Grass." "Lambs-tongue." British type. Native. Area, 1, 2, 3, 4, 5, 6, 7, 8. Range, 0–1,100 feet or higher. P. June–September.

Fields, waysides, and waste places. Very common. First record: *Linton, B.R.C. Rep.,* 1874, p. 84.

Abnormal forms bearing two or more spikes on one stalk, or having leaves at the base of the spikes occasionally occur.

Plantago maritima, L.—Sea Plantain. British type. Native. Area, 3, 4, 5, 6. Range, 0–50 feet. P. June–September.

Sea-coast, in muddy salt marshes, and river estuaries. Common. First record: *Ashfield, Fl. Prest.,* 1858. " Lea Marsh."

Plantago Coronopus, L.—Buck's Horn Plantain. British type. Native. Area, 3, 4, 5. Range, 0–100 feet. A. or B. July–August.

Sea-shores, and dry sandy or clayey banks near the sea. Frequent, but less so than the preceding species. First record: *Ashfield, Fl. Prest.,* 1858.

N. 3. Near Overton, *Wi.* Thurshouse Sands, *Wh.*
W. 4. Coast near Knott End, *Wi.* Glasson, *Wh.*
    5. Ashton Marsh, *Ashfield,* 1858. Blackpool and Fleetwood, *Wh.*

Var. *ceratophyllum,* Rap.

W. 5. Abundant on clay banks between Blackpool and Little Bispham, 1896, always on the marly soil of the drift cliffs, never on the sand, *Wh.*

Littorella uniflora, Asch.—Shore-weed. British type. Native. Area, 4, 5. Range, only seen at 70 feet. P. July.

Margins of pools and slow streams. Rare. First record: *Wilson, Journ. Bot.,* 1900, p. 45.

W. 4. and 5. Margin of canal near Garstang, 1891, *Wi.*

ILLECEBRACEÆ.

Scleranthus annuus, L.—Knapwell. British type. Native? Area, 7. Range, only seen at 130 feet. A. July–September.

Sandy fields. Very rare. First record: The present one.

E. 7. Field near Sandholme, Garstang. Not seen recently, *Wi.*

We know of only two localities for this plant in West, viz.: glaciated rocks south of Galgate, and between Ansdell and Wrea Green, *Wi.* In North it ascends to 880 feet on Dalton Crag; and in East to 920 feet in Foxdale.

Teucrium Chamædrys, L. — Wall Germander. Alien. First record: The present one.

In a field near Silverdale, which was for sale for building ground. *W.K.* in letter to *L.P.*

Sideritis montana, L.—Alien. First record: The present one.

W. 5. A few examples on the sandhills near St. Annes, 1906, *C.B.* (sp.)

Ajuga reptans, L.—Common Bugle. British type. Native. Area, 1, 2, 3, 4, 5, 6, 7, 8. Range, 0–800 feet or higher. P. May–June.

Damp woods and grassy places. Very common. First record: *Linton, B.R.C. Rep.,* 1874, p. 84.

It ascends to 800 feet on Leck Fell.

PLANTAGINACEÆ.

Plantago major, L. — Great Plantain. "Rat-tail Dock." Fruiting spike = "Rat-tails." British type. Native. Area, 1, 2, 3, 4, 5, 6, 7, 8. Range, 0–1,100 feet. P. June–August.

First record: *Linton, B.R.C. Rep.,* 1874. p. 84.

It ascends to 1,100 feet on Leck Fell, and to 860 feet in Hindburndale.

Var. *intermedia* (Gilib.)—Frequent in dry places.

W. 4. Canal towing path between Glasson and Galgate, *Wh.*
    5. Between Lytham and St. Annes (*Herb. E. S. Marshall*). Fleetwood, *Wh.* Near Uncle Tom's Cabin, Blackpool, *Wh.*
E. 8. Roadside near Thornley, 1884, *Lewis.*

Plantago media, L.—Hoary Plantain. English type. Native. Area, 1, 2, 3, 5, 7, 8. Range, 0–500 feet or higher. P. June–September.

Dry pastures and grassy banks. Common on limestone, rare elsewhere. First record: *Watson, Top. Bot.,* 1883.

N. Common throughout the division.
W. 5. Ashton, *Fl. Prest. N.*
E. 7. Canal bank, Garstang, *Wi.*
    8. Hacking Boat, *Fl. St.* Pastures by the Ribble above Preston, *Wi.* Near Grimsargh, *A.A.D.*

AMARANTHACEÆ.

Amaranthus blitum, L.—Wild Amaranth. Alien. First record: *Wheldon, Journ. Bot.,* 1901, p. 25.

W. 5. On ballast, Preston Dockyard, growing with *Ambrosia trifida,* 1900, *Wh.*

CHENOPODIACEÆ.

Chenopodium album, L. — White Goosefoot. "Fat Hen." British type. Colonist. Area, 1, 2, 3, 4, 5, 6, 7, 8. Range, 0–780 feet. A. July–September.

Cultivated ground and waste places. Very common. First record: *Linton, B.R.C. Rep.,* 1874, p. 84.

All three forms occur, var. *incanum,* Mcq, var. *viride,* Syme, and var. *paganum,* Reich.

Chenopodium polyspermum, L.—Many-seeded Goosefoot. Alien. First record: The present one.

N. 3. Between Heaton and Overton, 1906, *E.C.*

Chenopodium leptophyllum, Nutt.—Alien. First record: The present one.

W. 5. Waste ground, formerly sandhills, near St. Annes, 1906, *C.B.* (sp. !)
A native of lake and sea shores in North America.

Chenopodium anthelminticum, L.—Alien. First record: The present one.

W. 5. With the preceding at St. Annes, 1906, *C.B.* (sp. !)
This and other aliens near St. Annes occurred when the ground was disturbed for building purposes. The seed was probably introduced with fowl corn, as portions of the sandhills are occasionally railed off as poultry enclosures, *C.B.*

Chenopodium opulifolium, Schrad. — Alien. First record: *Wheldon, Journ. Bot.,* 1900, p. 45.

W. 5. Ballast heap near Preston, 1898, and on an embankment on the marsh near Fleetwood Docks, *Wh.* St. Annes, 1906, *C.B.* (sp. !)

Chenopodium serotinum, L. (*C. ficifolium,* Sm.)—Alien. First record: *Wheldon, Journ. Bot.,* 1901, p. 25.

W. 5. With the next on waste ground, Fleetwood Docks, 1900, *Wh.* On ground, formerly sandhills, being turned up for building, St. Annes, 1906, *C.B.*

Chenopodium murale, L.—Alien. First record: *Syme, Top. Bot.,* 1883 (locality not known).

W. 5. Wyre Docks, Fleetwood, 1900, *Wh.*

Chenopodium urbicum, L. — Alien. First record : *Wheldon, Journ. Bot.,* 1900, p. 45.

    W. 3. Casually on a newly-made road near Morecambe, *Wh.*

       4. On a manure heap near Preesall, 1899, *Wh.*

Chenopodium rubrum, L. — Red Goosefoot. English type. Native. Area, 4, 5, 8. Range, 0–100 feet. A. July–September.

River estuaries, and waste ground near the sea. First record : *Wheldon, Journ. Bot.,* 1900, p. 45.

    W. 4. Embankment by the roadside between Preesall and Knott End, *Wh.* Glasson Dock, two plants, *Wh.*

       5. River bank below Preston, 1899, *Wh.* Between Lytham and Guide House Inn, *S. & T.* Lytham, *Fl. Prest. N.* Wyre Dock, *Wh.*

    E. 8. Ribble bank above Preston, *Wh.*

Chenopodium Bonus-Henricus, L. — Good King Henry. British type. Denizen. Area, 1, 2, 3, 4, 5, 6, 7, 8. Range, 0–700 feet. P. June–August,

Roadsides and waste places, but generally in the neighbourhood of houses. Not unfrequent. First record : *Linton, B.R.C. Rep.,* 1874, p. 84.

Beta vulgaris, L.—Beet. Alien. First record : *Wilson, Journ. Bot.,* 1902.

    N. 3. Casual on railbanks near Silverdale, *Wi.*

Beta maritima, L.—Sea Beet. British type. Native. Area, 5. Range, 0–25 feet. P. July–September.

Waste ground near the seashore. Very rare. First record : *Flora of Preston and Neighbourhood,* 1903.

    W. 5. Ribble banks near Lytham, 1901, *A.A.D.* (sp.).

Atriplex littoralis, L. — Grass-leaved Sea Orache. British type. Native. Area, 4, 5. Range, 0–25 feet. A. July–September.

Muddy seashores and salt marshes. Not uncommon. First record : *Ashfield, Fl. Prest.,* 1860.

    W. 4. Rawcliffe and Hambleton Marshes, *Pearson.* Pilling and Knott End, *Wi.* Preesall Marsh, *Wh.*

       5. Between Stena and Fleetwood, *Ashfield,* 1860. Abundant on Fleetwood Marsh, *Wh.*

Atriplex patula, L.—Narrow-leaved Orache. British type. Native. Area, 1, 2, 3, 4, 5, 6, 7. 8. Range, 0–600 feet or higher. A. July–October.

Cultivated and waste ground. Common. First record : *Linton, B.R.C. Rep.,* 1874, p. 84.

Atriplex erecta, Huds.—Halberd-leaved Orache. British type. Native. Area, 1, 2, 3, 4, 5, 6, 7, 8. Range, 0–550 feet. A. July–October.

In similar situations to the preceding. Common. First record : *Lewis, B.R.C. Rep.,* 1884, p. 19.

    It was observed at 550 feet in a patch of cultivated land in Lower Hindburn.

Atriplex hastata, L. — Spear-leaved Orache. British type. Native. Area, 1, 2, 3, 4. 5, 6, 8. Range, 0–120 feet. A. July–August.

Cultivated ground, about manure heaps, and waste places near the sea. Frequent, but less so than either of the preceding. First record : *Linton, B.R.C. Rep.,* 1874, p. 84.

Atriplex deltoidea, Bab.—Triangular-leaved Orache. British type. Native. Area, 3, 4, 5. Range, 0–50 feet. A. July–October.

Rich waste ground near the sea. Rare. First record : *Wheldon, Journ. Bot.,* 1900, p. 45.

    W. 5. Norbreck near Blackpool, and near Preston Docks, *Wh.*

Var. *prostrata,* Bab.—Seashores and river estuaries.

    N. 3. Morecambe and Heysham, *Wh.*

       4. Wyre banks near Preesall, *Wh.*

       5. Fleetwood and Preston, *Wh.*

Atriplex Babingtonii, Woods.—Babington's Orache. British type. Native. Area, 3, 4, 5. Range, 0–50 feet. A. July–October.

Sea-coast. Frequent. First record : *Wheldon, Journ. Bot.,* 1900.

    N. 3. Near Morecambe, *Wh.* Silverdale Cove, *L.P.* Sparingly on shingle near Hest Bank, *Wh.* Overton, *Wi.*

    W. 4. Frequent about Pilling, *Wi.* Preesall Marsh, *Wh.*

       5. Little Bispham, *Wh.*

Var. *virescens,* Lange.

    W. 4. Wyre Mouth near Knott End, and shore near Glasson, *Wh.*

Atriplex laciniata, L.—Frosted Orache. British type. Native. Area, 3, 4, 5. Range, 0–50 feet. A. July–September.

Sandy or shingly shores. On several parts of the coast. First record : *Ashfield, Fl. Prest.,* 1858.

    N. 3. Silverdale, *Wi.*

    W. 4. Knott End, *Ashfield,* 1858. Still there and at Pilling, *Wi.*

       5. Between Naze Point and Lytham, *Ashfield.* Ansdell ! *S. & T.* Fleetwood, and near Little Bispham, and South Shore, Blackpool, *Wh.*

Obione portulacoides, Moq. — Sea Purslane. English type. Native. Area, 3, 4, 5. Range, 0–25 feet. P. August–October.

Seashores and muddy salt marshes. Not common, but locally abundant. First record : *Ashfield, Fl. Prest.,* 1860.

    N. 3. Bolton-le-Sands, and between Far Naze and Sunderland Point, *Wi.* Salt Marsh near the Ferry, Basil Point, *Wi.*

    W. 4. Knott End, *Wi.* Wyre banks and Salt Marsh near Preesall, *Wh.*

       5. Between Skippool and the sea, *Ashfield,* 1860. Still abundant, especially on the marsh near Fleetwood, *Wh.* Between Lytham and the Guide House Inn ! *S. & T.*

Salicornia europæa, L.—Glasswort. British type. Native. Area, 3, 4, 5. Range, 0–10 feet. A. August–September.

Muddy seashores and salt marshes. Common. First record : *Ashfield, Fl. Prest.,* 1858. " Ashton Marsh."

    Called " Samphire " locally and used as a pickle. In some parts the beds are carefully preserved and protected by railings.

Suæda maritima, Dum. — Sea Blite. British type. Native. Area, 3, 4, 5. Range, 0–15 feet. A. July–September.

Muddy salt marshes, and shores of tidal rivers. Rather common. First record : *Ashfield, Fl. Prest.,* 1858. " Ashton Marsh and thence to Lytham."

Var. *procumbens,* Syme.

    W. 4. Salt Marsh near Glasson, *Wh.*

Salsola Kali, L.—Saltwort. British type. Native. Area, 3, 4, 5. Range, 0–25 feet. A. July–October.

Seashores on sand or shingle. Not common. First record : *Ashfield, Fl. Prest.,* 1858.

    N. 3. Between Far Naze and Sunderland Point, *Wi.*

    W. 4. Near Pilling, *Wi.* , Glasson, *Wh.*

       5. Lytham ! *Ashfield,* 1858. St. Annes, *Wh.* Between Blackpool and Fleetwood, *Wi.* Fairhaven, *A.A.D.*

POLYGONACEÆ.

Polygonum Convolvulus, L.—Black Bindweed. British type. Colonist. Area, 1, 2, 3, 4, 5, 6, 7, 8. Range, 0–600 feet. A. July–September.

Waste ground, cornfields, etc. Frequent, except in East. First record : *Linton, B.R.C. Rep.,* 1874, p. 84.

Polygonum aviculare, L.—Knot-grass. British type. Native. Area, 1, 2, 3, 4, 5, 6, 7, 8. Range, 0–850 feet or higher. A. May–October.

Fields, roadsides, and waste ground. A very common weed. First record : *Linton, B.R.C. Rep.,* 1874, p. 84.

    The varieties *agrestinum,* Jord., *vulgatum,* Syme, and *rurivagum,* Jord., are all frequent. A specimen from railbanks, Longridge, 1901 (*Wh.*), was referred somewhat doubtfully by Mr. J. G. Baker to var. *microspermum,* Jord. The var. *littorale,* Link, is rare, and occurs on the sea-coast near Morecambe and Fleetwood (*Wh.*). It is sometimes mistaken for the next species, and is probably specifically distinct.

Polygonum Roberti, Lois.—Sea Knot-grass. British type. Native. Area, 3, 4, 5 Range, 0–25 feet. A. July–September.

Sandy or shingly shores. Rare. First record : *Linton, B.R.C. Rep.,* 1874, p. 84.

    N. 3. Near Heysham Harbour.

    W. 4. Pilling and Knott End, *Wi.*

       5. Fleetwood, *Wh.* Ansdell ! *S. & T.* Near Blackpool, *Wi.* Lytham, *Fl. Prest. N.*

    Erysiphe communis (Wallr.), occurs on its leaves occasionally, as well as on those of the preceding species.

Polygonum Hydropiper, L.—Water Pepper. British type. Native. Area, 1, 2, 3, 4, 5, 6, 7, 8. Range, 0–900 feet. A. August–September.

Ditches and watery places. Common. First record : *Linton, B.R.C. Rep.,* 1874, p. 84.

    It ascends to 900 feet on Whitmoor.

Polygonum Persicaria, L.—Persicary. " Red-knees," in the Fylde district—from the swollen red nodes of the stem. British type. Native. Area, 1, 2, 3, 4, 5, 6, 7, 8. Range, 0–900 feet. A. July–September.

Cultivated and waste ground. Very common. First record : *Linton, B.R.C. Rep.,* 1874, p. 84.

    It ascends to 900 feet in Roeburndale.

Polygonum lapathifolium, L.—Pale Persicary. British type. Native. Area, 3, 4, 5, 6, 7, 8. Range, o–500 feet. A. July–September.

Cultivated fields, waste ground, and damp places. Common. First record: *Linton, B.R.C. Rep.,* 1874, p. 84.

No records for districts 1 and 2, but certain to occur.

Polygonum amphibium, L.—Water Persicary. British type. Native. Area, 1, 2, 3, 4, 5, 6, 7, 8. Range, o–350 feet. P. July–September.

Ponds, canals, and ditches. Common, especially in the Preston and Kendal Canal. First record: *Linton, B.R.C. Rep.,* 1874, p. 84.

The var. *terrestre,* Leers, is less frequent than the type. It occurs near Little Bispham and Preston, *Wh.*

Polygonum Bistorta, L.—Snake-weed. Bistort. "Patience Dock." British type. Native. Area, 1, 2, 3, 4, 5, 6, 7, 8. Range, o–500 feet. P. June–July.

Damp meadows. Frequent. First record: *Linton, B.R.C. Rep.,* 1874, p. 84.

Polygonum Fagopyrum, L.—Buck-wheat. Alien. First record: *Wilson, B.R.C. Rep.,* 1883, p. 248.

E. 7. Occurs in cornfields near Garstang, year after year, *Wi.*

Rumex conglomeratus, Murr.—Sharp Dock. British type. Native. Area, 1, 2, 3, 4, 5, 6, 7, 8. Range, o–400 feet or higher. P. July–August.

Waste ground, roadsides, and damp places. Common. First record: *Linton, B.R.C. Rep.,* 1874, p. 84.

Rumex sanguineus, L.—*var. viridis,* Sibth.—British type. Native. Area, 1, 2, 3, 4, 5, 6, 7, 8. Range, o–600 feet. P. July–August.

Woods and damp shady places. Frequent. First record: *Lees, B.R.C. Rep.,* 1882, p. 137. "Wennington."

We have never seen the typical red-veined plant in West Lancashire.

Rumex maritimus, L.—Golden Dock. English type. Native. Area, 5. Range, o–25 feet. P. or B. July–August.

Marshes near the sea. Rare. First record: *Ashfield, Fl. Prest.,* 1860.

W. 5. Between Naze Point and Lytham, *Ashfield,* 1860. It occurs also on the opposite bank of the Ribble in South Lancashire.

Rumex obtusifolius, L.—Broad-leaved Dock. British type. Native. Area, 1, 2, 3. 4, 5, 5, 7. 8. Range, o–900 feet. P. July–September.

Roadsides, cultivated and waste ground. Very comn on. First record: *Linton, B.R.C. Rep.,* 1874, p. 84.

It ascends to 900 feet in Hindburndale.

Rumex crispus, L. — Curled Dock. British type. Native. Area, 1, 2, 3, 4, 5, 6, 7, 8. Range, o–600 feet. P. July–September.

Roadsides, fields, and waste places. Very common. First record: *Linton, B.R.C. Rep.,* 1874, p. 84.

Var. *trigranulatus,* Syme.—Waste ground about estuaries and docks. A doubtful native.

N. 3. Near Heysham Docks.
W. 4. Knott End, *S. & T.*
5. Abundant on Fleetwood Marsh, and more sparingly on waste land near Preston Docks, *Wh.*

Rumex acutus, L.—Meadow Dock. English type. Native. Area, 8. Range, only seen at about 450 feet. P. June–August.

Ditches and marshy places. Rare. First record: *Wheldon, Journ. Bot.,* 1900, p. 45.

E. 8. Near Knowl Green, 1899, *Wh.* Said to be a hybrid between *R. crispus* and *R. obtusifolius,* but it produces seed freely.

Rumex aquaticus, L.—Grainless Dock. Scottish type. Native. Area, 2, 8. Range, 150–350 feet. P. July–August.

Damp places, ditches, and riverbanks. Rare. First record: *Fl. St.,* 1891, p. 32.

N. 2. Bank of the Lune below Kirkby Lonsdale.
E. 8. Crowshaw Reservoir, *Fl. St.,* 1891. Still there, 1905!

Rumex Hydrolapathum, Huds.—Great Water Dock. English type. Native. Area, 3, 4. Range, o–25 feet. P. July–August.

S

Ditches and marshes in the low country. Rare. First record: *Wilson, B.R.C. Rep.,* 1884, p. 19.

N. 3. Abundant on Storrs Moss, *Wi.*
W. 4. Ditches between Cockerham and Preesall, 1883, *Wi.* In a pit near the proposed Knott End Railway terminus, *Wh.*

Rumex Acetosa, L.—Sorrel. "Sour Dock." British type. Native. Area, 1, 2, 3, 4, 5, 6, 7, 8. Range, o–1.930 feet. P. June–July.

Meadows and damp grassy places. Very common. First record: *Top. Bot.,* 1883.

The rusty purple fungus, *Ustilago Kuchniana,* Wolff, occurs not uncommonly upon this species.

Rumex Acetosella, L.—Sheep's Sorrel. British type. Native. Area, 1, 2, 3, 4, 5, 6, 7, 8. Range, o–2,040 feet. P. June–August.

Dry pastures, banks, and heathy places. Common. First record: *Top. Bot.,* 1883.

ARISTOLOCHIACEÆ.

Asarum europæum, L. — Asarabacca. Alien. First record: *Turner and Dillwyn, Botanist's Guide.*

W. 5. Several woods in Lancashire ; near Preston, *Botanist's Guide.* "Mr. Pearson says he has been informed that this plant is found at Barton," *Ashfield,* 1860. Ribby Hall Wood near Kirkham, *A.A.D.*

Aristolochia Clematitis, L.—Motherwort. Alien. First record: *Lucas,* in *Ashfield's Fl. Prest.,* 1860.

W. 5. Field at Warton on the road to Lytham, *Mr. Lucas.*

THYMELACEÆ.

Daphne Laureola, L.—Spurge Laurel. English type. Native. Area, 3. Range, 150–300 feet. Sh. March–April.

Woods and bushy places on limestone. Very rare. First record: *Petty, The Naturalist,* 1902.

N. 3. Near Silverdale, 1897, *L.P.* Seen also by one of the Authors in two localities near Silverdale, in one of which it occurs in fair quantity.

EUPHORBIACEÆ.

Euphorbia Helioscopia, L. — Wart-weed. Sun Spurge. British type. Colonist. Area 1, 2, 3, 4, 5, 6, 7, 8. Range, o–550 feet. A. July–September.

Cultivated land. Common. First record: *Linton, B.R.C. Rep.,* 1874, p. 84.

Ascends in East to 550 feet in Lower Hindburndale.

Euphorbia Cyparissias, L. — Cypress Spurge. Alien. First record: *Melvill, B.E.C. Rep.,* 1897.

Abundant in a grassy field near Lytham Vicarage, *J. C. Melvill.* Still abundant, for many yards the predominant plant, and with it *Brassica monensis* and *Diplotaxis muralis, Wh.*

Euphorbia Paralias, L.—Sea Spurge. Atlantic type. Native. Area, 5. Range, o–30 feet. P. August–November.

Sand-dunes. Rare. First record: *Ashfield, Fl. Prest.,* 1858.

W. 5. Between Lytham and Blackpool, *Ashfield,* 1858. Still there! It usually grows in loose blown sand associated with *Psamma, Cynoglossum,* and *Senecio Jacobæa.* In South Lancashire, *E. portlandica* is often a member of the group, but it is very rare with us.

Euphorbia portlandica, L.—Portland Spurge. Atlantic type. Native. Area, 5. Range, o–30 feet. P. June–August.

Sand-dunes. Very rare. First record: *Flora of Preston and Neighbourhood,* 1903, p. 38.

W. 5. St. Annes to Blackpool, *Fl. Prest. F.*

Euphorbia exigua, L. — Dwarf Spurge. English type. Colonist. Area, 2, 3, 4, 6. Range, o–200 feet. A. July–September.

Cultivated land, especially in cornfields. Rather rare. First record: *Linton, B.R.C. Rep.,* 1874, p. 84.

N. 2. Between Nether Kellet and Bolton-le-Sands.
3. Cornfield near Silverdale, *Wi.* Bottoms Lane, *L.P.* Shore shingle, Bolton-le-Sands, *Wi.*
W. 4. Cabus, near Garstang.
E. 6. Near Galgate, *Wi.*

Euphorbia Peplus, L. — Petty Spurge. Garden Spurge. "Chicken-weed." British type. Colonist. Area, 1, 2, 3, 4, 5, 6, 7, 8. Range, o–250 feet. A. July–September.

Cultivated fields and gardens. Common. First record: *Linton, B.R.C. Rep.,* 1874, p. 84.

Euphorbia Lathyris, L.—Caper Spurge. Alien. First record: The present one.

> W. 4. Near St. Michael's, *P.J.H.*
> 5. Near Preston, *H.B.*

Mercurialis annua, L.—Annual Dog's Mercury. Alien. First record: *Salmon and Thompson, Journ. Bot.,* 1902, p. 296.

> W. 4. Near Knott End, 1902, *S. & T.*

Mercurialis perennis, L.—Perennial Dog's Mercury. British type. Native. Area, 1, 2, 3, 4, 5, 6, 7, 8. Range, 0–1,160 feet. P. March–May.

Woods, hedgebanks, and shady places. Common. First record: *Linton, B.R.C. Rep.,* 1874, p. 84.

> It ascends to 1,160 feet in the shelter of the pot-holes on Leck Fell.
> A very curious autumn-flowering form of this plant was found by Mr. F. J. George, of Chorley, under a hedge at the foot of Frenchwood Hill, Preston, but he tells us that through cutting down the hedge when enlarging the park, the plant is now extinct. Specimens were sent by Mr. George to the herbarium at Kew.

URTICACEÆ.

Ulmus glabra, Huds.— Wych Elm. British type. Native. Area, 1, 2, 3, 4, 5, 6, 7, 8. Range, 0–1,060 feet. T. March–April.

Woods and hedges. Common. First record: *Top. Bot.,* 1883.

> It ascends to 1,060 feet in the main gully of Mallowdale Fell, Roeburndale.

Ulmus campestris, L.—Common Elm. English type. Denizen. Area, 1, 2, 4. Range, 0–300 feet. T. March–April.

Hedges and parks. Not very common, and not often planted; frequent and very fine near Over Burrow. First record: *Linton, B.R.C. Rep.,* 1874, p. 84.

Humulus Lupulus, L.—Hop. English type. Denizen. Area, 1, 2, 3, 4, 5, 6, 7, 8. Range, 0–500 feet. P. July–August.

In hedges, but always near houses. Not uncommon. First record: *Ashfield, Fl. Prest.,* 1858. "Between Broughton and Myerscough."

Urtica dioica, L.—Nettle. British type. Native. Area, 1, 2, 3, 4, 5, 6, 7, 8. Range, 0–2,030 feet. P. June–September.

Roadsides, waste places, and woods. Very common. First record: *Linton, B.R.C. Rep.,* 1874, p. 84.

Var. *angustifolia,* Blytt. Shady woods. Rare.

> N. 2. Wood by the Lune between Caton and Aughton.
> E. 7. Damas Gill near Dolphinholme, and wood near Abbeystead, *Wh.*
> 8. Knowl Green, 1899, *Wh.*

Urtica urens, L.—Small Nettle. British type. Native. Area, 1, 2, 3, 4, 5, 6, 7, 8. Range, 0–300 feet. A. June–September.

Waste ground and roadsides, generally near houses. Frequent, but' much less common than the preceding. First record: *Linton, B.R.C. Rep.,* 1874, p. 84.

Parietaria ramiflora, Moench.—Wall Pellitory. British type. Native. Area, 1, 2, 3, 4. Range, 0–250 feet. P. July–September. Old walls and rocks. Rare. First record: *Jenkinson, Brit. Pl.,* 1775, p. 238.

> N. 1. Near Hornby on the Leck side of the Wenning, *Wi.*
> 3. Cliffs about Silverdale, *Ashfield.* Still there, and in the Heysham Peninsula at Middleton, *Wi.*
> W. 4. On the walls of Lancaster Castle, *Jenkinson,* 1775. Walls at Wardless, *Ashfield.* Cockerham Churchyard, *E.C.*

MYRICACEÆ.

Myrica Gale, L.—Sweet Gale. "Fleffwood." British type. Native. Area, 2, 3, 4, 7. Range, 0–690 feet. Sh. May.

Peat bogs and swampy places. Becoming rare. First record: *Ashfield, Fl. Prest.,* 1858.

> N. 2. Arkholme Moor, and swamps near Over Kellet and Nether Kellet, *Wi.* Lords Lot Wood.
> 3. Haweswater Tarn and Thrang Moss, *Wi.* Heysham Moss.
> W. 4. Pilling Moss, *Ashfield,* 1858. Abundant on Cockerham Moss, *Wi.*
> E. 7. Bank of the Wyre near Tarnbrook, and on the south slope of Wardstone above the village, at 690 feet.

CUPULIFERÆ.

Betula verrucosa, Ehrh.—White Birch. British type. Native. Area, 1, 2, 3, 6, 7, 8. Range, 0–400 feet. T. May.

Woods and heaths. Frequent amongst the hills. First record: *Wilson, Journ. Bot.,* 1900, p. 46. "Middlebarrow Wood."

Betula alba, L. (*B. glutinosa,* Fries.).—Common Birch. British type. Native. Area, 1, 2, 3, 4, 5, 6. Range, 0–1,090 feet. Sh. or T. May.

Woods, heaths, and damp moors. More common than the preceding. First record: (for the aggregate species), *Linton,* 1874 (for the segregate), *Wilson, B.R.C. Rep.,* 1887, p. 105. "Bleasdale."

> It ascends in North to 1,080 feet in Navel Pot, Leck Fell; and in East to 1,090 feet on the south fork of Foxdale Beck.
> The var. *pubescens,* Ehrh. is the commoner form.

Alnus glutinosa, Gaert. — Alder. "Owler." British type. Native. Area, 1, 2, 3, 4, 5, 6, 7, 8. Range, 0–1,050 feet. T. March–April.

Damp woods, marshes, and streamsides. Common, especially in East and West. First record: *Linton, B.R.C. Rep.,* 1874.

> It ascends to 1,050 feet in the main gully of Mallowdale Fell, Roeburndale.

Carpinus Betulus, L.—Hornbeam. Alien.

Occasionally planted in woods and hedgerows, but nowhere indigenous.

Corylus Avellana, L.—Hazel. Nut-bush. "Lamb's tails" or "Lady's fingers"—the catkins. British type. Native. Area, 1, 2, 3, 4, 5, 6, 7, 8. Range, 0–1,160 feet. T. February–April.

Woods, thickets, and hedgerows. Very common, especially on the limestone of North. First record: *Linton, B.R.C. Rep.,* 1874, p. 84.

> It ascends in North to 1,160 feet on Leck Fell.

Quercus Robur, L.—Oak. British type. Native. Area, 1, 2, 3, 4, 5, 6, 7, 8. Range, 0–1,000 feet. T. May–June.

Woods and rocky places amongst the hills, and occasionally planted in hedgerows. Common. First record: *Linton, B.R.C. Rep.,* 1874, p. 84.

Var. *pedunculata,* Ehrh.—Area, 1, 2, 3, 4, 5, 6, 7, 8.

> Much the most frequent of the three forms.

Var. *intermedia,* Don.—Area, 8. Rare.

> E. 8. Near Hurst Green.

Var. *sessiliflora,* Salisb.—Area, 1, 2, 3, 6, 7, 8.

> N. 1. Near Nether Burrow.
> 2. Near Halton, *Wh.*
> 3. Warton Crag, *Wh.*
> E. 6. Near Wray, 1899.
> 7. Garstang and Lower Bleasdale.
> 8. Hodder Valley, *Wi.*

Fagus sylvatica, L.—Beech. English type. Denizen. Area, 1, 2, 3, 4, 5, 6, 7, 8. Range, 0–600 feet. T. May.

Woods, parks, and hedgerows. Common in the low country, but does not occur in the aboriginal woods of the hilly districts. First record: *Linton, B.R.C. Rep.,* 1874, p. 84.

AMENTIFERÆ.

Salix pentandra, L. — Sweet Bay Willow. Scottish type. Native. Area, 1, 2, 3, 5, 6, 7, 8. Range, 50–780 feet. T. June.

Frequent in some parts of North and East; very rare in West. First record: *Lewis, B.R.C. Rep.,* 1884, p. 20.

> N. 1. Near Leck, *Wi.* Ease Gill.
> 2. Frequent, *Wi.*
> 3. Near Silverdale, and in a swampy field east of Yealand, *Wi.*
> W. 5. Near Inskip, *Wi.*
> E. 6. Lower Hindburn, Quernmore, and Clougha, *Wi.*
> 7. Bleasdale, *Wi.*
> 8. Between Alston and Thornley, 1882, *Lewis!* Frequent in this district.

Salix decipiens, Hoffm.—White Welsh or Varnished Willow. British type. Area, 7. Range, only seen at 440 feet. T. May.

Damp thickets by streams. Rare. First record: The present one.

> E. 7. Bank of the Tarnbrook Wyre near Lower Emmetts, July, 1906.
> Of our example, Mr. E. F. Linton states that it is the usual form of the Midland Counties, which differs slightly from the south country form. He thinks it is far from certain that this is of the hybrid origin usually attributed to it, *S. triandra × fragilis. S. fragilis* grows in the immediate vicinity of our plant, but we have not observed *S. triandra* there, nor in the vice-county, which certainly tends to support Mr. Linton's contention.

Salix fragilis, L. — Crack Willow. British type. Native probably. Area, 1, 2, 3, 4, 5, 6, 7, 8. Range, 0–470 feet. T. April–May.

River banks and marshes. Frequent, but in many of its stations doubtfully indigenous. First record: *Wilson, B.R.C. Rep.*, 1887, p. 105. "Halton."

Willows generally are known by the country folk as "Wythins" in West Lancashire.

Salix alba, L. — White Willow. British type. Native, or perhaps only denizen. Area, 1, 2, 3, 4, 5, 6, 7, 8. Range, 0–440 feet. T. May.

Moist woods and hedgerows. Frequent, but often planted. First record: *Wilson, B.R.C. Rep.*, 1887, p. 105. "Near Garstang."

Salix cinerea, L.—Grey Sallow. British type. Native. Area, 1, 2, 3, 4, 5, 6, 7, 8. Range, 0–1,280 feet. T. April.

Moist woods, thickets, and hedgerows. Common. First record: *Top. Bot.*, 1883.

Ascends in North to 1,070 feet on Leck Fell, and in East to 1,280 feet in the main gully of Mallowdale Fell.

Salix aurita, L.—Round-eared Sallow. British type. Native. Area, 1, 2, 3. 5, 6, 7, 8. Range, 0–1,500 feet. Sh. April–May.

Moist heathy places, especially in upland districts. Common. First record: *Linton, Summary of Comital Distribution*, 1878.

It ascends to 1,500 feet in Dale Gill, Botton Head Fell.

Salix lutescens, A. Kern.=*S. aurita × cinerea.*

 E. 8. Wood on the banks of the Wyre near Emmetts with both parents, and on Lea Fell in the same vicinity.

Salix caprea, L.—Goat Sallow. "Palms," "Geslings"=goslings —the catkins. British type. Area, 1, 2, 3, 4, 5, 6, 7, 8. Range, 0–1,100 feet. T. April.

Woods and hedges in drier situations than either of the preceding. Common. First record: *Linton, Summary of Comital Distribution*, 1878.

Ascends in North to 1,100 feet on Leck Fell, and in East to about 900 feet in Roeburndale. *Melampsora farinosa* (Pers.), occurs as an orange powder on the undersides of the leaves of both this and the two preceding species very frequently.

Salix repens, L. — Dwarf Willow. British type. Native. Area, 1, 2, 3, 5, 6, 7, 8. Range, 0–700 feet. Sh. May.

Damp sandy ground and moist heaths. Frequent. First record: *Wilson, B.R.C. Rep.* 1884, p. 20. "Between Wennington and Burton."

The form known as *S. argentea,* Sm. is abundant on the coast sandhills.

Salix phylicifolia, L. — Tea-leaved Willow. Scottish type. Native. Area, 1, 6, 7, 8. Range, 600–100 feet. Sh. May.

River banks and streamsides in upland districts. Rare. First record: *Fl. St.,* 1891, p. 34.

 N. 1. Bank of the Lune near Tuns all, *Wi.* Lower Ease Gill.
 E. 6. Between Botton and Wray, *Wi.*
  7. Near Marshaw, *Wi.*
  8. Ribble bank above Jumbles; Higher Bridge Island, *Fl. St.*

Salix nigricans, Sm.—Dark broad-leaved Willow. Scottish type. Native. Area, 1, 7, 8. Range, 150–500 feet. Sh. April.

Rocky banks of streams amongst the hills. Rare. First record: *Fl. St.,* 1891, p. 34.

 N. 1. Between Kirkby Lonsdale and Nether Burrow, *Wi.* Lower Ease Gill.
  7. Near Lower Emmetts in Over Wyresdale.
 E. 8. Rocks at the head of Bolton Roughs, 1891, *Fl. St.*

Salix viminalis, L.—Common Osier. British type. Native. Area, 1, 2, 3, 4, 5, 6, 7, 8. Range, 0–470 feet. T. April–May.

Marshy places and river banks. Not very common. First record: *Wilson, B.R.C. Rep.,* 1887, p. 105.

 N. 1. Near Leck and Melling, *Wi.*
  2. Arkholme, *Wi.*
  3. Between Overton and Middleton, *Wh.*
 W. 4, 5. Both banks of the Wyre near Garstang; and near Myerscough, *Wi.*
 E. 6. Near Caton, *Wh.* Near Salter, Roeburndale, at 470 feet.
  7. Near Tarnbrook.
  8. Near Ribchester, *Wh.*

Salix viminalis × S. caprea (S. Smithiana, Willd.).

 N. 1. Side of Leck Beck, *L.P.*
 E. 8. Crowshaw reservoir; Thornley, etc., *Fl. St.*

Salix purpurea, L.—Purple Willow. British type. Native. Area, 1, 2, 3, 4, 5, 6, 7, 8. Range, 0–450 feet. Sh. April–May.

Low meadows bordering rivers and ditches. Not common. First record: *Top. Bot.*, 1883.

 N. 1. Ease Gill, *Wh.* & *Wi.* Near Wennington and Kirkby Lonsdale; and by the Greeta near Cantsfield, *Wi.* Leck Beck near Nether Burrow, and by the Lune near Tunstall, abundant, *Wi.*
  2. Near Arkholme, *Wi.*
 W. 4. Near Churchtown, *Wi.*
  5. Wyre banks near Catterall, *Wi.*
 E. 6. Common by the River Wenning, and by the Lune near Halton, *Wi.*
  7. Wyre banks near Garstang, *Wi.*
  8. River banks above Jumbles, and Higher Bridge Island, *Fl. St.* Ribble banks above Alston, *Wh.*

[*Salix herbacea,* L.—We are informed by Mr. W. West that this occurs in gritstone crevices on the east side of Greygarth Fell just beyond our boundary. It may possibly be found on the Lancashire side!]

Populus alba, L. — White Poplar. Abele. English type. Denizen. Area, 5, 7. Range, 0–75 feet. T. April.

Woods and streamsides in the low country. Not common. First record: The present one?

Populus tremula, L.—Aspen. British type. Native. Area. 1, 2, 3, 5, 6, 7, (8). Range, 0–890 feet. T. April.

Woods and hedges. Scattered thinly over the division; not always wild. First record: *Petty, The Naturalist*, 1902, p. 49.

 N. 1. Near Wennington; and several bushes by the Roman road near Cantsfield, *Wi.*
  2. Near Arkholme, *Wh.* & *Wi.* Near Nether Kellet, *Wi.*
  3. Eaves Wood, and other woods at Silverdale, 1902, *L.P.* Heysham, *Wi.* Near Carnforth. Ascent of Warton Crag from Warton; and in hedge by the canal near Hest Bank, *Wh.*
 W. 5. Lea near Preston, *H.B.* Near Myerscough, *Wi.*
 E. 6. On rocks by the stream in Foxdale at 890 feet.
  7. Near Tarnbrook.
  8. Near Ribchester, planted.

Populus nigra, L.—Black Poplar. Alien.

Woods, hedgerows, and streamsides. Not uncommon, but always planted. First record: *Petty, The Naturalist*, 1893, p. 100. "Banks of Leck Beck."

EMPETRACEÆ.

Empetrum nigrum, L.—Crowberry. Scottish type. Native. Area, 1, 2, 4, 6, 7, 8. Range, 25–1,920 feet. Sh. April–May.

Peaty moors, especially on the millstone grit. Abundant on the fells of East; becoming very rare in West. First record: *Ashfield, Fl. Prest.*, 1858.

 N. 1. Greygarth Fell at 1,920 feet.
  2. Whittington Moor, *Wi.* & *Wh.* Near Docker, *Wi.*
 W. 4. Cockerham Moss, *Wi.*
 E. 6, 7, 8. Longridge and Beacon Fells, *Ashfield*, 1858. Still there, and common on all the moors of this division, ascending to the summit of Wardstone.

CERATOPHYLLACEÆ.

Ceratophyllum demersum, L. — Hornwort. English type. Native. Area, 8. Range, only seen at about 125 feet. P. July–August.

Ponds and slow-flowing streams. Very rare. First record: *Fl. St.,* 1891, p. 35.

 E. 8. In the Hodder at Seed Holme Nook, 1887, *Fl. St.*

CONIFERÆ.

Juniperus communis, L. — Juniper. British type. Native. Area, 2, 3, 6, 7. Range, 0–1,400 feet.

Limestone hills, scars, and pavements of North and rocky heathery cloughs amongst the gritstone fells of East. Common in some parts of North; less frequent in East; not found in West. First record: *Ashfield, Bot. Chron.*, 1864.

 N. 2. Dalton Crag, *Wh.* & *Wi.* Arkholme Moor, *Wi.*
  3. Dry hills and roadsides in the Silverdale district, abundant, *Ashfield*, 1864. Still plentiful there!
 E. 6. Mallowdale, Upper Roeburndale, Hindburn, Udale, and Foxdale, *Wi.* Foot of Black Moor near Cougha, *Wh.*
  7. Tarnbrook Fell, *Joseph Pye.*

Taxus baccata, L.—Yew. English type. Native. Area, 2 3. Range, 0–840 feet. T. April.

Truly indigenous on the limestone scars and in the aboriginal woods of North, where it is abundant. Unknown in East and West, except as an occasionally planted tree. First record: *Ashfield, Bot. Chron.*, 1864.

N. 2. Dalton Crag, ascending to 840 feet.
  3. Woods about Silverdale, very plentiful, *Ashfield*, 1864. Still forms a striking feature in the scenery of this district.

Pinus sylvestris, L.—Scotch Fir. Scottish type. Denizen. Area, 1, 2, 3, 4. 5. 6. 7, 8. Range, 0–1,300 feet. T. May–June.

Woods and hedgerows. Probably all the existing trees have been planted. Frequent.

Larix europæa, L.—Larch. Alien.

Frequently planted, but nowhere wild. On the south end of the White side of Tarnbrook Fell a plantation of larches shown on the ordnance map at 1,290 feet has been almost completely blown down ; several trees are still standing but killed. At 1,150 feet they appear to be uninjured.

CLASS II.—MONOCOTYLEDONES.

HYDROCHARIDEÆ.

Elodea canadensis, Mich.—American Water Weed. Water-thyme. English type. Denizen. Area, 1, 2, 3, 4, 5, 6, 7. Range, 0–150 feet. P. June–September.

Canals, ponds, and ditches. Frequent, and especially abundant in the Preston and Kendal canal. First record : *Ashfield, Fl. Prest.* "Canal north of Preston, abundant, especially near Nateby Hall," 1866.

Hydrocharis Morsus-ranæ, L. — Frog-bit. English type. Native. Area, 4, 8. Range, 0–150 feet. P. July–August.

Ditches in the low country. Rare. First record : *Ashfield, Fl. Prest.,* 1858.

W. 4. Ditches between Cockerham Church and the sea, *Ashfield*, 1860. Still there, *Wi.*
E. 8. West side of Ribbleton Moor, *Ashfield*, 1858. Probably extinct now.

Stratoites Aloides, L.—Water Aloe. Water Soldier. English type. Denizen. Area, 3, 5, (7). Range, 0–160 feet. P. July.

Ponds and ditches. Rare. First record : *Ashfield, Fl. Prest.,* 1858.

N. 3. Bank Well, Silverdale, *J. C. Melvill.* Still there.
W. 5. Pit on Moor Hall Estate, north of Preston, near footpath to Cadeley Mill, *Ashfield*, 1858. Pond north-west of Preston ; near South Shore, Blackpool.
E. 7. In a pond near Garstang (introduced), *Wi.*

ORCHIDACEÆ.

Neottia nidus-avis, Rich.—Bird's Nest Orchis. British type. Native. Area, 8. Range, 100–250 feet. P. June.

Damp woods. Very rare. First record : *Fl., St.,* 1891, p. 35

E. 8. Middleton Woods, Goosnargh, 1870, not seen since, *R. Standen.* Bolton Roughs, *Fl. St.,* 1891. Tunbrook Wood, *H.B.* Red Scar, *Fl. Prest. N.*

Listera cordata, R. Br. — Mountain Tway-blade. Scottish type. Native. Area, 6, 7, 8. Range, 1,100–1,500 feet. P. June.

Peat bogs and wet heathy places on the higher moors of East. Rare. First record : *Wilson, B.R.C. Rep.,* 1884, p. 20.

E. 6. Littledale Fell, Udale.
  7. South-west shoulder of Wardstone, at 1,500 feet, 1883, *Wi.* Moor above middle of Gavell's Clough, Tarnbrook Fell at 1,400 feet.
  8. Leagram Fells, *Fl. St.*

Listera ovata, R. Br.—Tway-blade. British Type. Native. Area, 1, 2, 3, 4, 5, 6, 7, 8. Range, 0–880 feet. P. June.

Woods, banks, and shady places. Frequent. First record : *Ashfield, Fl. Prest.,* 1858. "Fulwood and Tunbrook Wood."

It ascends in North to 880 feet on Dalton Crag.

Spiranthes spiralis, Koch.—Lady's Tresses. English type. Native. Area, 3. Range, 0–400 feet. P. August–September.

Meadows and sandy fields. Very rare. First record : *Jenkinson, Brit. Pl.,* 1775, p. 222.

N. 3. On Yealand Common, but not very plentifully ; above John Jenkinson's Wood, and by the side of Grisedale Wall, Yealand, *Jenkinson,* 1775. Silverdale, frequent, *Ashfield.* On a lawn near Silverdale, *Rev. J. A. Barnes.*

Epipactis Helleborine, Crantz. — Broad-leaved Helleborine. British type. Native. Area, 1, 2, 3, 4, 5, 6, 7, 8. Range, 0–700 feet. P. July–August.

Woods and shady banks. Frequent, especially in North. First record : *Jenkinson, Brit. Pl.,* 1775, p. 224. "In Cringle-barrow Wood, amongst the rocks, plentifully !"

Epipactis atrorubens, Schultes.—Purple Helleborine. Local type. Native. Area, 2, 3. Range, 100–500 feet. P. July.

Limestone rocks and "scree" slopes. Rare. First record : *Wilson, Journ. Bot.,* 1900, p. 46.

N. 3. Warton Crag and Gatebarrow Wood, 1892, and other woods near Silverdale, *Wi.* Seen recently also in this district by *L.P.* and *W.K.*
  2. Wood near Over Kellet, *Wi.*

Epipactis longifolia, All. *(E. palustris,* Crantz.). — Marsh Helleborine. English type. Native. Area, 3, 4, 5, 8. Range, 0–500 feet. P. July.

Marshes amongst the coast sandhills and boggy ground on limestone. Rare, but locally abundant. First record : *Jenkinson, Brit. Pl.,* 1775, p. 225.

N. 3. In a meadow near Yealand, *Jenkinson,* 1775. Near Haweswater, *Wi.*
W. 4. Near Nateby, *E.C.*
  5. Lytham, *Ashfield.* Still abundant, although decreasing yearly, in damp hollows on the sandhills.
E 8. Leagram, *Fl. St.*

Orchis pyramidalis, L. — Pyramidal Orchis. Germanic type. Native. Area, 3, 5. Range ? P. July–August.

Pastures and banks on limestone or sandy soil. Very rare. First record : The present one.

N. 3. { Yealand and Lytham neighbourhoods, *R. E. Leach.* No localities
W. 5. { given, but specimen sent to one of the Authors by Mr. Leach.

Orchis ustulata, L.—Burnt Orchis. Germanic type. Native. Area, 2, 3. Range, ?–400 feet. P. June–July.

Limestone pastures. Rare. First record : *Simpson, Top. Bot.,* 1883.

N. 2. Over Kellet neighbourhood, 1903, *Wi.*
  3. Silverdale, *Miss K. Pickard.*

Orchis morio, L.—Green-winged Orchis. English type. Native. Area, 2, 3, 4, 5, 6, 8. Range, 0–500 feet. P. May–June.

Pastures, meadows, and grassy hills ; usually on limestone or clay. Frequent in some parts of North, local in East and West, but abundant where it occurs. First record : *Ashfield, Fl. Prest.,* 1858.

N. 2. Kit Bill Wood near Over Kellet, *Wi.*
  3. Frequent in the limestone area ; and more sparingly on the sandstone formation near Heysham.

W. 4. Abundant in meadows at Pilling, *Wi.*
  5. Several places in Cottam and Lea, *Ashfield*, 1858. Still there, *H.B.*
E. 6. Meadows near the Crook of Lune, *Wi.*
  8. Thornley Quarry near Chipping, *W. Clitheroe.*

Orchis mascula, L.—Early Purple Orchis. British type. Native. Area, 1, 2, 3, 4, 5, 6, 7, 8. Range, 0–1,100 feet. P. April–June.

Woods, copses, and grassy banks. Common. First record : *Ashfield, Fl. Prest.,* 1858. "Near Salwick Brook, Wood-plumpton."

Ascends to 1,100 feet on Leck Fell.
This and other species of *Orchis* are known as "Crowfoot" in various parts of West Lancashire.

Orchis incarnata, L.—Narrow-leaved Marsh Orchis. British type. Native. Area, 2, 3, 5, 7. Range, 0–250 feet. P. June.

Marshes, and in boggy, sandy ground amongst the coast sand dunes. Rare, but locally abundant. First record : *Wilson, B.R.C. Rep.,* 1884, p. 20.

N. 2. Bogs near Docker, Dunnald Mill Hole, and Nether Kellet.
  3. Margin of Haweswater Tarn, *Wi.*
W. 5. Abundant on the sandhills about St. Annes, 1876, *Wi.* Still occurs but in much reduced numbers.
E. 7. Near Wheatland Hall, between Scorton and Dolphinholme, *Wi.*

Orchis latifolia, L.—Marsh Orchis. British type. Native. Area, 2, 3, 4. Range, 0–250 feet. P. June–July.

Marshes and wet meadows. Rare. First record : For the aggregate species, *Ashfield, Fl. Prest.,* 1858. "Near Lytham," probably the preceding. For *O. eu-latifolia, Wilson, B.R.C. Rep.,* 1883, p. 249. "Margin of Hawswater Tarn."

N. 2. Gressingham Moor, and between Caton and Kellet, *Wi.*
  3. Near Burton, *Wh.* & *Wi.* Near Silverdale and Yealand, *Wi.*
W. 4. Cockerham and Pilling, *Wi.* Near Nateby, *S. Wi.*

Orchis maculata, L.—Spotted Orchis. "Crowfoot." British type. Native. Area, 1, 2, 3, 4, 5, 6, 7, 8. Range, 0–1,100 feet. P. June.

Damp meadows and woods. Common, and especially so in the upland heathy pastures of North and East. First record : *Linton, B.R.C. Rep.,* 1874, p. 85.

Ascends in North to 1,000 feet on Leck Fell, and in East to 920 feet at the Trough of Bowland, and 1,100 feet on Mallowdale Fell. In the more hilly districts it has frequently paler, and sometimes white flowers.

Ophrys muscifera, Huds.—Fly Orchis. English type. Native. Area, 2, 3. Range, 50–400 feet. P. May–June.

Woods and rough grassy or rocky ground on the limestone of North. Rare, and becoming more so. First record: *Jenkinson, Brit. Pl.*, 1775, p. 223.

N. 2. and 3. Recorded from various localities in the districts of Silverdale, Yealand, and Kellet, by Jenkinson, Grindon, Ashfield, Williams, and others. Still occurs, and is fairly plentiful locally, *Wi.*

Gymnadenia conopsea, R. Br.—Fragrant Orchis. British type. Native. Area, 1, 2, 3, 7, 8. Range, 0–600 feet. P. July.

Pastures and grassy banks, principally on a calcareous soil. Common on the limestone of North; rare in East; no records for West. First record: *Jenkinson, Brit. Pl.,* 1775, p. 220.

N. 1. Left bank of the Greeta near Wrayton, *Wi.*
2. Nether Kellet.
3. "In Sinderbarrow meadows, plentifully, within the liberties of Yealand Redman," *Jenkinson*, 1775. Frequent near Silverdale, *Ashfield*. Still frequent there! Near Leighton Beck; Gatebarrow and Bolton-le-Sands, *Wi.*
E. 7. Field near Claughton, *Wi.*
8. Bolton Roughs; Leagram, *Fl. St.*

Habenaria bifolia, R. Br.—Lesser Butterfly Orchis. British type. Native. Area, 2, 3, 6, 7, 8. Range, 0–500 feet or higher. P. June–July.

Upland pastures. Rather rare. First record: *Ashfield, Fl. Prest.*, 1858.

N. 2. Whittington Moor and near the Redwell Inn, *Wh.* & *Wi.* Gressingham, and Kellet Seeds, *Wi.*
3. Silverdale and Thrang End, *Wi.* Near Borwick.
E. 6. Lancaster Moor.
7. Claughton, *P.J.H.*
8. Chaigley, *Fl. St.*

Habenaria montana, Dur.—Butterfly Orchis. British type. Native. Area, 1, 2, 3, 6, 7, 8. Range, 0–750 feet. P. June–July.

Woods, meadows, and pastures. Frequent amongst the hills. First record: *Wilson, B.R.C. Rep.,* 1887, p. 106.

N. 1. Near Wennington, *Wi.*
2. Near Nether Kellet.
3. Near Challan Hall, etc., Silverdale, *L.P.*
E. 6. Near Ellel, *Wi.* In all the dales, ascending to 750 feet on Haylot Fell.
7. Bleasdale, common, *Wi.* Wyresdale, frequent.
8. Near Stonyhurst, common, *Fl. St.* Kemple End, *Wh.*

Habenaria viridis, R. Br. — Frog Orchis. British type. Native. Area, 1, 2, 3, 6, 7, 8. Range, 0–700 feet. P. June–July.

Meadows, pastures, and banks. Frequent. First record: *Ashfield, Bot. Chron.,* 1864.

N. 1. Near Wennington, *Wi.*
2. Kellet Seeds and Dalton Crag, *Wi.*
3. The Trough, Trowbarrow, *Ashfield,* 1864. Frequent in the Silverdale district, *Wi.*
E. 6. Near Ellel, *Wi.*
7. Near Garstang, *Wi.* Lea Fell, at 550 feet.
8. Forty-acre Field, Stonyhurst, *Fl. St.* Near Leagram, *Wi.*

IRIDACEÆ.

Iris Pseudacorus, L. — Yellow Flag or Iris. "Daggers." British type. Native. Area, 1, 2, 3, 4, 5, 6, 7, 8. Range, 0–400 feet. P. June.

Swampy fields, watery places, and pondsides in the low country. Common. First record: *Ashfield, Fl. Prest.,* 1858. "Brockholes Wood."

Sisyrinchium anceps.—Alien. First record: *Beesley,* 1903.

Railbank near Preston, 1900, *H.B.* 'sp.) A native of North America.

Crocus officinalis, Huds. (*C. vernus,* All.).—Purple Crocus. Denizen. Area, 3. Range, only occurs at about 25 feet. P. March. First record: The present one?

N. 3. Meadows near Torrisholme, abundant, 1885, *Wi.*

Crocus nudiflorus, Sm.—Alien. First record: The present one?

E. 8. Near Goosnargh, *R. Standen.*

AMARYLLIDACEÆ.

Narcissus Pseudo-narcissus, L.—Daffodil. Lent Lily. English type. Native. Area, 2, 3, 4, 5, 6, 7, 8. Range, 0–400 feet. P. March–April.

T

Woods, meadows, and streamsides. Not uncommon, and often truly wild, although no doubt introduced in many of its localities. First record: *Ashfield, Fl. Prest.,* 1860.

N. 2. Docker, *Wi.*
3. Silverdale and Thrang End, *Wi.*
W. 4. Tarnacre, *S. Wi.*
5. Near St. Michael's, *Ashfield.* Near Salwick, *A.A.D.*
E. 6. Near Low Gill, Lower Hindburndale.
7. Woods and banks near the Wyre, Calder, and Brock, especially abundant near Dolphinholme.
8. Turner Fold, Kemple End and Paper Mill Wood, *Fl. St.*

Narcissus poeticus, L.—Pheasant's Eye. Alien.

W. 4. Meadow in the township of Old Rawcliffe, *James Pearson,* in *Science Gossip.*

Galanthus nivalis, L.—Snowdrop. English type. Denizen. Area, 1, 4, 5, 7, 8. Range, 0–400 feet. P. February–April.

Banks of streams, woods, meadows, and old orchards, often appearing quite wild. Not unfrequent. First record: *Ashfield, Fl. Prest.,* 1858. "Hedgebank four miles from Preston on the Lancaster Road."

LILIACEÆ.

Asparagus officinalis, L.—Asparagus. Atlantic type. Denizen. Area, 5. Range, 0–25 feet. P. August.

Naturalized on the sandhills, where it occurs as the *var. hortensis* only, and has no claim to be regarded as a native. First record: *Ashfield, Fl. Prest.,* 1858. "Lytham." Still occurs sporadically on the sandhills from time to time!

Polygonatum multiflorum, All.—Solomon's Seal. English type. Native. Area, 3, 4, 5, 7, 8. Range, 50–300 feet. P. June.

Woods and thickets, native on limestone, but probably introduced elsewhere. Rare. First record: *Jenkinson, Brit. Pl.,* 1775, p. 70.

N. 3. In the Throng Wood and in the Hining Wood within the township of Warton, *Jenkinson,* 1775. Wood near Yealand Storrs, *Wi.*
W. 4. Near Nateby, *Wi.*
5. Copp, *P.J.H.* Wood at Ribby Hall near Kirkham, *A.A.D.*
E. 7. Wood at Brock, *A.A.D.*
8. Roadside east of Inglewhite, *Wi.* Goosnargh, *P.J.H.*

Polygonatum officinale, All. — Angular Solomon's Seal. English type. Native. Area, 2, 3. Range, 250–880 feet. P. June.

Crevices of limestone "pavement," and bushy limestone scars. Rare. First record: *Wilson, B.R.C. Rep.,* 1884, p. 48.

N. 2. Dalton Crag, 1883, and since; abundant, *Wi.* Ascends to 880 feet
3. Waterslack Wood, *L.P.* [Still grows in Ashfield's old locality in Middlebarrow Wood, but beyond our limits, *Wi.*] Gatebarrow Wood, *Wi.*

Convallaria majalis, L.—Lily of the Valley. Germanic type. Native. Area, 2, 3 (5), (7). Range, 50–870 feet. P. May.

Rocky woods and scars on the limestone of North, where it is frequent and in some localities abundant. Not found truly wild in East or West. First record: *Jenkinson, Brit. Pl.,* 1775, p. 70.

N. 2. Dalton Crag, ascending to 870 feet, *Wh.* & *Wi.* Kellet Seeds, *Wi.*
3. Amongst the rocks near Leighton Furnace; in the Flatwood, and in Mr. Townley's woods, plentifully, *Jenkinson,* 1775. Hills between Yealand and Leighton Beck; Gatebarrow and Middlebarrow Woods, *Wi.* Castlebarrow; Waterslack; and Hagg Woods, *L.P.* Criagle-barrow, *Wi.*
W. 5. Wood near Kirkham, *A.A.D.*
E. 7. Reported doubtfully by Ashfield from this district where it is unlikely to be more than a garden escape.

Allium vineale, L. — Crow Garlic. English type. Native. Area, 1, 2, 3, 4, 5, 6. Range, 0–50 feet. P. July.

Streamsides and banks. Rare. First record: *Wilson, Journ. Bot.,* 1900, p. 46.

N. 1. Leck district, *L.P.*
2. and 3. The *var. compactum* is abundant on the banks of the Keer, near Warton, extending into both districts 2 and 3, *Wi.*
W. 4. Coast bank at Cockerham, *Mrs. A. Wilson.*
5. Salwick Brook, Lea, *H.B.*
E. 6. Near Lancaster, 1896, *Wi.*

Allium oleraceum, L.—var. *complanatum,* Fr.—Field Garlic. Germanic type. Native. Area, 1. Range, only seen at about 120 feet. P. August.

Rare. First record: *Wilson, Journ. Bot.,* 1902, p. 349.

N. 1. Left bank of the Greeta near Wrayton, 1901, *Wi.*

Allium ursinum, L.—Wood Garlic. Ramsons. "Stinking Onion." British type. Native. Area, 1, 2, 3, 4, 5, 6, 7, 8. Range, 0–1,160 feet. P. May.

Woods, thickets, and shady banks. Very common, especially

in some parts of East and North. First record: *Linton, B.R.C. Rep.,* 1874, p. 85.

It ascends to 1,160 feet on Leck Fell. The fungus, *Peronospora Schleideni,* Ung., sometimes appears upon its leaves.

[*Allium Scorodoprasum,* L.—Sand Leek. Incognit.

Dr. Chas. Leigh, in his *Nat. Histy. of Lancs., Chesh., and the Peak,* p. 956, 1700, says, " *Rocamboes* grow in the meadows near Preston and make an agreeable sawce." Pryor refers " *Rocamboes* " to *Allium scorodoprasum.* This plant occurs on the banks of the Mersey and may possibly have been formerly found on the banks of the Ribble, although we have been unable to obtain any confirmation of Leigh's statement.]

Endymion non-scriptum, Garcke.—Wild Hyacinth. Bluebell. "Crowfoot." British type. Native. Area, 1, 2, 3, 4, 5, 6, 7, 8. Range, 0–1,200 feet. P. April–May.

Moist woods and shady banks. Common ; very abundant in the cloughs of North and East. First record: *Linton, B.R.C. Rep.,* 1874, p. 85.

Ascends to 1,190 feet on Ireby Fell, and to 1,200 feet on Fairsnape Fell. In Crag Wood, near Clougha, the leaves were noticed to be affected by the fungus, *Uromyces concentrica,* Lév., *Wh.*

Ornithogalum umbellatum, L.—Star of Bethlehem. "Jack-go-to-bed-at-noon." English type. Denizen. Area, 3, 4, 5, 7. Range, 0–100 feet. P. May.

Woods and grass hedgebanks. Rare. First record : *Jenkinson, Brit. Pl.,* 1775, p. 68.

N. 3. " I found this near Leighton Hall," *Jenkinson,* 1775.
W. 4. Near Cartford Bridge, *P.J.H.*
5. Thickets near Cottam Hall, *Ashfield.* Still there, *A.A.D.*
E. 7. Bank of the Wyre between Garstang and Scorton, *Wi.*

Tulipa sylvestris, L.—Wild Tulip. Alien? First record: *Fl. Prest. N.,* 1903, p. 42.

E. 5. Near Whittingham, 1870 (approx.). Now lost, *R. Standen,* in litt. to J. C. Melvill.

Gagea lutea, Gawl.—Yellow Star of Bethlehem. Intermediate type. Native. Area, 8. Range, only seen at about 140 feet. P. March–April.

Damp thickets and shady places, generally near streams. Very rare. First record: *Fl. St.,* 1891, p. 37.

E. 8. Ree Deep, one plant, *Fl. St.* (several plants on the Yorkshire side near Lower Hodder Bridge, *Fl. St.*).

Fritillaria Meleagris, L.—Fritillary. Alien. First record : *Ashfield, Fl. Prest.,* 1860.

W. 4. Meadow near Stainhall, *Ashfield.*

Asphodelus fistulosus, L.—Alien. First record : The present one.

W. 5. With other aliens near St. Annes, 1906, *C.B.* (sp.).

A native of the Greek Islands and the Mediterranean countries.

Narthecium ossifragum, Huds.—Bog Asphodel. British type. Native. Area, 1, 2, 3, 4, 6, 7, 8. Range, 25–1,730 feet. P. July.

Peat bogs in the low country, and swampy upland pastures. Common amongst the fells of East, but becoming very rare on the lowland mosses. First record: *Jenkinson, Brit. Pl.,* 1775, p. 69.

N. 1. Lower Ease Gill, *Wi.*
2. Whittington and Arkholme Moors, and bogs near Docker and Over Kellet, *Wi.*
3. On the Whitemoss, plentifully near Yealand, *Jenkinson.* 1775. Heysham Moss.
W. 4. Cockerham Moss.
E. 6, 7, 8. Abundant in suitable situations, ascending to 1,730 feet on Wardstone.

Paris quadrifolia, L.—Herb Paris. British type. Native. Area, 1, 2, 3, 4, 6, 7, 8. Range, 0–1,025 feet. P. May–June.

Moist woods and thickets. Frequent, especially on the limestone. First record: *Jenkinson, Brit. Pl.,* 1775, p. 82. " In all the woods about Yealands, Lancs., some with five or six leaves."

It ascends to 1,025 feet in the shelter of limestone crevices on Leck Fell, *Wi.* On limestone it often occurs with five or more leaves.

DIOSCOREACEÆ.

Tamus communis, L.—Black Bryony. English type. Native. Area, 1, 2, 3, 4, 5, 6, 7, 8. Range, 0–400 feet. P. May–June.

Hedges and thickets. Common in North, less frequent elsewhere. First record: *Ashfield, Fl. Prest.,* 1860. " Tunbrook Wood."

It ascends in North to 400 feet on Kellet Seeds.

JUNCACEÆ.

Juncus bufonius, L.—Toad Rush. British type. Native. Area, 1, 2, 3, 4, 5, 6, 7, 8. Range, 0–1,000 feet. A. July–August.

Swampy ground and damp places by roadsides. Very common. First record : *Linton, B.R.C. Rep.,* 1874, p. 85.

It ascends to 1,000 feet in Hindburndale.

Juncus squarrosus, L.—Heath Rush. British type. Native. Area, 1, 2, 3, 6, 7, 8. Range, 25–2,030 feet. P. June–July.

Moorland pastures and heaths. Common in East, more rare in North, except on Leck Fell, where it is abundant. No records for West. First record: *Linton, B.R.C. Rep.,* 1874, p. 85.

Juncus compressus, Jacq.—Round-fruited Rush. British type. Native. Area, 5, 8. Range, 0 to above 500 feet. P. July–August.

Marshy places. Rare. First record: *Fl. St.,* 1891, p. 37.

W. 5. Shore between Lytham and the Guide House Inn, *S. & T.*
E. 8. Longridge Fell above Chad's Well, *Fl. St.,* 1891.

Juncus Gerardi, Lois.—Mud Rush. British type. Native. Area, 3, 4, 5. Range, 0–25 feet. P. June–August.

Salt marshes and shores of tidal rivers. Common. First record: *Watson, Top. Bot.,* 1883.

Juncus inflexus, L.—Hard Rush. English type. Native. Area, 1, 2, 3, 4, 5, 6, 7, 8. Range, 0–600 feet or higher. P. July–August.

Marshy pastures and heaths, especially on clay soil. Very common. First record : *Top. Bot.,* 1883.

Juncus diffusus, Hoppe. (*J. effusus* × *glaucus*).—English type. Native. Area, 3, 8. Range, 0–400 feet. P. July–August.

In similar situations to the next species. Rare. First record : *Fl. St.,* 1891, p. 37.

N. 3. Marshy pasture near Bare, *F.A.L.*
E. 8. Hudd Lee Farm near Stonyhurst, 1891, *Fl. St.*

Juncus effusus, L.—Soft Rush. British type. Native. Area, 1, 2, 3, 4, 5, 6, 7, 8. Range, 0–1,520 feet or higher. P. July–August.

Pool sides and swampy places. Very common. First record : *Linton, B.R.C. Rep.,* 1874, p. 85.

It ascends to 1,520 feet on Tarnbrook Fell.

Juncus conglomeratus, L.—Common Rush. British type. Native. Area, 1, 2, 3, 4, 5, 6, 7, 8. Range, 0–1,900 feet. P. July–August.

Damp pastures and marshy ground. Very common. First record: *Linton, B.R.C. Rep.,* 1874, p. 85.

Juncus maritimus, Lam.—Sea Rush. British type. Native. Area, 3, 4. 5. Range, 0–20 feet. P. July–August.

Salt marshes and muddy seashores. Not uncommon. First record: *Ashfield, Fl. Prest.,* 1860.

N. 3. Between Silverdale and Carnforth, and in marshes near Overton and Middleton, *Wi.* Basil Point, *Wi.*
W. 4. Side of the Wyre at the Shard, Wardless, and other places in that locality ! *Ashfield,* 1860. Knott End, *Wh.* Glasson and Cockerham, *Wi.*
5. Fleetwood Marsh, *Wh.* Near Lytham ! *S. & T.* Skippool, *Wh.*

Juncus bulbosus, L.—Lesser Jointed Rush. British type. Native. Area, 1, 2, 4, 5, 6, 7, 8. Range, 0–1,000 feet or higher. P. July–September.

Boggy ground, moors, and swamps ; sometimes floating. Common. First record: *Linton, B.R.C. Rep.,* 1874, p. 85.

It ascends to 1,000 feet on Ireby Fell, in North.

Juncus obtusiflorus, Ehrh.—Blunt-flowered Rush. English type. Native. Area, 3. Range, 0–30 feet. P. July–August.

Marshy places and tarn sides. Rare. First record : *F. A. Lees, The Naturalist,* 1899, p. 302.

N. 3. In a bed of *Arundo Phragmites,* by a brook near the L. & N.W. Railway, north of Bare, 1899, *F.A.L.* Hawes Water ! *J.F.P.* Between Carnforth and Bolton-le-Sands, *Wi.* Ditch between Carnforth Station and the shore, *Wh.*

Juncus articulatus, L.—Shining-fruited Rush. British type. Native. Area, 1, 2, 3, 4, 5, 6, 7, 8. Range, 0–1,490 feet. P. July–August.

Marshes and swampy ground, ditch sides, and damp moors. Common. First record: *Linton, B.R.C. Rep.,* 1874, p. 85.

It ascends to 1,490 feet on Tarnbrook Fell.

Juncus acutiflorus, Ehrh.—Sharp-flowered Rush. British type. Native. Area, 1, 2, 3, 4, 6, 7, 8. Range, 0–1,000 feet or higher. P. July–August.

Marshes and swampy fields. First record: *Linton, B.R.C. Rep.*, 1874, p. 85.

It ascends to 1,000 feet by the Tatham Beck, Hindburndale.

Luzula pilosa (O. Kuntze.).—Hairy Wood-Rush. British type. Native. Area, 1, 2, 3, 4, 6, 7, 8. Range, 75–1,480 feet. P. April–May.

Woodland banks and thickets. Common amongst the hills. First record: *Ashfield, Fl. Prest.,* 1858. "Redscar."

It ascends to 1,480 feet on Tarnbrook Fell.

Luzula sylvatica, Gaud.—Great Wood-Rush. British type. Native. Area, 1, 2, 3, 6, 7, 8. Range, 30–1,600 feet. P. May–June.

Wooded cloughs and heathy banks. Frequent amongst the hills of East. First record: *Ashfield, Fl. Prest.,* 1858.

N. 1.　By the Lune near Kirkby Lonsdale, *Wi.* Ease Gill, ascending to 1,350 feet; and in the Gorge of the Greeta.

2.　Wash Dub Wood, and between Caton and Aughton.

3.　Near Heysham, *Wi.*

E. 6.　Hindburn and Roeburndale, common, *Wi.* Near Grassward Hall, *Wh.* Near Mill Houses, attacked by *Puccinia oblongata*, Link.

7.　Nickey Nook, *Ashfield,* 1858. Abundant in some parts of Wyresdale and Bleasdale, ascending to 1,600 feet on Tarnbrook Fell, *Wi.*

8.　Banks of the Loud, *Wi.*

Luzula campestris, DC.— Field Wood-Rush. British type. Native. Area, 1, 2, 3, 4, 5, 6, 7, 8. Range, 0–1,900 feet. P. April–May.

Pastures and dry grassy places. Very common. First record: *Ashfield, Fl. Prest.,* 1858. "Near Preston Cemetery."

It ascends in North to 1,900 feet on Greygarth Fell, and in East to 1,600 feet or higher on Tarnbrook Fell.

Luzula multiflora, Lej.—Many-headed Wood-Rush. British type. Native. Area, 1, 2, 3, 4, 6, 7, 8. Range, 25–2,040 feet. P. June.

Boggy heaths, moors and upland pastures. Frequent in North and East, but becoming rare in West owing to drainage and cultivation. First record: *Ashfield Fl. Prest.,* 1860. "Nickey Nook and Beacon Fell."

The states known as var. *umbellata,* and var. *congesta,* are both frequent. It ascends in North to 2,040 feet on Greygarth Fell.

TYPHACEAE.

Typha latifolia, L. — Bulrush. Reed-Mace. "Hardheads." British-type. Native. Area, 1, 3, 4, 5, 6, 7, 8. Range, 0–400 feet. P. July.

Ponds, marshes, canals and mill-dams. Common, except in the dales. First record: *Ashfield, Fl. Prest.,* 1858. "Between Kirkham and Marton"!

Typha angustifolia, L.—Lesser Reed-Mace. English type. Native. Area, 4, 5, 7, 8. Range, 0–250 feet. P. July.

Ponds and watery places in the low country. Hardly frequent, and much rarer than the preceding. First record: *Ashfield, Fl. Prest.,* 1858.

W. 4.　Rawcliffe Moss, *Pearson.* Ponds near Cabus, *Wi.*

5.　Between Kirkham and Marton, 1858, and near Marton Mere, *Ashfield.* Pit near Lea, *H.B.* Pit near Copp Church, *P.J.H.*

E. 7.　Canal near Garstang, *Wh.* & *W.* Near Claughton, *E.C.*

8.　Pond near Goosnargh, *S. Wi.*

Sparganium erectum, L.—Branched Bur-reed. British type. Native. Area, 1, 2, 3, 4, 5, 6, 7, 8. Range, 0–600 feet. P. July–August.

Ditches, slow streams and ponds. Very common. First record: *Linton, B.R.C. Rep.,* 1874, p. 85.

Dr. F. A. Lees states that this is called by boys at Bare, "Wiskers" or "Whiskers."

Sparganium neglectum, Beeby.— British type? Native. Area, 5. Range, 0–? P. July–August.

Ponds and ditches. Probably frequent in the low country. First record: *Salmon and Thompson, Journ. Bot.,* 1902, p. 295.

W. 5.　Pond between Lytham and the Guide House Inn, *S. & T.* Ponds near Marton and Staining, *Wh.*

Sparganium simplex, Huds.—Unbranched Bur-reed. British type. Native. Area, 3, 4, 5, 6, 7, 8. Range, 0–400 feet. P. July–August.

Marshes, ditches, and slow streams. Frequent in parts of West; more rare in other districts. First record; *Ashfield, Fl. Prest.,* 1858. "Moorhall Estate."

Sparganium minimum, Fries.—Floating Bur-reed. British type. Native. Area, 2, 3. Range, 0–160 feet. P. August–September.

Tarns and marshy pools. Rare. First record: *Wilson, B.R.C. Rep.,* 1884, p. 22.

N. 2.　Between Carnforth and Nether Kellet.

3.　Little Hawes Water, 1883, and in a marsh near Borwick, *Wi.* Bank Well, Silverdale, *L.P.,* also *W.K.*

ARACEAE.

Arum maculatum, L.—Lords and Ladies. Cuckoo Pint. "Ramp Bobs." "Baboramps." "Parson-in-the-Pulpit." "Bulls and Cows." English type. Native. Area, 1, 2, 3, 4, 5, 6, 7, 8. Range, 0–1,200 feet. P. April–May.

Copses, hedgebanks, and shady places. Common. First record: *Top. Bot.,* 1883.

It ascends to 1,200 feet on the limestone of Leck Fell. The spotting of the leaves does not appear to depend upon the nature of the soil or situation, spotless and spotted plants frequently growing adjacent to each other.

Acorus Calamus, L.—Sweet Flag. English type. Native. Area, 4, 5, 6, 7, 8. Range, 0–150 feet.

In numerous places along the Preston and Kendal Canal, more rarely in ponds. First record: *Ashfield, Fl. Prest.,* 1858. "Plentiful in a pit near Preston Cemetery, flowering in considerable quantity."

LEMNACEAE.

Lemna trisulca, L.—Ivy-leaved Duckweed. English type. Native. Area, 2, 3, 4, 5, 6, 7. Range, 0–150 feet. A. July–August.

Ponds, ditches, and canals. Frequent. First record: *Lawson,* in Ray's Corresp., 1718. "Hederula aquatica Ger. In ditches between Warton and Carnforth."

Lemna minor, L.—Lesser Duckweed. "Jenny Green-teeth," in both West and South Lancashire. British type. Native. Area, 1, 2, 3, 4, 5, 6, 7, 8. Range, 0–400 feet. A. June–July.

Stagnant ponds and ditches. Very common in the low country. First record: *Linton, B.R.C. Rep.,* 1874, p. 85.

Lemna gibba, L.—Gibbous Duckweed. English type. Native. Area, 3, 5, 8. Range, 0–180 feet. A. July–August.

Floating on still pools. Rare. First record: *Ashfield, Fl. Prest.,* 1858.

N. 3.　Pond near Overton.

W. 5.　Pond between Little Marton and Marton Mere, *Wh.* Little Plumpton, *Wi.* Rossall, *Wh.*

E. 8.　Ribbleton Moor, *Ashfield,* 1858.

Lemna polyrhiza, L.—Greater Duckweed. English type. Native. Area, 4, 5. Range, 0–25 feet. A. Has not been found in flower in Britain.

Slow streams. Very rare. First record: *Ashfield, Fl. Prest.,* 1860.

W. 4.　Canal between Glasson and Galgate, *Wh.*

5.　Canal at Preston, *Ashfield,* 1860.

ALISMACEAE.

Alisma Plantago, L.—Water Plantain. British type. Native. Area, 1, 2, 3, 4, 5, 6, 7, 8. Range, 0–400 feet or higher. P. July–August.

Ponds, ditches, and slow streams. Common. First record: *Linton, B.R.C. Rep.,* 1874, p. 85.

Alisma ranunculoides, L.—Small Water Plantain. British type. Native. Area, 2, 3, 4, 5, 6, 7, 8. Range, 0–300 feet. P. July–August.

Ponds and ditches. Frequent in the low country. First record: *Pearson,* in *Ashfield's Fl. Prest.,* 1858. "Canal near Garstang"!

Sagittaria sagittifolia, L. — Arrowhead. English type. Denizen or Native. Area, 5. Range, only known at 75 feet. P. July–September.

Canals and ditches. Rare. First record: *Linton, B.R.C. Rep.,* 1875, p. 85.

W. 5.　Canal near Preston, *E.F.L.* Canal, Ashton "Summit," *H.B.*

Butomus umbellatus, L. — Flowering Rush. English type. Native or Denizen. Area, 4, 5, 7. Range, only known at 75 feet. P. June–August.

Frequent in the Preston and Kendal Canal, in the portion south of Garstang, more rarely further north. First record: *Ashfield, Fl. Prest.,* 1858. "Canal near Preston."

NAIADACEÆ.

Triglochin palustre, L.—Arrow-grass. British type. Native. Area, 1, 3, 4, 5, 6, 7, 8. Range, 0–1,250 feet. P. July.

Marshy places. Not uncommon. First record: *Ashfield, Fl. Prest.,* 1858. "Nickey Nook."

It ascends to 1,250 feet on Salter Fell, Roeburndale.

Triglochin maritimum, L.—Sea Arrow-grass. British tyye. Native. Area, 3, 4, 5. Range, 0–25 feet. P. June–August.

Muddy sea shores and salt marshes, Frequent. First record: *Ashfield, Fl. Prest.,* 1858. "Ashton Marsh."!

Potamogeton natans, L.—Broad-leaved Pondweed. British type. Native. Area, 1, 2, 3, 4, 5, 6, 7, 8. Range, 0–600 feet. P. July–August.

Ponds. Very common in the low country. First record: *Top. Bot.,* 1883.

Potamogeton polygonifolius, Pourr. — Oblong-leaved Pondweed. British type. Native. Area, 1, 2, 3, 6, 7, 8. Range, 200–1,490 feet. P. July.

Frequent in pools, ditches, and sluggish streams, especially in the dales and on the fells. First record: *Bailey, B.R.C. Rep.,* 1883, p. 197.

Ascends in North to 730 feet on Ireby Fell, and in East to 1,490 feet on the south-west side of Wardstone.

Potamogeton alpinus, Balb.—Reddish Pondweed. British type. Native. Area, 4, 5, 7, 8. Range, 0–700 feet. P. August.

Ponds, ditches, and slow streams. Infrequent. First record: *Linton, B.R.C. Rep.,* 1874, p. 85.

W. 4. Canal near Lancaster; in the Cocker near Cockerham; and in a ditch near Winmarleigh, *Wi.*
  5. Pit near Barton, *Wi.*
E. 7. Barnacre, and ponds near Admarsh, Bleasdale, *Wi.* Pond near Catshaw.
  8. Pit near Fulwood, 1873, *E.F.L.* Reservoir near Grimsargh, *Wh.*

Potamogeton heterophyllus, Schreb.—British type. Native. Area, 8. Range, ? P. June–July.

Ponds and ditches. Very rare. First record: *Bailey, Top. Bot.,* 1883.

E. 8. Stonyhurst Ponds, *Fl. St.*

Potamogeton perfoliatus, L.—Perfoliate Pondweed. British type. Native. Area, 2, 3, 4, 5, 6, 7, 8. Range, 0–200 feet or higher. P. July–August.

Ponds, rivers, and streams in the low country. Abundant. First record: *Ashfield, Fl. Prest.,* 1858. "Canal above Preston."

Potamogeton crispus, L.—Curled Pondweed. British type. Native. Area, 1, 2, 3, 4, 5, 6, 7, 8. Range, 0–200 feet or higher. P. July–August.

Ditches, rivers, canals, and ponds. Common. First record: *Ashfield, Fl. Prest.,* 1858. "Canal near Preston"!

It is common in the canal, as also in the rivers Greeta, Lune, Wyre, Hodder, and Ribble, often accompanied by the preceding.

Potamogeton densus, L.—Opposite-leaved Pondweed. English type. Native. Area, 1, 2, 4, 5, 6, 7. Range, 0–150 feet. P. June–July.

Canals, rivers, and streams in the low country. Frequent. First record: *Ashfield, Fl. Prest.,* 1858. "Canal above Preston"!

Frequent in the canal, and also in the river Lune.

Potamogeton obtusifolius, M. & K.—Grassy Pondweed. English type. Native. Area, 5, 6, 7, 8. Range, 0–250 feet. P. June–July.

Ponds in the low country. Rare. First record: *Wheldon, Journ. Bot.,* 1902, p. 349.

W. 5. Catterall Old Mill-dam, *Wi.* Ditches near Marton Mere, *Wh.*
E. 6. Mill-dam, near Quernmore (*var. fluvialis,* L. & M.).
  7. Barnacre, *Wi.*
  8. Grimsargh Reservoir, *Wh.* (*var. fluvialis,* L. & M.).

Mr. Arthur Bennett referred some of our specimens to the *var. fluvialis,* Lange and Mortensen. Mr. Linton informs us that the station on which his Top. Bot. record was based is in South Lancashire.

Potamogeton Friesii, Rupr.—English type. Native. Area, 4. Range, only seen at about 60 feet. P. June–July.

Ponds and slow streams. Rare. First record: The present one.

W. 4. Near Winmarleigh, July, 1900, *Wh.* & *Wi.* (teste, A. Fryer).

Potamogeton pusillus, L.—Small Pondweed. British type. Native. Area, 1, 2, 3, 4, 5, 6, 7, 8. Range, 0–600 feet. P. July–August.

Ponds, ditches, and slow streams. Not uncommon. First record: *Ashfield, Fl. Prest.,* 1858. "Ashton Marsh"! "Preston Canal"!

Ascends to about 650 feet in a pond near Catshaw, in East.

Potamogeton pectinatus, L. — Fennel-leaved Pondweed. British type. Native. Area, 3, 6, 8. Range, 0–200 feet. P. June–July.

Rivers and ponds, especially near the sea. First record: *Linton, B.R.C. Rep.,* 1874, p. 85.

N. 3. Ditches between Carnforth and Bolton-le-Sands, *Wh.*
W. 5. In the old river bed at Ashton, near Preston, and pond near Fleetwood, *Wh.*
E. 6. In the Lune between Lancaster and the canal aqueduct, 1903, *Wi.*
  8. In the Ribble above Preston, *Wh.* Grimsargh, *Fl. Prest. N.*

Ruppia rostellata, Koch.—Tassel Pondweed. British type. Native. Area, 3, 4, 5. Range, 0–25 feet. P. July–September.

Brackish ditches and pools. Rare. First record: *Wilson, Journ. Bot.,* 1900, p. 46.

N. 3. Ditches near Bolton-le-Sands, 1893, and near Overton, *Wi.*
W. 4. Pond near Glasson, *Wh.*
  5. Freckleton Marsh, *Wh.*

Zannichellia palustris, L.—Horned Pondweed. British type. Native. Area, 2, 3, 4, 5, 6. Range, 0–100 feet. P. June–August.

Ponds and ditches in the low country. First record: *Wilson, Journ. Bot.,* 1900, p. 46.

N. 2. Canal near Carnforth, 1892, and near Skerton, *Wi.*
  3. Pool near Wood Well, Silverdale, *L.P.* Abundant near Overton, *Wi.*
W. 4. Ellel Grange and canal near Stodday, *Wi.*
  5. Freckleton Marsh, *Wh.* Old river bed at Ashton, *H.B.*
E. 6. Canal near Lancaster, *Wh.*

Var. *pedunculata,* Reich.—Brackish ponds on the seashore.

W. 4. Salt marsh between Glasson and Cockersand, *Wh.*
  5. Fleetwood marsh, *Wh.*

CYPERACEÆ.

Eleocharis palustris, R. & S.—Creeping Spike-rush. British type. Native. Area, 1, 2, 3, 4, 5, 6, 7, 8. Range, 0–500 feet. P. June–July.

Ponds, slow streams, and marshy places. Common. First record: *Top. Bot.,* 1883.

Eleocharis uniglumis, Schultes.—Link's Spike-rush. British type. Native. Area, 3, 5. Range, 0–20 feet. P. July–September.

Wet sandy places near the sea. Rare. First record: *Chadwick, Top. Bot.,* 1883.

N. 3. Damp sandy places on the coast near Middleton, and by the seashore between the Keer estuary and Bolton-le-Sands.
W. 5. St. Annes, 1880, *H. Searle* (in Herb. Wheldon). Lytham, *J. C. Melvill.*

Eleocharis multicaulis, Sm. — Many-stemmed Spike-rush. British type. Native. Area, 8. Range, only seen at 300 feet. P. July–August.

Marshy places on limestone soil. First record: *Wilson, Journ. Bot.,* 1901, p. 25 (*Top. Bot.,* 1883, with a query).

E. 8. Boggy ground (limestone) in Thornley Quarry, near Chipping, 1900, *Wi.*

Scirpus pauciflorus, Lightf.—Chocolate-headed Spike-rush. British type. Native. Area, 1, 2, 3, 7. Range, 0–950 feet. P. June–July.

Boggy and marshy ground on heaths and on the seashore. Rare, but locally abundant. First record: *Wilson, Journ. Bot.,* 1900, p. 46.

N. 1. Lower Ease Gill, on the north side of Leck Beck, and in a second locality near, *Wi.*
  2. Bog east of Dunnald Mill Hole.
  3. Salt marsh near Bolton-le-Sands, 1892, *Wi.*
E. 7. Roadside north-west of Harrisend Fell, *Wi.*

Scirpus cæspitosus, L.—Scaly-stemmed Spike-rush. "Deer's-hair." British type. Native. Area, 1, 2, 4, 6, 7, 8. Range, 25–1,920 feet. P June.

Turfy moors and peat bogs. Abundant on the fells of East and North, and formerly so on the mosses of West, but now becoming rarer there owing to the bogs being reclaimed. First record: *Ashfield, Fl. Prest.,* 1858. "Longridge Fell."

It ascends in North to 1,920 feet on Greygarth Fell.

Scirpus fluitans, L. — Floating Spike-rush. British type. Native. Area, 3. Range, 0–20 feet. P. July.

Swamps and ditches. Rare. First record: *Wheldon, Journ. Bot.*, 1900, p. 46.

N. 3. Ditches near Morecambe, 1899, *Wh.* Ditch near Bare station, *Wi.* A specimen exists in the Motley Herbarium labelled " Poulton-le-Sands, 1840 " (*Naturalist*, 1902).

Scirpus setaceus, L.—Bristle-like Club-rush. British type. Native. Area, 1, 2, 5, 6, 7, 8. Range, 0–860 feet or higher. P. July–August.

Wet sandy or peaty ground. Frequent amongst the hills, rare in the low country. First record: *Linton, B.R.C. Rep.*, 1874, p. 85.

N. 1. Near Cantsfield ; and in the Gorge of the Greeta, *Wi.*
2. Arkholme Moor, *Wi.*
W. 5. Canal bank near Ashton, *H.B.*
E. 6, 7, 8. Frequent.
It ascends to 860 feet in Hindburndale.

Scirpus lacustris, L.—Club-rush. Bulrush. British type. Native. Area, 2, 3, 4, 5. Range, 0–75 feet. P. July–August.

Tarns, ponds, and marshes in the low country. Rather rare. First record: *Lees, The Naturalist*, 1899.

N. 2. Two large beds in the river Keer, *Wi.*
3. Behind Bare Village, 1899, *F.A.L.* Margin of both the Haweswater Tarns, *Wi.* Ditch on Silverdale Moss Road, *L.P.* Near Borwick, *Wi.*
W. 4. Pond near Nateby, *Wi.*
5. Pond near Sowerby, *Wi.* Marton Mere, *Wh.*

Scirpus Tabernæmontani, Gmel.—Glaucous Bulrush. English type. Native. Area, 3. Range, 0–15 feet. P. July–August.

Brackish ditches and pools near the seashore. Very rare, but locally plentiful. First record: *Wilson, Journ. Bot.*, 1900, p. 46.

N. 3. Abundant in ditches, and sparingly on the shore near Bolton-le-Sands, 1893, *Wi.* Ditches near the shore south of Carnforth, *Wh.* Abundant also in a ditch near Overton, *Wi.*
It also occurs just beyond our borders in a ditch at the north-east end of Middlebarrow Wood, *Wi.*

Scirpus maritimus, L.—Sea Club-rush. British type. Native. Area, 3, 4, 5. Range, 0–15 feet. P. July–August.

Pools in salt marshes and by tidal rivers. Not uncommon. First record: *Ashfield, Fl. Prest.*, 1858. " Between Naze Point and Lytham, and on Ashton Marsh "!

Var. *compactus,* Koch. (*conglobatus,* S. Gray).

N. 3. Near Ovangle, *Wi.* Shore north of Carnforth, *Wh.*
W. 4. About the mouth of the Wyre with the type, 1900, *Wh.*
5. Near the Lake, Ansdell, *Wh*

Scirpus sylvaticus, L.—Wood Club-rush. British type. Native. Area, 1, 2, 5, 6, 7, 8. Range, 0–340 feet. P. June–July.

Sandy, marshy ground, generally by rivers. Rather rare. First record: *Ashfield, Fl. Prest.*, 1860.

N. 1. Lune banks near Melling ; between Kirkby Lonsdale and Nether Burrow ; and exceedingly abundant in marshes near Tunstall, *Wi.*
2. North bank of the Lune above Lancaster, *Wh.* Between Caton and Aughton, *Wh. & Wi.* Very fine near Arkholme, *Wi.*
W. 5. By Salwick Brook, Lea, *Ashfield*, 1860. Banks of both Wyre and Calder near Catterall, and in Sowerby Marshes, *Wi.* Stavens Pool, near Inskip, *P.J.H.*
E. 6. South bank of the Lune near Lancaster, and abundant in marsh north-east of Caton, *Wi.*
7. Meadows near the Wyre above Garstang, *Wi.*
8. Near Loud Lower Bridge, *Wh.* Near Barton Mill, and banks of the Hodder, near Whitewell, *Wi.*

Blysmus compressus, Panz.—English type. Native. Area, 1. Range, 100–110 feet. P. June–July.

Damp pastures and boggy places Very rare. First record: *Wilson, Journ. Bot.*, 1902, p. 349.

N. 1. Left bank of the Greeta near Wrayton, 1901 ; and where it joins the Lune below Tunstall, very fine and plentiful, *Wi.*

Blysmus rufus, Link.—Scottish type. Native. Area, 3, 5. Range, 0–20 feet. P. July.

Salt marshes and damp sands by the sea. Very rare. First record: *Wilson, Journ. Bot.*, 1900, p 46.

N. 3. Salt marsh, near Bolton-le-Sands, abundant, 1892, *Wi.* Salt marsh below Heald Brow, Silverdale, *Wi.*
W. 5. Near Lytham, *J. C. Melvill.*

Eriophorum vaginatum, L.—Hare's-tail Cotton-grass. British type. Native. Area, 1, 2, 3, 4, 6, 7, 8. Range, 25–2,030 feet. P. May–June.

Damp heaths. Frequent in East, rare in West. First record: *Martyn, Catalogus Cantabrigiensis,* 1763.

N. 1. Leck Fell, *L.P.* Greygarth Fell.
2. Bog near Docker, *Wi.*
3. Heysham Moss.

U

W. 4. " On Pillin Moss," *Martyn.* Cockerham Moss.
E. 6, 7, 8. Frequent on the fells, forming the principal vegetation of the wet summit plateaux.

Eriophorum polystachion, L.—Common Cotton-grass. "Moss-crop." British type. Native. Area, 1, 2, 3, 4, 6, 7. 8. Range, 25–2,030 feet. P. May–June.

Boggy and swampy places. Frequent in East, rare in West. First record: *Ashfield, Fl. Prest.,* 1858.

N. 1. Greygarth Fell, *Wh. & Wi.* Near Old Wennington, *Wi.*
2. Near Docker, *Wi.* Lords Lot Wood.
3. Hawes Water Moss, *L.P.* Storrs Moss, *L.P.* Heysham Moss.
W. 4. Cockerham Moss.
E. 6, 7, 8. Common on the moors of this division.

Rynchospora alba, Vahl.—White Beak-rush. British type. Native. Area, 3, 4, 7. Range, 20–650 feet. P. July.

Deep, spongy places, in very wet peat bogs, amongst sphagnum. Very rare. First record: *Wilson, Journ. Bot.*, 1900, p. 46.

N. 3. Heysham Moss.
W. 4. Cockerham Moss, 1877, *Wi.* Had become very scarce there in 1901.
E. 7. Tarnbrook Fell at 650 feet.

Schœnus nigricans, L.—Black Bog-rush. British type. Native. Area, 3. Range, 0–300 feet. P. June–July.

Wet sandy or peaty places on limestone. Rare. First record: *Ashfield, Bot. Chron.*, 1864.

N. 3. Plentiful near the Tarn, Silverdale, *Ashfield*, 1864. Still there ! Castle Barrow, *L.P.* Boggy ground near Leighton Beck, *Wi.* Estuary of the Keer near Carnforth, *Wi.*

Cladium Mariscus, Br.—Twig-rush. English type. Native. Area, 3. Range, 20–25 feet. P. July.

Boggy or swampy margins of tarns. Rare, but locally plentiful. First record: *Ashfield, Bot. Chron.*, 1864.

N. 3. Margin of Great Tarn, Silverdale, *Ashfield*, 1864. Still there, and also at the Little Tarn !

Carex dioica, L.—Separate-headed Sedge. Scottish type. Native. Area, 2, 3, 6, 7 (8). Range, 50–1,250 feet. P. June.

Spongy bogs. Rare. First record: *Ashfield, Fl. Prest.*, 1858.

N. 2. Bog near Docker.
3. Marshy ground by the Leighton Beck.
E. 6. Littledale Fell, Udale, *Wh. & Wi.* Salter Fell at 1,250 feet, *Wi.*
7. South side of Blaze Moss ; Gavells Clough ; and Grizedale near Abbeystead. Marshaw Fell, and Tarnbrook Fell, ascending in this district to 1,150 feet.
[8. Ribbleton Moor, *Ashfield*, 1858. Locality now drained.]

Carex pulicaris, L.—Flea Sedge. British type. Native. Area, 1, 2, 3, 6, 7, 8. Range, 0–1,500 feet P. June.

Boggy moors and heaths. Very common on the moorland tracts. First record: *Top. Bot.*, 1883

It ascends to 1,500 feet on Tarnbrook Fell in East, and to 900 feet to Ireby Fell in North.

Carex disticha, Huds.—Soft Brown Sedge. English type. Native. Area, 1, 2, 3, 4, 5, 8. Range, 0–600 feet. P. June.

Ditches and marshy places. Not common. First record: *Ashfield, Fl. Prest.*, 1860.

N. 1. Boggy field near the Gorge of the Greeta, *Wi.*
2. Bog near Docker, *Wi.*
3. Swamp near Borwick, *Wi.*
W. 4. Ditch at Cockerham, *Wi.*
5. Canal side above Preston, and near Naze Point, *Ashfield*, 1860.
E. 8. Kemple End and Stonyhurst Park, *Fr. St.*

Carex arenaria, L.—Sand Sedge. British type. Native. Area, 3, 4, 5. Range, 0–50 feet. P. June.

Sandy seashores. Common amongst the sandhills of West and North. First record: *Ashfield, Fl. Prest.*, 1858. "Lytham sandhills."

It was observed to be the host of *Puccinia scœleriana*, P. & M., near Lytham in 1903.

Carex diandra, Schrank. (*C. teretiuscula*, Good.).—Lesser Panicled Sedge. British type. Native. Area, 2. Range, only found at 200 feet. P. June.

Marshes. Very rare. First record: The Authors, *Journ. Bot.*, 1902, p. 349.

N. 2. Bog near Docker, abundant, June 1901.

Carex paniculata, L.—Panicled Sedge. British type. Native. Area, 2, 3, 4, 5, 6, 7, 8. Range, 0–1,200 feet. P. June.

Swampy thickets, by streams, and in watery places. Frequent. First record: *Wilson, B.R.C. Rep.*, 1883, p. 249. " By the canal near Garstang."

It ascends to 1,200 feet on Tarnbrook Fell.

Carex vulpina, L.—Fox Sedge. British type. Native. Area, 2, 3, 4, 5, 6, 7, 8. Range, 0–150 feet. P. June.

Ditches, poolsides, and marshes. Not uncommon in the low country. First record: *Bailey, B.R.C. Rep.*, 1875.

N. 2. Canal near Borwick and Bolton-le-Sands, *Wi.*
3. Silverdale, *Bailey*, 1875. Between Hest Bank and Bolton-le-Sands, *Wh.* Rather common in this district.
W. 4 and 5. Frequent.
E. 6. Canal near Lancaster.
7. Canal near Garstang, *Wi.*
8. Near Barton Mill, *Wi.*

Var. *nemorosa,* Lumn.

W. 4. Little Bispham, *Wh.*
5. By the canal between Galgate and Glasson, *Wh.*
Extreme forms of this variety look very distinct, but there are numerous intermediate states connecting it with the type.

Carex muricata, L.—Greater Prickly Sedge. British type. Native. Area, 1, 2, 3, 5, 7, 8. Range, 0–250 feet. P. June.

Dry hedgebanks and roadsides on calcareous or gravelly soils. Rather rare. First record: *Wilson, B.R.C. Rep.*, 1883, p. 249.

N. 1. Near Wrayton.
2. Near Whittington, and between Carnforth and Over Kellet, *Wi.* Near Borwick and Docker.
3. Near Warton and Yealand Redmayne, *Wi.*
W. 5. Catterall Lane near Garstang, *Wi.* Coast between Lytham and Freckleton, and old river bed near Preston, *A.A.D.*
E. 7. Near Garstang, *Wi.*
8. Roadside near Longridge, *Wh.*

Carex echinata, Murr.—Lesser Prickly Sedge. British type. Native. Area, 1, 2, 3, 4, 6, 7, 8. Range, 0–2,000 feet. P. May–June.

Boggy moors and swamps amongst the hills. Common. First record: *Ashfield, Fl. Prest.*, 1860. " Beacon Fell."

It ascends to 2,000 feet on Greygarth Fell, and to 1,600 feet or higher on Tarnbrook Fell.

Carex remota, L. — Remote-spiked Sedge. British type. Native. Area, 1, 2, 3, 4, 5, 6, 7, 8. Range, 0–890 feet. P. June.

Damp shady places, woods, and banks of ditches and pools. Common. First record: *Ashfield, Fl. Prest.*, 1860. " Near Preston Cemetery, and Nickey Nook."

It ascends to 890 feet in Foxdale.

Carex canescens, L.—White Sedge. British type. Native. Area, 1, 2, 4, 6, 7. Range, 25–1,450 feet. P. June–July.

Peat bogs, and peaty ditch sides. Not common. First record: *Wilson, B.R.C. Rep.*, 1883, p. 249.

N. 1. Ease Gill at 1,050 feet.
2. Whittington Moor, *Wi.* Bog near Docker and Lords Lot Wood.
W. 4. Cockerham Moss, 1884, *Wi.*
E. 6. Botton Head Fell at 1,450 feet, *Wi.*
7. Blaze Moss above Marshaw ; both sides of Tarnbrook Fell ; and Lower Bleasdale, *Wi.*

Carex leporina, L.—Oval-spiked Sedge. British type. Native. Area, 1, 2, 3, 4, 5, 6, 7, 8. Range, 0–1,900 feet. P. June–July.

Damp ill-drained fields and marshy ground, especially on clay. Common. First record: *Linton, B.R.C. Rep.*, 1874, p. 85.

It ascends to 1,900 feet on Greygarth Fell.

Carex elata, All.—Tufted Sedge. English type. Native. Area, 2, 3. Range, 0–400 feet. P. June.

Bogs and marshes. First record: *Wilson, Journ. Bot.*, 1905, p. 95.

N. 2. Between Carnforth and Nether Kellet, and in a swamp near Henridden at 400 feet.
3. Ditch side near Silverdale Moss, June, 1904, *Wi.*

Var. *turfosa,* Ar. Benn.—

N. 3. Marsh between Borwick and Yealand, June, 1905, *Wh.* & *Wi.* First gathered here in 1888, and supposed to be a form of *C. Goodenowii,* Gay, *Wi.*

Carex acuta, L.—Slender-spiked Sedge. British type. Native. Area, 1, 5, 6. Range, 40–100 feet. P. June.

Marshes and watery places. Rare and decreasing. First record: *Ashfield, Fl. Prest.*, 1858. " Frequent," but no locality given.

N. 1. Very abundant in marshes near Tunstall, *Wi.*
E. 6. Bank of Lune a mile above Caton ; also between Crook of Lune and Halton, *Wi.*, and lower down between Skerton and Halton, *Wh.*
W. 5. Barton, *Fl. Prest. N.* Near Myerscough, 1907, *Wi.*

Carex rigida, Good.—Rigid Sedge. Highland type. Native. Area, 1. Range, 1,950–2,000 feet. P. June–July.

Exposed mountain slopes. Very rare. First record: *Wilson, Journ. Bot.*, 1902, p. 349.

W. 1. Greygarth Fell at 2,000 feet, July, 1901, *Wi.* (Confirmed by A. Bennett).

Carex Goodenowii, J. Gay.—Common Sedge. British type. Native. Area, 1, 2, 3, 4, 5, 6, 7, 8. Range, 0–2,030 feet. P. June.

Poolsides, marshes, and moors. Common. First record: *Linton, B.R.C. Rep.*, 1874, p. 85.

It ascends to near the summit of Greygarth Fell.

Carex flacca, Schreb.—Glaucous Heath Sedge. British type. Native. Area, 1, 2, 3, 4, 5, 6, 7, 8. Range, 0–2,030 feet. P. May.

Damp pastures and marshy places. Very common. First record: *Top. Bot.*, 1883.

Carex limosa, L.—Mud Sedge. Scottish type. Native. Area, 4. Range, only found at 25 feet. P. June.

Deep spongy peat bogs, in very wet places amongst sphagnum. Very rare. First record: *Wilson, B.R.C. Rep.*, 1883, p. 249.

W. 4. Amongst sphagnum in the wettest parts of Cockerham Moss, 1881, *Wi.*

Carex digitata, L.—Fingered Sedge. English type. Native. Area, 3. Range, 150–250 feet. P. May–June.

Dry rocky woods and sheltered banks on limestone. Very rare. First record: *Sidebotham,* in *Baker's Fl. of Lake Dist.*, 1885, p. 223.

N. 3. Abundant in Silverdale where it was found by Mr. Sidebotham, *Baker*, 1885. Still there, 1888, and in Gatebarrow Wood, Cringlebarrow Wood and Eaves Wood, 1904, *Wi.*

Carex pilulifera, L. — Round-headed Sedge. British type. Native. Area, 1, 2, 6, 7. Range, 100–1,450 feet. P. May–June.

Mountain pastures and grassy slopes amongst the hills. Rather uncommon generally. First record: *Wilson, B.R.C. Rep.*, 1884, p. 24.

N. 1. Ease Gill, on grit.
2. On Dalton Crag at 880 feet.
E. 6. Roeburndale, 1884, *Wi.* Between Hornby and Lower Salter.
7. At 1,450 feet on the south side of Tarnbrook Fell ; near Bay Horse ; and Hazelhurst, *Wi.*

Carex verna, Chaix.—Spring Sedge. British type. Native. Area, 1, 2, 3, 4, 5, 6, 7, 8. Range, 0–1,100 feet or higher. P. April–May.

Dry pastures, heaths, and grassy places. Common. First record: *Linton, B.R.C. Rep.*, 1874, p. 85.

It ascends to 1,100 feet on Leck Fell.

Carex pallescens, L.—Pale Sedge. British type. Native. Area, 1, 2, 6, 7, 8. Range, 0–890 feet. P. June.

Damp banks, borders of thickets, and grassy places, generally on clay soil. Common in East, rare in North and West. First record: *Wilson, B.R.C. Rep.*, 1883, p. 249. " Damp fields near Garstang."

It ascends to 890 feet in Foxdale.

Carex panicea, L.—Pink-leaved Sedge. British type. Native. Area, 1, 2, 3, 4, 5, 6, 7, 8. Range, 0–920 feet or higher. P. May–June.

Damp heaths and swampy ground. Common, especially amongst the hills. First record: *Bailey, B.R.C. Rep.*, 1875. " Silverdale."

It ascends to 930 feet in the Trough of Bowland, and will probably be found to occur considerably higher on some of the fells.

Carex pendula, Huds.—Great Pendulous Sedge. British type. Native. Area, 1, 2, 5, 6, 7, 8. Range, 50–700 feet. P. June.

Shady banks and woods, generally on the sides of damp cloughs amongst the hills. Not common. First record: *Linton, B.R.C. Rep.*, 1874, p. 85.

N. 1. Gorge of the Greeta.
2. Woods by the Lune between Caton and Aughton.
W. 5. In private grounds near St. Michael's-on-Wyre, perhaps accidentally introduced, *P.J.H.*

E. 6. Gully above, and west of Lower Salter Bridge, and near Caton, *Wi.* Crook of Lune. Near Low Gill, Lower Hindburndale, on dripping scars by the river.

7. Grizedale, and Calder and Brock Valleys, *Wi.*

8. Ree Deep, *Fl. St.* Tunbrook Wood, *A.A.D.* Wood below Whitewell.

The favourite situation for this species is on wet shale banks, or clayey oozy places beneath them, in the wooded cloughs which cut through millstone grit strata. These shales are somewhat calcareous, and support such plants as *C. pendula, Festuca sylvatica, Hypnum commutatum, Barbula tophacea, Eucladium verticillatum,* &c.

Carex sylvatica, Huds.—Wood Sedge. British type. Native. Area, 1, 2, 3, 4, 6, 7, 8. Range, 0–900 feet. P. May–June.

Woods and thickets. Frequent. First record: *Ashfield, Fl. Prest.,* 1858. "Tunbrook Wood."

It ascends to 900 feet near Ease Gill Kirk.

Carex helodes, Link. (*C. lævigata,* Sm.).—Smooth Sedge. British type. Native. Area, 7. Range, 300–350 feet. P. June.

Marshy woods and moist shady places. Rare. First record: *Wilson, Journ. Bot.,* 1900, p. 46.

E. 7. Woods by the Calder near Garstang, 1888, *Wi.*

Carex binervis, Sm.—Green-ribbed Sedge. British type. Native. Area, 1, 2, 6, 7, 8. Range, 200–1,500 feet. P. June.

Dry heaths and stony mountain pastures. Not uncommon in East. First record: *Fl. St.,* 1891, p. 40.

N. 1. Lower Ease Gill, *Wi.*

2. Arkholme Moor, *Wi.* Whittington Moor.

E. 6. Udale, *Wh.* & *Wi.* Botton Head Fell at 1,500 feet, *Wi.*

7. Near Tarnbrook, and at 1,400 feet on the south side of Tarnbrook Fell, *Wi.* Wardstone, and Calder Valley above Oakenclough.

8. Kemple End, *Fl. St.,* 1891. Slope of Fairsnape Fell towards Chipping, *Wh.* Dewhurst Clough, White Stone Clough at 1,450 feet, and near Dinkling Green.

Carex distans, L.—Distant-spiked Sedge. British type. Native. Area, 3, 4, 5. Range, 0–25 feet. P. June.

Salt marshes and waste ground near the sea. Frequent. First record: *Wilson, B.R.C. Rep.,* 1883, p. 249.

N. 3. Keer estuary and Salt marsh near Silverdale, *Wi.* Basil Point, *Wh.*

W. 4. Wyre estuary, and near Cockerham and Glasson, *Wi.* Salt marsh near Preesall, *Wh.*

5. Ansdell, *S. & T.* Ribble estuary, *Wi.*

Carex Hornschuchiana, Hoppe.—Tawny Sedge. British type. Native. Area, 1, 2, 6, 7, 8. Range, 200–1,250 feet. P. June.

Peaty upland marshes and boggy sides of rills. Not common. First record: *Wilson, Journ. Bot.,* 1900, p. 46.

N. 1. Lower Ease Gill, and between Wrayton and Wennington, *Wi.*

2. Whittington Moor, *Wi.*

E. 6. Salter Fell at 1,250 feet, *Wi.* By the Roeburn below Wolfhole Crag; Clougha.

7. Fairsnape Fell, 1891; Lower Bleasdale; Marshaw and Tarnbrook Fells; and near Grizedale near Abbeystead at 1,150 feet.

8. Near Dinkling Green, *Wi.*

Carex extensa, Good.—Long-bracteated Sedge. British type. Native. Area, 3, 4, 5. Range, 0–25 feet. P. June.

Salt marshes. Not common generally, but locally abundant. First record: *Ashfield, Fl. Prest.,* 1866.

N. 3. Keer estuary, and near Sunderland Point, *Wi.* Basil Point, and shore between Carnforth and Bolton-le-Sands, *Wh.*

W. 4. Wyre bank between Skippool and Shard Ferry, *Ashfield,* 1864! Wyre estuary, *Wi.* Very fine near Coatwalls, *Wh.* Salt marsh near Glasson, *Wi.*

5. Freckleton and Lytham, *Wi.*

Carex flava, L.—Yellow Sedge. British type. Native. Area, 1, 2, 3, 5, 6, 7, 8. Range, 0–1,850 feet. P. June.

Wet pastures and boggy places. Common. First record: *Linton, B.R.C. Rep.,* 1874, p. 85.

It ascends to 1,850 feet on Greygarth Fell, and to 1,490 feet in the Great Clough of Tarnbrook Fell.

Carex Œderi, Retz.—Œder's Sedge. British type. Native. Area, 5. Range, 20–25 feet. P. June–July.

Boggy ground amongst the coast sandhills. Rare, and rapidly dying out. First record: *Wheldon, Journ. Bot.,* 1900, p. 46.

5. Sandhills between St. Annes and Lytham, 1897, *Wh.* (teste Kukenthal).

Carex hirta, L.—Hairy Sedge. British type. Native. Area, 1, 2, 3, 4, 5, 6, 7, 8. Range, 0–400 feet or higher. P. June–July.

Marshes, poolsides, and damp grassy places. Common. First record: *Linton, B.R.C. Rep.,* 1874, p. 85.

Var. *hirtæformis.*

W. Bank of the Wyre near Scorton, 1901, *H.B.* & *the Authors.*

The greatest altitude at which we have any note of this species is at 400 feet near Abbeystead in East.

Carex Pseudo-Cyperus, L.—Cyperus-like Sedge. English type. Native. Area, 3, 4, 8. Range, 0–200 feet. P. June.

Pondsides, swamps, and ditches. Not common. First record: *Ashfield, Fl. Prest.,* 1858.

N. 3. Little Hawes Water and Bank Well, Silverdale, *Wi.*

W. 4. Near Garstang, *Wi.*

E. 8. Higher Brockholes, *Ashfield,* 1858. Near Goosnargh, *Wi.* Pit near Whittingham Asylum, *Wh.*

Carex acutiformis, Ehrh.—Lesser Pond Sedge. British type. Native. Area, 1, 2, 3, 4, 5, 6, 7, 8. Range, 0–1,100 feet. P. June.

Ditches, marshes, and watery places. Common. First record: *Ashfield, Fl. Prest.,* 1858. "Frenchwood."

Ascends in East to 1,100 feet on Mallowdale Fell.

Carex riparia, Curt.—Greater Pond Sedge. British type. Native. Area, 3 (4). Range, 0–100 feet. P. June.

Broad, slow-flowing ditches, and marshes in the low country. Rare. First record: *Wilson, B.R.C. Rep.,* 1884, p. 25.

N. 3. Swamp near Borwick, *Wi.*

[W. 4. Ditch west of Lancaster, 1838, *Wi.* Now lost through building.]

Carex rostrata, Stokes.—Bottle Sedge. British type. Native. Area, 1, 2, 3, 4, 5, 6, 7, 8. Range, 0–1,000 feet. P. June.

Poolsides, marshes, and peaty bogs. Not uncommon, especially amongst the hills. First record: *Wilson, B.R.C. Rep.,* 1883, p. 249. "Near Garstang."

It ascends to 1,100 feet on Mallowdale Fell, in East.

Carex vesicaria, L.—Bladder Sedge. British type. Native. Area, 2, 3, 5. Range, 0–300 feet. P. June.

Margins of tarns and pools, and bogs in the low country. Rare. First record: *Wilson, B.R.C. Rep.,* 1884, p. 24.

N. 2. Near Dunnald Mill Hole.

3. Little Hawes Water near Silverdale, 1883; Storrs Moss; near Middle-barrow Quarry and other localities near Silverdale, *Wi.*

W. 5. Sowerby Marshes, abundant, *Wi.*

GRAMINEÆ.

Setaria viridis, Beauv.—Bristle Grass. Alien. First record: *Beesley, Journ. Bot.,* 1902, p. 349.

W. 4. Wardless, *H.B.* (sp.).

5. Waste ground on sandhills near St. Annes, 1906, *C.B.* (sp.).

Setaria glauca, Beauv.—Glaucous Bristle Grass. Alien. First record: *Beesley, Journ. Bot.,* 1902, p. 349.

W. 5. Near Ashton, 1900, *H.B.* ! With the preceding at St. Annes, *C.B.* (sp.).

Phalaris canariensis, L.—Canary Grass. Alien. Area, 3, 5, 7, 8. Not unfrequent on waste ground near houses and on rubbish heaps near railways and docks. First record: *Linton, B.R.C. Rep.,* 1874, p. 85.

Phalaris minor, Retz.—Alien. First record: *Wheldon, Journ. Bot.,* 1901, p. 26.

W. 5. Waste ground near Preston Docks, 1900, *Wh.*

Phalaris arundinacea, L.—Reed Grass. British type. Native. Area, 1, 2, 3, 4, 5, 6, 7, 8. Range, 0–580 feet. P. July.

Riversides, marshes, and watery places. Common. First record: *Linton, B.R.C. Rep.,* 1874, p. 85.

It ascends in East to 580 feet on Greenside, near Tarnbrook.

Anthoxanthum odoratum, L.—Sweet Vernal Grass. British type. Native. Area, 1, 2, 3, 4, 5, 6, 7, 8. Range, 0–2,030 feet. P. May–June.

Meadows, pastures, and grassy places. Very common. First record: *Linton, B.R.C. Rep.,* 1874, p. 85.

Anthoxanthum Puellii, Lecoq. & Lamot.—Alien. First record: *Wheldon, Journ. Bot.,* 1901, p 26.

W. 5. Waste ground near Preston Docks, 1900, *Wh.*

Alopecurus agrestis, L.—Slender Fox-tail Grass. English type. Colonist? Area? Range? A. June–August.

Very rare. First record: *Syme, Top. Bot.,* 1883.

It is not recorded by Ashfield, nor in *Fl. St.* or *Fl. Prest. N.* We have never seen it, and perhaps it is only an alien.

Alopecurus geniculatus, L.—Bent Fox-tail Grass. British type. Native. Area, 1, 2, 3, 4, 5, 6, 7, 8. Range, 0–900 feet or higher. P. June.

Poolsides and damp places, especially where water has stood in winter. Common. First record: *Linton, B.R.C. Rep.,* 1874, p. 85.

It ascends to 900 feet near the Tatham Beck.

Alopecurus pratensis, L.—Meadow Fox-tail Grass. British type. Native. Area, 1, 2, 3, 4, 5, 6, 7, 8. Range, 0–500 feet or higher. P. May–June.

Meadows and grassy places. Abundant. First record: *Wilson, B.R.C. Rep.*, 1883, p. 249. " Meadows at Garstang."

Milium effusum, L.—Millet Grass. British type. Native. Area, 1, 2, 3, 4, 6, 7, 8. Range, 0–780 feet. P. June.

Damp woods. Not very common. First record: *Petty, Naturalist*, 1902. (The *Top. Bot.* record was an error. *Linton, Journ. Bot.*, 1900, p. 37.)

    N. 1. Woods by the Greeta, *Wi.*
       2. Between Lancaster and Halton, *Wh.* Near Aughton, *Wi.* Wood between Over Kellet and Capernwray, *Wi.*
       3. Eaves Wood, 1900, *L.P.*
    W. 4. West of Churchtown near Garstang, *Wi.*
    E. 6. Trough Brook Gill, Clougha, at 780 feet, *Wi.*
       7. Near Claughton and Garstang, *Wi.*
       8. Tunbrook near Preston, *H.B.*

Phleum pratense, L.—Meadow Cat's-tail. Timothy Grass. British type. Native. Area, 1, 2, 3, 4, 5, 6, 7, 8. Range, 0–500 feet or higher. P. July.

Pastures, meadows, and grassy places. Very common. First record: *Top. Bot.*, 1883.

Phleum arenarium, L.—Sea Cat's-tail Grass. English type. Native. Area, 5. Range, 0–30 feet. P. June.

Sandhills near the sea. Rare. First record: *Linton, B.R.C. Rep.;* 1874, p. 85.

    W. 5. Lytham, 1874, *E.F.L.* Fairhaven, *P.J.H.* St. Annes! *H.B.*

Agrostis alba, L.—Marsh Bent Grass. British type. Native. Area, 1, 2, 3, 4, 5, 6, 7, 8. Range, 0–930 feet. P. July–August.

Damp ground and watery places. Very common. First record: *Top. Bot.*, 1883.

    It ascends to 930 feet in Hindburndale.

Var. *stolonifera*, L.—Common on the coast.

    Very fine on Fleetwood Marsh, *Wh.*

Var. *maritima*, Mey.—Rare on sandy shores.

    N. 3. Silverdale, 1901, *Wi.*
    W. 5. St. Annes, *Wi.*

Var. *coarctata*, Hoffm.—Frequent on the coast in damp waste ground and salt marshes.

    W. 4. Near Glasson, *Wh.*
       5. Wyre banks above, and seacoast south of Fleetwood, 1900, *Wh.* Ribblebanks near Preston Docks, *Wh.*

Agrostis tenuis, Sibth.—Bent Grass. Twitch Grass. British type. Native. Area, 1, 2, 3, 4, 5, 6, 7, 8. Range, 0–2,080 feet. P. July–August.

Pastures, heathy and grassy places. Very common. First record: *Linton, B.R.C. Rep.*, 1874, p. 85.

Var. *nigra*, With.—Cultivated fields; not common.

    W. 4. Field near Cockerham Moss, *S. & T.*
       5. Near Rossall, *Wh.*
    E. 8. Fields near Redscar, *Wh.*

Var. *pumila*, Lightf.—Grassy banks and upland pastures.

    N. 2. Pastures near Henidden at 450 feet.
    W. 5. Carrs Green Common, *Wi.*
    E. 6. West side of Caton Moor, *Wi.*
       7. Grizedale, near Garstang, *Wi.*

Calamagrostis epigejos, Roth.—Small-reed. English type. Native. Area, 3, 4. Range, 0–30 feet. P. July–August.

Damp woods, thickets, and swampy banks in the low country. Very rare. First record: *J. H. Rossall* in *B.R.C. Rep.* for 1874, p. 26.

    N. 3. Sea-coast between Far Naze and Sunderland Point, *Wi.* Banks between Basil Point and the Ferry House, *Wh.*
    E. 4. Bank facing the Lune between Aldcliffe and Stodday, near Lancaster, *J. H. Rossall.*

Apera Spica-venti, Beauv.—Wind-grass. Alien. First record: *Wheldon, Journ. Bot.*, 1901, p. 26.

    W. 5. Preston Docks, 1900, *Wh.* Ashton, *H.B.*

Ammophila arenaria, Link.—Marram. " Star Grass." British type. Native. Area, 3, 4, 5. Range, 0–50 feet. P. July–August.

Sandy seashores and sandhills. Common on the more sandy portions of the coast. First record: *Linton, B.R.C. Rep.*, 1874, p. 86.

Aira caryophyllea, L.—Silvery Hair-grass. British type. Native. Area, 1, 3, 4, 5, 6. Range, 0–600 feet. A. June–July.

Dry banks and fields on a gravelly or sandy soil. Not common. First record: *Linton, B.R.C. Rep.*, 1874, p. 86.

    N. 1. Dry sandy ground above the bank of Ireby Beck, at 600 feet, *Wi.*
       3. Overton, *Wi.* Banks by the river at Basil Point, *Wh.*
    W. 4. Moss Side, St. Michael's, *P.J.H.* River bank between Knott End and Preesall, *Wh.*
       5. Wyre banks above Fleetwood, *Wh.* St. Annes, *H.B.* Lytham, *Fl. Prest. N.*
    E. 6. Clougha at 600 feet, *Wi.*

Aira præcox, L.—Early Hair-grass. British type. Native. Area, 1, 2, 3, 4, 5, 6, 7, 8. Range, 0–890 feet or higher. A. May.

Dry banks. Not uncommon. First record: *Linton, B.R.C. Rep.*, 1874, p. 86.

    It ascends to 890 feet in Foxdale.

Aira cæspitosa, L.—Tufted Hair-grass. British type. Native. Area, 1, 2, 3, 4, 5, 6, 7, 8. Range, 0–1,920 feet. P. July.

Ditch sides, damp meadows, and swampy woods. Very common. First record: *Top. Bot.*, 1883.

    It ascends in North to 1,920 feet on Greygarth Fell.

Aira flexuosa, L.—Waved Hair-grass. British type. Native. Area, 1, 2, 3, 4, 6, 7, 8. Range, 0–2,030 feet. P. July.

Woodland banks and sandy heaths. Common in the hill districts of East; rarer in West and North. First record: *Wilson, B.R.C. Rep.*, 1887, p. 111. " Near Garstang."

Holcus mollis, L.—Creeping Soft-grass. British type. Native. Area, 1, 2, 3, 4, 5, 6, 7, 8. Range, 0–1,900 feet. P. July.

Woods and shady banks. Abundant in sandy soil amongst the gritstone hills of East; rarer elsewhere. First record: *Wilson, B.R.C. Rep.*, 1887, p. 111.

    It ascends to 1,900 feet on Greygarth Fell in North, and to 1,120 feet in Gavell's Clough in East.

Holcus lanatus, L.—Meadow Soft-grass. " Yorkshire Fog." British type. Native. Area, 1, 2, 3, 4, 5, 6, 7, 8. Range, 0–930 feet or higher. P. June.

Meadows, pastures, and grassy places. Very common. First record: *Linton, B.R.C. Rep.*, 1874, p. 86.

    It ascends to 930 feet in Hindburndale, and to 800 feet on Greenside in Wyresdale.

Trisetum flavescens, Beauv.—Yellow Oat-grass. English type. Native. Area, 1, 2, 3, 4, 5, 6, 7, 8. Range, 0–? P. June–July.

Dry pastures and grassy places. Common. First record: *Linton, B.R.C. Rep.*, 1874, p. 86.

Avena pubescens, Huds.—Downy Oat-grass. British type. Native. Area, 3. Range, 0–100 feet. P. June–July.

Dry sandy or limestone banks and walls. Rare. First record: *Petty, Naturalist*, 1902, p. 52.

    N. 3. On a wall in Silverdale village, *L.P.*, 1902. Sea-bank near Middleton, *Wi.*

Avena pratensis, L.—Narrow-leaved Oat-grass. British type. Native. Area, 2, 3. Range, 0–200 feet. P. July.

Dry limestone banks and fields. Rare. First record: *Petty, Naturalist*, 1902, p. 53.

    N. 2. Rocks by the Greeta, *Wi.*
       3. Near Silverdale Cove, 1902, *L.P.*

Avena fatua, L.—Wild Oat. Alien. First record: *Wheldon, Journ. Bot.*, 1902, p. 349.

    E. 8. Roadside near Ribchester, 1900, *Wh.*

Arrhenatherum elatius, Mert. & Koch. *(A. avenaceum*, Beauv.).—False Oat. British type. Native. Area, 1, 2, 3, 4, 5, 6, 7, 8. Range, 0–1,280 feet. P. June–July.

Hedgebanks, thickets, and pastures. Very common. First record: *Top. Bot.*, 1883.

    Ascends in North to 1,280 feet on Ireby Fell.

    The var. *nodosum*, Reichb., is frequent in drier situations.

Sieglingia decumbens, Bernh.—Heath Grass. British type. Native. Area, 1, 2, 3, 4, 5, 6, 7, 8. Range, 0–1,250 feet. P. July–August.

Dry sandy upland pastures. Frequent in the hilly districts. First record: *Wilson, B.R.C. Rep.*, 1883, p. 249. " Near Garstang."

    It ascends to 1,250 feet on Salter Fell, Roeburndale.

Phragmites communis, Trin.—Common Reed. British type. Native. Area, 1, 2, 3, 4, 5, 6. Range, 0–110 feet. P. July–August.

GRAMINEÆ.

Ditches, marshes, and pondsides in the low country. Common in West; rare in East; frequent in the western part of North. First record: *Linton, B.R.C. Rep.,* 1874, p. 85.

Observed to be the host of *Puccinia phragmitis* (Schum), near Stalmine, 1905.

Sesleria cœrulea, Ard.—Blue Mountain Grass. Highland type. Native. Area, 1, 2, 3, 8. Range, 0–880 feet. P. April–May.

Dry rocky limestone hills and scars. Very abundant on the limestone hills of North. First record: *Grindon, Naturalist,* 1864, or *Ashfield, Fl. Prest.,* 1864.

N. 1. Gorge of the Greeta.
2. Dalton Crag, up to 880 feet, *Wh.* & *Wi.* Near Over Kellet, and Kellet Seeds, *Wi.*
3. Abundant near Silverdale, *Ashfield,* 1864. Also *Grindon,* in *Naturalist,* 1864. Still very abundant there, and also near Leighton Beck, etc., *Wi.*
E. 8. Near Greystoneley.

Cynosurus cristatus, L.—Dog's-tail Grass. British type. Native. Area, 1, 2, 3, 4, 5, 6, 7, 8. Range, 0–800 feet or higher. P. June–August.

Meadows, pastures, or grassy places. Abundant. First record: *Linton, B.R.C. Rep.,* 1874, p. 86.

It ascends to 800 feet on Greenside, Wyresdale.

Cynosurus echinatus, L.—Prickly Dog's-tail Grass. Alien.

W. 5. Fleetwood Docks, 1902, *Wh.* Preston Docks, *A.A.D.*

Kœleria cristata, Pers.—Crested Hair-grass. British type. Native. Area, 2, 3, 8. Range, 0–480 feet. P. June–July.

Dry rocky pastures. Frequent on the limestone hills of North. First record: *Pinder, Top. Bot.,* 1883.

N. 2. Over Kellet, *Wi.*
3. Silverdale, *Rev. G. Pinder.* Still frequent there, ascending to 450 feet on Warton Crag!
E. 8. Limestone rocks near Dinkling Green.

Molinia cœrulea, Moench.—Blue Moor Grass. British type. Native. Area, 1, 2, 3, 4, 5, 6, 7, 8. Range, 0–1,430 feet. P. July.

Swampy heaths and peaty thickets. Frequent in East and North, and formerly so in West, but now rare there. First record: *Wilson, B.R.C. Rep.,* 1883, p. 249. "Moorlands near Bleasdale."

Ascends in East to 1,090 feet on the south fork of Foxdale Beck; and in North to 1,430 feet on Leck Fell.

Var. *breviramosa,* Parn.

W. 4. Cockerham Moss, 1902, *Jones* & *Wh*

Catabrosa aquatica, Beauv.—Water Whorl Grass. British type. Native. Area, 2, 3, 4, 5, 7. Range, 0–250 feet. P. June–July.

Pondsides and watery places. Infrequent. First record: *Wilson, Journ. Bot.,* 1900, p. 46.

N. 2. Near Priest Hutton, *Wi.*
3. Near Borwick, 1888, *Wi.*
W. 4. Near Nateby, *Wi.*
5. Ashton, near Preston, *H.B.*
E. 7. Canal near Garstang, *Wi.*

Melica montana, Huds. *(M. nutans,* L.). — Nodding Melic Grass. Scottish type. Native. Area, 2, 3. Range, 100–700 feet. P. June.

Rough bushy ground and rocky places on limestone. Rare, but locally plentiful. First record: *Wilson, Journ. Bot.,* 1900, p. 46.

N. 2. Dalton Crag at 700 feet.
3. Rough limestone ground in a wood near Silverdale, 1888, and in Gatebarrow Wood, *Wi.*

Melica nutans, L. *(M. uniflora,* Retz.).—Wood Melic Grass. British type. Native. Area, 1, 2, 3, 4, 5, 6, 7, 8. Range, 0–500 feet or higher. P. May–June.

Dry woodland banks and shady places. Common in the hill districts. First record: *Pearson* in *Fl. Prest.,* 1860. "Ravine above Greenhalgh Castle."

Dactylis glomerata, L. — Cock's-foot Grass. British type. Native. Area, 1, 2, 3, 4, 5, 6, 7, 8. Range, 0–920 feet or higher. P. June.

Meadows, pastures, and waste ground. Very common. First record: *Linton, B.R.C. Rep.,* 1874, p. 86.

It ascends in North to the summit of Dalton Crag, and in East to 920 feet in Hindburndale. The orange bands of the fungus, *Epichloe typhina* (Pers.), may be occasionally seen on its stems or leaf sheaths.

Briza media, L. — Trembling Grass. "Dothering Grass." British type. Native. Area, 1, 2, 3, 4, 5, 6, 7, 8. Range, 0–870 feet or higher. P. June–July.

V

Meadows, pastures, and grassy places. Common. First record: *Linton, B.R.C. Rep.,* 1874, p. 86.

It ascends to 870 feet on Dalton Crag, and to 600 feet on Greenside, Wyresdale.

Poa annua, L.—Annual Meadow Grass. British type. Native. Area, 1, 2, 3, 4, 5, 6, 7, 8. Range, 0–2,050 feet. A. January–December.

Cultivated fields, gardens, and waste places. Very common. First record: *Linton, B.R.C. Rep.,* 1874, p. 86.

Poa nemoralis, L. — Wood Meadow-grass. British type. Native. Area, 1, 2, 5, 6, 7, 8. Range, 0–1,120 feet. P. June–July.

Woods and bushy banks. Not common. First record: *Linton, B.R.C. Rep.,* 1874, p. 86.

N. 1. Cantsfield; Melling; Wrayton; Wennington, &c. Frequent in this division, ascending to 1,120 feet in a pot hole on Leck Fell, *Wi.*
2. North bank of Lune above Halton, *Wh.* Near Arkholme, *Wi.*
W. 5. Near St. Michael's.
E. 6. Aqueduct embankment by the Lune above Lancaster, *Wh.* Crook of Lune, *Wi.* Lower Hindburn, *Wi.*
7. Near Garstang, *Wi.*
8. Alston, *H.B.*

Poa compressa, L.—Flat-stemmed Meadow-grass. British type. Native. Area, 4, 5, 8. Range, 0–350 feet. P. July–September.

Dry banks, roadsides, and walls. Rare. First record: *Wheldon, Journ. Bot.,* 1900, p. 46.

W. 4. Plentiful with *Festuca rigida* near Glasson, *Wh.*
5. In the Churchyard at Copp, *P.J.H.* Near Fleetwood, *Wh.*
E. 8. Wall near Preston Wives, Longridge, 1899, *Wh.*

Poa pratensis, L. — Smooth Meadow-grass. British type. Native. Area, 1, 2, 3, 4, 5, 6, 7, 8. Range, 0–2,000 feet. P. June–July.

Meadows, banks, and grassy places. Very common. First record: *Linton, B.R.C. Rep.,* 1874, p. 86.

Poa trivialis, L.—Rough Meadow-grass. British type. Native. Area, 1, 2, 3, 4, 5, 6, 7, 8. Range, 0–1,280 feet. P. June–August.

Meadows and shady grassy places. Common. First record: *Linton, B.R.C. Rep.,* 1874, p. 86.

It ascends to 1,280 feet on Mallowdale Fell.

Glyceria fluitans, R. Br.—Floating Grass. British type. Native. Area, 1, 2, 3, 4, 5, 6, 7, 8. Range, 0–910 feet. P. June–July.

Marshes, ditches, and watery places. Very common. First record: *Linton, B.R.C. Rep.,* 1874, p. 86.

It ascends to 910 feet in Roeburndale.

Glyceria plicata, Fries.—Branched Floating Grass. English type. Native. Area, 1, 2, 3, 5, 8. Range, 0–250 feet. P. June–September.

Borders of ditches and ponds. Rare. First record: *Wheldon, Journ. Bot.,* 1901, p. 26.

N. 1. Marshy broads near Tunstall, *Wi.*
2. Near Aughton, *Wi.*
3. Ponds on the Heysham Peninsula near Overton.
W. 5. In the old river bed at Ashton, near Preston, *Wh.*
E. 8. Ditch between Grimsargh and Alston, *Wh.*

Glyceria aquatica, Wahlb. — Reed Meadow-grass. English type. Native. Area, 5. Range, only seen at 50 feet. P. July–August.

Marshy places and ponds. Rare. First record: *Linton, B.R.C. Rep.,* 1874, p. 86.

W. 5. Great Marton near Blackpool, 1899, *H.B.* (sp.!).

Sclerochloa maritima, Lindl. — Sea Meadow-grass. British type. Native. Area, 3, 4, 5. Range, 0–25 feet. P. June–July.

Salt marshes and mud-flats on the coast. Common. First record: *Carr, B.R.C. Rep.,* 1884, p. 25. "Tidal creek near Carnforth."!

Sclerochloa rigida, Link.—Hard Meadow-grass. British type. Native. Area, 2, 3, 4. Range, 0–300 feet. A. June.

Dry sandy ground and wall tops. Rare. First record: *Wilson, B.R.C. Rep.,* 1887, p. 112.

N. 2. Near Over Kellet, and near Carnforth, *Wi.*
3. Cringlebarrow (on limestone rocks); between Heald Brow and Silverdale Station, and near Carnforth, *Wi.* Near Heysham.
W. 4. Side of footpath near Glasson, plentiful, *Wh.* Near Bay Horse, *Wi.*

Festuca loliacea, Woods.—Dwarf Wheat-grass. English type. Native. Area, 3, 5. Range, 0–200 feet or higher. A. June–July.

Sandy ground near the sea. Rare. First record: *Ray's, Fasciculus*, 1688, p. 11.

N. 3. Near Bare, *Mr. Newton* (in Ray's Fasc., 1688). Near Heysham, 1902, *W.W.M.*

W. 5. Near Ingol and Cottam, *H.B.*

The Bare locality was the first recorded British station.

Festuca fasciculata, Forsk. (F. *uniglumis*, Sol.). — Single-glumed Fescue. English type. Native. Area, 5. Range, 0–30 feet. A. June.

Sandhills on the sea-coast, in the drier places. Rare. First record: *Wheldon, Journ. Bot.*, 1900, p. 46.

W. 5. Sandhills near St. Annes, 1897, *Wh.*

Festuca bromoides, L.—Slender Fescue. British type. Native. Area, 3, 4. Range, 0–100 feet. A. June.

Dry banks and waysides. Very rare. First record: *Rev. P. J. Hornby, Journ. Bot.*, 1902, p. 349.

N. 3. Wall below Grisdale Wood, 1906.

W. 4. Moss Side near St. Michael's, 1902, *P.J.H.*

Festuca ovina, L.—Sheep's Fescue. British type. Native. Area, 1, 2, 3, 4, 5, 6, 7, 8. Range, 0–2,050 feet. P. June–July.

Pastures and banks. Very common, especially in the hill country. First record: *Wilson, B.R.C. Rep.*, 1884, p. 26.

It ascends to 2,050 feet on Greygarth Fell. The var. *duriuscula*, Fr., is frequent, especially in the low country, where it is sometimes sown.

The var. *glauca*, Hack., occurs at Silverdale (*Wh.*), and probably elsewhere on the limestone hills of North.

Festuca rubra, L.—Creeping Fescue. British type. Native. Area, 2, 3, 4, 5, 8. Range, 0–350 feet. P. June–July.

Dry sandy ground, especially near the sea. Frequent and locally abundant. First record: *Watson, Top. Bot.*, 1883.

N. 2. Near Carnforth, *Wi.*

3. Heysham, *Wh.*

W. 4. Very fine by the Wyre opposite Preesall, *Wh.* Pilling, *H.B.*

5. Frequent along the coast.

E. 8. Near Higher Hodder Bridge, *Wh.*

Var. *juncea*, Hack. Sea shore near Fleetwood, July 1902, *Wh.* (teste *Hackel, B.E.C. Rep.*, 1903, p. 31.).

Festuca sylvatica, Vill. — Wood Fescue. Scottish type. Native. Area, 1, 7, 8. Range, 250–850 feet. P. July.

Shady and rocky banks amongst the hills, generally on limestone or calcareous shale. First record: *Wilson, Journ. Bot.*, 1900, p. 46.

N. 1. Ease Gill, near Leck, 1888, *Wi.*

E. 7. Dripping rocks by the Wyre above Dolphinholme, *Wi.*

8. Gorge of the Hodder below Whitewell, *Wh. & Wi.* Greystoneley Glen, *Wi.*

Festuca elatior, L.— Tall Fescue. British type. Native. Area, 1, 2, 4, 5, 6. Range, 0–600 feet. P. July.

River banks and damp places. Not very frequent. First record: *Wilson, Journ. Bot.*, 1900, p. 46.

N. 1. Ease Gill, 1888 ; Melling ; and Nether Burrow, *Wi.* Field near Collin Holme, *L.P.* Gorge of the Greeta.

2. Near Borwick and Arkholme, *Wi.* Between Caton and Aughton.

4. River bank near Cartford Bridge, *Wi.*

W. 5. Wyre banks near St. Michael's, *P.J.H.* Near Cottam, *H.B.* Old river bed near Ashton, *Wh.*

E. 6. Lune banks near Caton, *Wi.*

Festuca pratensis, Huds. — Field Fescue. British type. Native. Area, 1, 2, 3, 4, 5, 6, 7, 8. Range, 0–800 feet or higher. P. July.

Moist meadows and grassy places. Very common. First record: *Wilson, Journ. Bot.*, 1900, p. 47. "Near Cowkins, Lower Hindburn, 1887."

It ascends to 800 feet on Greenside, Wyresdale.

Festuca gigantea, Vill. — Tall Brome Grass. British type. Native. Area, 1, 2, 3, 4, 5, 6, 7, 8. Range, 0–400 feet or higher. P. July.

Damp woods and shady banks. Common. First record: *Lewis, B.R.C. Rep.*, 1884, p. 26. "Near Thornley."

Var. *triflora*, Sm.

N. 2. Wood by the Lune between Lancaster and Halton, 1899, *Wh.*

Bromus ramosus, Huds.—Rough Brome Grass. British type. Native. Area, 1, 2, 3, 4, 5, 6, 7, 8. Range, 0–400 feet or higher. P. July–August.

Hedgebanks, woods, and thickets. Common, especially in North. First record: *Wilson, B.R.C. Rep.*, 1887, p. 112. "Yealand."

Bromus sterilis, L. — Barren Brome Grass. British type. Native. Area, 1, 2, 3, 4, 5, 6, 7, 8. Range, 0–300 feet. A. June–July.

Dry hedgebanks, roadsides and waste ground. Frequent. First record: *Top. Bot.*, 1883.

Bromus maximus, Desf. — Great Brome Grass. Alien. First record : *Melvill & Bailey, Journ. Bot.*, 1904, p. 56.

W. 5. Sandhills near St. Annes, October 3rd, 1903, *C.B. & J.C.M.*

Serrafalcus mollis, Parl.—Goose Grass. Soft Brome Grass. British type. Native. Area, 1, 2, 3, 4, 5, 6, 7, 8. Range, 0–500 feet or higher. A. June–July.

Meadows, pastures, and waste places. Common. First record : *Top. Bot.*, 1883.

Brachypodium sylvaticum, Roem. & Schultes.—False Brome Grass. British type. Native. Area, 1, 2, 3, 4, 5, 6, 7, 8. Range, 0–880 feet. P. June–July.

Woodlands, banks, and rocky places. Common. First record : *Linton, B.R.C. Rep.*, 1874, p. 86.

It ascends to 880 feet on Dalton Crag.

Lolium perenne, L.—Perennial Rye Grass. Darnel. British type. Native. Area, 1, 2, 3, 4, 5, 6, 7, 8. Range, 0–860 feet or higher. P. June–July.

Pastures, grassy places, and waste ground. Abundant. First record : *Linton, B.R.C Rep.*, 1874, p. 86.

It ascends to 860 feet on Whitmoor.

Lolium italicum, Braun. — Italian Rye Grass. Alien. First record : *Petty, The Naturalist*, 1893.

N. 1. Near Leck, *L.P.*

W. 4. Fleetwood Docks, *Wh.* Fields near Winmarleigh.

Lolium temulentum, L.—Bastard Darnel. Alien. First record : *Beesley, Journ. Bot.*, 1902, p. 349.

W. 4. Near Winmarleigh, 1899, *H.B.* (sp.).

Triticum caninum, L.—Fibrous-rooted Couch Grass. British type. Native. Area, 1, 2, 3, 4, 5, 6, 7, 8. Range, 0–500 feet. P. July.

Bushy banks and woods. Not uncommon. First record : *Petty, The Naturalist*, 1893, p. 102. "Side of Leck Beck above Cowan Bridge."

Triticum repens, L.—Couch Grass. Black Twitch. British type. Native. Area, 1, 2, 3, 4, 5, 6, 7, 8. Range, 0–890 feet. P. July–August.

Hedgerows, banks, and waste places. Very common. First record : *Top. Bot.*, 1883.

It ascends to 890 feet in Foxdale.

Near Fleetwood, a form of this plant occurs in blown sand, which is referred by Messrs. Hackel and Druce to *A. repens × pungens*. It occurred also near Preesall, with both parents, June, 1900, *Wh.*

Triticum littorale, Host.—Sea Couch Grass. British type. Native. Area, 3, 4, 5. Range, 0–50 feet. P. July.

Sandy or muddy banks on the sea coast. First record : *Lees, The Naturalist*, 1899, p. 303.

N. 3. Near Bare, *F.A.L.*, 1899. Near Basil Point, *E.C.* !

W. 4. Near Glasson, and borders of Preesall salt marsh, *Wh.*

5. Fleetwood, *Wh.*

Triticum acutum, DC.—Sea Wheat. British type ? Native. Area, 3. Range, 0–50 feet. P. July–August.

Sandy seashores. Rare or overlooked. First record : The present one.

N. 3. Near Morecambe, 1899, *J. Beanland* (in Herb. C. E. Salmon) !

Triticum junceum, L.—Rushy Sea Wheat. British type. Native. Area, 3, 4, 5. Range, 0–50 feet. P. July–August.

Sandy or shingly seashores. Common. First record : *Linton, B.R.C. Rep.*, 1874, p. 86.

Lepturus filiformis, Trin.—Sea Hard Grass. English type. Native. Area, 3, 4, 5. Range, 0–25 feet. A. July.

Salt marshes and mud-flats on the coast. Not unfrequent. First record : *Syme, Top. Bot.*, 1883.

N. 3. Silverdale, Bolton-le-Sands and Ovangle, *Wi.* Coast between Middleton and Sunderland Point.

W. 4. Between Knott End and Pilling, *S. & T.* Cockerham coast, *Wi.*

5. Abundant on Ashton Marsh, *Wh.*

Nardus stricta, L.—Mat Grass. British type. Native. Area, 1, 2, 3, 5, 6, 7, 8. Range, 0–2,030 feet. P. July.

Heathy pastures and moors. Common in East and North; very rare in West, where we have only seen it on Carrs Green Common. First record: *Linton, B.R.C. Rep.*, 1874, p. 86.

Elymus arenarius, L.—Sea Lyme Grass. British type. Native. Area, 3, 5. Range, 0–30 feet. P. June-August.

Sandy seashores and banks of tidal rivers. Rare, and planted in *some* of its stations. First record: *Rev. P. J. Hornby, Journ. Bot.*, 1905, p. 95.

N. 3. Abundant on earth thrown up during the excavation of Heysham Harbour, *Wi.*

W. 5. Left bank of the Wyre above St. Michael's, 1903, *P.J.H.* Sandhills near St. Annes! *H.B.* Waste sandy ground near St. Annes Fire Station, several fine patches, *Wh.* Lytham, *Fl. Prest. N.*

Hordeum murinum, L.—Wall Barley. English type. Native? Area, 5. Range, 0–50 feet. P. June–August.

Waste places and at the foot of walls and buildings. Rare. First record: *Wheldon, Journ. Bot.*, 1900, p. 47.

W. 5. About Preston Docks on waste ground, and near the ferry and about the docks at Fleetwood, *Wh.* St. Annes, *H.B.*

[*Hordeum marinum*, Huds., recorded in *Journ. Bot.*, 1900, is probably an error. The record was made in 1883, and unfortunately no specimens were kept. As it has not been found since, it is bracketed until re-discovered.]

CRYPTOGAMEÆ.

FILICES.

Hymenophyllum peltatum, Desv. (*H. unilaterale*, Bory.).—Filmy Fern. Atlantic type. Native. Area, 6, 7. Range, 900–1,800 feet. P. July-August.

Wet, shady or mossy rocks in mountain glens, generally near streams. Very rare. First record: *Stabler, The Naturalist*, 1888.

E. 6. Clougha Scar, 1881, *G. Stabler.* Damp gritstone rocks on Wardstone.

7. Damp recesses of broken grit-rocks, Long Crag, Over Wyresdale.

Hymenophyllum tunbridgense, Sm.—Filmy Fern. Atlantic type. Native. Area, 6. Range, ? P. July–August.

In similar localities to the preceding. Very rare. First record: The present one.

E. 6. Clougha Scar, *G. Stabler* (in letter to the Authors).

Pteris aquilina, L.—Bracken. British type. Native. Area, 1, 2, 3, 4, 5, 6, 7, 8. Range, 0–1,500 feet. P. July–September.

Moorland banks, woods, barren pastures and commons. Very abundant in North and East, much less frequent in West. First record: *Linton, B.R.C. Rep.*, 1874, p. 86.

It ascends to 1,500 feet on Botton Head Fell, Mallowdale Fell, and Wardstone.

Cryptogramme crispa, R. Br.—Parsley Fern. Highland type. Native. Area, 1, 6, 7. Range 700–2,000 feet. P. August.

Stony mountain slopes on Silurian strata, and gritstone rocks at considerable elevations. Very rare. First record: *Woodward, Turner and Dillwyn's Botanists' Guide*, 1805.

N. 1. South-west side of Greygarth Fell at 2,000 feet; and very fine on Casterton Fell above Leck Eeck, *Wi.*

E. 6. Near Lancaster, *Woodward.* "Mr. Simpson informs me that on the moors near Lancaster, it grows at a very slight elevation above sea-level. Mr. W. Wilson found it in the same neighbourhood on the road to the Asylum," *Newman's Brit. Ferns.* Dale Gill and Antey Gill, Upper Hindburn ascending to 1,600 feet, *Wi.* Haylot Fell, Roeburndale.

7. Formerly on Nickey Nook, *A.B.* White side of Tarnbrook Fell, *Wi.*

Blechnum Spicant, With.—Hard Fern. British type. Native. Area, 1, 2, 3, 4, 6, 7, 8. Range, 50–1,800 feet. P. July–August.

Heathy banks, rocky woods, and cloughs amongst the gritstone hills. Frequent in East, less common in North and West. First record: *Ashfield, Fl. Prest.*, 1858. "Longridge Fell."

It ascends to 1,800 feet on Wardstone.

Asplenium Adiantum-nigrum, L.—Black Spleenwort. British type. Native. Area, 1, 2, 3, 4, 5, 6, 8. Range, 0–890 feet. P. July–August.

Rocks and walls, generally on slate or grit strata. Rare. First record: *Ashfield, Fl. Prest.*, 1858.

N. 1. Near Leck, *L.P.* Ease Gill, *Wi.* Near Whittington.

2. Near Newton, between Arkholme and Kirby Lonsdale, *Wi.*

N. 3. Near Silverdale Cove; and near the Post Office; and on cliffs near Gibraltar, *E. S. Pickard.* Trowbarrow, *L.P.* Formerly on sea-cliffs at Far Naze, *Wi.* Now destroyed by making Heysham Harbour.

W. 4. On bridges between Galgate and Glasson, and on a wall between Glasson and Lancaster, *Wh.*

5. Wyre banks near Skippool, *Pearson.* Near Lytham, *A.A.D.*

E. 6. On grit-rocks in Foxdale, very fine at 890 feet.

8. Goosnargh Churchyard walls, *Ashfield*, 1858. Leagram, *Fl. St.*

Asplenium marinum, L. — Sea Spleenwort. British type. Native. Area, 3. Range, 15-30 feet. P. August.

Sheltered caves and rock crevices on sea cliffs. Very rare. First record: *Lawson, Ray's Corresp.*, 1718, as *Filix marina anglica*, Park.

N. 3. "Under a shadowy sea-rock by Middleton near Lancaster," *Lawson*, 1688. "Mr. Simpson observed it abundantly upon rocks near Heysham," *Newman's Brit. Ferns.* Recorded also by *Ashfield.* Seen near Heysham in 1896, *Wi.* And by the Authors in 1906.

Asplenium viride, Huds.—Green Spleenwort. Highland type. Native. Area, 1, 3, 8. Range, ?–1,390 feet. P. July–August.

Damp rocky mountain gills on limestone. Rare, but locally abundant. First record: *Wilson, B.R.C. Rep.*, 1887, p. 113.

N. 1. Limestone rocks on Leck Fell and in Ease Gill, ascending to 1,390 feet, 1883, *Wi.*

3. Silverdale, *E. S. Pickard* (per *J.F.P.*).

E. 8. Leagram, *Fl. St.*

Asplenium Trichomanes, L.—Common Spleenwort. British type. Native. Area, 1, 2, 3, 4, 5, 6, 7, 8. Range, 0–1,250 feet. P. July–August.

Rocks and walls. Common in North; less frequent in the remaining divisions. First record: *Ashfield, Fl. Prest.*, 1858. "Bridge near Woodplumpton Brook."

It ascends to 1,250 feet on grit-rocks in Hindburndale.

Asplenium Ruta-muraria, L. — Wall Rue. British type. Native. Area, 1, 2, 3, 4, 5, 6, 7, 8. Range, 0–890 feet or higher. P. July–August.

Walls and rocks, mostly on—but not restricted to—limestone. Common in North; less plentiful elsewhere. First record: *Linton, B.R.C. Rep.*, 1874, p. 86.

It ascends to near the summit of Dalton Crag.

Athyrium Filix-fœmina, Roth.—Lady Fern. British type. Native. Area, 1, 2, 3, 4, 5, 6, 7, 8. Range, 0–1,520 feet. P. July-August.

Woods, hedgebanks, and shady places. Very common. First record: *Linton, B.R.C. Rep.*, 1874, p. 86.

Sub-sp. *A. incisum*, Newm.—Area, 1, 2, 3, 4, 5, 6, 7, 8. Frequent in damp sheltered woods and lanes. Ascending to 1,520 feet on Botton Head Fell. First record The present one.

Sub-sp. *A. rhæticum*, Roth.—Area, 2, 6, 7. Not unfrequent amongst the hills; in somewhat more exposed localities than the preceding. First record: *Wheldon, Journ. Bot.*, 1900.

N. 2. Near Over Kellet.

E. 6. Near Cowkins in Lower Hindburn, *Wi.*

7. Near Quernmore, 1899, and between Bay Horse and Five Lanes, *Wh.*

Ceterach officinarum, Willd. — Scaly Spleenwort. English type. Native. Area, 1, 3, 5, 8. Range, 0–600 feet. P. July.

Limestone rocks and walls. Rare. First record: *Simpson* in *M. & B. Nat. Print. Ferns*, 1859.

N. 1. Walls at Burrow, *Mr. Hindson* in *Newman's Brit. Ferns.*

3. Silverdale, *Simpson*, 1859. Seen there in 1902, *L.P.*

W. 5. In a locality near Garstang, *Wi.*

E. 8. Leagram, *Fl. St.* Near Chipping, *F. Clitheroe.*

Phyllitis Scolopendrium, Newm. (*S. vulgare*, Sym.).—Hart s Tongue. British type. Native. Area, 1, 2, 3, 4, 5, 6, 7, 8. Range, 0–875 feet or higher. P August.

Rocky woods and hedgebanks; especially abundant and luxuriant in the crevices of limestone pavement. Very common in North; rather uncommon in East and West. First record: *Linton, B.R.C. Rep.*, 1874, p. 86.

It ascends to 875 feet on Dalton Crag.

Cystopteris fragilis, Bernh.—Brittle Fern. Bladder Fern. British type. Native. Area, 1, 2, 3, 5, 6, 7, 8. Range, 0–1,390 feet. P. July–August.

Rocks and walls, generally, but not invariably on calcareous strata. Not common. First record: *Simpson, Moore & Braabury's Nature printed Ferns*, 1859.

N. 1. Ease Gill and Greygarth Fell, ascending to 1,390 feet, *Wi.*
2. Near Whittington, Dalton Crag, and Over Kellet, *Wi.*
3. Silverdale, Simpson, *Moore & Bradbury's Nature printed Ferns,* 1859. Warton and Yealand, *L.P.*
W. 5. Catforth, *Fl. Prest. N.*
E. 6. Foxdale, Roeburndale, and Dale Gill, Hindburn, *Wi.* Udale and near Pott Yeats, *Wh.*
7. [Formerly grew in Nickey Nook, but was rooted up, *A.B.*]. Rocky bank of the Wyre above Dolphinholme.
8. Bolton Roughs ; Leagram, *Fl. St.* Near Whitewell.

Var. *dentata,* Hook.

N. 3. Silverdale, *Moore & Bradbury,* 1859.
E. 8. Leagram, *Fl. St.*

Polystichum aculeatum, H. Schott.—Prickly Shield Fern. British type. Native. Area, 1, 2, 3, 6, 7, 8. Range, 0–1,420 feet. P. July–August.

Limestone scars and pavements, and occasionally in hedgebanks and thickets. Not very frequent, but locally abundant. First record : *Linton, B.R.C. Rep.,* 1874, p. 86.

Ascends to 1,420 feet in Upper Ease Gill, and to 890 feet in Foxdale.

Var. *lobatum,* Sw. — Less frequent than the preceding, but recorded from districts 1, 2, 8.

Polystichum angulare, Presl.—Willdenow's Fern. English type. Native. Area (5 or 7), 8. Range, ? P. July–August.

Thickets and shady banks. Very rare. First record : *Pearson, Fl. Prest.,* 1858.

E. 5 or 7. Plentiful in hedges between Garstang and Bowgrave, *Pearson,* 1858. Locality quite altered during the last 30 years—old hedgebanks and roadside ditches improved away, and the fern has disappeared with them, *Wi.*
8. Buck Banks, Leagram, *Fl. St.*

Lastrea montana, Moore. (*L. Oreopteris,* Presl.).—Sweet Mountain Fern. British type. Native. Area, 1, 2, 6, 7, 8. Range, 150–1,100 feet. P. August.

Damp rocky banks and heathy places amongst the hills. Common in North and East ; very rare, if found at all, in West. First record : *Pearson, Fl. Prest.,* 1860. " Ravine above Greenhalgh Castle."

It ascends to 1,100 feet on Mallowdale Fell.

Lastrea Filix-mas, Presl.—Male Fern. British type. Native. Area, 1, 2, 3, 4, 5, 6, 7, 8. Range, 0–1,500 feet. P. July–August.

Rocky woods, banks, and shady thickets. Very common. First record : *Linton, B.R.C. Rep.,* 1874, p. 86.

Ascends in North to 1,390 feet in Upper Ease Gill ; and in East to 1,500 feet on Botton Head Fell.

Var. *Borreri,* Newm.—Not unfrequent in the cloughs of East.

E. 6. Dale Gill, Hindburn, *Wi.* Clougha, *Wh.*
7. Fairsnape Fell, *Wh.*

Lastrea rigida, Presl.—Rigid Shield Fern. Local type. Native. Area, 1, 2, 3. Range, 150–1,000 feet. P. August.

Crevices of limestone pavement. Very scarce, except in one locality. First record : *Rev. J. Smythes, Phytologist,* 1843.

N. 1. Limestone scars on Leck Fell at 1,000 feet, *Wi.*
2. "Was found by the Rev. J. Smythes near the top lock of the Lancaster and Kendal Canal in Lancashire," *Phyt.* i., p. 478. Dalton Crag, *Wi.,* ascending to 870 feet.
3. Silverdale, *Moore & Bradbury's Nature printed Ferns,* 1859. Near Silverdale, 1878, *Wi.* Still there, 1902, *L.P.* Gatebarrow district, 1902, and since ; widely scattered, but not abundant, *Wi.*

Lastrea spinulosa, Presl.—Prickly Shield Fern. English type. Native. Area, 2, 3, 4, 8. Range, 0–500 feet. P. August.

First record : *Wilson, B.R.C. Rep.,* 1887, p. 113.

N. 2. Swamp near Over Kellet, 1886, and on Whittington Moor, and near Docker, *Wi.*
3. Thrang Moss, and in a swamp near Borwick, *Wi.* Heysham Moss.
W. 4. Cockerham Moss, *Wi.*
E. 8. Gaswood, Stonyhurst, *Fl. St.*

Lastrea aristata, Brit. & Rendle. (*L. dilatata,* Presl.).—Common Shield Fern. British type. Native. Area, 1, 2, 3, 4, 5, 6, 7, 8. Range, 0–1,950 feet. P. July–August.

Damp woods, hedgebanks, and shady rocky places. Common. First record : *Linton, B.R.C. Rep.,* 1874, p. 86.

Var. *nana,* Newm.—Amongst grit rocks on the higher fells of North and East. Rare.

N. 1. Greygarth Fell at 1,950 feet, *Wi.*
E. 6. North-west shoulder of Wardstone at 1,800 feet, 1902, *Wh. & Wi.* Dale Gill at 1,250 feet, *Wi.*
7. Hell Crag.

Polypodium vulgare, L.—Polypody. British type. Native. Area, 1, 2, 3, 4, 5, 6, 7, 8. Range, 0–1,520 feet. P. August.

Woodland banks, hedges, and old tree trunks. Common. First record : *Linton, B.R.C. Rep.,* 1874, p. 86.

It ascends in North to 1,520 feet upon rocks on Greygarth Fell, and in East to 1,250 feet in Dale Gill, Hindburn.

Polypodium Dryopteris, L.—Oak Fern. Scottish type. Native. Area, 1, 6, 7, 8. Range, 250–1,250 feet. P. July.

Rocky, wooded cloughs and gills in the hilly districts. Rather rare, but locally abundant. First record : *Ashfield, Fl. Prest.,* 1858.

N. 1. Gorge of the Greeta, plentiful.
E. 6. Near Botton Mill, and at 1,250 feet in Dale Gill, Hindburn, *Wi.* Middle Gill, Hindburn ; in the gully on Mallowdale Fell ; and in Foxdale, *Wh. & Wi.* Gully above Lower Salter Bridge, and by the Harter Beck, Roeburndale, *Wi.*
7. Nickey Nook, *Ashfield,* 1858. Wyre banks below Abbeystead.
8. Dean Brook ; Leagram ; Bolton Roughs, *Fl. St.*

Polypodium Robertianum, Hoffm. — Limestone Polypody. English type. Native. Area, 1, 2, 3, 7. Range, 100–1,200 feet. P. July.

Crevices of limestone "pavement" and stony "scree" slopes of limestone scars ; less frequently on banks of calcareous shale. Locally abundant. First record : *Wilson, B.R.C. Rep.,* 1884, p. 26.

N. 1. Leck Fell and Ease Gill, ascending to 1,200 feet, *Wi.*
2. Dalton Crag, 1883, and since, *Wi.*
3. Rocks behind the Tarn, Silverdale, *E. S. Pickard* (In *Herb., J.F.P.*).
E. 7. Banks of calcareous shale, Gorge of the Wyre.

Polypodium Phegopteris, L. — Beech Fern. Scottish type. Native. Area, 1, 6, 7, 8. Range, 250–1,450 feet. P. July.

In similar situations to the Oak Fern. Rather rare. First record : *Ashfield, Fl. Prest.,* 1858.

N. 1. Near Anneside, Leck Fell, *L.P.* Ease Gill, *Wi.* Gorge of the Greeta.
E. 6. Whiteray Gill ; hillside below White Moss, and below Botton Mill, Hindburn ; Udale ; and gully above Lower Salter Bridge, *Wi.* Head of Foxdale, *Wh. & Wi.* Trough Brook Gill, *Wi.*
7. Calder Valley near Garstang, *Wi.* Great Clough of Tarnbrook Fell at 1,450 feet ; south-west slope of Wardstone ; and on the White Side of Tarnbrook Fell. Side gully leading into Brock, Lower Bleasdale.
8. Brockholes Wood, *Ashfield,* 1858. Hodder Wood, Leagram, and Bolton Roughs, *Fl. St.*

Osmunda regalis, L.—Royal or Flowering Fern. British type. Native. Area, formerly grew in 3, 4, 5, 7, 8. Range, 0–550 feet. P. July.

Thickets on peat bogs and swampy ground in woods. Formerly frequent, now very rare ; almost extinct. First record : *Jenkinson, Brit. Pl.,* 1775, p. 241.

N. 3. [On the side of White Moss, Yealand ; and on the sides of Leighton Moss, *Jenkinson,* 1775.] Gone now from these localities ; but it still grows, according to Dr. Lees, in a station in this district.
W. 4. Winmarleigh Moss and Nateby Moss, *Pearson.* Cogie Hill, *P.J.H.,* is practically the same locality. [Formerly near Garstang, *Wi.*].
5. In two places near Inskip, *P.J.H.* [Near Catforth, *Ashfield.*]
E. 7. [Near Hawthornthwaite, Wyresdale, and Claughton near Garstang in two or more localities, now gone, *A.B.*].
8. Wood between Higher and Lower Brockholes, very plentifully and luxuriantly, *Ashfield.* Leagram, *Fl. St.*

Probably several more of the above records should now be bracketed as extinct. This fern was formerly so very abundant in Lancashire, that in certain parts its fronds were cut by farmers to line baskets of fruit and vegetables for the markets.

Ophioglossum vulgatum, L.—Adder's Tongue. British type. Native. Area, 2, 3, 5, 6, 7, 8. Range, 0–740 feet. P. June.

Damp pastures and open places in woods. Frequent. First record : *Jenkinson, Brit. Pl.,* 1775, p. 241.

N. 2. Henridden, very abundant, *Wi.*
3. " Found by one of my pupils on Yealand Common betwixt Yealand and Warton, under the high ridge of rocks plentifully," *Jenkinson,* 1775. Still there in 1903, and at Silverdale, *Wi.*
W. 5. St. Annes, *C.B.*
E. 6. Near Haylot at 740 feet, *Wi.*
7. Frequent about Garstang, *Wi.*
8. Longridge Fell. *Fl. Prest. N.*

Botrychium Lunaria, Sw.—Moonwort. British type. Native. Area, 2, 3, 7, 8. Range, 0–700 feet or higher. P. May–July.

Hill pastures. Less frequent than the preceding. First record : *Jenkinson, Brit. Pl.,* 1775, p. 241.

N. 2. Hilly pasture north-east of Dunnald Mill Hole.
3. " Under the ledge of rock between Yealand and Warton, and in the Flatwood, Yealand, *Jenkinson,* 1775." Still at Warton, *L.P.*
E. 7. Sullam, near Garstang, *Wi.*
8. Forty Acre Field, Stonyhurst, and Leagram, *Fl. St.* Longridge Fell, *Fl. Prest. N.*

EQUISETACEÆ.

Equisetum maximum, Lam.—Great Horse-tail. English type. Native. Area, 1, 4, 6, 7, 8. Range, 50–500 feet. P. April.

Damp woods and swampy thickets, especially amongst the hills. Frequent in East; rare elsewhere. First record: *Linton, B.R.C. Rep.,* 1874, p. 86.

   N. 1. Near Wennington, *Wi.* Woods by the Greeta.

   W. 4. Cabus.

   E. 6, 7, 8. Localities too numerous to mention, and locally very abundant.

Equisetum arvense, L. — Field Horse-tail. British type. Native. Area, 1, 2, 3, 4, 5, 6, 7, 8. Range, 0–1,420 feet. P. March–May.

Roadsides, banks, and waste ground; generally on stiff soils. Very common. First record: *Linton, B.R.C. Rep.,* 1874, p. 86.

   Ascends in North to 1,420 feet in Upper Ease Gill; and in East to 900 feet on Haylot Fell.

Equisetum sylvaticum, L.—Wood Horse-tail. British type. Native. Area, 1, 4, 6, 7, 8. Range, 0–910 feet. P. May.

Damp woods and bushy places. Frequent amongst the hills; ascending to 910 feet on Haylot Fell; rare in the low country. First record: *Ashfield, Fl. Prest.,* 1860.

   N. 1. Near Cantsfield.

   W. 4. Near Winmarleigh.

   E. 6, 7 and 8. Found in all three districts.

Equisetum palustre, L. — Marsh Horse-tail. British type. Native. Area, 1, 2, 3, 5, 6, 7, 8. Range, 0–1,000 feet or higher. P. July.

Damp places, marshes, and moors. Common. First record: *Linton, B.R.C. Rep.,* 1874, p. 86.

Var. *polystachium,* Auct.

   E. 8. Side of the old Seminary Pond, *Fl. St.*

Equisetum limosum, L.—Water Horse-tail. "Fishing-rods." "Toad-pipes." British type. Native. Area, 1, 2, 3, 4, 5, 6, 7, 8. Range, 0–400 feet or higher. P. July.

Ponds, slow streams, and marshes. Common. First record: *Linton, B.R.C. Rep.,* 1874, p. 86.

Equisetum hyemale, L.—Dutch Rush. Scottish type. Native. Area, 6. Range, 20–30 feet. P. August.

Damp thickets and shady banks of streams. Very rare. First record: *Wilson, Journ. Bot.,* 1900, p. 47.

   E. 6. Bank of the Lune near Halton, 1899, *Wi.*

Equisetum variegatum, Schleich. — Variegated Horse-tail. Scottish type. Native. Area, 5. Range, 0–25 feet. P. July–August.

Damp sandy ground on the coast. Very rare. First record: *Wheldon, Journ. Bot.,* 1900, p. 47.

   W. 5. Damp places on the sandhills between Lytham and St. Annes, and between St. Annes and South Shore, 1898, *Wh.*

LYCOPODIACEÆ.

Lycopodium Selago, L. — Fir Club-moss. Local name in Wyresdale, "Shepherd's Luck" British type. Native. Area, 1, 6, 7, 8. Range, 85–1,950 feet. P. April–May.

High stony moors and rocky places amongst the hills. Rather rare. First record: *Linton, B.R.C. Rep.,* 1874, p. 86.

   N. 1. Greygarth Fell at 1,950 feet, and Leck Fell, on grit; and in Ease Gill, on limestone, *Wi.*

   E. 6. Lythe Fell, Hindburn; *Wi.* Clougha, *Wh.* Haylot Fell.

   7. Very fine on rocks on both sides of Tarnbrook Fell; and at only 85 feet on Permian Red Sandstone in a railway cutting east of Garstang, *Wi.*

   8. Rocks by Dean Brook, *Fl. St.*

Lycopodium clavatum, L.—Common Club-moss. Stag's-horn. British type. Native. Area, 1, 6, 7. Range, 700–1,700 feet. P. July–September.

Bare stony slopes on the fells. Rare. First record: *Wilson, B.R.C. Rep.,* 1884, p. 27.

   N. 1. Leck Fell, 1½ miles from the Westmorland border, 1883, *Wi.*

   E. 6. Goodber Common, and by the Harter Beck, Roeburndale, *Wi.* Middle Gill, Hindburn.

   7. Catshaw Greave, and north side of Harrisend Fell in very small quantity, *Wi.* Greenside, near Tarnbrook, and Trough of Bowland.

Lycopodium alpinum, L.—Savin-leaved Club-moss. Highland type. Native. Area, 1. Range, 1,700–1,350 feet. P. August. High stony moorlands. Very rare. First record: *Wilson, B.R.C. Rep.,* 1884, p. 27.

   N. 1. Greygarth Fell and Upper Ease Gill, *Wi.*

W

Selaginella selaginoides, Link.—Lesser Club-moss. Highland type. Native. Area, 1, 3, 5, 7. Range, 0–1,600 feet. P. August.

Wet mossy places and margins of tarns and springs. First record: *Wilson, B.R.C. Rep.,* 1883, p. 249.

   N. 1. Greygarth Fell, Leck Fell and Ease Gill; and abundant by the Ireby Beck, *Wi.* Gorge of the Greeta.

   3. Hawes Water, 1881–2, *Wi.*

   W. 5. Damp hollows on the Lytham sandhills, 1877, *Wi.* Still there, *H.B.* (sp).

   E. 7. Near Damas Gill Head.

CHARACEÆ.

Chara fragilis, Desv.—British type. Native. Area, 1, 2, 3. Range, 0–250 feet. June–August.

Slow and stagnant waters. Rare. First record: *W. Wilson, B.R.C. Rep.,* 1877, p. 255. "Near Lancaster."

   N. 1. In a small backwater of Leck Beck below Cowan Bridge, *L.P.*

   2. Bog near Docker, and canal between Borwick and Burton, and near Lancaster, *Wi.*

   3. Wood Well, 1898, but not in 1901, *L.P.* Near Leighton Beck, *Wi.*

Var. *delicatulum* (Braun.) (*C. delicatula,* Braun.).

   N. 3. Clear slow-flowing stream between Borwick and Yealand, June, 1905.

Chara vulgaris, L.—British type. Native. Area, 3, 4, 5. 8. Range, 0–450 feet.

Ponds and ditches. First record: *Fl. St.,* 1891, p. 45.

   N. 3. Ditches near Silverdale Station, *Wi.*

   W. 5. Shore near St. Annes, and ditch between Blackpool and Marton, *Wh.*

   E. 8. Stonyhurst Infirmary Plantation, Crowshaw Reservoir, and quarry near Leagram Mill, *Fl. St.,* 1891. Thornley Quarry, *Wi.*

Var. *longibracteata,* Kutz.

   W. 4. Pond on the shore south of Glasson Dock, *Wh.*

   5. Pond in a potato field left side of Middle Road going from Blackpool to St. Annes, 1900, *Wh.*

Var. *papillata,* Wallr.

   N. 3. Wood Well, Silverdale, 1899, *Wh.*

Chara aspera, Willd.—British type. Native. Area, 4. Range, ? July–August.

Slow streams. Rare. First record: *Beesley, Wheldon,* and *Wilson, Journ. Bot.,* 1902, p. 350.

   W. 4. Canal near Cabus, July, 1901, *H.B., Wh.* & *Wi.*

Nitella glomerata, Chevall.—English type? Native. Area, 5, 6. Range, 0–70 feet.

Pools amongst the sandhills and stagnant waters. Rare. First record: *Wheldon, Journ. Bot.,* 1902, p. 350.

   W. 4. Pond near Glasson Dock, *Wh.*

   5. Shallow pools (where rain had stood and which must be often quite dry) on the sandhills, St. Annes, *Wh.*

   E. 6. Canal near Lancaster, 1900, *Wh.*

Nitella prolifera, Kutz.—English type? Native. Area, 4, 5. Range, only seen at 70 feet. July–August.

Slow streams. Rare. First record: The present one (referred to doubtfully in *Journ. Bot.,* 1902, p. 350).

   W. 4. Canal near Cabus, 1900.

   5. Canal near Brock, *Wi.*

Nitella opaca, Agardh.—British type. Native. Area, 4, 7. Range, 0–60 feet. May–June.

Ponds and ditches. Very rare. First record: *Wilson, Journ. Bot.,* 1905, p. 96.

   W. 4. Deep ditch between the river Conder and the Glasson Canal, not fruiting, but closely resembling our other examples of this species in habit, *Wi.*

   E. 7. The first vegetation to appear in a newly made pond near Garstang, June 1903, *Wi.*

The pond was situated in a grass field and had only been dug about 18 months. It was made as an experiment to see what water plants would appear in it, and was carefully railed off to prevent access of cattle. Besides the *Nitella,* five other species appeared, viz., *Alisma Plantago, Callitriche species, Glyceria fluitans, Juncus conglomeratus* and *J. articulatus.*

[**Nitella flexilis,** Agardh.

A handsome species, resembling this perhaps more than any other, but without fruit, occurred abundantly in Grizedale Reservoir, September, 1900. Messrs. Groves think it referable to this or to *N. opaca.*]

MUSCI.

SPHAGNALES.

Sphagnum fimbriatum, Wils. — Area, 2, 4, 6, 7. Range, 25–1,400 feet.

On the wetter parts of the peat mosses and in swampy places amongst the hills. Not common. First record: *Wilson, Journ. Bot.*, 1899. p. 467.

N. 2. Bog near the Redwell Inn.

W. 4. Cockerham Moss, the vars. *tenue*, Grav., and *compactum*, Warnst.

E. 6. Tatham Beck and Thrushgill Gully, Hindburn. The var. *robustum*, Braith., occurs on Thrushgill Fell.

7. Calder Valley above Oakenclough, 1898, *Wi.* Upper Grizedale. Damas Gill Head.

Sphagnum Russowii, Warnst.—Area, 6, 7. Range, 700–1,200 feet.

Bogs amongst the hills. Rare. First record: The Authors, *Journ. Bot.*, 1901. p. 295.

Var. *virescens*, Russ.

E. 6. Caton Moor, 1899.

Var. *rhodochroum*, Russ.

E. 6. Tatham Beck, 1899, *Wi.* Dale Gill, Hindburn.

7. Grizedale, near Abbystead.

Sphagnum Warnstorfii, Russ.—Area, 2, 4, 6, 7. Range, 25–800 feet or higher.

Wet birch swamps, etc. Rare. First record: *Wilson, Journ. Bot.*, 1901. p. 295.

Var. *versicolor*, Russ.

N. 2. Whittington Moor and bog near Docker, *Wi.*

E. 6. Udale.

7. Marshaw Fell, Wyresdale, 1900, *Wi.*

Var. *virescens*, Warnst.

N. 2. Bog near Docker, *Wi.*

Var. *flavescens*, Warnst.

W. 4. Cockerham Moss, 1900.

Sphagnum rubellum, Wils.—Area, 1, 2, 3, 4, 6, 7, 8. Range, 20–2,020 feet.

Moorland bogs. Common in East. First record: *Wilson, Journ. Bot.*, 1899, p. 467.

N. 1. Greygarth Fell (var. *versicolor*, Russ., and var. *viride*, Warnst.).

2. Whittington Moor (var. *purpurascens*, Warnst.).

3. Heysham Moss (var. *versicolor*, Russ., and var. *pallescens*, Warnst.).

W. 4. Cockerham Moss, 1898, *Wi.* (vars. *pallescens*, Warnst.; *purpurascens*, Warnst.; *flavum*, C. Jens. and *versicolor*, Russ.).

E. 6, 7 and 8. Frequent in all the divisions, principally as var. *rubrum*, Grav. (with the *forma robusta*, Warnst.), var. *purpurascens*, Warnst.; var. *versicolor*, Russ., and more rarely, as var. *viride*, Warnst.

Sphagnum fuscum, Klinggr.—Area, 4. Range, only seen at 25 feet.

Deep bogs, amongst *S. medium*. Very rare. First record: The Authors, *Journ. Bot.*, 1901.

W. 4. Cockerham Moss, 1900 (var. *fuscescens*, Warnst.). And var. *pallescens*, Russ., in the same vicinity 1903, *Jones & Wh.*

We have not yet found this on the moorlands associated with *S. rubellum*, which Mr. Horrell indicates as its usual habitat elsewhere.

Sphagnum acutifolium, R. & W.—Area, 1, 3, 4, 6, 7, 8. Range, 20–2,020 feet.

Wet peaty moorlands. Frequent. First record: The Authors, *Journ. Bot.*, 1901, p. 295.

The following varieties have been noted, localities for the rarer ones only being indicated :—Var. *flavo-rubellum*, Warnst. (Mallowdale Fell) ; var. *griseum*, Warnst. ; var. *pallescens*, Warnst. (with *forma robusta, subf. dasyclada*) ; var. *pallido-glaucescens*, Warnst. (Harrisend Fell) ; var. *purpurascens*, Warnst. (Peacock Hill and summit of Greygarth Fell) ; var. *rubrum*, Warnst. (Whitmoor and Clougha) ; var. *versicolor*, Warnst., *f. robusta* (Tarnbrook Fell) ; var. *viride*, Warnst., with *f. ano-orthoclada*, *subf. dasyclada*, and *f. gracilis, subf. drepanoclados*.

Sphagnum quinquefarium, Warnst.—Area, 1, 6, 7. Range, 600–2,020 feet.

Upland moors. Rare. First record: *Wheldon, Journ. Bot.*, 1901, p. 295.

N. 1. Near Ease Gill Kirk (var. *viride*, Warnst.) ; Greygarth Fell (vars. *fusco-flavum*, Warnst., and *roseum*, Warnst.). Leck Fell (var. *pallescens*, Warnst.).

E. 6. Mallowdale Fell, 1900, *Wh. & Wi.* Clougha, *Wh.*

7. Hawthornthwaite Greave (var. *viride*, Warnst.).

Shagnum subnitens, R. & W.—Area, 1, 2, 3, 4, 5, 6, 7, 8. Range, 20–1,700 feet or higher.

Bogs and dripping rocks by streams. Common. First record: The Authors, *Journ. Bot.*, 1901, p. 295.

This is the most abundant representative of the *acutifolia* in West Lancashire. The more frequent varieties are, *flavo-rubellum*, Warnst., *versicolor*, Warnst., *violascens*, Warnst., *obscurum*, Warnst., and *virescens*, Warnst. Less frequent are var. *flavescens*, Warnst. (Gressingham Moor and Clougha), var. *griseum*, Warnst. (Ease Gill, Bleasdale Fell, and Wardstone) ; and var. *pallescens*, Warnst. (Upper Grizedale and Haylot Fell).

Sphagnum squarrosum, Pers.—Area, 2, 3, 6, 7, 8. Range, 20–1,400 feet.

Margins of moorland pools and streams and near boggy springs. Frequent. First record: *Wheldon, Journ. Bot.*, 1899, p. 467. "Dolphinholme."

The vars. *spectabile*, Russ., and *subsquarrosum*, Russ., occur in numerous stations. The var. *imbricatum*, Schimp., which occurs in South Lancashire, has not yet been detected within our limits.

Sphagnum teres, Angstr.—Area, 4, 7. Range, 25–500 feet or higher.

Deep marshy bogs. Very rare. First record: *Wheldon, Journ. Bot.*, 1899, p. 467.

Var. *subsquarrosum*, Warnst.

E. 7. Dolphinholme, 1897, *Wh.*

Var. *squarrosulum*, Warnst.

4. Cockerham Moss.

Sphagnum riparium, Angstr.—Area, 4. Range, only seen at 25 feet.

Deep wet bogs. Very rare. First record: *H. Beesley and the Authors, Journ. Bot.*, 1902, p. 413. This, perhaps the most handsome British species, occurs abundantly, but over a very restricted area on Cockerham Moss, where it was discovered in 1901. It had not been recorded with certainty from any British locality prior to this, and was identified for us by Mr. Horrell.

Sphagnum cuspidatum, R. & W.—Area, 1, 2, 3, 4, 6, 7, 8. Range, 20–2,020 feet.

Submerged in moorland pools. Common. First record: *Wheldon, Journ. Bot.*, 1899, p. 468. "Longridge Fell."

The vars. *falcatum*, Russ., *submersum*, Schp., and *plumosum*, Nees., all occur in numerous localities.

Sphagnum trinitense, C. Mull.—Area, 4, 7, 8. Range, 25–800 feet and probably much higher.

Submerged in moorland pools. Rare. First record: The Authors, *Journ. Bot.*, 1901, p. 296.

W. 4. Cockerham Moss, 1900.

E. 7. Lower Bleasdale, *Wi.*

8. Longridge Fell, *Wh.* Moor near Lickhurst, *Wi.*

Sphagnum pulchrum, Warnst.—Area, 3, 4, 6, 7. Range, 20–1,200 feet.

Deep bogs. Rare. First record: *Wilson, Journ Bot.*, 1899, p. 468.

N. 3. Heysham Moss.

W. 4. Cockerham Moss.

E. 6. Upper Roeburndale, 1898, *Wi.*

7. Grizedale, near Abbeystead ; Tarnbrook Fell.

Sphagnum Torreyanum, Sull.—Area, 6. Range, uncertain.

Submerged in moorland pools. Rare. First record: The Authors, *Journ. Bot.*, 1902, p. 413.

E. 6. Mallowdale Fell at about 1,400 feet, September, 1900 (*teste Horrell.*)

Sphagnum obtusum, Warnst.—Area, 4. Range, only seen at 25 feet.

Deep bogs amongst birches. Rare. First record: The Authors in "The European Sphagnaceæ," E. C. Horrell, *Journ. Bot.*, 1900, p. 343.

W. 4. Cockerham Moss, 1900.

Sphagnum recurvum, R. & W.—Area, 1, 2, 3, 4, 5, 6, 7, 8. Range, 20–2,020 feet.

Marshy places on moors. Common. First record: The Authors in "The European Sphagnaceæ," E. C. Horrell, *Journ. Bot.*, 1900, p. 296. "Cockerham Moss." The var. *mucronatum*, Warnst., is the more common plant and occurs in every district.

Var. *amblyphyllum*, Warnst.

N. 1. Bog near Docker.

E. 6. Upper Roeburndale. Clougha, *Wh.*

7. Upper Grizedale, *Wi.*

8. Longridge Fell, *H.B.*

A curious small form occurs in running water on Mallowdale Fell, with rigid, brittle stems and branches, and more closely imbricate leaves.

Sphagnum molluscum, Bruch.—Area, 1, 3, 4, 6, 7. Range, 20–1,000 feet or higher.

Damp heaths and moorlands. Rare. First record: *Wilson, Journ. Bot.,* 1899, p. 467 (as *S. tenellum,* Ehrh.).

 N. 1. Ease Gill.
  3. Heysham Moss, fruiting freely.
 W. 4. Cockerham Moss (*f. compacta,* Warnst.).
 E. 6. Whitmoor, 1898, *Wi.*
  7. Tarnbrook Fell.

Sphagnum compactum, DC.—Area, 2, 6, 7, 8. Range, from about 400–1,000 feet.

Damp heaths and moorlands. Frequent. First record: *Wilson, Journ. Bot.,* 1899, p. 467. Vars. *imbricatum,* Warnst., and *subsquarrosum,* Warnst., occur.

 N. 2. Gressingham and Arkholme Moors, *Wi.*
 E. 6. Whitmoor, 1898, *Wi.* White Moss, Hindburn.
  7. Tarnbrook Fell, *Wi.* Catshaw Fell, *Wi.*
  8. Longridge Fell, *Wh.* (" a *hemi-isophyllous* form," Warnst.).

Sphagnum contortum, Limpr.—Area, 2, 6. Range, 200–1,200 feet or higher.

Very wet bogs amongst bushes, etc. Rare. First record: *Wilson, Journ. Bot.,* 1899, p. 467 (as *S. laricinum,* Spruce).

 N. 2. Bog near Docker.
 E. 6. Wolfhole Crag, 1898, *Wi.*

Sphagnum subsecundum, Limpr.—Area, 1, 2, 7, 8. Range, 400–2,020 feet.

Boggy moorlands. Rare. First record: *Wheldon, Journ. Bot.,* 1899, p. 467.

 N. 1. Summit of Greygarth Fell.
  2. Quarry near Over Kellet, *J. W. Hartley,* and the Authors.
 E. 7. Abbeystead Fell.
  8. Longridge Fell, 1898, *Wh.*

Sphagnum inundatum, Warnst.—Area, 2, 5, 6, 7, 8. Range, 25–1,200 feet or higher.

Submerged, or in very wet places on moorland bogs. Perhaps more frequent than the records indicate. First record: *Wilson, Journ. Bot.,* 1901, p. 296.

 N. 2. Lords Lot Wood, Arkholme Moor, 1900, and in Wash Dub Wood, *Wi.*
 W. 5. Near Catforth, *H.B.*
 E. 6. Clougha, *Wh.* Lythe Fell, Hindburn, *Wi.*
  7. Lower Bleasdale.
  8. Longridge Fell ! *H.B.*

Sphagnum Gravetii, Warnst.—Area, 6, 7, 8. Range, 100–1,200 feet or higher.

Submerged in moorland poo s. Rare. First record: *Wilson, Journ. Bot.,* 1902, p. 413.

 E. 6. Tatham Beck, Hindburn, 1899, *Wi.*
  7. West side of Harrisend Fell Lee Fell, and in the Great Clough of Tarnbrook Fell.
  8. Fulwood, *H.B.*

Sphagnum rufescens, Warnst.—Area, 1, 2, 3, 5, 6, 7, 8. Range, 25–2,020 feet.

Ditches, boggy places, and dripping rocks on the fells. Common. First record: The Authors, *Journ. Bot.,* 1901. " Whittington Moor and Longridge Fell."

  This is the most common species of the *Subsecunda* group, and is very variable.

Sphagnum aquatile, Warnst.—Area, 6, 7, 8. Range, 600–1,200 feet or higher ?

Pools and ditches on the fells. Rare. First record: The Authors, in " *The European Sphagnaceæ,*" E. C. Horrell, *Journ. Bot.,* 1900, p. 389.

 E. 6. Whitmoor.
  7. Tarnbrook Fell.
  8. Longridge Fell, 1898, *Wh.*

  The Whitmoor locality appears to be t he first British record for this plant.

Sphagnum crassicladum, Warnst.—Area, 6, 7, 8. Range, 300–1,700 feet.

Submerged in deep moorland pools and ditches. Common in East, occurring on nearly all the fells. First record: *Wheldon, Journ. Bot.,* 1901, p. 296. "Slope of Fairsnape Fell towards Chipping."

Sphagnum obesum, Warnst.—Area, 2. Range, only seen at about 370 feet.

Submerged in deep ponds. Very rare. First record: The Authors, *M.E.C. Report,* 1906, p. 210.

 N. 2. Pond in Lords Lot Wood, May, 1905 (*teste Horrell*).

Sphagnum turfaceum, Warnst.—Area, 2, 3, 4, 6, 8. Range, 25–1,000 feet or higher.

Lowland mosses and moorland bogs. Rare. First record: *Wheldon,* in *Horrell's* " *The European Sphagnaceæ,*" *Journ. Bot.,* 1900, p. 476.

 N. 2. Lords Lot Wood, Arkholme Moor, *Wi.*
  3. Heysham Moss.
 W. 4. Cockerham Moss.
 E. 6. Clougha, *Wh.*
  8. Longridge Fell, 1898, *Wh.*

Sphagnum medium, Limpr.—Area, 3, 4, 6, 7. Range, 20–1,600 feet or higher.

Peat bogs and wet moors. Not common, but locally abundant. First record: The Authors, in " *The European Sphagnaceæ,*" E. C. Horrell, 1900, p. 477.

 N. 3. Heysham Moss.
 W. 4. Cockerham Moss.
 E. 6. Upper Roeburndale, 1898, *Wi.* White Moss, and Tatham Moor, Hindburndale.
  7. Tarnbrook Fell.

  The vars. *roseum,* Warnst., *roseo-pallescens,* Warnst., and *glauco-purpurascens,* Warnst., all occur, and are connected by intermediate forms.

Sphagnum cymbifolium, Warnst.—Area, 1, 2, 3, 4, 5, 6, 7, 8. Range, 25–1,700 feet or higher.

Boggy woods and cloughs amongst the hills. Frequent and widely distributed, but much less common than the next species. First record: The Authors, *Journ. Bot.,* 1899, p. 467.

  The vars. *glaucescens,* Warnst., *pallescens,* Warnst., and *glauco-pallens,* Warnst., occur in several districts each. The var. *flavo-glaucescens,* Warnst. is plentiful at Damas Gill Head, and on Whittington Moor ; in the latter locality it is accompanied by var. *fusco-pallens,* Warnst., which is also reported by Mr. Beesley from near Catforth. The var. *fusco-glaucescens* occurs in the bog near Docker.

Sphagnum centrale, Jensen.*—Area, 2, 6, 7.

Deep bogs. Very rare. First record: *Wilson, Journ. Bot.,* 1902, p. 413.

 N. 2. Bog near Docker, 1899, *Wi.*
 E. 6. Upper Roeburndale. Bog at the foot of Clougha, very fine, *Wh.*
  7. Head of Damas Gill.

 *It was recorded in *Journ. Bot.* as *S. cymbifolium,* var. *carneum,* Warnst. As Warnstorf now includes this variety under *S. centrale,* Jens., the number of West Lancashire *Sphagna* given in the table on page 107 must be increased to 32.

Sphagnum papillosum, Lindb.—Area, 1, 2, 3, 4, 5, 6, 7, 8. Range, 20–2,000 feet.

On mosses and heaths. Common, generally occurring wherever Sphagna grow at all. First record: *Wheldon, Journ. Bot.,* 1899, p. 467. " Longridge Fell."

  The vars. *normale,* Warnst., and *sublæve,* Limpr., are almost equally distributed ; the latter is exceptionally luxuriant and abundant on a dripping bank in Udale, growing in large cushions 18 inches deep. The var. *conferta,* Lindb., is common in drier situatious.

ANDREÆALES.

ANDREÆACEÆ.

Andreæa petrophila, Ehrh.—Area, 1, 6, 7, 8. Range, 450–2,050 feet.

Mountain rocks of Silurian, Yoredale, and Millstone Grit. Somewhat rare. First record: *Wilson, Journ. Bot.,* 1899, p. 468.

 N. 1. On Silurian rocks in Ease Gill, 1898, *Wi.* Greygarth Fell, on Millstone Grit, at 2,050 feet.
 E. 6. Greenbank Fell and Middle Gill, Hindburn ; Mallowdale Fell and north-west shoulder of Wardstone.
  7. Catshaw Greave, *Wi.* Head of Grizedale near Abbeystead, south-west slope of Wardstone, and rocks in Gavells Clough.
  8. Whitestone Clough.

Andreæa Rothii, W. & M.—Area, 1, 6, 7, 8. Range, 700–1,800 feet.

Millstone Grit rocks on the higher fells. Rare. Fruiting. First record: The Authors, *Journ. Bot.,* 1901, p. 296.

 N. 1. On grit rocks above the upper fa ls in Ease Gill, *Wi.* Greygarth Fell at 1,750 feet.
 E. 6. Upper Roeburndale, 1899. Udale and Foxdale, *Wh.* & *Wi.* Little Moor Beck, Hindburn, *Wi.* Wardstone at 1,800 feet.
  7. Great Clough of Tarnbrook Fell, and Harrisend Fell, *Wi.* Hawthorn thwaite Greave ; and crags on the south side of Wardstone.
  8. Dewhurst Clough, Burnslack Fell.

Var. *falcata,* Lindb.—Area, 6, 7.

 E. 6. Udale, with fruit, *Wi.* Roeburndale, and Wardstone.
  7. Catshaw Greave, 1900, *Wi.* Ha wthornthwaite Greave, and on the south side of Wardstone at 1,600 feet. Very fine on Hell Crag at 1,490 feet.

Andreæa crassinervia, Bruch.—Area, 6, 7.   Range, 900–1,250 feet.

Millstone Grit rocks on the higher fells.   Very rare.   First record : The Authors, *Journ. Bot.*, 1901, p. 296.

> E. 6.   Rocks by the Roeburn, at the head of Roeburndale, October, 1899.
>
> 7.   Great Clough of Tarnbrook Fell.
>
> With us the *Andreæaceæ* occur principally on the horizontal or sub-horizontal surfaces of rocks, in a similar manner to, and often accompanied by, the *Gyrophoræ* and other lichens.   Their scarcity on the perpendicular sides of larger rocks may be due to their intolerance of water charged with smoke-impurities derived from the surfaces above them in such situations.   In the lower country the *Orthotrichaceæ*, and some allied plants, show a similar predilection for *horizontal* trunks and branches.

BRYALES.

POLYTRICHACEÆ.

Catharinea undulata, W. & M.—Area, 1, 2, 3, 4, 5, 6, 7, 8. Range, 0–1,000 feet or higher.

Damp woods and shady banks, especially on sand or clay. Very common.   Fruiting.   First record : The Authors, *Journ. Bot.*, 1899, p. 468.

Var. *minor*, W. & M.

> N. 3.   Under trees near the shore, Hawes Water, near Silverdale, 1906, *Wh.*

Catharinea crispa, James.—Area, 6, 7, 8.   Range, 400–1,100 feet.

Sandy and stony sides and beds of streams amongst the gritstone moors of East.   Locally abundant.   Male plants only. First record : *Wheldon, Journ. Bot.*, 1899, p. 468.

> E. 6.   Tatham Beck and Botton, Hindburn ; Mallowdale and Harter Beck, Roeburndale ; Foxdale and Udale.
>
> It ascends to the exceptional altitude of 1,100 feet on Haylot Fell.
>
> 7.   Tarnbrook and Marshaw, Wyresdale.   Damas Gill, *Wh.*   Upper Grizedale, *Wi.*   Shore of Abbeystead Reservoir at 400 feet.
>
> 8.   Longridge Fell, 1899, and at 400 feet near Kemple End, *Wh.*
>
> This moss is a typical submontane species, and has (compared with the rest of the *Polytrichaceæ*, except the distinctly lowland species) a narrow altitudinal range. It does not usually ascend far up on the fells, preferring the lower reaches of the cloughs, nor does it ever seem to follow the streams down into the low country.   It is most abundant between 500 and 800 feet.

> N. 1.   Ease Gill, 1898, and on Greygarth Fell at 1,700 feet, *Wi.*   Gorge of the Greeta.
>
> 2.   Wash Dub Wood, *Wi.*
>
> E. 6.   Burn Moor, Hindburn, and Foxdale, *Wi.*   Near Botton, and on Mallowdale Fell.
>
> 7.   Barnacre Moor, Catshaw Greave, and Fairsnape Fell, *Wi.*   Grizedale, *Wh.* & *Wi.*   Scorton, *H.B.*
>
> 8.   Longride Fell, *Wh.*

Var. *humile*, Wahl.

> 8.   Near the summit of Jeffrey Hill, Longridge, 1898, *Wh.*

Polytrichum alpinum, L.—Area, 1, 6, 7.   Range, 660–2,050 feet.

Turfy and heathery banks on mountains.   Rather rare. Fruiting.   First record : *Wilson, Journ. Bot.*, 1899, p. 468.

> N. 1.   At 2,050 feet on the ridge of Greygarth Fell, *Wi.*   Ireby Fell.
>
> 6.   Middle Gill and Dale Gill, Hindburn, *Wi.*   Foxdale.   Clougha Pike, *Wh.*
>
> E. 7.   Blaze Moss and Tarnbrook Fell, *Wi.*

Polytrichum piliferum, Schreb.—Area, 1, 2, 5, 6, 7, 8.   Range, 20–2,030 feet.

Dry stony or sandy heaths.   Common.   Fruiting.   First record : *Wilson, Journ. Bot.*, 1899, p. 468.   "Fairsnape Clough."

Polytrichum juniperinum, Willd.—Area, 1, 2, 3, 4, 5, 6, 7, 8. Range, 20–1,400 feet or higher.

Dry sandy or peaty ground and waste places.   Not uncommon.   Fruiting.   First record : *Wheldon, Journ. Bot.*, 1899, p. 468.   "St. Annes and Fleetwood."

Polytrichum strictum, Banks.—Area, 1, 2, 4, 6, 7, 8.   Range, 25–2,050 feet.

Damp peaty mountain moors and boggy lowland heaths. Locally abundant.   Fruiting.   First record : *Wilson, Journ. Bot.*, 1899, p. 468.   "Bleasdale Fell."

Polytrichum gracile, Dicks.—Area, 3, 4, 5.   Range, 20–75 feet.

Peat bogs and turfy banks in the low country.   Rare.   Fruiting.   First record : *Wilson, Journ. Bot.*, 1899, p. 468.

> N. 3.   Thrang Moss and Heysham Moss.
>
> 4.   Cockerham Moss, August, 1898, and Rawcliffe Moss, *Wi.*
>
> 5.   Carrs Green, *Wi.*
>
> The seta of this species varies from one to three inches in length.

Oligotrichum incurvum, Lindb.—Area, 1, 6, 7, 8.   Range, 2,020–400 feet.

Sandy or gravelly mountain slopes and stony sides of streams amongst the hills.   Not unfrequent in parts of East.   Fruit not seen.   First record : *Wheldon, Journ. Bot.*, 1899, p. 468.

> N. 1.   South-west slope of Greygarth Fell at 2,020 feet, *Wi.*
>
> E. 6.   Middle Gill (900 feet), and Tatham Fell, Hindburn ; Antley Gill on Botton Head Fell (1,600 feet), and on the Lancashire side of the Cross of Greet, *Wi.*   Caton Moor.
>
> 7.   Tarnbrook Fell and Trough of Bowland, *Wi.*
>
> 8.   Near Kemple End (about 400 feet), and on Jeffrey Hill (950 feet), Longridge Fell, 1899, *Wh.*

Polytrichum nanum, Neck.—Area, 1, 3.   Range, 100–580 feet.

Sandy heathy ground.   Rare.   First record : *Wilson, Journ. Bot.*, 1901, p. 297.

> N. 1.   Near Leck village, at 580 feet.   Bank of the Lune near Over Burrow.
>
> 3.   Gatebarrow Wood near Silverdale, April, 1901, and near Leighton Hall, *Wi.*   Abundant by a path leading to the summit of Warton Crag, ascending to 400 feet, *Wh.*
>
> Prof. Barker thought some of our specimens were referable to the var. *longiseta*, Hampe.   Most of our plants have setæ about an inch in length.   Long capsules may be found on short setæ, and *vice-versa*, and the characters of this variety derived from the fruit seem to be unstable.

Polytrichum aloides, Hedw.—Area, 1, 2, 3, 5, 6, 7, 8.   Range, 50–1,300 feet or higher.

Sandy or peaty banks, quarries, and steep sides of moorland streams, especially where overhung by grass or heather.   Common in the cloughs of East ; rarer elsewhere.   First record : *Wilson, Journ. Bot.*, 1899, p. 468.

> N. 1.   Leck Fell.
>
> 2.   Near Pedder Pots, *Wh.*
>
> 3.   Silverdale, 1898, *Wi.*
>
> W. 5.   Lea and Cottam, *H.B.*   Left bank of Wyre near Catterall, *Wi.*
>
> E. 6, 7 and 8.   Common.

Var. *Dicksoni*, Wallm.

> E. 6.   By the Lune, between Lancaster and Caton, 1900, *Wh.*

Polytrichum urnigerum, L. — Area, 1, 2, 6, 7, 8.   Range, 85–1,700 feet or higher.

Stony and gravelly mountain slopes.   Not common.   Fruiting freely.   First record : *Wilson, Journ. Bot.*, 1899, p. 468.

Polytrichum formosum, Hedw.—Area, 1, 2, 3, 4, 5, 6, 7, 8. Range, 50–1,500 feet.

Peaty woodland banks, especially amongst the hills.   Common. Fruiting.   First record : *Wheldon, Journ Bot.*, 1899, p. 468. "Dolphinholme."

Polytrichum commune, L.—Area, 1, 2, 3, 4, 6, 7, 8.   Range, 25–2,050 feet.

Peat bogs and moorland swamps.   Abundant in East and parts of North, and still occurs, but only rarely, in West.   First record : The Authors, *Journ. Bot.*, 1899.

BUXBAUMIACEÆ.

Diphyscium foliosum, Mohr.—Area, 1, 6.   Range, 550–1,800 feet.

Clefts of slate and gritstone rocks, and turfy banks amongst the hills.   Very rare.   First record : The Authors, *Journ. Bot.*, 1902, p. 414.

> N. 1.   Silurian rocks, Lower Ease Gill, fruiting, in two localities about a mile apart, *Wi.*   By the Cant Beck above Leck Hall at 760 feet, fruiting.
>
> E. 6.   Grit scars on the north-west side of Wardstone at 1,800 feet, sterile, June, 1902.   This was the var. *acutifolium*, Lindb.   Dale Gill, Hindburn, at 1,100 feet.

TETRAPHIDACEÆ.

Tetraphis pellucida, Hedw.—Area, 1, 3, 4, 6, 7, 8.   Range, 50–1,500 feet.

Turfy banks and rotting stumps in woods, and damp shady gritstone rocks.   Frequent amongst the hills of East, growing luxuriantly on some of the gritstone scars.   Fruit not seen.   First record : *Wheldon, Journ. Bot.*, 1899, p. 468.   "Kemple End, Longridge."

> N. 1.   Gorge of the Greeta near Wrayton.
>
> 3.   Thrang Moss, *Wi.*
>
> W. 4.   Rocks by the canal near Galgate, *Wi.*
>
> E. 6, 7, 8.   Common.

Tetraphis Browniana, Grev.—Area, 1, 2, 6, 7.   Range, 200–800 feet.

Damp sandstone rocks, generally near waterfalls, or on the banks of streams, in the shady cloughs amongst the hills, growing downwards from the roofs of hollows and caves. Not common. First record: *Stabler, The Naturalist,* 1896, p. 138.

N. 1. Gorge of the Greeta above Wrayton.
2. Near the waterfall in Wash Dub Wood.
E. 6. Trough Brook Ghyll, Clougha, 1881, *G. Stabler!* Near Botton, Hindburn, *Wh.* & *Wi.* Whiteray Beck on Lythe Fell, and on the Harter Beck, Roeburndale, *Wi.*
7. Rocks near the Wyre, above Dolphinholme, *Wi.*

DICRANACEÆ.

Archidium alternifolium, Schimp.—Area, 6, 7. Range, 200–300 feet or higher.

Damp, sandy, bare patches amongst grass by roadside ditches, and in cultivated fields. Very rare. Fruiting. First record: *Wheldon, Journ. Bot.,* 1906, p. 100.

E. 6. Crook of Lune, *Wh.*
7. Ditch side near Dolphinholme, 1901, *Wh.* Fallow field near Garstang, *Wi.*

Pleuridium axillare, Lindb.—Area, 1, 2, 5, 7. Range, 100–400 feet.

On the mud of dried up ponds, and damp sandy margins of ditches and roadsides. Rather rare. Fruiting. First record: The Authors, *Journ. Bot.,* 1901, p. 297.

N. 1. Fields by the Roman road near Cowan Bridge and near Over Burrow.
2. Bog near Kellet.
W. 5. Staining, near Blackpool, *Wh.*
E. 7. On the mud of a recently drained mill-dam, Calder Vale, Dec., 1900. Damp peaty margin of a pond near Garstang, *Wi.* Ditch side near Dolphinholme, *Wh.*

Pleuridium subulatum, Rab.—Area, 1, 2, 3, 4, 5, 6. Range, 50–550 feet.

Bare sandy banks, fields, quarries, and stony places. Not common. Fruiting. First record: *Wilson, Journ. Bot.,* 1899, p. 468.

N. 1. Near Leck, *Wi.* Ease Gill, and near Whittington.
2. Quarry near Kirkby Lonsdale. Between Leighton Hall and Warton Crag, *Wi.*
3. Silverdale, 1898, *Wi.*

W. 4. Pilling, *Wi.*
5. Lea and Cadley, *H.B.*
E. 6. Near Brookhouse, *Wh.* Ascending to 550 feet in Lower Hindburn.

Pleuridium alternifolium. Rab.—Area, 4, 5, 8. Range, 25–100 feet.

Damp sandy or clayey ground. Rare. Fruiting. First record: *Wheldon, Journ. Bot.,* 1899, p. 468.

W. 4. Near Galgate, *Wi.*
5. Near Blackpool, 1898, *Wh.* Cottam, *H.B.*
6. Crook of Lune, *Wh.*
E. 8. Haighton, *H.B.*

Ditrichum tortile, Hampe.—Area, 3. Range, ?

Bare sandy ground. Rare. First record: The present one.

N. 3. North side of Warton Crag in the lane leading to the summit from Warton, 1906, *Wh.*

Ditrichum homomallum, Hampe.—Area, 1, 2, 6, 7, 8. Range, 100–1,870 feet.

Damp sandy places. Common amongst the fells of East, rare elsewhere. Fruits occasionally. First record: *Wilson, Journ. Bot.,* 1899, p. 468.

N. 1. Ease Gill near Leck, *Wi.* Greygarth Fell at 1,870 feet.
2. Lane near Pedder Pots, *Wh.*
E. 6. Frequent, especially on shaly slopes in the cloughs. Fruiting in Dale Gill.
7. Frequent. Fruiting on Tarnbrook Fell and near Garstang, *Wi.*
8. Frequent. Fruiting near Kemple End, *Wh.*

Ditrichum flexicaule, Hampe.—Area, 1, 2, 3, 5. Range, 25–1,200 feet.

Limestone rocks and dry calcareous banks; more rarely on the sandhills. Common in North, very rare elsewhere. Not fruiting. First record: *Wilson, Journ. Bot.,* 1899, p. 469.

N. 1. Ease Gill and Leck Fell, *Wi.*
2. Near Henridden, *Wi.* Dalton Crag.
3. Gatebarrow, Silverdale, 1898, *Wi.* Hawes Water shore and Trowbarrow, *Wh.* Warton Crag, *Wi.*
W. 5. Sandhills at St. Annes, *H.B.!*

Var. *densum,* B. & S.

N. 2. Kellet Seeds and Dalton Crag, *Wi.*
3. Trowbarrow, 1899, *Wh.* Near Leighton Beck, *Wi.*

x

Swartzia montana, Lindb.—Area, 1, 6. Range, 50–600 feet.

Crevices of damp mountain rocks. Rare. Fruiting. First record: *Wilson, Journ. Bot.,* 1899, p. 469.

N. 1. Ease Gill, 1898, *Wi.* Gorge of the Greeta.
E. 6. Rocks by the Lune near Halton, *Wi.*

Seligeria pusilla, B. & S.—Area, 1, 6. Range, 100–1,550 feet.

Shady rocks and damp walls. Very rare. Fruiting. First record: *Wheldon, Journ. Bot.,* 1899, p. 469.

N. 1. Upper Ease Gill, *Wi.*
E. 6. Damp wall near Caton, 1898, *Wh.*

Seligeria recurvata, B. & S.—Area, 1, 6, 7, 8. Range, 100–1,100 feet.

Shady sandstone rocks and walls. Rare. Fruiting. First record: *Wheldon, Journ. Bot.,* 1899, p. 469.

N. 1. On a stone by the Lune below Kirkby Lonsdale Bridge.
E. 6. Wall near Caton, 1898, *Wh.* Hindburndale, at 1,100 feet.
7 and 8. In two localities on Parlick Pike, *Wi.*

Brachyodus trichodes, Fürnr.—Area, 6, 7, 8. Range, 900–1,250 feet.

Damp sandstone rocks amongst the hills. Rare. Fruiting. First record: The Authors, *Journ. Bot.,* 1901, p. 297.

E. 6. Gully west of Dale Beck, Greenbank Fell, Hindburn, 1899, *Wh.* & *Wi.* Claughton Moor, *Wi.*
7. On a stone in Gavells Clough, and in the Great Clough of Tarnbrook Fell. South-west side of Parlick Pike, *Wi.*
8. South-east gully, Parlick Pike, *Wi.*

Ceratodon purpureus, Brid.—Area, 1, 2, 3, 4, 5, 6, 7, 8. Range, 0–2,050 feet.

Fields, banks, heathy places, and roadsides. Very common. Fruiting. First record: The Authors, *Journ. Bot.,* 1899, p. 496.

Rhabdoweisia denticulata, B. & S.—Area, 6. Range, only found at about 1,800 feet.

Clefts of mountain rocks. Very rare. Fruiting. First record: *Wilson, Journ. Bot.,* 1905, p. 96.

E. 6. Grit rocks on the north-west side of Wardstone, 1904, *Wi.*

Dichodontium pellucidum, Schimp.—Area, 1, 2, 4, 6, 7, 8. Range, 100–2,050 feet.

Wet rocks and sandy detritus by moorland streams. Frequent in some parts of North; very common in East; very rare in West. Fruiting. First record: *Wilson, Journ. Bot.,* 1899, p. 469. "Near Garstang, January, 1878."

Var. *fagimontanum,* Schimp.

N. 1. Greygarth Fell at 2,050 feet, *Wi.*
W. 4. Near Galgate, *Wi.*
E. 8. Longridge Fell, *Wi.*
A singular form with short erecto-patent leaves, bearing many shortly stalked multilocular gemmæ in their upper axils, occurs in Upper Ease Gill at 1,500 feet, *Wi.*

Dichodontium flavescens, Lindb.—Area, 1, 2, 6, 7, 8. Range, 50–1,270 feet or higher.

Sandy and gravelly banks of streams. Not common. Fruit not seen. First record: *Hamilton, The Naturalist,* 1898, p. 28.

N. 1. Lower Ease Gill, near Leck, *Wi.* Middle Ease Gill at 1,270 feet. Gorge of the Greeta.
2. Near Halton, *Wh.*
E. 6. Near Lancaster, 1897, *W.P.H.* Near Caton, *Wh.* Rocks by the Roeburn, near Lower Salter, and near Botton, *Wi.*
7. By the Wyre between Abbeystead and Dolphinholme, *Wi.*
8. Hodder Valley, and by the Ribble near Ribchester, *Wh.* Near the junction of the Loud and Hodder, *Wi.*

Dicranella heteromalla, Schimp.—Area, 1, 2, 3, 4, 5, 6, 7, 8. Range, 0–1,950 feet.

Damp sandy and turfy banks. Common. Fruiting. First record: The Authors, *Journ. Bot.,* 1899, p. 469.

Var. *interrupta,* B. & S.

N. 1. Wall near Leck Hall, *Wi.*

Var. *sericea,* Schimp.

E. 6. Tatham Beck, Hindburn, 1899, *Wi.*

Dicranella cerviculata, Schimp.—Area, 1, 3, 4, 5, 6, 7, 8. Range, 25–1,850 feet.

Peaty banks and sides of cuttings on heaths. Frequent. Fruits, sometimes twice in the year. First record: *Wilson, Journ. Bot.,* 1899, p. 469.

N. 1. Greygarth Fell at 1,850 feet, *Wi.*
3. Thrang Moss, *Wi.*

W. 4. Cockerham Moss, 1898, *Wi.* Near Pilling, *Wh.* Rawcliffe Moss, *Wi.*
    5. Marton Moss, *H. B.*
E. 6. Head of Roeburndale, *Wi.* North-west slope of Wardstone at 1,330 feet.
    7. Lower Bleasdale and west side of the Trough of Bowland, *Wi.*
    8. Wet moor near Lickhurst, and on Parlick Pike, *Wi.*

Var. *pusilla*, Schimp.

E. 6. Dale Gill, Hindburndale, at 1,500 feet, *Wi.*
W. 4. With the type on Cockerham Moss, 1902, *Jones & Wh.*

Dicranella rufescens, Schimp.—Area, 1, 6, 7, 8. Range, 100–900 feet or higher.

On shale and clay banks by streams, especially in cloughs on the fells. Frequent amongst the hills of East. Fruiting. First record: *Wilson, Journ. Bot.*, 1901, p. 297.

N. 1. Ease Gill, and by the Cant Beck above Leck Hall.
E. 6. Over Salter, Roeburndale, 1898, *Wi.* Caton Moor, Greenbank Fell, and Tatham Beck.
    7. Calder Valley above Oakenclough; Catshaw Greave, Emmetts and Deer Clough, Over Wyresdale; Blaze Moss, and Trough of Bowland.
    8. Bank of Barton Brook, *Wi.*

Dicranella varia, Schimp.—Area, 1, 2, 3, 4, 5, 6, 7, 8. Range, 0–800 feet.

Damp clay fields and banks. Frequent. First record: *Wilson, Journ. Bot.*, 1899, p. 469. "Near Garstang, Jan., 1878."

Ascends in North to 800 feet by the Cant Beck above Leck Hall, and in East to 600 feet near Dinkling Green.

Dicranella Schreberi, Schimp. — Area, 1, 6 (?). Range, uncertain.

Damp clay banks and sides of ditches. Very rare. First record: *Hunt*, in *Braithwaite's British Moss Flora.* "Near Lancaster, 1865" (as *Anisothecium crispum*, Lindb.).

Var. *elata*, Schimp.

N. 1. By the Cant Beck above Leck Hall, at 750 feet, *September*, 1906.

Dicranella squarrosa, Schimp.—Area, 1, 2, 6, 7, 8. Range, 400–1,200 feet or higher.

Boggy places, sides of rills, and dripping rocks about waterfalls in moorland districts. Frequent in East. Fruits rarely. First record: *Wilson, Journ. Bot.*, 1899, p. 469.

N. 1. Ease Gill.
    2. Arkholme Moor.
E. 6. Hindburn, in fine fruit, *Wi.* Udale and Foxdale.
    7. Calder Valley above Oakenclough, 1898; Marshaw Fell, and Fairsnape Clough, *Wi.* Gavells Clough.
    8. Spring on the north side of Jeffrey Hill, *Wh.*

Blindia acuta, B. & S.—Area, 1, 6, 7. Range, 600–2,050 feet.

Damp rocks and banks of mountain streams. Very rare. First record: *Wilson, Journ. Bot.*, 1901, p. 297.

N. 1. Greygarth Fell at 2,050 feet, and on Silurian rocks in Ease Gill, *Wi.*
E. 6. Dale Gill, Hindburndale.
    7. Gavells Clough, Wyresdale, 1900, *Wi.*

Dicranoweisia cirrata, Lindb. — Area, 1, 2, 3, 4, 5, 6, 7, 8. Range, 0–2,050 feet.

Walls, rocks, dry banks, and on the trunks of trees. Common, especially in North and East, where it fruits freely. First record: *Hamilton, The Naturalist*, 1898, p. 28. "Near Lancaster."

Campylopus flexuosus, Brid.—Area, 1, 2, 3, 4, 6, 7, 8. Range, 20–1,950 feet.

Turfy ground in woods and on heaths and rocks. Common. Fruiting rarely. First record: *Wheldon, Journ. Bot.*, 1899, p. 469. "Longridge Fell."

Var. *uliginosus*, Ren.—In boggy places on the fells.

N. 1. Greygarth Fell at 1,950 feet, *Wi.*
E. 6. Clougha, *Wh.* Bog on Salter Fell.
    7. Calder Moor above Oakenclough; White Moss on Hawthornthwaite Fell at 1,500 feet; and Luddocks Fell, Upper Calder, *Wi.* Tarnbrook Fell, Wyresdale.
    8. North side of Longridge Fell, *Wh.*

Var. *paradoxus*, Husn.—Frequent on bare peat on the fells of East; once seen on a tree trunk!

N. 2. Very dwarf, on a tree trunk seven feet from the ground, Halton, 1898, *Wh.* (teste *H. N. Dixon*).
E. 6. Tatham Moor and Middle Gill, Hindburn; and on Thrushgill Fell, *Wi.* Mallowdale Fell and Clougha.
    7. Fairsnape Fell, *Wh.* Calder Moors, and Catshaw Greave, *Wi.* Grizedale, *H.B.* Damas Gill Head and Abbeystead Fell.
    8. Longridge Fell, 1898, *Wh.*

Var. *zonatus*, Limpr.—Amongst Silurian and gritstone rocks in mountainous situations.

N. 1. Silurian rocks in Lower Ease Gill, *Wi.*
E. 6. Clougha Pike, *Wh.* Mallowdale Fell.
    7. White side of Tarnbrook Fell, 1900, *Wi.* Gavells Clough.
    8. Whitestone Clough, *Wh.* & *Wi.* Wolf Fell, *Wi.*

Campylopus pyriformis, Brid. — Area, 1, 2, 3, 4, 6, 7, 8. Range, 0–1,600 feet or higher.

Peaty moorlands and heathy ground. Common. Fruit rare. First record: *Wilson, Journ. Bot.*, 1899, p. 469. "Bleasdale Fell, and near Garstang with fruit, January, 1878."

Campylopus fragilis, B. & S.—Area, 1, 2, 3, 6, 7, 8. Range, 40–1,300 feet.

Turfy banks and rocks. Frequent. Fruit not seen. First record: *Wheldon, Journ. Bot.*, 1899, p. 469. "Longridge Fell, July, 1898."

The form known as var. "*densus*," occurs occasionally. Near Low Gill, Hindburn, on Kellet Seeds (*Wi.*), and in Lower Ease Gill.

Campylopus atrovirens, De Not. — Area, 6, 7. Range, 1,000–1,700 feet.

Damp gritstone rocks and bogs in mountainous places. Rare. Fruit not seen. First record: *Wheldon, Journ. Bot.*, 1901, p. 297.

E. 6. Clougha Pike, 1900, *Wh.* Wardstone at 1,700 feet.
    7. Tarnbrook Fell at 1,300 feet.

Campylopus brevipilus, B. & S.—Area, 6. Range, only at about 1,000 feet?

Damp heathy moors. Very rare. First record: *Stabler, The Naturalist*, 1896–7, p. 279.

E. 6. Clougha, 1881, *G. Stabler.*

Dicranodontium longirostre, B. & S.—Area, 1, 6, 7. Range, 1,280–2,040 feet.

Deep hollows and crevices of gritstone rocks in shade; the variety in open damp peaty ground on the higher fells. Rare. First record: The Authors, *Journ. Bot.*, 1901, p. 297.

N. 1. Greygarth Fell.
E. 7. Hell Crag on Tarnbrook Fell at 1,400 feet, *September*, 1900.

Var. *alpinum*, Milde.—First record: *Wilson, Journ. Bot.*, 1899, p. 470.

N. 1. Greygarth Fell from 1,800 to 2,040 feet, locally plentiful, *Wi.*
E. 6. Thrushgill Fell and Salter Fell at 1,280 feet, 1898, and since, *Wi.*

Dicranum Bonjeani, De Not.—Area, 1, 2, 3, 5, 6, 7, 8. Range, 20–1,300 feet.

Boggy heaths and swamps. Frequent. Fruit not seen. First record: *Wilson, Journ. Bot.*, 1899, p. 470. "Storrs Moss, Silverdale, 1898."

Var. *rugifolium*, Boswell.

N. 2. Bog near Docker, *Wi.*
Less typical plants approaching this variety have also been found near Silverdale (*Wh.*), and near Fulwood (*H.B.*).

Dicranum scoparium, Hedw.—Area, 1, 2, 3, 4, 5, 6, 7, 8. Range, 20–2,050 feet.

Woods, heaths, rocky places, and tree trunks. Common. Fruiting occasionally. First record: The Authors, *Journ. Bot.*, 1899, p. 470.

Var. *paludosum*, Schimp.—In boggy places. Rare.

N. 2. Whittington Moor and near Dalton Hall, *Wi.*
    3. Bog below Thrang End Quarry, *Wh.*
E. 6. Bog at the foot of Clougha, *Wh.*

Var. *orthophyllum*, Brid.—Sand dunes and dry grassy heaths. Locally plentiful.

N. 2. Between Caton and Borwick, *Wi.*
    3. Near Thrang End, *Wh.*
W. 5. Lytham, St. Annes, and Fleetwood, *Wh.*
E. 6. Upper Roeburndale and Clougha.

Var. *ericetorum*, Corbiere.—Damp heaths. Frequent.

N. 1. Ease Gill, *Wi.*
    2. Kellet Seeds, *Wi.*
W. 4. Cockerham Moss, 1900.
E. 7. Clougha, *H. B.* & *Wh.* Gavells Clough.
    8. Longridge Fell, *Wh.*

Var. *spadiceum*, Boul. (including vars. *alpestre* and *turfosum*).— On exposed heaths and mountain summits.

N. 1. Greygarth Fell at 2,050 feet.
E. 6. Wolfhole Crag and head of Antley Gill, Botton Head Fell at 1,650 feet, *Wi.* Summit of Wardstone.
   7. Calder Valley above Oakenclough and Bleasdale Fell, *Wi.* Tarnbrook Fell.

Dicranum majus, Turn. — Area, 1, 2, 5, 6, 7, 8. Range, 50–1,000 feet.

Woods and shaded banks amongst the hills. Common in East and parts of North ; rarer elsewhere. Fruiting. First record : *Wheldon, Journ. Bot.,* 1899, p. 470. " Dolphinholme, March, 1897."

Dicranum fuscescens, Turn. — Area, 1, 6, 7, 8. Range, 400–2,030 feet.

Mountain rocks and banks on peaty gritstone moors. Frequent in parts of North and East. Occasionally fruiting. First record : *Stabler, The Naturalist,* 1896-7, p. 280.

N. 1. Leck Fell and Greygarth Fell in fruit.
E. 6. Frequent and very fine amongst the higher hills of this division ; fruiting on Mallowdale Fell. Clougha, 1881, *Stabler.* Fruiting there in 1902, *Wh.*
   7. Frequent in this division, and especially abundant in cloughs and on scars at the head of Wyresdale.
   8. Wolf Fell, in small quantity at 1,150 feet, *Wi.*

Var. *falcifolium,* Braithw.—Exposed gritstone scars. Rare.

N. 1. Greygarth Fell at 1,900 feet, *Wi.*
E. 6. Upper Roeburndale, 1898, and Dale Gill, Hindburn, *Wi.* Near the Old Town Quarry, Clougha, *Wh.*

Leucobryum glaucum, Schimp.—Area, 1, 2, 3, 4, 5, 6, 7, 8. Range, 20–1,300 feet or higher.

Swampy heaths, wet woodland banks, and damp rocks. Common. Fruit not seen. First record : *Wilson, Journ. Bot.,* 1899, p. 470. " Lower Bleasdale, May, 1898."

FISSIDENTACEÆ.

Fissidens exilis, Hedw.—Area, 2, 5. Range, 0–100 feet.
Bare earth on banks, shaded by vegetation. Rare. Fruiting. First record : *Wheldon, Journ. Bot.,* 1899, p. 470.

N. 2. Warton Crag, *Wh.*
W. 5. Near Blackpool, 1895, *Wh.* Lea and Catforth, *H.B.*

Fissidens crassipes, Wils.—Area, 1, 6. Range, 50–200 feet.
Stones in streams. Rare. First record : The Authors, *Journ. Bot.,* 1905, p. 96.

N. 1. Rocks in the Greeta between Burton-in-Lonsdale and Wrayton.
E. 6. Crook of Lune near Caton, July, 1904.

Fissidens osmundoides, Hedw.—Area, 1, 7. Range, 600–1,270 feet.

Damp mountain slopes and dripping rocks. Rare. Fruiting. First record : The Authors, *Journ. Bot.,* 1902, p. 414.

N. 1. Silurian rocks by a spring in Middle Ease Gill, in fruit, and on Greygarth Fell at 1,270 feet, with *Cinclidium.*
E. 7. Gavells Clough, on the White side of Tarnbrook Fell, in fruit, September, 1902.

Fissidens adiantoides, Hedw.—Area, 1, 2, 3, 4, 5, 6, 7, 8. Range, 0–2,050 feet.

Bogs and damp rocks. Frequent. Fruiting. First record : *Hamilton, The Naturalist,* 1898, p. 28. " Near Lancaster."

A form with rather more obscure cells and of compact habit, occurs frequently on dry limestone rocks in North, and is liable to be mistaken for the next species. It is probably *F. collinus*, Mitt., which however seems to be a very unstable species.

Fissidens decipiens, De Not.—Area, 1, 2, 3, 8. Range, 0–1,200 feet.

Damp limestone rocks and moist grassy banks. Rare. Fruit not seen. First record : *Wilson, Journ. Bot.,* 1899, p. 470.

N. 1. Ease Gill, 1898, *Wi.* Leck Fell.
   2. Dalton Crag.
   3. Silverdale ; and near Ings Point, *Wh.*
W. 8. Near Whitewell.

Fissidens taxifolius, Hedw.—Area, 1, 2, 3, 4, 5, 6, 7, 8. Range, 0–1,500 feet or higher.

Earthy banks and sides of ditches on a clay soil. Common. Fruiting. First record : *Wheldon, Journ. Bot.,* 1899, p. 470. " Silverdale, 1898."

Ascends in North to 1,500 feet in Ease Gill.

Fissidens minutulus, Sull.—Area, 3. Range, 0–200 feet.
Sandy ground in calcareous districts. Rare. Fruiting. First record : The present one.

N. 3. Heald Brow near Silverdale.
Of these specimens, Herr. G. Roth writes—" The Heald Brow Fissidens is as it were a *F. minutulus*, with the characters of *F. tamarindifolius*, in producing in-novations from the axils of the leaves, and also in developing male flowers with rhizoids in the axils of the lower leaves, as we l as on special stems. It has narrow cells 7–10 μ, while *F. pusillus* has these from 10–14 μ, and also shorter pericœtial leaves. One could name this plant *F. minutulus,* var. with the inflorescence of *F. tamarindifolius.*"

Fissidens viridulus, Wahl.—Area, 1, 2, 3, 4, 5. Range, 0–200 feet.

Clay banks by hedges and sides of ditches. Rare. Fruiting. First record : *Wheldon, Journ. Bot.,* 1899, p. 470.

N. 1. Lune bank below Kirkby Lonsdale, and near Tunstall.
   3. Hawes Water, and shore bank on south side of Keer estuary, *Wi.*
W. 4. Bank of the Wyre near Preesall, 1899, *Wh.* Nateby, *Wi.*
   5. Bank of the Wyre below Shard Bridge, *Wh.*

Var. *Lylei,* Wils.—First record : *Journ. Bot.,* 1901.

N. 2. Near Borwick, *Wi.*
W. 4. Near Garstang, 1900, *Wi.*

Fissidens incurvus, Stark.—Area, 8. Range, only seen at about 300 feet.

Clay banks. Rare. Fruiting. First record : *Wheldon, Journ. Bot.,* 1899, p. 470.

E. 8. Hedgebank near Stonyhurst, 1899, *Wh.*

Fissidens tamarindifolius, Wils. — Area, 3. Range, 0–100 feet.

Stiff sandy ground. Very rare. Fruiting. First record : The present one.

N. 3. Near Carnforth.

Fissidens bryoides, Hedw.—Area, 1, 2, 3, 4, 5, 6, 7, 8. Range, 0–1,550 feet.

Woods, fields, and damp shady banks. Common. Fruiting. First record : *Wilson, Journ. Bot.,* 1899, p. 470. " Sandholme, near Garstang, January, 1877."

GRIMMIACEÆ.

Grimmia apocarpa, Hedw.—Area, 1, 2, 3, 5, 6, 7, 8. Range, 0–2,080 feet.

Rocks and walls. Common in North and East, rare in West. Fruiting. First record : *Wilson, Journ. Bot.,* 1899, p. 470. " Silverdale, 1898."

It ascends to the wall on the summit ridge of Greygarth Fell.

Var. *rivularis,* W. & M. — Frequent in rivers and streams. Numerous records in Districts 1, 2, 6, 7, 8.

Var. *gracilis,* W. & M.

N. 1. By the Leck Beck, Ease Gill, 1898, *Wi.*

Grimmia maritima, Turn.—Area, 3. Range, 0–30 feet.

Grit rocks by the sea. Locally abundant. Fruiting. First record : *Wheldon, Journ. Bot.,* 1899, p. 471.

N. 3. Rocks about Lower Heysham, 1898, *Wh.* Middleton Rocks, 2½ miles from the above.

This is one of the few exclusively maritime species. It frequently grows where it is liable to be submerged by high tides—an ordeal which few mosses are able to survive, although the following are able to withstand with impunity occasional immersion for short periods in sea water :—*Trichostomum tophaceum, T. mutabile, Pottia Heimii, Bryum Marratii,* and *Amblystegium serpens,* var. *salinum.* The annual mosses *Pottia intermedia, P. littoralis,* and *P. truncatula* thrive near the sea, but their habitats are usually beyond tidal reach, and only occasionally are they found on fallen clods of earth within reach of the water. But salt laden breezes seem to have anything but a prejudicial effect on them, and they become rarer as we go further inland.

Grimmia pulvinata, Sm.—Area, 1, 2, 3, 4, 5, 6, 7, 8. Range, 0–2,080 feet.

Walls and rocks, finest and most abundant on the limestone of North. Common. Fruiting. First record : The Authors, *Journ. Bot.,* 1899, p. 471.

Grimmia Doniana, Sm.—Area, 1, 6, 8. Range, 1,000–2,060 feet.

High gritstone rocks and walls. Rare. Fruiting. First record : *Wilson, Journ. Bot.,* 1901, p. 297.

N. 1. Greygarth Fell, on a wall near the summit, and on Yoredale Grit rocks in a second locality about two miles distant, *Wi.*
E. 6. Clougha, *Wh.*
   8. Wolf Fell, *Wi.*

Rhacomitrium aciculare, Brid.—Area, 1, 6, 7, 8. Range, 300–1,100 feet or higher.

Stones and rocks in sub-alpine streams. Frequent. Fruiting. First record: *Wilson, Journ. Bot.*, 1899, p. 471. "Roeburndale, near Lower Salter, 1898."

Rhacomitrium protensum, Braun.—Area, 7. Range, only seen at about 300 feet.

Damp rocks. Very rare. Fruit not seen. First record: *Wheldon, Journ. Bot.*, 1901, p. 297.

E. 7.   Rocks by the Wyre near Dolphinholme, very sparingly, 1901, *Wh.*

Rhacomitrium fasciculare, Brid. — Area, 1, 2, 3, 6, 7, 8. Range, 100–2,080 feet.

Subalpine rocks. Common in North and East; not seen elsewhere. First record: *Wheldon, Journ. Bot.*, 1899, p. 471. "Longridge Fell, 1898."

In District 3 we have seen it in one locality only, on a Silurian boulder near Borwick, *Wi.*

Rhacomitrium heterostichum, Brid. — Area, 1, 2, 6, 7, 8. Range, 100–2,050 feet.

Rocks and walls in hilly districts. Less common than the preceding. Fruiting. First record: *Wilson, Journ. Bot.*, 1899, p. 471. "Between Hornby and Lower Salter, 1898."

Var. *alopecurum*, Hub.

N. 1.   Ease Gill and Greygarth Fell, *Wi.*
E. 6.   Upper Roeburndale, *Wi.*

Var. *gracilescens*, B. & S.

N. 1.   Greygarth Fell, at 2,050 feet, *Wi.*
E. 7.   Catshaw Greave, *Wi.*

Rhacomitrium lanuginosum, Brid.—Area, 1, 2, 3, 5, 6, 7, 8. Range, 300–2,050 feet.

Stony heaths and mountain rocks and walls. Frequent amongst the hills; rare elsewhere. Rarely fruiting. First record: *Wilson, Journ. Bot.*, 1899, p. 471. "Lower Salter, 1898."

Rhacomitrium canescens, Brid. — Area, 1, 2, 5, 6, 7, 8. Range, 0–1,850 feet.

Stony heaths and sandy roadsides on the fells. Frequent. Fruits rarely. First record: *Wilson, Journ. Bot.*, 1899, p. 471.

N. 1.   Ease Gill and Greygarth Fell, *Wi.*   Leck Fell, in fruit.
   2.   Arkholme Moor, *Wi.*
W. 5.   On sandstone blocks in the embankment of the Ribble below Preston Docks, in small dark tufts, very depauperate, *Wh.*
E. 6.   Clougha, *Wh.*   Tatham Moor and near Wray, *Wi.*   Lower Salter, *Wh.* & *Wi.*   Upper Roeburndale (*forma epilosa*), *Wi.*
   7.   Tarnbrook; and east of Peacock Hill, Bleasdale, *Wh.* & *Wi.*   Marshaw, *Wi.*
   8.   Longridge Fell, *Wh.*   Lickhurst, *Wi.*

Var. *ericoides*, B. & S.

N. 1.   Ease Gill, 1898, *Wi.*

Campylostelium saxicola, B. & S.—Area, 1, 6. Range, only seen at about 200 feet.

Siliceous rocks and stones in damp shady places. Very rare. Fruiting. First record: The Authors, *Journ. Bot.*, 1906, p. 100.

N. 1.   Gorge of the Greeta, May, 1905.
E. 6.   Rocks by the Hindburn, near Mill Houses.

Ptychomitrium polyphyllum, Fürnr.—Area, 1, 2, 6, 7, 8. Range, 150–2,050 feet.

Rocks and walls in upland districts. Not common. Fruiting. First record: *Wilson, Journ. Bot.*, 1899, p. 471.

N. 1.   Ease Gill, *Wh.* & *Wi.*   Greygarth Fell and near Wennington Hall, *Wi.*
   2.   Whittington Moor and on a wall near Arkholme.
E. 6.   North side of Clougha, *Wh.*   Foxdale.
   7.   Garstang, 1877, *Wi.*   Damas Gill, *Wh.*   Scorton, *Wi.*
   8.   Near Dilworth Reservoir, Longridge, and on walls on Longridge Fell, *Wh.*

TORTULACEÆ.

Acaulon muticum, C.M.—Area, 5. Range, 50–100 feet.

Bare sandy ground near the sea. Very rare. Fruiting. First record: *H. Beesley, Journ. Bot.*, 1902, p. 414 (as *A. mediterraneum*, Limpr.).

W. 5.   Bank near Bispham, February, 1901, *H.B.* (sp.).

The specimens from Mr. Beesley, are similar to the slender form from the Isle of Man, which has been called *A. Holtii* and *A. mediterraneum*, but which Mr. Dixon regards as a mere form of *A. muticum*. The latter, as it occurs in South Lancashire, has, however, a rather different habit, and Mr. Holt's plant seems to be entitled to varietal rank.

Phascum cuspidatum, Schreb.—Area, 1, 2, 3, 4, 5, 6, 7, 8. Range, 0–500 feet or higher.

Fallow fields, banks, and bare earth. Frequent in the low country. Fruiting. First record: *Wheldon, Journ. Bot.*, 1899, p. 471. "Lytham, 1898."

Var. *piliferum*, B. & S.

N. 2.   Quarry near Kirkby Lonsdale.

Pottia recta, Mitt.—Area, 2, 3, 4. Range, 0–500 feet.

Bare earth by roadsides, hedgebanks, and fallow fields, on limestone or clay soil. Rather rare. Fruiting. First record: *Wilson, Journ. Bot.*, 1899, p. 471.

N. 2.   Near Henridden and Over Kellet, and between Kellet Seeds and Bolton-le-Sands, *Wi.*
   3.   Near Torrisholme; near Silverdale Church; between Silverdale and Carnforth; and near Warton, *Wi.*   Near Leighton Beck and by the canal, Yealand, *Wi.*   Not uncommon in this district.
W. 4.   Roadside near Garstang, 1899, *Wi.*

Pottia Heimii, Fürnr.—Area, 3, 4, 5. Range, 0–75 feet.

Muddy or sandy ground near the sea. Frequent. Fruiting. First record: *Yates, Journ. Bot.*, 1899, p, 471.

N. 3.   Ovangle and Overton, and on mud-capped walls on Oxcliffe Marsh near Lancaster, *Wh.*   Coast near Middleton.   Salt marsh near Heald Brow, Silverdale.
W. 4.   Pilling and Preesall, *Wh.*
   5.   Lytham, 1892, *Yates* (in *Herb. Miss Armitage*).   St. Annes; Fleetwood; and banks of the Wyre below Shard Bridge, *Wh.*

Pottia truncatula, Lindb.—Area, 1, 2, 3, 4, 5, 6, 7, 8. Range, 0–720 feet.

Fallow fields, and bare earth on hedgebanks, and by roadsides. Common. Fruiting. First record: *Wheldon, Journ. Bot.*, 1899, p. 471. "Lytham, 1898."

It ascends in North to 500 feet, and in East to 720 feet in Lower Hindburn. In the latter locality it occurs by a moorland cartroad, perhaps brought up from the low country on cart wheels or the feet of animals. It is a rare plant in the uncultivated moorland districts.

Pottia intermedia, Fürnr.—Area, 3, 4, 5, 7, 8. Range, 0–300 feet.

In similar situations to the preceding species, but much less common. Fruiting. First record: *Wheldon, Journ. Bot.*, 1901, p. 297.

N. 3.   Middleton and Overton, *Wi.*   Silverdale and Carnforth, *Wh.*
W. 4.   Wyre bank above Cartford Bridge, *Wi.*   Near Wardless, growing on an old tree trunk lying in the salt marsh, *Wh.*
   5.   Lytham, 1900, *Wh.*
E. 7.   Near Dolphinholme, *Wi.*
   8.   Tunbrook Wood, *H.B.*

Pottia littoralis, Mitt.—Area, 4, 5. Range, 0–20 feet.

Muddy or sandy places on sea shores where rarely reached by the tide. Rare. Fruiting. First record: *Wheldon, Journ. Bot.*, 1899, p. 471.

N. 3.   Salt marsh near Heald Brow, Silverdale.
W. 4.   Knott End, 1899, *Wh.*
   5.   Fairfield, *Wh.*

Pottia minutula, Fürnr.—Area, 2, 3, 4, 5, 6, 8. Range, 0–500 feet.

Clay banks and fallow fields. Rare. Fruiting. First record: *Wheldon, Journ. Bot.*, 1899, p. 471.

N. 2.   Near Henridden, and between Kellet Seeds and Bolton-le-Sands, *Wi.*
   3.   Ings Point and Silverdale, *Wh.*   Canal bank near Yealand, *Wi.*
W. 4.   Near Preesall, *Wh.*
   5.   Lytham Rifle Range, 1898, and near Staining, *Wh.*
E. 6.   Between Caton and Lancaster, *Wh.*
   8.   Tunbrook Wood, *H.B.*   Chipping, *Wi.*

Pottia lanceolata, C.M.—Area, 2, 3, 4. Range, 0–250 feet.

Dry calcareous banks. Rather rare. Fruiting. First record: *Wilson, Journ. Bot.*, 1901, p. 297.

N. 2.   Between Kellet Seeds and Bolton-le-Sands, *Wi.*
   3.   Near Silverdale Railway Station, 1901, and near Thrang End, *Wi.*   Near Woodwell, etc., Silverdale.   Near Carnforth, *Wh.*
W. 4.   Knott End, *H.B.* (sp.).

Tortula aloides, De Not.—Area, 1, 3, 4, 5, 6, 8. Range, 0–800 feet or higher.

On bare clay banks. Rare. Fruiting. First record: *Wheldon, Journ. Bot.*, 1899, p. 471.

N. 1.   Wennington.
   3.   Silverdale, and in the quarry near Jenny Brown's Point.
W. 4.   Wyre bank near Coat Walls, *Wh.*
   5.   Lytham, 1898, *Wh.*
E. 6.   Tatham Beck, Hindburn, *Wi.*
   8.   Tootle Heights near Longridge, *Wh.*

Tortula ambigua, Angstr.—Area, 2, 3, 4, 7.   Range, 0–400 feet.
Walls and banks.   Rare.   Fruiting.   First record: *Wilson,
Journ. Bot.*, 1899, p. 471.

> N. 2.   Near Borwick, *Wi.*
>   3.   Silverdale, and shore bank south of Keer estuary, *Wi.*   Warton Crag.
>   Wall near Heysham.
> W. 4.   Between Pilling and Cockerham, 1898. *Wi.*
> E. 7.   Barnacre, near Garstang, *Wi.*   Scorton and Grizedale, *H.B.*

Tortula muralis, Hedw.—Area, 1, 2, 3, 4, 5, 6, 7, 8.   Range,
0–1,500 feet.
Walls and stones.   Common.   Fruiting.   First record: The
Authors, *Journ. Bot.*, 1899, p. 471.

Var. *rupestris.*—Area, 1, 2, 3, 4, 8.   Very common on walls and
rocks in limestone districts ; rarely on mortar elsewhere, as
near Glasson, the only record for District 4.

Tortula subulata, Hedw.—Area, 1, 2, 3, 4, 5, 6, 7, 8.   Range,
0–1,200 feet.
Sandy banks.   Frequent.   Fruiting.   First record: *Wheldon,
Journ. Bot.*, 1899, p. 471.   " St. Annes sandhills, 1898."

Var. *subinermis*, Wils.—Tree roots by streams.

> N. 1.   Tree roots by the Lune near Over Burrow, with *Leskea polycarpa*, 1906.

Tortula mutica, Lindb.—Area, 1, 2, 3, 4, 6, 8.   Range, 50–200
feet.
Roots of trees by rivers, especially where liable to be occasionally submerged during floods.   Rather rare.   Fruit not seen.
First record: *Wheldon, Journ. Bot.*, 1899, p. 471.

> N. 1.   On willows near Tunstall and Over Burrow.
>   2.   By the Lune below Kirkby Lonsdale, and between Caton and Aughton.
>   By the Keer near Carnforth.
>   3.   Tree roots by the Keer, *Wi.*
> W. 4.   Tree roots by the Conder, between Glasson and Galgate, *Wh.*
> E. 6.   By the Lune between Caton and Lancaster, *Wh.*
>   8.   Near Lower Hodder Bridge, and by the Ribble below Stonyhurst,
>   1898, *Wh.*

Tortula lævipila, Schwaeg.—Area, 1, 2, 3, 4, 6, 7, 8.   Range,
25–600 feet.
Trunks of trees.   Frequent in North, rather rare and usually
small and ill-developed in the other divisions.   Fruits freely in

favourable localities, but is often barren.   First record: *Wheldon,
Journ. Bot.*, 1899, p. 471.   " Halton, fruiting, 1898."

Tortula intermedia, Berk.—Area, 1, 2, 3, 8.   Range, 0–2,050
feet.
Limestone walls and rocks.   Common in North, more sparingly
about Leagram and Whitewell in East.   Not seen in West.
Fruiting.   First record: *Wilson, Journ. Bot.*, 1899, p. 472.
" Silverdale, 1898."

Tortula ruralis, Ehrh.—Area, 2, 3, 8.   Range, 0–500 feet.
Walls in limestone districts.   Very rare.   Fruit not seen.
First record: *Wheldon, Journ. Bot.*, 1899, p. 472.

> N. 2.   Near Whittington, *Wi.*
>   3.   Silverdale, 1898, *Wh.*
> E. 8.   Near Whitewell, *Wh.*

Tortula ruraliformis, Dixon.—Area, 3, 4, 5.   Range, 0–30 feet.
Sandy ground near the sea.   Frequent.   Fruit not seen.
First record: *Wheldon, Journ. Bot.*, 1899, p. 472.   " St. Annes
sandhills, 1898."

> N. 3.   Near Middleton, and at Hest Bank and Bolton-le-Sands, *Wi.*
> W. 4.   Sparingly and very dwarf, near Knott End, *Wh.*   Near Pilling, *Wi.*
>   5.   Frequent from Lytham to Fleetwood, 1898, *Wh.*

Tortula papillosa, Wils.—Area, 3.   Range, 0–200 feet ?
Trunks of trees, and on branches and trunks of hawthorns in
hedgerows.   Very rare.   Not fruiting.   First record: *John
Nowell in Braithwaite's British Moss Flora,* p. 223.

> N. 3.   Silverdale, *John Nowell.*   Sparingly on a hawthorn near Heysham, *Wh.*

Barbula lurida, Lindb.—Area, 2, 5, 8.   Range, 50–200 feet.
Rocks, banks, and posts near streams.   Rare.   Not fruiting.
First record: *G. A. Holt, Journ. Bot.*, 1899, p. 472.

> N. 2.   By the Lune near Caton, *Wh.*
> W. 5.   Near Ashton, *H.B.* (sp.).
> E. 8.   By the Ribble opposite Stonyhurst, 1886, *Holt* (sp.).   Redscar, near
>   Preston, *Wh.*

Barbula rubella, Mitt.—Area, 1, 2, 3, 4, 5, 6, 7, 8.   Range,
0–2,050 feet.
Shaded banks, rocks, and walls.   Common.   Fruiting.   First
record: *F. C. King, Journ. Bot.*, 1899, p. 472.   " Longridge,
1881 (sp.)."

Y

Var. *dentata*, Braithw.

> N. 1.   Greygarth Fell at 2,050 feet, *Wi.*
> W. 5.   St. Annes, *H.B.*
> E. 8.   Longridge Fell, *Wh.*

Barbula tophacea, Mitt.—Area, 1, 2, 3, 4, 5, 6, 7, 8.   Range,
0–1,000 feet.
Dripping rocks amongst the hills, and wet clay banks in the
low country, especially near the sea.   Frequent.   Fruiting.   First
record: *Wheldon, Journ. Bot.*, 1899, p. 472.   " Blackpool, 1898."

The var. *acutifolia*, Schimp., occurs occasionally, but is so commonly
accompanied by forms connecting it with the type, that it seems to be hardly entitled
to varietal rank.

Barbula fallax, Hedw.—Area, 1, 2, 3, 4, 5, 6, 7, 8.   Range,
0–1,000 feet or higher.
Clay banks, roadsides, and waste stony ground.   Frequent.
Fruiting.   First record: *Wheldon, Journ. Bot.*, 1899, p. 472.
" Lindeth, 1898."

Var. *brevifolia*, Schultz.—Stony places and road-ides.

> N. 2.   Near Borwick and Over Kellet, *Wi.*
>   3.   Middlebarrow, 1899, *Wi.*   Silverdale, *Wh.*
> E. 6.   Clougha Quarry, *Wh.*   Near Caton, *Wh.*
>   8.   Near Chipping, *Wh.*

Barbula recurvifolia, Schimp.—Area, 2, 3.   Range, 0–650 feet.
Earthy banks in limestone districts.   Rare.   Not fruiting.
First record: *Wilson, Journ. Bot.*, 1899, p. 472.

> N. 2.   Dalton Crag and Henridden, *Wi.*   Over Kellet, *Wh.*   Highfield, near
>   Carnforth, *J. W. Hartley, Wh. & Wi.*
>   3.   Silverdale and Warton, *Wi.*   Thrang End, *Wh.*   Quarry near Jenny
>   Brown's Point.

Barbula spadicea, Mitt.—Area, 1, 2, 6, 7, 8.   Range, 50–700
feet.
Wet rocks and stony margins of streams and rivers.   Rare.
Fruiting.   First record: *Wheldon, Journ. Bot.*, 1899, p. 472.

> N. 1.   Ease Gill, *Wi.*
>   2.   Banks of the Lune near Caton, *Wh.*   Near Arkholme, *Wi.*
> E. 6.   Rocks by the Hindburn near Mill Houses.
>   7.   Banks of the Brock ; and of the Wyre near Abbeystead, *Wh.*
>   8.   Banks of the Hodder, and of the Ribble near Redscar, *Wh.*   In fine
>   fruit near the junction of the Hodder and Loud, *Wi.*

Barbula rigidula, Mitt.—Area 1, 2, 3, 4, 6, 8.   Range, 0–2,050
feet.
Rocks and walls.   Frequent, especially in calcareous
districts.   Fruiting.   First record: *Hamilton, The Naturalist,*
1898, p. 28.

> N. 1.   Frequent, ascending to 2,050 feet on Greygarth Fell, *Wi.*
>   2.   Near Halton, *Wh.*   Henridden, *Wi.*   Over Kellet, in fruit.
>   3.   Frequent.
> W. 4.   Wall near Galgate, *Wi.*
> E. 6.   Near Lancaster, 1897, *W.P.H*   Caton, *Wh.*   Crook of Lune.
>   8.   Near Ribchester, and near Whitewell.

Barbula cylindrica, Schimp.—Area, 1, 2, 3, 4, 6, 7, 8.   Range,
0–500 feet.
Walls and banks.   Frequent.   Fruiting occasionally.   First
record: *Wheldon, Journ. Bot.*, 1899, p. 472.

> N. 1.   Near Kirkby Lonsdale Station, in fruit ; and near Melling, *Wi.*
>   2.   Near Whittington, *Wh. & Wi.*   Over Kellet, *Wi.*
>   3.   Between Heaton and Overton, *Wh.*
> W. 4.   Near Pilling, and on a wall near Ellel Grange, *Wi.*
> E. 6.   Caton, 1898, *Wh.*
>   7.   Garstang and Brock Bottom, *Wi.*
>   8.   Longridge, *Wh.*

Barbula vinealis, Brid.—Area. 1, 2, 3, 4, 5, 6, 7, 8.   Range,
0–500 feet.
Rocks, walls, and sand dunes.   Not common.   Fruit not
seen.   First record: *Wheldon, Journ. Bot.*, 1899, p. 472.

> N. 1.   Near Wennington, *Wi.*
>   2.   Wall near Carnforth.
>   3.   Walls near Torrisholme, *Wi.*
> W. 4.   Knott End, *Wh.*
>   5.   St. Annes, 1898, Lytham, and South Shore, *Wh.*
> E. 6.   Near Wray.
>   7.   Grizedale, near Garstang, and walls near Catterall, *Wi.*
>   8.   Tootle Heights, *Wh.*

Barbula sinuosa, Braithw.—Area, 2, 3.   Range, 0–400 feet.
Limestone rocks and walls.   Rare.   Fruit not seen.   First
record: *Wheldon, Journ. Bot.*, 1902, p. 414.

> N. 2.   Near Over Kellet, *J. W. Hartley, Wh. & Wi.*   Warton Crag, *Wh.*
>   3.   Silverdale, 1901, in two or three localities, *Wh.*   Gatebarrow Wood, *Wi.*

Barbula Hornschuchiana, Schultz. — Area, 3, 5. Range, 0–200 feet.

Sandy ground near the sea. Very rare. Not seen in fruit. First record : *Wheldon, Journ. Bot.*, 1901, p. 297.

N. 3. Thrang End, and shore-bank on the south side of the Keer estuary, *Wi.*
W. 5. Near South Shore, Blackpool, 1898, *Wh.*

Barbula revoluta, Brid.—Area, 1, 2, 3, 4, 5, 6, 7, 8. Range, 0–700 feet.

Limestone walls, and on mortar. Very abundant in the calcareous districts, frequent elsewhere. Fruiting. First record : *Wheldon, Journ. Bot.*, 1899, p. 472. "Hodder Valley, 1898.

Barbula convoluta, Hedw.—Area, 1, 2, 3, 4, 5, 6, 8. Range, 0–500 feet.

Roadsides, walls, waste ground, and sandhills. Common. Fruiting only occasionally. First record : *Hamilton, The Naturalist*, 1898, p. 28. "Near Lancaster, 1897."

Forms approaching the var. *Sardoa, B. & S.*, are seen occasionally, as between Stodday and Galgate, (*Wi.*), and near Mitton, (*Wh.*).

Barbula unguiculata, Hedw.—Area, 1, 2, 3, 4, 5, 6, 7, 8. Range, 0–1,700 feet.

Waste ground and banks on a clay soil ; more rarely on rocks and walls. Common. Fruiting freely. First record : *Hamilton, The Naturalist*, 1898, p. 472.

It ascends to 1,700 feet on Greygarth Fell.

Var. *apiculata,* Hedw.

E. 6. Near Galgate, *Wi.*
  7. Near Garstang, *Wi.*

Var. *cuspidata,* B. & S.

N. 1. With *Polytrichum nanum* by the road above Leck at 550 feet.

Leptodontium flexifolium, Hampe.—Area, 1, 6, 7, 8. Range, 500–1,850 feet.

Peaty ground and bare stony places, generally on mountains. Not common. Fruit not seen. First record : *Hamilton, The Naturalist*, 1898, p. 28.

N. 1. Greygarth Fell at 1,850 feet.
E. 6. Near Lancaster, 1897, *W.P.H.* Clougha Quarry, *Wh.* Head of Roeburndale, *Wi.* Greenbank Fell and Marshaw Fell.
  7. Gully off Damas Gill, *Wh.* White side of Tarnbrook Fell.
  8. Near Knowl Green, *Wh.*

Weisia crispa, Mitt.—Area, 3. Range, only seen at about 350 feet.

Bare ground. Very rare or overlooked. Fruit not seen. First record : The present one.

N. 3. North side of Warton Crag, 1906.

Weisia squarrosa, C.M.—Area, 4. Range, 0–50 feet.

Clayey fallows and bare places on grassy banks. Very rare. Fruiting. First record : *Wheldon, Journ. Bot.*, 1899, p. 472.

W. 4. Coat Walls, near Preesall, 1899, *Wh.*

Weisia microstoma, C.M.—Area, 1, 2, 3, 4, 8. Range, 0–750 feet.

Stony places and earth amongst limestone rocks, and on ant-hills and crumbling banks on a calcareous soil. Frequent in the limestone districts, very rare in others. Fruiting. First record : *F. C. King, Journ. Bot.*, 1899, p. 472.

N. 1. Leck Fell.
  2. Dalton Crag, *Wi.* Near Henridden.
  3. Frequent in the Silverdale district. Near Carnforth.
W. 4. Preesall, *Wh.*
E. 8. Near Whitewell, 1880, *F. C. King* (in *Herb. Wh.*). Limestone rocks near Dinkling Green, *Wi.*

Weisia crispata, C.M.—Area, 1, 2, 3. Range, 0–1,000 feet.

Calcareous rocks, banks, and walls. Locally abundant. Fruiting. First record : *Wilson, Journ. Bot.*, 1899, p. 375 ("Weisia crispata in Britain." By H. N. Dixon).

N. 1. Ease Gill.
  2. Borwick, *Wi.* Dalton Crag.
  3. In many places about Silverdale, Warton, Yealand, and district.

Weisia viridula, Hedw.—Area, 1, 2, 3, 4, 5, 6, 7, 8. Range, 0–600 feet.

Hedgebanks and sandy fields. Common. Fruiting. First record : The Authors, *Journ. Bot.*, 1899, p. 473.

It ascends in North to 600 feet above Leck.

Var. *amblyodon,* B. & S.

W. 4. Near Cockersand Abbey, *Wi.*
E. 6. Bank of Dale Beck near Brookhouse, Caton, *Wh.*
  8. Near Mitton, *Wh.*

Var. *densifolia,* C.M.

E. 8. Abundant about the mouth of a cave in the limestone near Dinkling Green.

A form with slightly oblique capsules occurs with the type on the shore near Overton. The habit and texture of the var. *densifolia* seem to ally it as closely with *W. tortilis* as with *W. viridula.*

Weisia tenuis, C.M.—Area, 1. Range, only seen at 180 feet.

Sandstone rocks by streams. Very rare. First record : *Wilson, Journ. Bot.*, 1905, p. 96.

N. 1. Gorge of the Greeta between Burton-in-Lonsdale and Wrayton, 1904, *Wi.*

Weisia rupestris, C.M.—Area, 1, 2, 6, 7, 8. Range, 160–1,300 feet.

Wet limestone rocks and dripping calcareous banks. Locally abundant. Fruiting. First record : *Wilson, Journ. Bot.*, 1899, p. 473.

N. 1. Greygarth Fell, Ease Gill, and near Wennington, *Wi.* Gorge of the Greeta above Wrayton.
  2. Dalton Crag, *Wi.*
E. 6. Between Hornby and Lower Salter, 1898 ; by the Harter Beck, and the Roeburn ; and in Hindburndale, *Wi.* Udale ; Caton Moor ; and Crook of Lune.
  7. By the Wyre above Dolphinholme.
  8. Gorge of the Hodder near Whitewell.

A dwarf form with pellucid leaves (var. *humilis,* Ingham), occurs near the Crook of Lune with the type.

Weisia verticillata, Brid.—Area, 1, 2, 3, 6, 7, 8. Range, 20–500 feet.

Dripping rocks, especially those of calcareous shale. Rather rare. Fruiting occasionally. First record : *Wilson, Journ. Bot.*, 1899, p. 473.

N. 1. Gorge of the Greeta, in fruit.
  2. By the stream in Wash Dub Wood.
  3. Wet bank facing the shore between Carnforth and Bolton-le-Sands, *Wi.* In one or two places on the ridge of rocks extending from Woodwell to Gibraltar, near Silverdale. On the north side of Warton Crag.
E. 6. Crook of Lune.

E. 7. By the Brock near Garstang, 1898, *Wi.* By the Wyre below Abbey-stead, in fruit.
  8. Gorge of the Hodder below Whitewell, in fruit, and in a glen near Greystoneley.

Trichostomum crispulum, Bruch.—Area, 1, 2, 3, 8. Range, 0–750 feet.

Calcareous rocks and banks. Frequent in the limestone districts ; very rare elsewhere. Always sterile. First record : *Wheldon, Journ. Bot.*, 1899, p. 473.

N. 1. Gorge of the Greeta, *Wi.* Middle Ease Gill.
  2. Dalton Crag.
  3. Several localities near Silverdale and Yealand, 1898, *Wh.* Sea embankment between Carnforth and Bolton-le-Sands, *Wh.*
E. 8. Above Greystoneley, *Wh.* Near Whitewell.

Trichostomum mutabile, Bruch.—Area, 1, 2, 3, 6, 8. Range, 0–500 feet.

Rocks and banks, especially in calcareous districts or near the sea. Rare. Sterile. First record : *Hamilton, The Naturalist,* 1898, p. 28.

N. 1. Middle Ease Gill.
  2. Near Over Kellet, *J. W. Hartley, Wh. & Wi.*
  3. Silverdale, Lindeth, and near Woodwell, *Wh.* Rocks in Silverdale Cove, *Wi.*
E. 6. Near Lancaster, 1897, *W.P.H.*
  8. Near Greystoneley, *Wi.*

Var. *cophocarpum,* Schimp.

N. 1. Middle Ease Gill.
  2. Kellet Seeds, *Wi.*
  3. Silverdale, 1898, *Wh.*

Var. *littorale,* Dixon (*Trichostomum littorale,* Mitt.).

N. 1. Lower Ease Gill, *Wi.*
  3. Silverdale, 1899, *Wh.* Basil Point, *Wh.*

Trichostomum flavovirens, Bruch.—Area, 3, 4, 5. Range, 0–40 feet.

Sandy ground near the sea. Rare. Always sterile. First record : *J. Cash, Journ. Bot.*, 1899, p. 473.

N. 3. Near Heysham and at Basil Point, *Wh.* Between Middleton and Sunderland Point.
W. 4. Near Preesall, *Wh.* Between Pilling and Knott End, *Wi.*
  5. St. Annes ! 1884, *Cash.* Lytham and on the bank of the Wyre below Shard Bridge, *Wh.*

Trichostomum nitidum, Schimp.—Area, 1, 2, 3, 8. Range, 50–1,250 feet.

Limestone rocks and walls. Frequent in limestone districts, especially in North. Sterile. First record : *Wheldon. Journ. Bot.,* 1899, p. 473.

    N. 1. Ease Gill and Leck Fell, *Wi.*
       2. Dalton Crag.
       3. Silverdale, 1898, *Wh.* Near Hawes Water, *Wi.*
    W. 8. Between Leagram and Whitewell, *Wh.* Near Dinkling Green.

Trichostomum tortuosum, Dixon.—Area, 1, 2, 3, 6, 8. Range, 0–2,050 feet.

Abundant on limestone rocks and walls, especially in North ; occasionally in other districts on sandstone. Fruit not seen. First record : *Hamilton, The Naturalist,* 1898, p. 28.

    N. 1. Ease Gill and Greygarth Fell at 2,050 feet.
       2. Dalton Crag and near Whittington.
       3. Abundant throughout the Silverdale and Carnforth districts.
    E. 6. Lancaster, 1897, *W.P.H.* Hindburn and Roeburndale, *Wi.*
       8. Leagram, *Wh.* Near Whitewell.

Var. *fragilifolium,* Dixon.

    N. 1. Greygarth Fell at 2,050 feet, and in Ease Gill.
       2. Over Kellet, *Wi.* Dalton Crag.
       3. Frequent on exposed rocks.
    E. 6. Udale.
       8. Greystoneley, *Wh.* Wolf Fell on grit-rocks, *Wi.*

Pleurochæte squarrosa, Lindb.—Area, 3. Range, 10–100 feet.

Bare, stony, limestone banks near the sea. Very rare. Sterile. First record : *Wheldon, Journ. Bot.,* 1901, p. 297.

    N. 3. Open clearing in wood between Silverdale Church and Station, October, 1900, *Wh.* Also on a grassy slope below the sea cliffs north of Silverdale Cove, *Wi.*

Cinclidotus fontinaloides, P. Beauv.—Area, 1, 2, 3, 6, 8. Range, 50–750 feet.

Stones and rocks in rivers and streams. Frequent, especially amongst the hills. Fruiting. First record : *Hamilton, The Naturalist,* 1898, p. 28. " In the Lune above Lancaster, 1897."

Limestone rocks. Very rare. First record : The Authors, *Journ. Bot.,* 1905. p. 96.

    E. 8. Rocks between Greystoneley and Whitewell, 1903.

Zygodon conoideus, Hook. & Tayl.—Area, 1, 2, 3, 8. Range, 150–300 feet.

Trunks of trees. Rare. Fruits rarely. First record : The Authors, *Journ. Bot.,* 1902, p. 414.

    N. 1. Gorge of the Greeta, *Wi.*
       2. Trees by the Lune below Kirkby Lonsdale, 1901, and in Wash Dub Wood. Highfield, near Carnforth.
       3. On trees on the north side of Warton Crag.
    E. 8. On elders in the damp wood by the Hodder below Whitewell, fruiting abundantly, 1903. The elders were literally covered, both stem and branches, with this species and *Aulacomnium androgynum.*

ORTHOTRICHACEÆ.

Ulota Drummondii, Brid.—Area, 2. Range, only seen at about 200 feet.

Branches of trees, mixed with the next species. Very rare. First record : The Authors, *Journ. Bot.,* 1902, p. 414.

    N. 2. Wash Dub Wood, June, 1901.

Ulota Bruchii, Hornsch.—Area, 1, 2, 6, 7. Range, 150–1,050 feet.

Branches of trees ; more rarely on their trunks. Fruiting. Locally frequent in some of the sub-divisions ; quite unknown in others. First record : *Wilson, Journ. Bot.,* 1899, p. 509.

    N. 1. On an ash tree in a pot hole on Leck Fell, 1898, *Wi.* Ease Gill, *Wh.* & *Wi.* Wood by the Lune between Kirkby Lonsdale and Nether Burrow, *Wi.*
       2. Trees in Wash Dub Wood and in hedgerows near Whittington, *Wh.* & *Wi.* Dalton Crag and near Docker, *Wi.*
    E. 6. Lower Salter, and in the main gully of Mallowdale Fell at about 700 feet, *Wh.* & *Wi.* Trees by the falls of the Hindburn above Mill Houses, *Wi.* Middle Gill and Botton, *Wh.* & *Wi.*
       7. Gorge of the Wyre below Abbeystead, very sparingly.

Ulota crispa, Brid —Area, 1, 2, 3. Range, 100–1,000 feet.

Tree trunks and branches. Very rare. Fruiting. **First** record : *Wilson, Journ. Bot.,* 1905, p. 96.

ENCALYPTACEÆ.

Encalypta vulgaris, Hedw. — Area, 2, 3, 4, 5, 8. Range, 0–300 feet.

Dry rocks, walls, and banks, especially on calcareous soils. Infrequent. Fruiting. First record : *Wilson, Journ. Bot.,* 1899, p. 473.

    N. 2. Over Kellet, *S. Wi.* Borwick and Arkholme, *Wi.*
       3. Numerous places about Silverdale, 1898, Yealand and Warton, *Wh.* & *Wi.* Bolton-le-Sands, *Wi.*
    W. 4. Near Pilling and Cockersand Abbey, *Wi.* Knott End, *H.B.*
       5. St. Annes, *H.B.* Ansdell, *Fl. Prest. N.*
    E. 8. By the Hodder above Kemple End, *Wh.*

Encalypta streptocarpa, Hedw.—Area, 1, 2, 3, 4, 5, 6, 7, 8. Range, 0–2,050 feet.

Rocks and walls, mostly on limestone, but occasionally on sandstone. Common. Fruit very rare. First record : *Hamilton, The Naturalist,* 1898, p. 28. " Near Lancaster."

    E. 6. Fruiting sparingly on a damp wall near Caton, *Wh.*

Zygodon Mougeotii, B. & S.—Area, 1, 6, 7. Range, 500–1,000 feet.

Damp rocks by streams amongst the hills. Rare. Fruit not seen. First record : *Wilson, Journ. Bot.,* 1899, p. 473.

    N. 1. Lower Ease Gill, *Wi.*
    E. 6. Middle Gill, 1899, and Dale Gill, Hindburn ; and near Lower Salter, Roeburndale, *Wi.* Foxda e.
       7. By the Wyre between Abbeystead and Dolphinholme, on both sides of the river.

Zygodon viridissimus, R. Brown.—Area, 1, 2, 3, 4, 8. Range, 0–600 feet.

Tree trunks and old stumps ; rarely on walls and rocks. Frequent. Fruit not seen. First record : *Wheldon, Journ. Bot.,* 1899, p. 473.

    N. 1. Melling, *Wi.* Ease Gill ; and on willows near Tunstall.
       2. Dalton Hall and Borwick, *Wi.*
       3. Bare, and near Heysham, 1898, *Wh.* Middlebarrow, *Wi.*
    W. 4. Cockerham, *Wi.* Between Glasson and Galgate, *Wh.*
    E. 8. Near Stonyhurst, and near Longridge, *Wh.* Thornley, *Wi.*

Zygodon Stirtoni, Schimp.—Area, 8. Range, only seen at 500 feet.

    N. 1. Near Cowan Bridge. In Middle Ease Gill at 1,000 feet on hazels.
       2. Dalton Crag, ascending to 820 feet, on elder.
       3. Cringlebarrow, 1903, and near Yealand, *Wi.* Heald Brow, very sparingly.

Var. *intermedia,* Braithw.

    N. 2. With the type on Dalton Crag, 1904, *Wi.*

Ulota phyllantha, Brid.—Area, 2, 3. Range, 0–300 feet.

Trees near the sea. Very rare. Not fruiting. First record : The Authors, *Journ. Bot.,* 1905, p. 96.

    N. 3. Elders between Heaton and Overton, in small quantity and very stunted, *Wh.*
    N. 2. On trees near Carnforth, 1904, *J. W. Hartley, Wh.* & *Wi.* On an ash between Over Kellet and Capernwray, *Wi.*

Orthotrichum rupestre, Schleich.—Area, 1, 3. Range, 100–1,000 feet.

Rocks and walls. Rare. Fruiting. First record : *Wheldon, Journ. Bot.,* 1899, p. 509.

    N. 1. Ease Gill, near Leck, *Wi.*
       3. Silverdale, 1898, *Wh.*

Orthotrichum anomalum, Hedw. Var. *saxatile,* Milde — Area, 1, 2, 3, 8. Range, 0–2,080 feet.

Limestone rocks and walls. Abundant on the scar limestone. Fruiting freely. First record : *Wilson, Journ. Bot.,* 1899, p. 509. " Near Silverdale, 1898."

    We have not seen *O. anomalum* (type) in West Lancashire.

Orthotrichum cupulatum, Hoffm.—Area, 1, 2, 3, 7, 8. Range, 0–2,050 feet.

Limestone rocks and walls. Frequent. More widely spread, but less abundant than the last. Fruiting. First record : *Wilson, Journ. Bot.,* 1899, p. 509.

    N. 1. Greygarth Fell ; Ease Gill ; and near Cowan Bridge.
       2. Over Kellet, *Wi.* Dalton Crag.
       3. Silverdale, 1898, and Storrs Moss, *Wi.* Near Hawes Water, *Wh.*
    E. 7. Near Dolphinholme, *Wh.*
       8. Near Stonyhurst and Chipping, *Wh.*

Var. *nudum,* Braith.

    N. 1. Stones in Leck Beck, and tree roots by the Lune near Nether Burrow, *Wi.*
    E. 8. Banks of the Hodder, *Dr. J. B. Wood.* This was probably on the Lancashire side, as it is abundant there near Whitewell !

Orthotrichum leiocarpum, B. & S.—Area, 1, 2, 3, 6. Range, 150–550 feet.

On elder, hazel, and ash. Rare. Fruiting. First record : *Wilson, Journ. Bot.,* 1901, p. 297.

> N. 1. Near Wennington, *Wi.*
> 2. Dalton Crag, and on an ash tree between Over Kellet and Capernwray, *Wi.*
> 3. Between Silverdale Station and the village, 1901, *Wi.*
> E. 6. Lower Hindburn, at 550 feet.

Orthotrichum Lyellii, Hook. & Tayl.—Area, 1, 2, 3, 4, 6. Range, 50–400 feet.

Trunks of trees. Frequent in some parts of North ; very rare elsewhere. Fruit not seen. First record : *Wilson, Journ Bot.,* 1901, p. 297.

> N. 1. Near Melling ; Cantsfield ; and Burrow, *Wi.* Near Kirkby Lonsdale Station, and near Ireby.
> 2. Whittington, 1899, and near Dalton Houses, *Wi.* Near Burton ; and by the river below Kirkby Lonsdale.
> 3. Thrang Moss, and trees near Leighton Beck, *Wi.*
> W. 4. Near Bay Horse, *Wi.*
> E. 6. Near Wray, *Wi.*

Orthotrichum affine, Schrad.—Area, 1, 2, 3, 4, 6, 7, 8. Range, 100–500 feet.

Trunks of trees, more rarely on stones. Rather frequent in some parts of North, rare elsewhere. Fruiting. First record : *Wheldon, Journ. Bot.,* 1899. p. 509.

Var. *rivale,* Wils.

> E. 8. Tree roots by the Hodder near Mitton, *Wh.*

Orthotrichum rivulare, Turn.—Area, 1, 6, 8. Range, 50–150 feet.

Rocks and tree roots by streams. Rare. Fruiting. First record : *Hamilton, The Naturalist,* 1898, p. 28.

> N. 1. By the Lune between Nether Burrow and Kirkby Lonsdale, *Wi.*
> E. 6. Artle Beck, Caton, 1897, *W.P.H.* Crook of Lune, *Wh.*
> 8. Above Red Scar, and near Mitton, *Wh.*

Orthotrichum stramineum, Hornsch. — Area, 1, 2, 3, 8. Range, 50–500 feet.

Trees, principally in the limestone districts. Frequent locally in North ; very rare in the other divisions. Fruiting. First record : *Wilson, Journ. Bot.,* 1899, p. 509.

> N. 1. Near Kirkby Lonsdale Station, *Wi.* Springs Wood in Lower Ease Gill. Near Cowan Bridge, and Ireby.
> 2. Near Over Kellet, Capernwray, and Whittington, *Wi.* Holme Area below Kirkby Lonsdale Bridge.
> 3. In hedges near Yealand and Silverdale, *Wi.*
> E. 8. Cockleach near Longridge, and on oaks near Chipping, *Wh.*

Orthotrichum tenellum, Bruch.—Area, 2, 7. Range, 50–200 feet.

On the trunks of trees and horizontal branches in hedgerows. Very rare. Fruiting. First record : *Wilson, Journ. Bot.,* 1901, p. 297.

> N. 2. On an ash tree near Arkholme, *Wi.*
> E. 7. In small quantity in a hedge near Garstang, 1900, *Wi.*

Orthotrichum pulchellum, Wils.—Area, 1. Range, 400–500 feet.

Trees in woods. Very rare. Fruiting. First record : The Authors, *Journ. Bot.,* 1905, p. 96.

> N. 1. On elders in Springs Wood, Lower Ease Gill, 1904.

Orthotrichum diaphanum, Schrad.—Area, 1, 2, 3, 4, 5, 6, 7, 8. Range, 0–500 feet.

In crevices of the bark of trees, chiefly elder, hawthorn, and oak ; more rarely on stones. Widely distributed, but hardly common. First record : *Wheldon, Journ. Bot.,* 1899, p. 509. "Near Lower Hodder Bridge, 1898."

SPLACHNACEÆ.

Splachnum sphæricum, L.—Area, 1, 6, 7. Range, 1,200 (or lower)–1,900 feet.

On dung or remains of carcases of animals on the fells. Rather rare. Fruiting. First record : The Authors, *Journ. Bot.,* 1901, p. 297.

> N. 1. Greygarth Fell at 1,900 feet, *Wi.*
> E. 6. Wolfhole Crag and Botton Head Fell, *Wh.* & *Wi.* Salter Fell, *Wi.*
> 7. Slope of Wardstone towards Tarnbrook Fell at 1,600 feet. At the head of Great Clough in two localities, and Grizedale, near Garstang.

Tetraplodon mnioides, B. & S.—Area, 1, 4, 6, 7. Range, 25–1,700 feet.

On the remains of animals, or where they have lain, on moorlands. Rare. Fruiting. First record : *Wilson, Journ. Bot.,* 1899, p. 510.

> N. 1. On the remains of a sheep on Greygarth Fell at 1,460 feet.
> W. 4. Cockerham Moss, 1881, *Wi.*
> E. 6. On a dead sheep, Botton Head Fell at 1,700 feet, *Wi.* Clougha Scar, *Wh.*
> 7. North side of Marshaw Fell, on the bones of a sheep, *Wi.*

FUNARIACEÆ.

Discelium nudum, Brid.—Area, 6, 7, 8. Range, 500–1,050 feet.

Steep banks of clay or shale, chiefly near streams. Rare, but locally abundant in some of the dales. Fruiting. First record : *Wilson, Journ. Bot.,* 1901, p. 297.

> E. 6. Tatham Beck, Hindburn, 1899, *Wi.* Middle Gill, Hindburn ; Udale, and Caton Moor.
> 7. Tarnbrook Fell, Lower Emmetts, and Abbeystead Fell.
> 8. Parlick Pike, *Wi.*

Ephemerum serratum, Hampe.—Area, 3, 5. Range, 0–350 feet.

Fallow fields and bare earth. Rare. Fruiting. First record : *Wheldon, Journ. Bot.,* 1899, p. 510.

> N. 3. North side of Warton Crag, *Wi.* Mole hills near Hawes Water, *Wh.*
> W. 5. Lytham, near the Rifle Range, 1891, *Wh.*

Physcomitrium pyriforme, Brid.—Area, 2, 5, 7, 8. Range, 25–550 feet.

Clay banks and sides of roads and ditches. Infrequent. Fruiting. First record : *H. Beesley, Journ. Bot.,* 1901, p. 297.

> N. 2. Between Over Kellet and the Redwell Inn, *Wh.* Near Docker, *Wi.* Near Whittington. Near Dunnald Mill Hole.
> W. 5. Catforth, and Ashton near Preston, 1900, *H.B.* Banks of ditches inland from Ansdell, *Wi.*
> E. 7. Near Marshaw, *W.W.M.* ; Barnacre, *Wi.*
> 8. Near Kemple End, *Wh.*

Funaria ericetorum, Dixon.—Area, 3. Range, 100–400 feet.

Damp, bare, stony ground. Very rare. Fruiting. First record : The present one.

> N. 3. Near Grisdale Wood. North side of Warton Crag, 1906, *Wh.* & *Wi.* Also near the summit of Warton Crag, *Wh.*

Funaria calcarea, Wahl.—Area, 2, 3. Range, 50–500 feet.

Bare earth amongst limestone rocks. Locally abundant. Fruiting. First record : *Wilson, Journ. Bot.,* 1899, p. 510.

> N. 2. Near Henridden.
> 3. Silverdale, 1898, *Wi.* ; Trowbarrow, near Hawes Water ; Thrang End ; Yealand ; and very plentiful on Warton Crag, *Wh.*
> Often associated with *Weisia crispata, W. microstoma, Barbula recurvifolia, Encalypta vulgaris,* and *Hypnum chrysophyllum.*

Funaria hygrometrica, Sibth.—Area, 1, 2, 3, 4, 5, 6, 7, 8. Range, 0–2,050 feet.

Waste ground, banks, walls, and heaths. Abundant. Fruiting. First record : The Authors, *Journ. Bot.,* 1899, p. 510.

MEESIACEÆ.

Amblyodon dealbatus, P. Beauv.—Area, 1, 5, 6, 7. Range, 15–1,200 feet.

Springs on the hills and wet peaty places amongst the sandhills. Very rare. Fruiting. First record : *Stabler, The Naturalist,* 1897, p. 213.

> N. 1. Ireby Fell at 900 feet, *Wi.*
> W. 5. Between Lytham and St. Annes, 1898, *Wh.*
> E. 6. West side of Clougha, 1881, *Stabler.*
> 7. Moor just above the foot of Gavells Clough, Tarnbrook Fell, 1900, *Wi.*

Meesia trichoides, Spruce.—Area, 5. Range, only at about 20 feet.

Wet peaty hollows amongst the sandhills. Rare and rapidly becoming extinct through drainage. First record : *Wheldon, Journ. Bot.,* 1899, p. 510.

> W. 5. Between Lytham and St. Annes, 1898, *Wh.*
> This, like *Hypnum Wilsoni,* is in imminent danger of early extermination.

Aulacomnium palustre, Schwaeg.—Area, 1, 2, 3, 4, 5, 6, 7, 8. Range, 25–2,050 feet.

Wet heaths, bogs, and about moorland springs. Fruiting occasionally. First record : *Wilson, Journ. Bot.*, 1899, p. 510. " Storrs Moss, 1898, and Borwick, with fruit."

Aulacomnium androgynum, Schwaeg.—Area. 4, 5, 8. Range, 25–300 feet.

Banks and rotting tree trunks. Rare. Fruit not seen, but pseudopodia with gemmæ produced abundantly. First record : *H. Beesley, Journ. Bot.*, 1901, p. 298.

W. 4. Knott End, *H.B.* (sp.).
5. Lea, 1900, *H.B.* (sp.).
E. 8. Abundant on old elders near the Hodder two miles below Whitewell, *Wh. & Wi.*

BARTRAMIACEÆ.

Bartramia Œderi, Swartz.—Area, 1. Range, 600–1,500 feet.

Calcareous mountain rocks. Rare. Fruiting. First record : *Wilson, Journ. Bot.*, 1899, p. 510.

N. 1. Plentiful in Ease Gill near Leck, 1898, *Wi.* Leck Fell.

Bartramia ithyphylla, Brid.—Area, 1, 6. Range, 400–1,500 feet.

Mountain rocks and banks. Rare. Fruiting. First record : *Wilson, Journ. Bot.*, 1899, p. 510.

N. 1. Lower Ease Gill, 1898, *Wi.* Greygarth Fell.
E. 6. Mallowdale Fell Clough.

Bartramia pomiformis, Hedw.—Area, 1, 2, 4, 6, 7. Range, 15–700 feet.

Sandy or turfy banks and clefts of rocks. Not common. Fruiting. First record : *Wilson, Journ. Bot.*, 1899, p. 510.

N. 1. Ease Gill.
2. Near Skerton, *Wh.* Near Arkholme.
W. 4. Near Pilling, *Wi.*
E. 6. Caton Moor and Udale.
7. Sandholme, Garstang, 1898, *Wi.* Near Ellel Crag, *Wh.*

Philonotis fontana, Brid.—Area, 1, 3, 6, 7, 8. Range, 40–1,850 feet.

Peaty places about mountain springs and rill sides. Common amongst the hills ; rare elsewhere. First record : *Wheldon, Journ. Bot.*, 1899, p. 510. " Longridge Fell."

Forms approaching var. *ampliretis*, Dixon, occur, and others with something of the habit of *P. adpressa*, but not sufficiently marked to warrant their being placed on record. A very slender delicate form occurs on Warton Crag, which may be var. *pumila*, Dixon.

Philonotis calcarea, Schimp.—Area, 7, 8. Range, 600–1 270 feet.

Bogs and springs on the fells. Rather rare. First record : The Authors, *Journ. Bot.*, 1902 p. 414.

N. 1. By the Cant Beck above Leck Hall, and in Middle Ease Gill at 1,270 feet.
E. 7. Springs on the White side of Tarnbrook Fell and on Marshaw Fell with male flowers, 1902. Calder Valley above Oakenclough.
8. Wet moor near Lickhurst, *Wi.*

Breutelia arcuata, Schimp.—Area, 1, 3, 6, 7. Range, 220–1,200 feet.

Damp, stony and heathy mountain slopes. Frequent amorgst the hills. Not seen fruiting. First record : *Wilson, Journ. Bot.*, 1899, p. 510.

N. 1. Ease Gill and Ireby Fell, *Wi.*
3. Middlebarrow, *Wi.*
E. 6. Dale Beck, Hindburn ; Foxdale ; and Harter Beck, Roeburndale, *Wi.* Near Crag Wood, Clougha, *Wh.*
7. Gavells Clough and Marshaw Fell : north side of Harrisend Fell ; and Upper Grizedale, *Wi.*

BRYACEÆ.

Leptobryum pyriforme, Wils.—Area, 5. Range, 0–100 feet.

Sandy ground amongst short grass. Rare. First record : *Wheldon, Journ. Bot.*, 1899, p. 510.

W. 5. Sparingly on the Lytham sandhills, in fruit, 1898, *Wh.* Old river bank, Preston, *Fl. Prest. N.* Hedgebanks inland from Ansdell, abundant, *Wi.*

Webera elongata, Schwaeg.—Area, 1, 3, 7. Range, 50–600 feet.

Banks and clefts of rocks Very rare. Fruiting. First record : The Authors, *Journ. Bot.*, 1899, p. 510.

N. 1. In small quantity on Silurian rocks in Ease Gill, 1899.
3. Near Silverdale, *H.B.* (sp.).
E. 7. Near Garstang, *H.B.* (teste Dr. *Braithwaite*).

z

Webera nutans, Hedw.—Area, 1, 2, 3, 4, 5, 6, 7, 8. Range, 20–2,080 feet.

Peaty or sandy soil, on banks, heaths, and in woods. Very common. Fruiting. First record : The Authors, *Journ. Bot.*, 1899, p. 510. " Greygarth Fell."

Var. *longiseta*, B. & S.—Area, 1, 3, 4, 6, 7, 8. Frequent on damp heaths, ascending to the summit of Greygarth Fell.

Var. *bicolor*, Schimp.—Area, 1. Exposed mountain summits. Rare.

N. 1. Summit of Greygarth Fell at 2,080 feet, *Wi.*

Var. *elata*, Stalle.—A large tufted plant, which fruits but seldom, occurs on gritstone scars on Clougha and elsewhere. It agrees well with authentic examples of this variety.

Webera annotina, Schwaeg.—Area, 3, 5, 6, 7, 8. Range, 200–800 feet.

Sandstone quarries and stony or sandy waste ground and roadsides. Fruit not seen. Rare. First record : *Wheldon, Journ. Bot.*, 1899, p. 510.

N. 3. Carnforth, *H.B.*
W. 5. Near Garstang, *Wi.*
E. 6. Between Caton and Lower Salter, and on Tatham Moor.
7. Barnacre Moor near Garstang, *Wi.* Clougha, *Wh.*
8. Tootle Heights Quarry, 1898, *Wh.* Longridge Fell.

The above details represent the distribution of the aggregate species. The range of the segregates has not been fully ascertained, but the following occur :—

W. annotina, Hedw.—Area, 6, 7, 8. Range, ?–800 feet. To this sub-species belong the records given above for Tatham Moor, Clougha, and Tootle Heights, and probably some of the others.

W. erecta, Roth.—Area, 6, 7. Range, 200–800 feet. Damp sandy roadsides. First record : The Authors, *Journ. Bot.*, 1902, p. 415.

E. 6. With the type, abundant by roadside runnels on Tatham Moor, Hindburn, 1902.
7. Near Nicky Nook, *Wi.*

Webera carnea, Schimp.—Area, 1, 3, 4, 5, 6, 8. Range, 0–550 feet.

Damp clay banks. Not common. Fruiting. First record : *Hamilton, The Naturalist*, 1898, p. 28.

N. 1. Bank of the Lune near Over Burrow.
3. Thurshouse Sands, *Wh.*
W. 4. Near Preesall, *Wh.*
5. Near Blackpool and Lytham, *Wh.* Ashton, *H.B.*
E. 6. Near Lancaster, 1897, *W.P.F.* Caton, *Wh.*
8. Longridge, *Wh.* Near Dinkling Green at 550 feet.

Webera albicans, Schimp.—Area, 1, 2, 3, 4, 5, 6, 7, 8. Range, 0–1,980 feet.

Damp shady banks by streams and on wet rocks by mountain rills, etc. Frequent, especially in East. Occasionally fruiting. First record : *Wilson, Journ. Bot.*, 1899, p. 510. " By the Brook near Garstang, 1898."

Plagiobryum Zierii, Lindb.—Area, 1, 8. Range, 600–1,500 feet.

Damp mountain limestone rocks. Very rare, but locally abundant. Fruiting. First record : *Wilson, Journ. Bot.*, 1899, p. 510.

N. 1. Upper Ease Gill near Leck, in plenty at 1,500 feet, 1898, *Wi.* Young fruit seen in 1905.
E. 8. Near Whitewell, *Wi.*

Bryum filiforme, Dicks.—Area, 1. Range, only seen at 600 feet.

Wet mountain rocks. Very rare. Fruit not seen. First record : *Wilson, Journ. Bot.*, 1902, p. 414.

N. 1. On Silurian rocks, Lower Ease Gill, 1902, *Wi.*

Bryum pendulum, Schimp.—Area, 2, 3, 4, 5, 7. Range, 0–300 feet.

Sandy ground and walls. Common on the coast sandhills ; not frequent in other parts. Fruiting. First record : *Wheldon, Journ. Bot.*, 1899, p. 511.

N. 2. Near Lancaster and Over Kellet, *Wi.*
3. Roadside near Heysham Dock, *Wi.* Carnforth, *Wh.*
W. 4. Cabus near Garstang, *Wi.* Canal bridge near Winmarleigh.
5. Blackpool, 1895, Lytham, St. Annes, and Fleetwood, *Wh.*
7. Grizedale, near Garstang, *Wi.*

Bryum Warneum, Bland.—Area, 5. Range, only at about 20 feet.

Damp hollows in the sandhills, especially where water has stood during the winter. Rare. Fruiting. First record : *Wheldon, Journ. Bot.,* 1899, p. 511.

W. 5. St. Annes, July, 1898, *Wh.*

Bryum Marratii, Wils.—Area, 3. Range, 0–10 feet.

Damp sandy or muddy ground on the sea coast and margins of pools on the salt marshes. Very rare. First record : *Wilson, Journ. Bot.,* 1906.

N. 3. Salt marsh, south side of Keer estuary, November, 1905, *Wi.* Salt marsh between Middleton and Sunderland Point.

Bryum lacustre, Brid.—Area, 5. Range, only at about 20 feet.

With *B. Warneum,* and like it becoming much rarer through "improvements." Fruiting. First record : *Wheldon, Journ. Bot.,* 1899, p. 511.

W. 5. St. Annes, 1898, and since, *Wh.*

Bryum inclinatum, Bland.—Area, 1, 2, 3, 4, 5, 7, 8. Range, 0–2,000 feet.

Dry heaths, banks, and walls ; widely scattered over the vice-county, but nowhere very abundant. Fruiting. First record : *Wheldon, Journ. Bot.,* 1899, p. 511. "St. Annes, Lytham, and Preesall."

Ascends from the sea-coast near Middleton to 2,000 feet on Greygarth Fell.

Bryum uliginosum, B. & S.—Area, 5. Range, only at about 20 feet.

Damp sandy ground on the coast. Very rare and perhaps now extinct. Fruiting. First record : *Wheldon, Journ. Bot.,* 1899, p. 511.

W. 5. St. Annes, in fruit, 1899, *Wh.*

Bryum pallens, Swartz.—Area, 1, 2, 3, 4, 5, 6, 7, 8. Range, 0–2,040 feet.

Damp ground and banks of streams. Very common amongst the hills ; frequent in suitable situations elsewhere. First record : *Wilson, Journ. Bot.,* 1899, p. 511. "Silverdale, 1898."

The curious form formerly known as *B. origanum,* Bosw., which is not now admitted to be distinct, occurs near Silverdale, *Wh.* (teste *Dixon*), and near Dinkling

Green, *Wi.* It is a very marked form, with the young shoots of a claret colour, and leaves nearly or quite destitute of a border. It appears to be restricted to limestone soils and is perhaps worthy of varietal rank.

Bryum pseudotriquetrum, Schwaeg. — Area, 1, 5, 6, 7, 8. Range, 20–1,700 feet or higher.

Boggy ground amongst the hills. Fruiting occasionally. First record : *Wilson, Journ. Bot.,* 1899, p. 511. "Calder Valley above Oakenclough, 1898."

Var. *compactum,* B. & S.

W. 5. Damp sandy ground near St. Annes, in fruit, *Wh.*

Bryum bimum, Schreb. — Area, 1, 2, 3, 6, 7, 8. Range, 25–1,500 feet or higher.

Sphagnum bogs and damp rocks by pools and streams. Frequent ? Fruiting. First record : *Wheldon, Journ. Bot.,* 1899, p. 511.

N. 1. Ease Gill.
2. Bog near Docker, *Wi.*
3. Hawes Water, 1898, *Wh.* Thrang Moss.
E. 6. Lower Salter and Wolfhole Crag, *Wi.*
7. Lee Fell.
8. Longridge Fell, *Wh.*

Bryum affine, Lindb.—Area, 1, 6, 8. Range, 400–2,000 feet.

Damp walls and rocks. Rare. Fruiting. First record : *Wheldon, Journ. Bot.,* 1899, p. 511.

N. 1. Greygarth Fell, 1898, *Wh.*
6. Near Botton, Hindburn.
E. 8. Longridge Fell, 1898, *Wh.*

Bryum intermedium, Brid.—Area, 3, 5, 7, 8. Range, 0–800 feet.

Walls, banks, and damp ground where liable to inundation in winter, especially on clay soil. Not very frequent. Fruiting. First record : *Wheldon, Journ. Bot.,* 1899, p. 511.

N. 3. Ings Point, between Silverdale and Carnforth, *Wh.* Near Heysham Harbour.
W. 5. Sandhills at Lytham and St. Annes, 1898, *Wh.* Ashton, *H.B.*
E. 7. Calder Valley above Oakenclough, *Wi.*
8. Fulwood and Longridge, *Wh.*

Bryum cæspiticium, L.—Area, 1, 2, 3, 4, 5, 6, 7, 8. Range, 0–2,050 feet.

Walls, rocks, and dry banks. Common. Fruiting. First record : *Hamilton, The Naturalist,* 1898, p. 28. "Near Lancaster."

Bryum capillare, L.—Area, 1, 2, 3, 4, 5, 6, 7, 8. Range, 0–2,050 feet.

Walls, rocks, and trunks of trees. Very common. Fruiting. First record : *Hamilton, The Naturalist,* 1898, p. 28. "Near Lancaster."

Var. *macrocarpum,* Hübn.—Area, 1, 2, 3, 6, 7, 8. Frequent on rocks and walls, especially in hilly districts.

Var. *flaccidum,* B. & S.—Rare. Only seen on limestone rocks.

N. 3. Wood near Silverdale, *Wi.*
E. 8. Limestone rocks in Fence Wood near Dinkling Green.

Var. *rosulatum,* Mitt.—Area, 2, 3, 6, 7, 8. Not infrequent on the trunks of trees.

Bryum erythrocarpum, Schwaeg.—Area, 3. Range, 0–200 feet.

Bare earth on a sandy soil. Rare. Fruit not seen. First record : The present one.

N. 3. Near Yealand Redmayne, and on the sea coast near Carnforth, in each instance in small quantity, mixed with other mosses, and producing abundance of its characteristic compound crimson gemmæ.

Bryum atropurpureum, W. & M.—Area, 3, 4, 5, 8. Range, 0–400 feet.

Banks, bare earthy places, and waste ground. Not common. Fruiting. First record : *Wheldon, Journ. Bot.,* 1899, p. 511.

N. 3. Silverdale, Trowbarrow, and Thurshouse Sands, *Wh.*
W. 4. Near Galgate, *Wi.* Between Preesall and Pilling, *Wh.*
5. St. Annes, 1898, *Wh.*
E. 8. Roadside near Chipping, *Wh.*

Bryum murale, Wils.—Area, 1, 3, 5. Range, 0–250 feet.

Limestone rocks and on the mortar of walls. Rare. Fruiting. First record : *Wheldon, Journ. Bot.,* 1899, p. 511.

N. 1. On mortar of a wall near Tunstall ; and on a newly-mortared bridge between Kirkby Lonsdale Station and Cowan Bridge.
3. Several places about Silverdale, on limestone rocks and banks, *Wh.*
W. 5. On a newly-mortared wall round Fleetwood Barracks, very fine in 1899, *Wh.*

Bryum alpinum, Huds.—Area, 1, 6, 7. Range, 500–1,200 feet.

Wet mountain heaths and rocks. Rare. Fruit not seen. First record : *Wilson, Journ. Bot.,* 1899, p. 511.

N. 1. Silurian rocks in Ease Gill near Leck, 1898, *Wi.*
E. 6. Clougha, on grit rocks, *Wh.*
7. Tarnbrook Fell, *Wi.* Marshaw Fell, *P.J.H.* and *The Authors.*

Bryum argenteum, L.—Area, 1, 2, 3, 4, 5, 6, 7, 8. Range, 0–2,000 feet.

Waste ground and sides of footpaths. Very common. Fruiting. First record : The Authors, *Journ. Bot.,* 1899, p. 511.

Var. *lanatum,* B. & S.

Highfield, near Carnforth, *J. W. Hartley, Wh.* & *Wi.*

Bryum roseum, Schreb.—Area, 3, 5. Range, 20–120 feet.

Shady, bushy ground on sand or limestone. Rare. Not seen fruiting. First record : *Wheldon, Journ. Bot.,* 1899, p. 511.

N. 3. Silverdale, *Wi.*
W. 5. Amongst *Salix repens* near Fairhaven, 1898, *Wh.*

Cinclidium stygium, Swartz.—Area, 1. Range, only seen at about 1,270 feet.

Boggy rill-heads and wet spongy places on the fells. Very rare. Fruit not seen. First record : The present one.

N. 1. On the slope of Greygarth Fell towards Ease Gill, October, 1906.

Mnium affine, Bland.—Area, 1, 3, 4, 6, 7, 8. Range, 10–2,030 feet.

Damp woods and bogs. Frequent. Not seen in fruit. First record : *Wheldon, Journ. Bot.,* 1899, p. 511.

N. 1. By the Lune near Kirkby Lonsdale, and at 2,030 feet on Greygarth Fell, *Wi.* Gorge of the Greeta near Wrayton.
3. Near Burton, *Wh.* & *Wi.* Damp place in the wood near Silverdale Church, *Wh.* Marshy place near the shore below Heald Brow, *Wi.*
E. 6. Heights Wood near Clougha, *Wh.* Dale Gill, *Wi.*
7. Calder Valley, *Wi.*
8. Spade Mill Reservoir, 1898, and above Stonyhurst, *Wh.*

Var. *elatum*, B. & S.

N. 3. Borwick, 1898. *Wi.*
W. 4. Near Cabus.
E. 7. Calder Valley above Oakenclough; and Brock Bottom, *Wi.* Near Garstang, *H.B.* (5 or 7 ?).

Mnium cuspidatum, Hedw.—Area, 3, 5, 8. Range, 20–600 feet.

Damp sandy ground or damp rocks by streams. Rare. Fruiting. First record: *Wheldon, Journ. Bot.*, 1899, p. 511.

N. 3. Silverdale, in fruit, *Wh.*
W. 5. Near Fairhaven, *Wh.*
E. 8. Near Kemple End, 1898, *Wh.* Fence Wood, near Dinkling Green at 600 feet.

Mnium rostratum, Schrad.—Area, 1, 2, 3, 5, 6, 7, 8. Range, 20–2,050 feet.

Shady banks and damp rocks. Frequent, in some districts common. Fruiting occasionally. First record: *Hamilton, The Naturalist*, 1898, p. 28. "Near Lancaster, 1897."

Mnium undulatum, L.—Area, 1, 2, 3, 4, 5, 6, 7, 8. Range, 0–1,480 feet.

Shady banks and woods. Common. Fruiting very rarely. First record: *Wheldon, Journ. Bot.*, 1899, p. 512. "Near Ribchester, 1898."

Fruiting at Silverdale and in Heald Brow Wood, *Wi.* It ascends to 1,480 feet in Upper Ease Gill.

Mnium hornum, L.—Area, 1, 2, 3, 4, 5, 6, 7, 8. Range, 0–2,050 feet.

Banks, rocks, tree roots, etc., in woods and damp shady places. Very common, especially in East, where it is very luxuriant and fruits freely. First record: The Authors, *Journ. Bot.*, 1899, p. 512.

Mnium serratum, Schrad.—Area, 1, 6, 8. Range, 300–2,030 feet.

Rocks by streams in hilly districts. Rather rare, but locally plentiful. Fruiting. First record: *Wilson, Journ. Bot.*, 1899, p. 512.

N. 1. Ease Gill, and Gorge of the Greeta, *Wh. & Wi.* Greygarth Fell at 2,030 feet, *Wi.*
E. 6. Dale Gill, Hindburn, 1899, *Wi.*
8. Longridge, *H.B.* Fruiting near Whitewell, *Wh. & Wi.* By the Brock near Garstang, *Wi.*

Mnium orthorrhynchum, B. & S.—Area, 1. Range, only seen at about 1,400 feet.

Limestone rocks amongst the hills. Very rare. Fruiting. First record: *Wilson, Journ. Bot.*, 1899, p. 512.

N 1. Ease Gill, with fruit, 1899, *Wi.*

Mnium stellare, Reich.—Area, 1, 2, 3, 7, 8. Range, 75–250 feet.

Shady banks and woods. Rare. Fruit not seen. First record: *Wilson, Journ. Bot.*, 1899, p. 512.

N. 1. Near Melling, *Wi.*
2. Kellet Seeds, *Wi.*
3. Silverdale, *Wi.*
E. 7. By the Brock, 1898, *Wi.*
8. Hodder banks near Kemple End, *Wh.*

Mnium punctatum, L.—Area, 1, 2, 3, 4, 5, 6, 7, 8. Range, 0–1,500 feet or higher.

Damp shady banks, rocks, and walls, in bogs and by streams. Very common, except in the Fylde. Fruiting. First record: *Wilson, Journ. Bot.*, 1899, p. 512. "By the Brock near Garstang, 1898."

Mnium subglobosum, B. & S.—Area, 6, 7, 8. Range, 700–1,300 feet.

Peat bogs and springs on the fells. Frequent in some parts of East, very rare, if at all, in the other divisions. Fruiting. First record: *Wilson, Journ. Bot.*, 1899, p. 512.

E. 6. Goodber Common, *Wi.* Dale Gill, Hindburn.
7. Fairsnape Clough, in fruit, 1898; and Calder Valley above Oakenclough, *Wi.* Gavells Clough; Blaze Moss; and Grizedale near Abbeystead.
8. Fell near Burnslack, *Wi.*

FONTINALACEÆ.

Fontinalis antipyretica, L.—Area, 1, 2, 3, 4, 5, 6, 7, 8. Range, 25–1,200 feet.

Attached to stones or wood in streams and ponds. Common. It ascends to 1,200 feet on Bottom Head Fell. Fruit not seen. First record: *Wheldon, Journ. Bot.*, 1899, p. 512. "River Ribble, 1898."

Var. *gracilis*, Schimp.—Streams amongst the hills.

N. 1. Leck Beck, 1898, *Wi.* Ireby Fell.
E. 6. Udale; Roeburndale; and Crook of Lune, *Wh. & Wi.* Hindburn, *Wi.*
7. Near Clougha, *Wh.*
8. Burnslack Beck, *Wi.*

Fontinalis squamosa, L.—Area, 1, 6, 7. Range, 600–1,000 feet or higher.

Rapid mountain streams. Rare. Fruit not seen. First record: The Authors, *Journ. Bot.*, 1899, p. 512.

N. 1. Leck Beck, Ease Gill, *Wi.*
E. 6. Udale, 1899.
7. Marshaw Fell.

NECKERACEÆ.

Neckera crispa, Hedw.—Area, 1, 2, 3, 8. Range, 0–1,400 feet.

Limestone rocks and walls. Common in North, more sparingly in the limestone district near Whitewell. First record: *Wilson, Journ. Bot.*, 1899, p. 512. "Ease Gill, 1898."

Var. *falcata*, Boul., has the same range as the type, but is rather less common. It is a mere form passing into the type and depends on exposure to strong sunlight for its peculiar features.

Neckera pumila, Hedw.—Area, 3. Range, only seen at about 200 feet ?

Tree trunks. Very rare. Not seen in fruit. First record: *H. Beesley, Journ. Bot.*, 1905, p. 96.

Var. *Phillipeana*, Milde. (the type does not occur with us).

N. 3. On a tree near Silverdale, 1904, *H.B.* (sp.).

Neckera complanata, Hübn.—Area, 1, 2, 3, 4, 5, 6, 7, 8. Range, 0–1,000 feet or higher.

Rocks, walls, banks, and trunks of trees. Common. Fruiting, but not very commonly. First record: *Wheldon, Journ. Bot.*, 1899, p. 512. "Stonyhurst, 1898, and Caton."

Homalia trichomanoides, Brid.—Area, 1, 2, 3, 4, 5, 6, 7, 8. Range, 0–700 feet.

Shaded rocks and trunks of trees. Frequent. Fruits more frequently than the preceding. First record: *Wilson, Journ. Bot.*, 1899, p. 512. "Gatebarrow, 1898, and Borwick."

HOOKERIACEÆ.

Pterygophyllum lucens, Brid.—Area, 1, 2, 6, 7, 8. Range, 100–1,500 feet.

Moist woodland banks and shady glens, generally on shale or clay. Not uncommon, especially in East. Fruiting. First record: *Wheldon, Journ. Bot.*, 1899, p. 512.

N. 1. Ease Gill, ascending to 1,500 feet, *Wi.* Gorge of the Greeta, near Wrayton.
2. By the Lune between Caton and Aughton.
E. 6. Harter Beck and Lower Salter, *Wi.* Heights Wood, *Wh.* Near Botton, Hindburn.
7. Calder Woods; Barnacre; Wyresdale, and Brock Valley, *Wi.*
8. Near Upper Hodder Bridge, 1898, *Wh.* Longridge Fell, *H.B.*

LEUCODONTACEÆ.

Leucodon sciuroides, Schwaeg.—Area, 1, 2, 5. Range, 50–400 feet.

Trunks of trees and occasionally on walls. Rare. Not fruiting. First record: *Wheldon, Journ. Bot.*, 1899, p. 512.

N. 1. On a wall near Ireby, *Wi.*
2. Tree trunk at Halton, 1898, *Wh.* Near Borwick and Whittington, or trees, *Wi.*
W. 5. Weeton, near Blackpool, *H.B.*

Pterogonium gracile, Swartz.—Area, 2. Range, only seen at about 150 feet.

Trunks of trees. Very rare. Not fruiting. First record: The Authors, *Journ. Bot.*, 1902, p. 415.

N. 2. Holme Area below Kirkby Lonsdale, 1901.

Antitrichia curtipendula, Brid.—Area, 1, 2. Range, 200–450 feet.

Rocks and walls. Very rare. Not seen in fruit. First record: The Authors, *Journ. Bot.*, 1902, p. 415.

N. 1. Wall near Leck, *Wi.*
2. On a wall in Wash Dub Wood, 1901.

Porotrichum alopecurum, Mitt.—Area, 1, 2, 3, 4, 6, 7, 8. Range, 0–1,100 feet or higher.

Damp walls, rocks, and hedgebanks, and often especially luxuriant in crevices of limestone pavement. Common, particularly in North. Fruit rare. First record: *Wilson, Journ. Bot.*, 1899, p. 512. "Silverdale, 1898."

Fruiting on grit-rocks, Calder Valley near Garstang, and on limestone in Leapers Wood, Kellet Seeds, *Wi.*, and in Ease Gill.

LESKEACEÆ.

Leskea polycarpa, Ehrh.—Area, 1, 2, 4, 5. 6, 7, 8. Range, 0–300 feet.

Tree roots by lowland streams. Frequent. First record: *Wheldon, Journ. Bot.*, 1899, p. 512. "Near Ribchester, 1899, and Caton."

A large lax form (var. *paludosa*, Schimp.) occurs occasionally, but does not appear to differ in any essential particular from the type.

Anomodon viticulosus, H. & T.—Area, 1, 2, 3, 8. Range, 0–700 feet.

Common on rocks, walls, and tree roots on calcareous soils. Fruiting sparingly. First record: *Wilson, Journ. Bot.*, 1899, p. 512. "Silverdale, 1898, Borwick, etc."

Heterocladium heteropterum, B. & S.—Area, 1, 2, 6, 7. Range, 250–900 feet.

Damp rocky glens and near waterfalls. Rare. Fruit not seen. First record: The Authors, *Journ. Bot.*, 1901, p. 298.

N. 1. Lower Ease Gill, *Wi.* Gorge of the Greeta.
    2. By the fall in Wash Dub Wood.
E. 6. Waterfall near Botton, Hindburn, 1899, *Wh.* & *Wi.* Harter Beck, Roeburndale, *Wi.* Clougha, *Wh.* Troughbrook Ghyll, *Wi.*
    7. Calder Valley near Garstang, and rocks by the Wyre near Dolphinholme, *Wi.*

Thuidium tamariscinum, B. & S.—Area, 1, 2, 3, 4, 5, 6, 7, 8. Range, 0–2,050 feet.

Shady woods and banks. Common. Fruits occasionally. First record: *Wheldon, Journ. Bot.*, 1899, p. 512. "Dolphinholme, 1897."

Thuidium delicatulum, Mitt.—Area, 1, 6. Range, 400–500 feet.

Damp sub-alpine woods. Very rare. Fruit not seen. First record: *Wilson, Journ. Bot.*, 1905, p. 96.

N. 1. Wet wood in Lower Ease Gill, July, 1903, *Wi.*
E. 6. By the Harter Beck, Roeburndale.

Thuidium recognitum, Lindb.—Area, 2, 3. Range, 0–700 feet.

Banks and shaded rocks on calcareous soil. Rare. Fruit not seen. First record: *Wilson, Journ. Bot.*, 1899, p. 512.

N. 2. Dalton Crag, *Wh.* & *Wi.* Kellet Seeds, *Wh.*
    3. Silverdale, 1898, *Wi.* Trowbarrow and near Yealand, *Wh.* Flat, bare ground near the salt marsh, south side of Keer estuary, *Wi.*

HYPNACEÆ.

Climacium dendroides, W. & M.—Area, 1, 2, 3, 4, 5, 6, 7, 8. Range, 15–2,080 feet.

Bogs and marshes. Frequent. Fruit not seen. First record: *Yates, Journ. Bot.*, 1899, p. 513. "Fulwood, 1892."

A small condensed state (*forma depauperata*, Boul.) occurs on the sandhills and on dry ground in the limestone districts, but in both situations it appears to be connected with the type by intermediate forms. Its peculiarities are apparently produced by the frequent droughts to which it is subjected in the habitats named.

Cylindrothecium concinnum, Schimp.—Area, 2, 3. Range, 100–400 feet.

Dry, grassy calcareous hills and banks. Rare. Not fruiting. First record: *Wheldon, Journ. Bot.*, 1901, p. 298.

N. 2. Roadside, south-west of Dalton Crag, *Wi.*
    3. Silverdale, 1900, *Wh.*

Orthothecium intricatum, B. & S.—Area, 1, 2, 3. Range, 50–1,500 feet.

Damp limestone rocks amongst the hills. Very rare. Not fruiting. First record: *Wilson, Journ. Bot.*, 1899, p. 513.

N. 1. Upper Ease Gill, in plenty at about 1,500 feet with *Plagiobryum Zierii*, 1898, and in a pot-hole on Leck Fell, *Wi.* Middle Ease Gill.
    2. Kellet Seeds, very sparingly, *Wi.*
    3. Wall, Leighton Beck, near Silverdale, *Wi.*

Camptothecium sericeum, Kindb.—Area, 1, 2, 3, 4, 5, 6, 7, 8. Range, 0–2,050 feet.

Rocks, walls, and trunks of trees. Common; especially abundant and luxuriant on the limestone. Fruiting. First record: *Hamilton, The Naturalist*, 1898, p. 28. "Near Lancaster, 1897."

Camptothecium lutescens, B. & S.—Area, 1, 2, 3, 5. Range, 0–800 feet.

Gravelly banks on calcareous soil and on the coast sandhills. Frequent. Fruit not seen. First record: *Wheldon, Journ. Bot.*, 1899, p. 513.

N. 1. Ease Gill, *Wi.*
    2. Dalton Crag; Kellet Seeds; and near Whittington, *Wi.*
    3. Silverdale; Warton; and near Leighton Beck, *Wi.* Near Hawes Water and Yealand, *Wi.*
W. 5. Sandhills near St. Annes, 1898, *Wh.*

Brachythecium glareosum, B. & S.—Area, 3, 5, 7, 8. Range, 0–300 feet.

Rocks, banks, and woods, especially on calcareous or clay soils. Rare. Fruit not seen. First record: *Wilson, Journ. Bot.*, 1899, p. 513.

N. 3. Silverdale, *Wh.* Warton; Borwick; and near Hawes Water, *Wi.*
W. 5. Salwick, *H.B.*
E. 7. Garstang, 1898, *Wi.*
    8. Quarry near Chipping, *Wi.*

Brachythecium albicans, B. & S.—Area, 3, 4, 5, 7. Range, 0–375 feet.

Dry, sandy or stony ground. Locally abundant, especially on the coast sandhills. Fruit very rare. First record: *Wheldon, Journ. Bot.*, 1899, p. 513.

N. 3. Wood Well and Heald Brow, near Silverdale, *Wh.* Bolton-le-Sands; Hest Bank; coast south of Heysham; and near Middleton, *Wi.*
W. 4. Abundant on the sea-coast near Pilling, *Wi.* Knott End, *Wh.* Between Cockersand and Glasson, *Wi.*
    5. Lytham, in fruit, 1898, *Wh.* St. Annes; South Shore; Blackpool; and near Fleetwood, *Wh.*
E. 7. Roadside, Upper Claughton, near Garstang, *Wi.*

Brachythecium salebrosum, B. & S.—Area, 3, 5. Range, 0–100 feet.

Stones, sandy banks, and tree roots. Rare. Fruiting. First record: *H. Beesley, Journ. Bot.*, 1901, p. 298.

N. 3. Near Leighton Beck, *Wi.*
    5. Ashton, near Preston, 1900, *H.B.* Lea, *Fl. Prest. N.*

Brachythecium Mildeanum, Schimp. — Area, 5. Range, 15–25 feet.

Damp sandy or peaty places on the sandhills. Rare.

Fruiting. First record: *Wheldon, Journ. Bot.*, 1899, p. 513 (as *B. salebrosum* var. *palustre*, Schimp.).

W. 5. St. Annes, 1898; near Fairhaven; and near Fleetwood, *Wh.*

Brachythecium rutabulum, B. & S.—Area, 1, 2, 3, 4, 5, 6, 7, 8. Range, 0–1,000 (?) feet.

Trees, rocks, walls, or banks, especially on a sandy or clayey soil. Common in damp shady places. Fruiting. First record: *Wilson, Journ. Bot.*, 1899, p. 513. "Garstang, December, 1876."

Var. *longisetum*, B. & S.—Marshy ground amongst rushes, etc.

N. 2. Near Borwick, *Wi.*
W. 5. Salwick, *H.B.*
E. 8. Longridge, 1898, *Wh.*

Var. *robustum*, B. & S.—Wet grassy places and margins of ponds.

N. 2. Pedder Pots, *Wh.*
    3. Thrang Moss.
W. 5. Bispham, 1898, and very fine by a pond on the sandhills, St. Annes, *Wh.* Cadley, *H.B.*
E. 6. Hindburn, *Wi.*

Var. *densum*, Schimp.—Tree roots and dry rocks.

N. 3. Near Woodwell, Silverdale, *Wh.*

A small form of a light green or yellow colour is plentiful under dwarf *Salix repens* on the coast sandhills, and was recorded in *Journ. Bot.*, 1899, p. 513, as var. *plumulosum*, B. & S., but Mr. Dixon considers that it is not that variety.

Brachythecium rivulare, B. & S.—Area, 1, 2, 3, 5, 6, 7, 8. Range, 40–1,200 feet or higher.

Damp rocks and stones at the margins of streams and in moorland rills. Common in East, less common in North, and very rare in West. Rarely fruiting. First record: *Wheldon, Journ. Bot.*, 1899, p. 513.

N. 1. Ease Gill.
    2. Near Docker.
    3. Leighton Beck, *Wi.*
W. 5. Ashton and Cottam, *H.B.*
E. 6, 7, 8. Common.

Var. *latifolium*, Husn.

E. 6. Near Botton, Hindburn, 1899. A curious form, referred to this variety by Mr. Bagnall.

A form very near var. *chrysophyllum*, Bagn., occurs near Lower Salter, *Wi.*

Brachythecium velutinum, B. & S.—Area, 1, 2, 3, 5, 6, 7, 8. Range, 0–1,000 (?) feet.

Tree trunks, rocks, walls, and banks. Common. Fruiting. First record: *Wheldon, Journ. Bot.,* 1899, p. 513. " Near Ribchester, 1898."

The var. *intricatum,* Hedw., is not uncommon, and the var. *prælongum,* B. & S., is reported from Lytham by H. Beesley (teste *Dr. Braithwaite*).

Brachythecium populeum, B. & S.—Area, 1, 2, 3, 4, 5, 6, 7, 8. Range, 0–2,050 feet.

Stones, walls, and banks. Frequent. Fruiting. First record: *Wilson, Journ. Bot.,* 1899, p. 513. " Garstang, January, 1877."

Brachythecium plumosum, B. & S.—Area, 1, 2, 6, 7, 8. Range, 100–2,050 feet.

Damp sandstone rocks and stones, especially by streams. Common amongst the hills. Fruiting. First record: *Hamilton, The Naturalist,* 1898, p. 28. " Near Lancaster, 1877."

The var. *homomallum,* B. & S., is frequent in streams amongst the hills of East.

Brachythecium purum, Dixon.—Area, 1, 2, 3, 4, 5, 6, 7, 8. Range, 0–1,000 (?) feet.

Banks and grassy places. Common. Fruit rare. First record: *Wheldon, Journ. Bot.,* 1899, p. 513. " Dolphinholme, 1897."

With fruit on Warton Crag, *Wi.*

Hyocomium flagellare, B. & S.—Area, 1, 2, 3, 6, 7, 8. Range, 50–1,400 feet or higher.

Wet rocks by streams in hilly districts. Common in East and parts of North. Fruit not seen. First record: *Wilson, Journ. Bot.,* 1899, p. 513. " In the Brock, Fairsnape Clough, 1898."

Eurynchium piliferum, B. & S.—Area, 1, 2, 3, 4, 5, 6, 7, 8. Range, 0–500 feet or higher.

Woods and damp shady banks. Frequent. Rarely fruiting. First record: *Wheldon, Journ. Bot.,* 1899, p. 513. " Yealand, with fruit, 1898."

Eurynchium crassinervium, B. & S.—Area, 1, 2, 3, 6 8. Range 0–500 (?) feet.

Damp shaded rocks and stones, more rarely on tree roots. Rare. Not fruiting. First record: *G. A. Holt, Journ. Bot.,* 1899, p. 513.

N. 1. By the Greeta, Wrayton, *Wi.*
　　2. Quarry between Kirkby Lonsdale and Whittington, *Wi.*
　　3. Wall between Heysham and Morecambe ; between Warton and Yealand ; and wood at Wood Well near Silverdale, *Wh.* Middlebarrow Wood, *Wi.*
E. 6. Between Caton and Lancaster, *Wh.*
　　8. Hodder Banks, 1886, *G. A. Holt.* Still there, 1898, *Wh.*

Eurynchium speciosum, Schimp.—Area, 5. Range, only seen at 75 feet.

Damp wall by water. Very rare. Fruit not seen. First record: *H. Beesley, Journ. Bot.,* 1902, p. 415.

W. 5. Canal side, Ashton, near Preston, 1901, *H.B.*

Eurynchium prælongum, Hobk.—Area, 1, 2, 3, 4, 5, 6, 7 8. Range, 0–1,500 feet or higher ?

Woods, banks, and hedgerows, especially on a stiff soil. Very common. Fruiting. First record: *Wilson, Journ. Bot.,* 1899, p. 514. " Near Garstang, 1876."

The var. *Stokesii,* B. & S., is frequent in woods, especially in the hilly districts.

Eurynchium Swartzii, Hobk.—Area, 1, 2, 3, 4, 5, 6, 7, 8. Range, 0–1,000 feet ?

Fields, banks, and grassy places, especially on calcareous soil, sometimes on clay. Frequent, but not nearly so common as the preceding. Fruit not seen. First record: *Hamilton, The Naturalist,* 1898, p. 28. " Near Lancaster, 1897."

A dark green broad-leaved form occurs on limestone rocks near Dinkling Green (var. *atrovirens,* B. & S.).

Eurynchium abbreviatum, Schimp.—Area, 1, 2, 3. Range, 0–500 feet.

Shady woods, on earth or damp stones. Rare. Fruit not seen. First record: *Wheldon, Journ. Bot.,* 1901, p. 298.

N. 1. Wall near Leck, *Wi.*
　　2. Near Whittington, *Wi.*
　　3. Silverdale, 1900, *Wh.* Thrang End, *Wi.*
A 1

Eurynchium pumilum, Schimp.—Area, 1, 2, 6, 8. Range, 100–600 feet.

Damp shaded rocks. Rare. Fruit not seen. First record: *Wheldon, Journ. Bot.,* 1899, p. 514.

N. 1. Ease Gill.
　　2. Wash Dub Wood.
E. 6. Caton, 1898, *Wh.*
　　8. Near Whitewell.

Eurynchium Teesdalei, Schimp.—Area, 8. Range, 600 feet.

Pot-holes, caves, and damp, shaded, limestone rocks. Rare. First record: The present one.

E. 8. Pot-hole in Fence Wood, and cave near Dinkling Green, April, 1907.

Eurynchium tenellum, Milde.—Area, 1, 2, 3, 6, 7. Range, 0–1,700 feet.

Damp, shaded limestone rocks. Not uncommon. Fruiting. First record: *Wheldon, Journ. Bot.,* 1901, p. 298.

N. 1. Middle Ease Gill and near Wennington.
　　2. Wall near Halton, *Wi.* Near Dunnald Mill Hole, *Wi.*
　　3. Near Silverdale Cove ; Wood Well, etc., 1900, *Wh.* Cringlebarrow, *Wi.* Warton Crag, *Wh.*
E. 6. Wardstone.
　　7. Near Scorton, *H.B.*

Eurynchium myosuroides, Schimp.—Area, 1, 2, 3, 4, 5, 6, 7, 8. Range, 0–900 feet.

Rocks, banks, and tree roots. Frequent. Rarely fruiting. First record: *Wheldon, Journ. Bot.,* 1899, p. 514. " Near Kemple End, 1898."

The *forma filescens,* Ren., occurs on elder trunks near Heaton in the Heysham Peninsula.

Eurynchium myurum, Dixon.—Area, 1, 2, 3, 5, 6, 7, 8. Range, 0–700 feet.

Roots of trees, rocks, and banks in shady places. Frequent in calcareous districts, less common elsewhere. Fruiting sparingly. First record: *Wheldon, Journ. Bot.,* 1899, p. 513. " Caton, 1898, and Hodder Valley."

Eurynchium striatum, B. & S.—Area, 1, 2, 3, 4, 5, 6, 7, 8. Range, 0–700 feet or higher.

Grassy banks, rocks in woods, and shady hedgerows. Common. Fruiting. First record: *Wilson, Journ. Bot.,* 1899, p. 514. " Near Garstang, 1877."

Eurynchium rusciforme, Milde.—Area, 1, 2, , 5, 6, 7, 8. Range, 25–1,200 feet.

Stones and rocks in streams. Common, especially amongst the hills. Fruiting. First record: *F. C. King, Journ. Bot.,* 1899, p. 514. " Brock, 1886 " (sp.).

Var. *atlanticum,* Brid.

E. 6. Near Botton, Hindburn.
　　8. Longridge Fell, *Wh.*

Var. *inundatum,* Brid.

E. 7. Grizedale, near Garstang, 1898, *Wi.*

Var. *alopecuroides,* Brid.—In rapid moorland streams.

E. 6. Dale Gill, Hindburn, 1902.
　　8. Dripping rocks by the Hodder, Whitewell, 1903, *Wi.*

Eurynchium murale, Milde.—Area, 1, 2, 3, 4, 5, 6, 7, 8. Range, 0–2,000 feet.

Damp, shaded rocks and walls. Frequent. Fruiting. First record: *Wheldon, Journ. Bot.,* 1899, p. 514. " Near Preston, fruiting, 1897."

Var. *julaceum,* Schimp.—On drier rocks, especially limestone.

N. 1. Near Cowan Bridge.
　　2. Near Carnforth, 1900, *Wh.*
　　3. Near Silverdale, *Wh.*
E. 8. Exposed limestone rocks, opposite Whitewell.

Var. *complanatum,* B. & S.—Damp rocks and stones by streams.

N. 3. Near Silverdale, *Wh.*
E. 7. Brock Valley, 1898, with fruit, *Wi.*
　　8. Near Whitewell.

Eurynchium confertum, Milde.—Area, 1, 2, 3, 4, 5, 6, 7, 8. Range, 0–2,030 feet.

Stones, walls, tree trunks, and banks. Very common. Fruiting. First record: *Wilson, Journ. Bot.,* 1899, p. 514. " Near Garstang, 1876."

Eurynchium megapolitanum, Milde. — Area, 5. Range, 15–30 feet.

Dry banks on the sandhills. Rare. Fruiting. First record: *Wheldon, Journ. Bot.,* 1899, p. 514.

> W. 5. St. Annes and near Fairhaven, in fruit, 1898, *Wh.*

Plagiothecium depressum, Dixon.—Area, 1, 6, 8. Range, 400–1,300 feet.

Shaded rocks. Very rare. Not fruiting. First record: *Wilson, Journ. Bot.,* 1899, p. 514.

> N. 1. Middle Ease Gill, *Wi.*
> E. 6. Upper Roeburndale, 1898, *Wi.*
> 8. Near Knowl Green, on sandstone, *Wh.*

Plagiothecium elegans, Sull.—Area, 1, 2, 4, 5, 6, 7, 8. Range, 0–1,950 feet.

Sandstone rocks and peaty soil on moors and in woods. Frequent, especially on the moorlands of East. Fruit not seen. First record: *Wheldon, Journ. Bot.,* 1899, p. 514. " Longridge Fell, 1898."

Var. *collinum*, Wils.

> N. 2. Grit-rocks in Kellet Park Wood.
> E. 6. Clougha Scar, *Wh.*

Plagiothecium pulchellum, B. & S.—Area, 6. Range, only seen at about 1,000 feet.

Amongst other mosses in crevices of gritstone rocks. Very rare. Sterile. First record: The Authors, *Journ, Bot.,* 1899, p. 514.

> E. 6. Clougha Pike, 1899.

Plagiothecium denticulatum, B. & S.—Area, 1, 2, 3, 4, 5, 6, 7, 8. Range, 0–1,400 feet or higher.

Banks, rocks, and rotting tree trunks in shady places and woods. Common. Fruiting. First record: *Wilson, Journ. Bot.,* 1899, p. 514. " Near Garstang, 1877."

Var. *majus*, Boul.—Area, 1, 2, 4, 6, 7, 8. Frequent, especially amongst the hills, where it is as common as the type.

Var. *aptychus*, Spr.—Rare and not always fruiting, and on this account some of the records are perhaps not free from doubt.

> E. 6. Clougha, *Wh.*
> 7. Fairsnape Clough, *Wi.* Between Bay Horse and Ellel, in fruit, *Wh.*

Plagiothecium sylvaticum, B. & S.—Area, 1, 2, 5, 6, 7, 8. Range, 50–600 feet or higher.

Shaded banks and in woods. Rather rare. Fruit not seen. First record: *Wilson, Journ. Bot.,* 1899, p. 514.

> N. 1. Springs Wood, Lower Ease Gill.
> 2. Between Halton and Lancaster, *Wi.*
> W. 5. Salwick, *H.B.* (sp.).
> E. 6. Between Lancaster and Caton, *Wh.*
> 7. Calder Woods near Garstang, 1898, and Barnacre, *Wi.*
> 8. Redscar and Knowl Green, *Wh.* Near Whitewell.

Plagiothecium undulatum, B. & S.—Area, 1, 2, 5, 6, 7, 8. Range, 50–2,050 feet.

Damp woodland banks and shaded heathy slopes amongst the hills. Common in East. Fruiting frequently and sometimes very profusely in damp sheltered cloughs. First record: *Wheldon, Journ. Bot.,* 1899, p. 514. " Dolphinholme, 1897."

Amblystegium compactum, Aust.—Area, 2. Range, only seen at about 250 feet.

Dark hollows in limestone rocks. Very rare. Sterile. First record: The Authors, *Journ. Bot.,* 1906, p. 101.

> N. 2. In the cave known as Dunnald Mill Hole, 1905.

Amblystegium serpens, B. & S.—Area, 1, 2, 3, 4, 5, 6, 7, 8. Range, 0–1,000 (?) feet.

On banks and old decaying stumps. Common. Fruiting. First record: *Hamilton, The Naturalist,* 1898, p. 28. " Near Lancaster, 1897."

Var. *salinum*, Carr.—Sandy or clayey ground near the sea.

> N. 3. Salt marsh on south side of Keer estuary, *Wi.*
> W. 5. Sandhills near Fairhaven and South Shore, 1899, and on fallen clods of earth occasionally washed by the tide below Shard Bridge, *Wh.*

Amblystegium Juratzkanum, Schimp.—Area, 1, 3, 4, 5, 8. Range, 0–300 feet.

Damp walls, tree roots, and rotting wood in wet places. Rather rare. Fruiting. First record: *Wheldon, Journ. Bot.,* 1901, p. 298.

> N. 1. Willows by the Lune near Tunstall.
> 3. Between Ovangle and Heaton, *Wh.* Near Middleton.
> W. 4. Canal side between Galgate and Glasson, 1900, *Wh.*
> 5. Near Lee, *H.B.* (sp.). Near Myerscough, *Wi.* On a rotting tree trunk in a pit near Staining, *Wh.*
> E. 8. Old log in a ditch near Chaigley, *Wh.*

Amblystegium irriguum, B. & S.—Area, 3, 6, 8. Range, 25–100 feet or higher.

Stones and rocks in streams. Rare. Fruit not seen. First record: *Hamilton, The Naturalist,* 1898, p. 28.

> N. 3. Stones by Leighton Beck, *Wi.*
> E. 6. Near Lancaster, 1897, *W.P.H.* Crook of Lune.
> 8. Tunbrook Wood, *Mr. George* (sp.)

Amblystegium fluviatile, B. & S.—Area, 1, 3. Range, 25–700 feet.

Stones in streams. Rare. Sterile. First record: *Wilson, Journ. Bot.,* 1901, p. 298.

> N. 1. River Lune below Kirkby Lonsdale, and in the Leck Beck, Ease Gill, *Wi.*
> 3. In the Leighton Beck, 1900, *Wi.*

Amblystegium filicinum, De Not.—Area, 1, 2, 3, 4, 5, 6, 7, 8. Range, 0–1,200 feet or higher.

Swampy ground, streamsides and damp rocks. Common. Fruiting rarely, and then only in the smaller creeping forms found in drier localities. First record: *Wheldon, Journ. Bot.,* 1899, p. 514. " Near Ribchester, 1898."

Var. *Whiteheadii*, Wheldon. Wet sandy places.

> W. 5. St. Annes, 1898, *Wh.*
> E. 8. Near Ribbleton, *H.B.*

Var. *gracilescens*, Schimp.

> N. 3. Hawes Water, *Wh.*
> E. 6. Rocks by the Hindburn near Mill Houses, *Wi.*

Var. *trichodes*, Brid.

> N. 2. Near Whittington and Arkholme, *Wi.*
> 3. Coast banks between Silverdale and Ings Point, 1900, *Wh.*
> E. 8. Limestone cave near Dinkling Green, *Wi.*

Hypnum riparium, L.—Area, 2, 3, 4, 5, 6, 7, 8. Range, 0–250 feet.

Marshes and margins of ditches and ponds, on tree roots, posts, and stones. Frequent. Fruiting. First record: *F. C. King, Journ. Bot.,* 1899, p. 514. " Ashton-on-Ribble, 1881 " (sp.).

Var. *longifolium*, Schimp.—Peaty ground and rotting wood by ponds and ditches, where subject to inundation.

> N. 3. By ditches between Morecambe and Snatchems, 1899, *Wh.* Pool in old quarry east of Heysham Harbour, *Wi.* In the Conder between Glasson and Galgate, *Wh.*
> E. 7. Pond at Claughton ; and near Sandholme, Garstang, *Wi.*
>
> A robust form near var. *longifolium*, Schimp., but with more closely imbricate and less markedly distichous leaves occurs in a ditch between Overton and Middleton. It is perhaps referable to the var. *elongatum*, Schimp., *Wh.*

Hypnum elodes, Spruce.—Area, 2, 3, 5. Range, 15–290 feet.

Boggy places on calcareous or sandy soil. Rare. Fruit not seen. First record: *Wheldon, Journ. Bot.,* 1899, p. 515.

> N. 2. Bog near Dunnald Mill Hole.
> 3. Shore of Hawes Water Tarn, Silverdale, *Wi.* In a damp hollow of the limestone pavement, Gatebarrow Wood.
> W. 5. Sandhills, St. Annes, 1898, and in a ditch near Fairhaven Lake, *Wh.*

Hypnum polygamum, Schimp.—Area, 5, 8. Range, 15–180 feet.

Marshy places on a sandy soil. Fruiting freely. Local, but plentiful on the coast sandhills. First record: *Wheldon, Journ. Bot.,* 1899, p. 515.

> W. 5. Sandhills at St. Annes ; Fairhaven ; and South Shore ; fruiting, 1898, *Wh.*
> E. 8. Ribbleton, near Preston, *H.B.*

Var. *stagnatum*, Wils.

> With the type at St. Annes, *Wh.*

Hypnum stellatum, Schreb.—Area, 1, 2, 3, 5, 6, 7, 8. Range, 20–1,200 feet.

Bogs and marshes, especially in hilly districts. Frequent. Fruit not seen. First record: *Wheldon, Journ. Bot.,* 1899, p. 515. " Upper Hodder Bridge and Longridge Fell."

Var. *protensum*, Röhl.—Damp rocks and walls.

N. 3. Gatebarrow, near Silverdale, *Wi.*
E. 6. Lancaster, 1897, *W.P.H.* Wall near Caton, *Wh.*
   8. Wall near Tootle Heights ; near Mitton ; and on stones near Chipping, *Wh.*

Hypnum chrysophyllum, Brid.—Area, 1, 2, 3, 8. Range, 0–900 feet.

Calcareous rocks and banks. Frequent in North. Fruit not seen. First record : *Wheldon, Journ. Bot.*, 1899, p. 515.

N. 1. Middle Ease Gill and near Wennington.
   2. Near Carnforth and on Dalton Crag, *Wi.*
   3. Frequent in the Silverdale district.
E. 8. Above Higher Hodder Bridge, 1898, *Wh.*

Hypnum aduncum, Hedw.—Area, 2, 3, 4, 5, 7, 8. Range, 0–100 feet.

Marshes and pools in the low country. Frequent. Fruit not seen. First record : *Wheldon, Journ. Bot.*, 1899, p. 515. " Near Preston, 1897."

Group *typicum*, Ren.—Rare. Chiefly near the sea coast about St. Annes, where it is becoming very scarce now. The var. *gracilescens*, Schimp., and forms *falcata*, Ren., and *diversifolia*, Ren., occur there. The var. *aquaticum*, Sanio, was found by Mr. Beesley near Catforth.

Group *Kneiffii*, Ren.—Frequent and very variable.

Var. *polycarpon*, Bland.

N. 3. Between Morecambe and Snatchems, *Wh.*

Var. *attenuatum*, Boul.

W. 5. Very fine and typical, near Catterall, Garstang, *Wi.*

Var. *intermedium*, Schimp.

N. 3. Thrang Moss, *Wi.* Pond near Silverdale, *Wh.*
   4. Between Cockerham and Cockersand, *Wi.*
   5. Ashton (*forma laxifolia*, Schimp.), *H.B.* Carrs Green, *Wi.*
E. 8. Near Preston, *Wh.*

Group *pseudofluitans*, Ren.—Rare.

N. 3. Thrang Moss.
W. 5. Pool on the sandhills near St. Annes, 1898, *Wh.* Near Ashton and Catforth, *H.B.*

Most of our plants in this group come under var. *paternum*, Sanio.

A form which very closely approaches var. *flexile*, Ren., occurs at St. Annes, and is especially fine in a pool and in a ditch by the new Fairhaven Marine Lake, *Wh.*

Hypnum Sendtneri, Schimp.—Area, 5. Range, only at about 15 feet.

Does not appear to ascend to the fells of the hill districts in Lancashire. Very rare and rapidly disappearing. Fruit not seen. First record : *Wheldon, Journ. Bot.*, 1899, p. 515.

W. 5. Sandhills near St. Annes, 1898, and near the Marine Lake at Fairhaven, 1905, *Wh.*

Hypnum lycopodioides, Schwaeg.—Area, 5. Range, only at about 15 feet.

Boggy places on the sandhills. Very rare. Sterile. First record : *Wheldon, Journ. Bot.*, 1899, p. 515.

W. 5. With the last at St. Annes, 1898, *Wh.*

Seen as late as 1906, in one place only, and will probably become extinct in a very short time.

Hypnum Wilsoni, Schimp.—Area, 5. Range, only at about 15 feet.

Boggy places and pools on the sandhills. Rare. Sterile. First record : *Wheldon, Journ. Bot.*, 1899, p. 515.

W. 5. St. Annes, and with it the var. *hamatum*, Schimp., 1898, and following years, *Wh.*

Not seen in 1906, the slack in which it grew having apparently been drained and covered with fresh turf sods for making a golf green.

Hypnum uncinatum, Hedw.—Area, 1, 2, 5, 6, 7. Range, 50–2,050 feet.

Moist rocks and banks amongst the hills. Rather rare. Fruiting. First record : *Wilson, Journ. Bot.*, 1899, p. 515.

N. 1. Ease Gill, 1898 ; and (var. *plumosum*, Schimp.) on Greygarth Fell, *Wi.*
   2. Near Over Kellet, *Wi.*
W. 5. Ashton, near Preston (var. *plumosum*, Schimp.), *H.B.*
E. 6. Udale, *Wh.* & *Wi.* Gully west of Lower Salter, *Wi.*
   7. On Sullom Side, *Wi.*

Hypnum fluitans, L.—Area, 1, 2, 4, 6, 7, 8. Range, 20–2,050 feet.

Wet moors and in pools and rills on a peaty soil, both in the lowland mosses and amongst the hills. Common in East ; more rare elsewhere. Fruiting. First record : *Wilson, Journ. Bot.*, 1899, p. 515.

Var. *Jeanbernati,* Ren.—Area, 6, 7. Frequent on the fells.

E. 6. Mallowdale Fell and White Moss, Hindburn, with the *f. condensata*, Sanio, *Wh.* & *Wi.* Tatham Moor, *Wi.*
   7. Both sides of Tarnbrook Fell ; Wardstone ; Harrisend Fell ; Bleasdale Fell (*f. tenella*, Ren.), *Wh.* & *Wi.* Catshaw Fell and Hawthornthwaite Fell, *Wi.*

Var. *elatum,* Ren. & Arnell.—Area, 4.

W. 4. A form resembling this occurs on Cockerham Moss, but is not quite typical.

Var. *gracile,* Boul.—Area, 1, 4, 6, 7, 8. Mosses and moorlands, ranging from Cockerham Moss to the summit of Greygarth Fell.

Var. *atlanticum,* Ren.—Area, 1, 6, 7, 8. Frequent in the hilly districts of North and East ; ascending to 1,500 feet or higher.

Var. *Payoti,* Ren.—Area, 6. Very rare. Moorland pools.

E. 6. In a deep pool on Greenbank Fell, 1899.

Var. *setiforme,* Ren.—Area, 6. Very rare. Moorland pools.

E. 6. Goodber Common, 1902, *Wi.*

Var. *falcatum,* Schimp.—Area, 6, 7, 8. Not infrequent on the fells, where it fruits freely.

Var. *ovale,* Ren.—Area, 1. Rare. Moorland bogs.

N. 1. Greygarth Fell at 1,800 feet, *Wi.*

Hypnum exannulatum, Gumb.—Area, 1, 2, 5, 6, 7, 8. Range, 15–1,900 feet.

In similar situations to the last. Rarely fruiting. Frequent in East, much rarer in the other divisions. First record : *Wilson, Journ. Bot.*, 1899.

Var. *pinnatum,* Boul.—Area, 1, 2, 6, 7, 8. Frequent, especially in East.

The *forms acuta*, Sanio, and *stenophylloides*, Ren., are frequent ; the *f. polyclada*, Ren., is rare, having only been seen on Mallowdale Fell, where it is very fine ; and the *f. montana*, Ren., has only been noticed on Goodber Common. A curious and prettily plumose plant, having the auricles of this group, but the habit of green forms of the var. *falcifolium* occurs at the foot of Clougha, near Crag Wood, and was doubtfully referred by Mons. Renauld to var. *molluscum*, Sanio, which however that authority does not now regard as anything more than an accidental variation unworthy of separation. It is a mere state of *forma stenophylloides*, Ren.

* For the original description of these forms see " The North of England Harpidia," *The Naturalist*, 1902, p. 65.

Var. *purpurascens,* Schimp.—Area, 7. Rare, and not so characteristic as in more mountainous districts.

E. 7. Calder Valley above Oakenclough, 1898, and on Marshaw Fell, *Wi.*

The plant figured under this name by Renauld in *Class. Harpid. Muscologia Gallica*, has been named var. *nivale*, Ren., and does not occur in West Lancashire. It has a more alpine distribution than has the var. *purpurascens*.

Var. *brachydictyon,* Ren.—Area, 6, 8. Rare.

E. 6. White Moss, Hindburn, *Wi.*
   8. Longridge Fell, 1898, *Wh.*

Var. *falcifolium,* Ren.—Area, 1, 2, 5, 7. Rare.

N. 1. Ireby Fell.
   2. Ditches on Arkholme Moor.
W. 5. St. Annes, 1898, *Wh.* Catforth, *H.B.*
E. 7. Marshaw Wyre above Marshaw (*f. viridis*, Boul.), *Wi.* Over Wyresdale (*f. inundata*, Ren.), *Wh.* & *Wi.* Between Greenbank and Dolphinholme, *Wh.*

Hypnum vernicosum, Lindb.—Area, 2. Range, only seen at about 200 feet.

Deep bogs. Very rare. Not seen in fruit. First record : *Wilson, Journ. Bot.*, 1901, p. 299.

N. 2. Bog near Docker, and growing near it in another part of the same bog, the var. *majus*, Boul., the latter very luxuriant and typical, *Wi.*

Hypnum revolvens, Swartz.—Area, 1, 2, 3, 5, 6, 7, 8. Range, 15–1,400 feet or higher.

Bogs and marshes, and especially about springs on the fells. Frequent in East. Fruiting occasionally. First record : *Wheldon, Journ. Bot.*, 1899, p. 515.

This species varies greatly in size, colour, and general habit. None of its varieties, however, seem to be stable. The var. *Cossoni*, Ren., occurs in several localities, and the *f. falcata*, Sanio, in Udale ; but other abnormal states occur which are probably as worthy of separate names as these, and it does not seem to be desirable to attempt to define them, as has been done by Sanio.

Hypnum intermedium, Lindb.—Area, 3, 6, 8. Range, 50–1,200 feet.

Bogs. Very rare. Not seen in fruit. First record : The Authors, *Journ. Bot.*, 1899, p. 515.

N. 3. By Leighton Beck, *Wi.*
E. 6. Udale.
   8. Boggy moor near Dinkling Green, Whitewell.

Hypnum commutatum, Hedw.—Area, 1, 2, 3, 6, 7, 8. Range, 25–1,500 feet or higher.

Wet calcareous banks, dripping rocks, and by springs amongst the hills. Common. Fruiting occasionally. First record: *Hamilton, The Naturalist,* 1898, p. 28. " Near Lancaster, 1897."

It ascends to 1,500 feet in Ease Gill.

Hypnum falcatum, Brid.—Area, 1, 3, 6, 7, 8. Range, 25–1,850 feet.

In similar situations to the preceding. Frequent. Fruiting. First record: *Wheldon, Journ. Bot.,* 1899, p. 515. " Hawes Water, 1898."

Var. *virescens,* Schimp.—On wet calcareous rocks.

E. 8. On tufa in the Gorge of the Hodder below Whitewell, 1903.

Hypnum cupressiforme, L.—Area, 1, 2, 3, 4, 5, 6, 7, 8. Range, 0–2,050 feet.

Stones, walls, tree trunks, and on grassy or heathy banks. Very common. Fruiting. First record: *Wilson, Journ. Bot.,* 1899, p. 516. " Garstang, 1877."

Var. *filiforme,* Brid.—Area, 1, 2, 3, 6, 7, 8. On the trunks of trees. Frequent. Very fine in Lower Ease Gill.

Var. *mamillatum,* Brid.—Area, 3, 8.

N. 3. Silverdale, *Wh.*
E. 8. Kemple End, July, 1896, *Wh.*

Var. *tectorum,* Brid.—Area, 1, 2, 3, 5, 6, 8.

N. 1. Near Cantsfield.
2. Kellet Seeds, *Wi.*
3. Several places about Silverdale ; and near Heysham.
W. 5. Sandhills at Lytham and St. Annes, especially about rabbit warrens, *Wh.*
E. 6. Near Caton, *Wh.*
8. Limestone hills west of Whitewell.

Var. *ericetorum,* B. & S.—Area, 1, 2, 3, 4, 5, 6, 7, 8. Common in heathy or peaty places.

Var. *elatum,* B. & S.—Area, 3.

N. 3. Near Hawes Water, *Wi.* Not typical, but apparently nearer this than any other of the named varieties.

Hypnum resupinatum, Wils.—Area, 1, 2, 5, 6, 7, 8. Range, 0–1,000 feet or higher.

Tree trunks and walls. Frequent. Fruits occasionally. First record: *Wheldon, Journ. Bot.,* 1899, p. 516.

N. 1. Near Burrow, *Wi.* Very fine in Ease Gill, in fruit.
2. Highfield, near Carnforth, *Wi.*
W. 5. Lea and Ashton, in fruit, *H.B.*
E. 6. Caton and Hornby, *Wh.*
7. Garstang and Grizedale, in fruit, *Wi.*
8. Mitton, 1896, *Wh.* Near Whitewell, in fruit.

Hypnum Patientiæ, Lindb.—Area, 1, 2, 3, 6, 7, 8. Range, 100–800 feet or higher.

Sandy or gravelly places and muddy roadsides. Widely, but sparingly, distributed. Not seen in fruit. First record: *Wilson, Journ. Bot.,* 1899, p. 516.

N. 1. Ease Gill, *Wi.* Near Nether Burrow.
2. Roadside near Lords Lot Wood and Arkholme Moor, *Wi.*
3. Warton, *Wi.*
E. 6. Tatham Moor ; and near Botton, Hindburn, *Wi.* Caton Moor ; Ellel.
7. Garstang, 1878, and Lower Bleasdale, *Wi.* Near Emmetts.
8. Roadside near the source of the Loud, *Wi.*

Hypnum molluscum, Hedw.—Area, 1, 2, 3, 4, 5, 6, 7, 8. Range, 0–2,030 feet.

Limestone rocks and banks, wet oozy places amongst the fells, and occasionally on banks in the low country. Abundant in North, frequent in East, rare in West. Fruiting. First record: *Hamilton, The Naturalist,* 1898, p. 28. " Near Lancaster, 1897."

Hypnum palustre, Huds.—Area, 1, 2, 5, 6, 7, 8. Range, 0–1,500 feet.

Stones and rocks in or by rivers and streams. Common amongst the hills. Fruiting. First record: *Wheldon, Journ. Bot.,* 1899, p. 516. " Kemple End, 1896."

Var. *hamulosum,* B. & S.

N. 1. Gorge of the Greeta.
E. 6. Crook of Lune.
7. Damas Gill, *Wh.*

Var. *subsphæricarpon,* B. & S.

E. 7. By the Brock, *Wi.*
8. By the Hodder and Ribble, *Wh.*

Hypnum ochraceum, Turn.—Area, 1, 6, 7, 8. Range, 400–1,500 feet or higher.

In rills and streams on the fells. Frequent. Fruit not seen. First record: *Wilson, Journ. Bot.,* 1899, p. 516. " Calder Valley above Oakenclough, 1898."

Var. *complanatum,* Milde.

E. 6. Tatham Beck, *Wi.*

Var. *uncinatum,* Milde.—A slender form resembling *Hypnum aduncum.*

E. 6. Marsh near the Trough Ghyll Brook, Clougha, *Beesley* & *Wh.*

Var. *flaccidum,* Milde.

E. 6. Hindburn, 1899.
7. Marshaw Fell.
This species varies very greatly in size, colour, and direction of the leaves.

Hypnum scorpioides, L.—Area, 1, 2, 3, 7. Range, 50–1,270 feet.

Oozy places and springs on the fells ; more rarely in lowland bogs. Rather rare. Fruiting rarely. First record: *Wilson, Journ. Bot.,* 1901, p. 299.

N. 1. Lower Ease Gill, *Wi.* Greygarth Fell at 1,270 feet.
2. Bog near Docker, *Wi.* Bog east of Dunnald Mill Hole.
3. Bog near Leighton Beck, 1899, *Wi.*
E. 7. Marshaw Fell, *Wi.* Blaze Moss, in fruit, and Tarnbrook Fell.

Hypnum stramineum, Dicks.—Area, 1, 5, 6, 7, 8. Range, 250–1,400 feet or higher.

Moorland swamps and peaty ground by the side of rills amongst the hills. Common in East. Fruit not seen. First record: *Wilson, Journ. Bot.,* 1899, p. 516. " Calder Valley above Oakenclough, 1898."

Hypnum cordifolium, Hedw.—Area, 2, 3, 4, 5, 7, 8. Range, 15–350 feet.

Ponds and reedy marshes, chiefly in the low country. Frequent. Fruiting occasionally. First record: *Wheldon, Journ. Bot.,* 1899, p. 516. " Near Ribchester, 1898."

Hypnum giganteum, Schimp.—Area, 1, 2, 5. Range, 15–800 feet.

Bogs. Very rare. Rarely fruiting. First record: *Wilson, Journ. Bot.,* 1901, p. 299.

N. 1. Lower Ease Gill, *Wi.*
2. Bog near Docker, in fruit, 1900, *Wi.* Bog between Carnforth and Nether Kellet.
W. 5. Sandhills, St. Annes, 1899, *Wi.* Now gone ?

Hypnum sarmentosum, Wahl.—Area, 1, 6, 7. Range, 650–800 feet ; not seen higher, but very likely to occur.

Mountain bogs and rills. Very rare. First record: *Wilson, Journ. Bot.,* 1901, p. 299.

N. 1. Rill on the Silurian rocks in Ease Gill, *Wi.*
E. 6. Udale, *Wi.*
7. Marshaw Fell, Wyresdale, at only 650 feet, *Wi.*

Hypnum cuspidatum, L.—Area, 1, 2, 3, 4, 5, 6, 7, 8. Range, 0–2,040 feet.

Marshes and wet places by pools and ditches. Very common. Fruits occasionally. First record: *Wheldon, Journ. Bot.,* 1899, p. 516. " Near Ribchester, 1898."

A large aquatic form with spreading leaves, which agrees with a plant sent us by Mr. Holt labelled *forma submersa,* occurs in ponds near Garstang, *Wi.*

Hypnum Schreberi, Willd.—Area, 1, 2, 3, 4, 5, 6, 7, 8. Range, 50–1,900 feet.

Heaths, woods, and upland pastures. Common. Fruit not seen. First record: *Wheldon, Journ. Bot.,* 1899, p 516. " Longridge Fell, 1896."

Hylocomium splendens, B. & S.—Area, 1, 2, 3, 5, 6, 7, 8. Range, 0–2,030 feet.

Heaths, banks, and woods. Common in North ; less frequent in East and West. Fruit not seen. First record: *Wheldon, Journ. Bot.,* 1899, p. 516. " Hodder Valley, 1896."

Hylocomium brevirostre, B. & S.—Area, 1, 2, 3, 8. Range, 75–700 feet.

Woods and shady banks. Rare. Fruiting occasionally. First record: *Wheldon, Journ. Bot.,* 1899, p. 516.

N. 1. Gorge of the Greeta above Wrayton, in fruit.
2. By the brook in Wash Dub Wood and on Dalton Crag, *Wi.*

N. 3.   Trowbarrow, 1898, *Wh.*   Gatebarrow Wood, and near Leighton Beck, *Wi.*

E. 8.   Tunbrook Wood, *H.B.*   Near Whitewell.

Hylocomium loreum, B. & S.—Area, 1, 2, 3, 6, 7, 8.   Range, 150–2,030 feet.

Upland woods and moors.   Abundant on the fells of North and East.   Fruit not seen.   First record: *Wilson, Journ. Bot., 1899, p. 516.*   " Calder Valley above Oakenclough, 1898."

Hylocomium squarrosum, B. & S.—Area, 1, 2, 3, 4, 5, 6, 7, 8. Range, 0–2,030 feet.

Damp banks and grassy places.   Common.   Fruiting very rarely.   First record: *Wheldon, Journ. Bot., 1899, p. 516.* " Dolphinholme, 1897."

Fruiting by roadside near Winmarleigh (Dist. 4), 1906, *Wi.*   It ascends to 2,030 feet on Greygarth Fell.

Hylocomium triquetrum, B. & S.—Area, 1, 2, 3, 5, 6, 7, 8. Range, 0–1,480 feet.

Woods and banks.   Frequent, especially in North ; rarer amongst the fells of East.   Not seen fruiting.   First record: *Wheldon, Journ. Bot., 1899, p. 516.*   " Stonyhurst, 1896."

It ascends to 870 feet on Dalton Crag and to 1,480 feet in Ease Gill.

Hylocomium rugosum, De Not.—Area, 2, 3.   Range, 50–800 feet.

Dry limestone rocks and banks.   Locally abundant.   Not fruiting.   First record: *Wilson, Journ. Bot., 1899, p. 516.*

N. 2.   Dalton Crag, ascending to 800 feet.

3.   Thrang End, 1898, *Wi.*   Trowbarrow ; and near Hawes Water ; Yealand, *Wh.*   Near Leighton Beck, *Wi.*

HEPATICÆ.

MARCHANTIALES.

RICCIACEÆ.

Riccia glauca, L.—Area, 1, 2, 3, 4, 5.   Range, 0–200 feet.

Fallow fields and ditch sides, especially on a clay soil. Locally common.   First record: *Wheldon, Journ. Bot., 1906, p. 102.*

N. 1.   Roman road near Cowan Bridge and fields near Over Burrow.

2.   Fields near Borwick and Over Kellet.

3.   Thrang End, near Silverdale, *Wi.*

W. 4.   Near Pilling, *Wi.*

5.   Field near Marton Mere, by the footpath to Staining, September, 1905, *Wh.*

Var. *minima*, Pears.

N. 1.   Roman road near Cowan Bridge.

3.   Fields near Yealand Redmayne, *Wi.*

W. 4.   Several localities near Pilling, *Wi.*

Riccia Lescuriana, Aust. (*R glaucescens*, Carr.)—Area, 2, 3. Range, 100–700 feet.

In hollows of scar limestone " pavement " where water stands after rain.   Very rare.   First record: The Authors, *Journ. Bot., 1905, p. 96.*

N. 2.   Dalton Crag, *Wi.*

3.   Gatebarrow Wood, near Silverdale, September, 1904, and since.

Riccia sorocarpa, Bisch.—Area, 1, 3   Range, 0–200 feet or higher.

Bare ground by footpaths, etc., on a limestone soil.   Rare. First record: *Wilson, Journ. Bot., 1906, p. 102.*

N. 3.   Edge of footpath on hill near Silverdale Station, 1905, and on the north side of Warton Crag, near Leighton Hall, *Wi.*   Rabbit holes and wall tops on Heald Brow near Silverdale.

E. 4.   Fields near Nateby with *R. glauca*, *Wi.*

Riccia fluitans, L.—Area, 5, 8.   Range, 0–300 feet.

Floating on ponds amongst *Lemna*, or creeping over muc at their margins.   Rare.   First record: *Wheldon, Journ. Bot., 1901, p. 299.*

B·1

W. 5.   Pit near Weeton, *Wi.*

E. 8.   Abundant in a pond in a field by the roadside between Whittingham Asylum and Longridge, July, 1900, *Wh.*   Between Goosnargh and Barton, *H.B.*

MARCHANTIACEÆ.

Reboulia hemisphærica, Raddi.—Area, 1, 2, 3, 4, 8.   Range, 0–1,200 feet.

Calcareous rocks and banks.   Frequent in North.   First record: The Authors, *Journ. Bot., 1899, p. 518.*

N. 1.   Leck Fell, *Wi.*   Ease Gill, in fine fruit, ascending to 1,200 feet.

2.   Dalton Crag, 1899, and near Borwick and Carnforth.

3.   Middlebarrow and other places about Silverdale, *Wi.*

W. 4.   Hedgebanks near Pilling, *Wi.*

E. 8.   Near Whitewell, fruiting.

Conocephalum conicum, Dum.—Area, 1, 2, 3, 4, 5, 6, 7, 8. Range, 0–1,500 feet.

Wet rocks and damp, shady banks, especially by streams. Very common.   First record: *Wheldon, Journ. Bot., 1899, p. 518.* " Silverdale."

This is very much more abundant than any other species of the *Marchantiaceæ* in West Lancashire.   It ascends to 1,500 feet in Ease Gill.

Lunularia cruciata, Dum.—Area, 2, 3, 4, 5, 7, 8.   Range, 0–400 feet.

Shaded walls, banks, and garden paths, and sometimes in greenhouses.   Rare.   First record: *Wheldon, Journ. Bot., 1902, p. 416.*

N. 2.   Wall between Halton and Slyne, 1902, *Wh.*

3.   Apparently quite native on limestone banks near Silverdale, *Wi.*

W. 4.   Hedgebank near Pilling, *Wi.*

5.   Near Myerscough, *Wi.*

E. 7.   Garden ground near Garstang ; abundant, *Wi.*

8.   By a ditch near Hurst Green.

It is quite possible that this plant is only a colonist in some parts of West Lancashire.   Most of the records are those of habitats in the vicinity of houses or gardens.   The same remarks would apply to its distribution in South Lancashire.

Preissia commutata, Nees.—Area, 1, 2, 3, 5, 6, 7, 8.   Range, 0–1,250 feet or higher.

Shady rocks, especially near streams, and damp, sandy ground

near the sea.   Thinly but widely distributed.   First record: *Wheldon, Journ. Bot., 1899, p 518.*

N. 1.   Ease Gill, Ireby Beck, and Gorge of the Greeta above Wrayton.

2.   Streamside in Wash Dub Wood.

3.   Margin of Hawes Water Tarn and in Gatebarrow Wood near Silverdale, *Wi.*

W. 5.   Sandhills near St. Annes, 1898, *Wh.*

E. 6.   Rocks by the Lune between Caton and Halton, *Wi.*

7.   Nickey Nook and Gavells Clough, *Wi.*   Rocks by the Wyre above Dolphinholme.

8.   Longridge Fell and near Whitewell, *Wh.*

Marchantia polymorpha, L —Area, 1, 2, 3, 4, 5, 6, 7, 8. Range, 0–1,350 feet or higher.

Damp earth, walls and rocks, or in bogs.   Not unfrequent. First record: The Authors, *Journ. Bot., 1899, p. 518.*

A luxuriant form occurs on stones in the stream in Kellet Park Wood.

JUNGERMANNIALES.

JUNGERMANNIACEÆ, ANACROGYNÆ.

Aneura pinguis, Dum.—Area, 1, 2, 3, 5, 6, 7, 8.   Range, 0–1,400 feet.

Oozy banks, amongst mosses in bogs, and by rills and springs on the fells.   Common.   First record: *Wheldon, Journ. Bot., 1899, p. 518.*

This species is exceedingly variable in size and habit.   It is sometimes very robust and densely tufted on the fells, in wet bogs amongst *Sphagnum.*   In other situations it is procumbent and has broader and shorter fronds.   On the sandhills near St. Annes a form occurs but little more robust than the next species. This, however, is not worth varietal distinction, as under cultivation it increases greatly in size and cannot be distinguished then from the ordinary plant.

Aneura multifida, Dum.—Area, 7, 8.   Range, 250–1,000 feet or higher.

Wet, loamy places in woods, sides of ditches, and amongst moss and other hepatics at the roots of rushes in boggy places. Frequent ?   First record: *Wheldon, Journ. Bot., 1899, p. 518.*

E. 7.   North side of Harrisend Fell, *Wi.*   Calder Valley above Oakenclough.

8.   Stream near Stonyhurst and near Loud Lower Bridge, *Wh.*   Near Dinkling Green.

Aneura sinuata, Dum.—Area, 6.　Range, 600–800 feet.

Wet rocks in shady places, by waterfalls, etc.　Very rare. First record : The Authors, *Journ. Bot.*, 1899, p. 518.

E. 6.　Udale, 1899.
8.　Near Dinkling Green.

Aneura latifrons, Lindb.—Area, 7.　Range, only seen at about 300 feet.

Damp, turfy ground.　Very rare.　First record : The present one.

E. 7.　Calder Woods, near Garstang, 1900.

Metzgeria furcata, Dum.—Area, 1, 2, 3, 6, 8.　Range, 0–1,200 feet.

Trunks of trees and occasionally on rocks.　Common.　First record : *Wheldon, Journ. Bot.*, 1899, p. 518.　"Silverdale and Yealand."

Metzgeria conjugata, Lindb.—Area, 1, 3, 8.　Range, 100–850 feet or higher.

Damp, shaded rocks.　Rare.　First record : *Wheldon, Journ. Bot.*, 1899, p. 518.

N. 1.　Gorge of the Greeta and Ease Gill.
W. 3.　Thrang End near Silverdale, 1898, *Wh.*
E. 8.　Pot-hole in Fence Wood near Dinkling Green.

Metzgeria pubescens, Raddi.—Area, 1, 2, 3.　Range, 0–1,000 feet or higher.

Rocks, stones, and tree trunks in calcareous districts. Frequent in some parts of North.　First record : *Wilson, Journ. Bot.*, 1899, p. 518.

N. 1.　Near Leck and at the Dry Foss, etc., Ease Gill, 1898, *Wi.*
2.　Over Kellet and Kellet Seeds, *Wi.*
3.　Silverdale, *Wh.*

Pellia endiviæfolia, Dum.—Area, 2, 5, 6, 7, 8.　Range, 0–1,200 feet.

Wet banks and rocks.　Somewhat rare.　First record : *Wheldon, Journ. Bot.*, 1899, p. 518.

N. 2.　Bank of the Lune, near Arkholme, *Wi.*
W. 5.　St. Annes, 1898, *Wh.*

E. 6.　Hindburn, *Wi.*
7.　Fairsnape Clough, *Wi.*
8.　Longridge Fell, *Wh.*

Pellia Neesiana, Limpr.—Area, 1, 6, 7.　Range, 600–1,890 feet.

Wet rocks amongst the hills.　Frequent.　First record : *Wilson, Journ. Bot.*, 1901, p. 416.

N. 1.　Greygarth Fell at 1,890 feet.
E. 6.　Whiteray Gill, Hindburn, 1900, *Wi.*　By the Little Moor Beck, Lythe Fell, and in Dale Gill.
7.　Gavells Clough and near the Marshaw Wyre above Marshaw.

Pellia epiphylla, Dum.—Area, 1, 2, 3, 4, 5, 6, 7, 8.　Range, 0–1,400 feet or higher.

Wet, shady rocks, and banks of ditches and streams. Common. First record : *Wheldon, Journ. Bot.*, 1899, p. 518.　"Blackpool."

Blasia pusilla, L.—Area, 3, 6, 7, 8.　Range, 0–900 feet.

Moist heaths and wet, clayey banks, especially by moorland streams.　Rare.　First record : The Authors, *Journ. Bot.*, 1901, p. 299.

N. 3.　Wet bank on the coast south of the Keer estuary, *Wi.*
E. 6.　Caton Moor, 1900, and near Ivah, Hindburn.　By the Tatham Beck, *Wi.*
7.　Abbeystead Fell.
8.　Near Hurst Green.

Fossombronia pusilla, L.—Area, 1, 3, 4, 7.　Range, 25–220 feet.

Sandy banks and fallow fields.　Rather rare, perhaps overlooked.　First record : The present one.

N. 1.　Roman road near Cowan Bridge.
3.　Fields near Yealand Redmayne, *Wi.*
W. 4.　Two localities near Pilling, 1905, *Wi.*　Between Glasson and Galgate, on bank of the Conder, *Wh.*　Near Nateby, *Wi.*
E. 7.　Fields below Bruna Hill, Garstang, *Wi.*

Fossombronia cæspitiformis, De Not.—Area, 3.　Range, 100–370 feet.

Sandy ground and bare earth amongst rocks on a limestone soil.　Very rare.　First record : *Wilson, Journ. Bot.*, 1906, p. 102.

N. 3.　North side of Warton Crag, near Leighton Hall, where it was associated with *Riccia sorocarpa*, 1905, *Wi.*　Also on the footpath up the crag from Warton village, *Wh.*

JUNGERMANNIACEÆ, ACROGYNÆ.

Marsupella emarginata, Dum.—Area, 6, 7, 8.　Range, 500–1,600 feet.

Wet rocks and stones by moorland streams.　Rather rare. First record : *Wheldon, Journ. Bot.*, 1899.

E. 6.　Foxdale Head ; Udale, and by the Tatham Beck, Hindburn.
7.　Gavells Clough, *Wi.*　Great Clough of Tarnbrook Fell, and near Marshaw in the Trough of Bowland.
8.　Longridge Fell, 1898, *Wh.*

Var. *aquatica,* Lindb.

E. 6.　In the Foxdale Beck.

Nardia compressa, Gray.—Area, 6, 7, 8.　Range, 700–1,550 feet.

By springs and on rocks in the beds of mountain streams. Frequent in East.　First record : *Wheldon, Journ. Bot.*, 1901, p. 299.

E. 6.　Haylot Fell ; banks of the Roeburn below Wolfhole Crag, and in Hindburndale.
7.　Great Clough of Tarnbrook Fell.
8.　Longridge Fell, 1898, *Wh.*

Nardia scalaris, Gray.—Area, 1, 2, 3, 4, 5, 6, 7, 8.　Range, 25–1,900 feet.

Damp banks, heaths, and rocks.　Common.　First record : *Wheldon, Journ. Bot.*, 1899.　"Jeffrey Hill, Longridge."

Nardia hyalina, Carr.—Area, 6, 7, 8.　Range, 200–1,200 feet.

Wet rocks and banks, near streams.　Rare.　First record : The Authors, *Journ. Bot.*, 1906, p. 101.

E. 6.　Greenbank Fell and by the river Hindburn.
7.　Banks of the Calder near Garstang, *Wi.*　Great Clough of Tarnbrook Fell.
8.　Near Kemple End with *Catharinea crispa*, 1898, *Wh.*

Nardia obovata, Carr.—Area, 1, 3, 6, 7, 8.　Range, 25–1,550 feet.

Wet, peaty ground and about springs on the fells.　Frequent, especially in East.　First record : *Wheldon, Journ. Bot.*, 1899, p. 518.

N. 1.　Ease Gill, near Leck, *Wi.*
3.　Heysham Moss.

E. 6.　Udale ; Thrush Gill Fell, *Wi.*
7.　Plateau of Tarnbrook Fell.
8.　Longridge Fell, 1898, *Wh.*

Aplozia crenulata, Dum.—Area, 1, 3, 6, 7, 8.　Range, 25–1,300 feet.

Wet, peaty ground, and damp, sandy places by roadsides. Frequent, especially amongst the hills in East.　First record : *Wheldon, Journ. Bot.*, 1899, p. 518.　"Longridge Fell."

Var. *gracillima* (Sm.).

N. 1.　Roadside above Leck Hall at 550 feet.
3.　Heysham Moss.
E. 7.　Catshaw Greave ; Brock Valley.

Aplozia sphærocarpa, Dum.—Area, 1, 6, 7, 8.　Range, 700–1,550 feet.

Wet, clayey banks of streams and damp soil in woods. Rather rare.　First record : *Wheldon, Journ. Bot.*, 1899, p. 518.

N. 1.　Ease Gill.
E. 6.　Dale Gill and Middle Gill, Hindburn, *Wi.*　Lythe Fell.
7.　Tarnbrook Fell.
8.　Kemple End, Longridge Fell, 1898, *Wh.*

Aplozia cordifolia, Dum.—Area, 1, 6, 7, 8.　Range, 600–1,200 feet.

Rocks in and by rills on the fells.　Rare.　First record : *Wheldon, Journ. Bot.*, 1899, p. 518.

N. 1.　Above Ease Gill Kirk, *Wi.*　Near Leck.
E. 6.　Udale ; and Dale Gill, Hindburn.
7.　Great Clough of Tarnbrook Fell ; very fine.
8.　Longridge Fell, sparingly, 1898, *Wh.*

Lophozia inflata, Howe.—Area, 1, 2, 4, 6, 7, 8.　Range, 400–2,000 feet.

Floating in bog-pools or forming wide patches on wet, peaty soil.　Common, especially on the fells.　First record : *Wheldon, Journ. Bot.*, 1899, p. 518.　"Longridge Fell."

Var. *heterostipa,* Macv.

N. 1.　Greygarth Fell, *H. Fisher (Victoria History of Lancashire,* p. 77).

Lophozia bantriensis, Hook.—Area, 1, 8. Range, only seen at about 910 feet.

In large patches around moorland springs. Rare or overlooked. First record : The present one.

N. 1. Ireby Fell, 1906, growing with *Amblyodon dealbatus* and *Selaginella selaginoides*, *Wi.* (teste *Macvicar*).

Var. *Muelleri*, Nees.—Damp limestone rocks by streams.

E. 8. Limestone rocks by the Hodder near Whitewell, April, 1907.

Lophozia turbinata, Steph.—Area, 1. Range, 100-200 feet.

Damp, calcareous rocks or banks. Rare. First record : The Authors, *Journ. Bot.*, 1905, p. 96.

N. 1. Gorge of the Greeta near Wennington, 1904.
3. Lane near Silverdale Station and plentiful in old mine shafts on the north side of Warton Crag.
The closely allied *L. badensis*, Schiffn., has not yet been detected in our district.

Lophozia ventricosa, Dum.—Area, 1, 2, 3, 4, 6, 7, 8. Range, 50-1,900 feet.

Damp rocks and banks in woods and on moors, and occasionally in bogs amongst *Sphagnum*. Common. First record : *Wheldon, Journ. Bot.*, 1899. " Above Kemple End."

Var. *porphyroleuca*, Limpr.

E. 6. On rotting sticks in Udale.

Lophozia bicrenata, Dum.—Area, 1, 3. Range, 200-400 feet.

About the mouths of rabbit-holes and on ant-hills on a sandy soil. Rare. First record : The present one.

N. 1. Mud-capped wall near Nether Burrow.
3. Sandy ground over limestone, north side of Warton Crag, 1906.

Lophozia excisa, Dum.—Area, 5, 6, 8. Range, 50-850 feet or higher.

Amongst mosses on peaty moorland banks. Rare. First record : The Authors, *Journ. Bot.*, 1899, p. 518.

W. 5. Bank of Wyre near Catterall, *Wi.*
E. 6. Udale, 1898.
8. Near Kemple End, Longridge Fell, 1898, *Wh.*

Lophozia incisa, Dum.—Area, 6. Range, 500-930 feet or higher.

Amongst mosses on moorland banks and rocks. Rare. First record : *Wheldon Journ. Bot.*, 1902, p. 416.

E. 6. Clougha Pike, 1899, *Wh.* Lythe Fell at 930 feet.

Lophozia quinquedentata, Cogn. — Area, 1, 6. Range, 500-2,050 feet.

Amongst mosses on siliceous rocks. Rare. First record : *Wilson, Journ. Bot.*, 1899, p. 518.

N. 1. Ease Gill, near Leck, 1898, and in a pot-hole on Leck Fell *Wi.* Greygarth Fell.
E. 6. Clougha, *Wh.*

Lophozia barbata, Dum.—Area, 1, 3, 6, 7, 8. Range, 100-1,950 feet.

On bare ground, rocks, and banks in intricate patches or mixed with mosses and *Cladoniæ*, especially on a peaty soil. Frequent amongst the hills. First record : *Wheldon, Journ. Bot.*, 1899. p. 518. " Longridge Fell."

Lophozia Flœrkii, Schiffn.—Area, 1, 6, 7, 8. Range, 500-1,950 feet.

Rocks and banks on the gritstone moors. Common in East and parts of North. First record : *Wheldon, Journ. Bot.*, 1899, p. 518. " Longridge Fell." Much the commonest form of the *Barbata* series in West Lancashire.

Var. *Naumanniana*, Nees.—On bare crumbling soil in shady places.

E. 6. Plentiful in Heights Wood at the foot of Clougha, *Wh.* Salter Fell, Roeburndale, *Wi.*

Lophozia Baueriana, Schiffr.—Area, 6, 8. Range, 600-1,400 feet or higher.

On elevated gritstone scars. Rare. First record : *Wheldon* in Macvicar's " *Notes on British Hepaticæ*," *Journ. Bot.*, 1907, p. 259.

E. 6. Windy Clough, Clougha, October, 1899, *Wh.*
8. Wolf Fell, *Wi.*

Mr. Macvicar *(Journ. Bot.*, 1907, p. 253) says : " Among the many English specimens of this group which I have examined, there is only the West Lancashire plant which I would consider as belonging to this species."

Lophozia gracilis, Steph.—Area, 1, 3, 6, 7, 8. Range, 50-1,950 feet or higher.

Amongst loose gritstone rocks on the fells. Locally abundant. First record : *Stabler, The Naturalist*, 1898, p. 342.

N. 1. Amongst *Dicranum fuscescens* on Greygarth Fell ; Ease Gill, *Wi.*
3. Grit rocks near Heysham, *Wi.*
E. 6. Clougha, 1881, *Stabler.* Still abundant there and in Udale. Thrush Gill Fell, Hindburn, *Wi.*
7. Catshaw Greave, *Wi.* White side of Tarnbrook Fell and on Long Crag.
8. Wolf Fell, *Wi.* Whitestone Clough.

Lophozia atlantica, Schiffn.—Area, 6, 7. Range, 600-1,900 feet.

In similar localities to the preceding. Only recently recognised and probably more frequent than the records indicate. First record : The present one.

E. 6. Udale, June, 1898.
7. White side of Tarnbrook Fell, *Wi.*
Determined for us, as in the case of many other of our rarer species, by Mr. S. M. Macvicar.

Sphenolobus minutus, Steph. — Area, 1, 6, 7. Range, 1,000-1,950 feet.

Exposed gritstone scars on the fells. Rare. First record : The Authors, *Journ. Bot.*, 1899, p. 518.

N. 1. Greygarth Fell, *Wi.*
E. 6. Clougha, 1899, *Wh.* & *Wi.* Thrush Gill Fell, Hindburn, *Wi.*
7. Hell Crag and Great Clough of Tarnbrook Fell ; Marshaw Fell.
With us it is often associated with *Lophozia gracilis, Lepidozia pinnata, Mylia Taylori, Scapania gracilis,* and *Dicranum fuscescens*.

Plagiochila spinulosa, Dum.—Area, 1, 2, 6. Range, 150-1,950 feet.

Shady rocks and banks in large cushions or scattered amongst mosses. Rare. First record : *Wilson, Journ. Bot.*, 1899, p. 518.

N. 1. Ease Gill near Leck, 1898, *Wi.* Greygarth Fell, accompanied by the var. *inermis*, Carr.
2. Amongst mosses (especially *Neckera complanata*) on a wall near Halton, *Wh.*
E. 6. Sparingly amongst *Lepidozia pinnata* on Clougha, *Wh.*

Plagiochila asplenioides, Dum.—Area, 1, 2, 3, 4, 6, 7, 8. Range, 0-2,040 feet.

On rocks and shady banks in extensive tufts or creeping

amongst mosses. Common. First record : *Wheldon, Journ. Bot.*, 1899, p. 518.

Var. *major*, Nees.—Frequent, especially in the wooded cloughs amongst the hills.

Var. *Dillenii* (Tayl.) — Common, especially in the limestone districts of North. In East on limestone rocks near Whitewell and Dinkling Green.

Var. *humilis*, Lindb.—Apparently very rare.

The only record is Upper Ease Gill at 1,300 feet, 1907, *Wi.*

Mylia Taylori, Gray.—Area, 1, 6, 7, 8. Range, 1,000-2,040 feet.

Heathy moorland gullies and exposed gritstone rocks on the fells. Abundant in some parts of East ; absent from the low country. First record : *Wilson, Journ. Bot.*, 1899, p. 518. " Bottom Head Fell."

Very frequently associated with *Bazzania trilobata* and *Dicranum fuscescens*. Finest and best developed on high gritstone scars that are frequently enveloped in mist, but are, nevertheless, well drained. It ascends to 2,040 feet on Greygarth Fell.

Mylia anomala, Gray.—Area, 1, 3, 4, 6, 7. Range, 20-1,400 feet.

Wet moorland bogs, often amongst *Sphagnum*, with lower vertical range than the preceding. Rare, and likely to become more so. First record : *Wilson, Journ. Bot.*, 1899, p. 518.

N. 1. Ireby Fell, *Wi.* Ease Gill.
3. Heysham Moss.
W. 4. Cockerham Moss, 1898, *Wi.*
E. 6. White Moss, Hindburn, *Wh.* & *Wi.* Salter Fell, Roeburndale, *Wi.*
7. Grizedale, near Abbeystead.

Lophocolea bidentata, Dum.—Area, 1, 2, 3, 4, 5, 6, 7, 8. Range, 0-1,500 feet.

Shady banks and damp, grassy places. Very common. First record : *Wheldon, Journ. Bot.*, 1899, p. 517. " Silverdale."

Var. *rivularis*, Raddi.—Abundant on Heysham Moss (District 3), where it was first discovered by the Authors in 1905. It had not previously been found in Britain.

For a description of this variety see *Journ. Bot.*, 1906, p. 101. It occurs in extensive green tufts, with the stems sometimes densely cæspitose, in wet places amongst *Sphagnum*. We are indebted to Herr. C. Warnstorf and Mr. Macvicar for its determination.

Lophocolea alata, Mitt.—Area, 6. Range, only seen at about 100 feet.

Bare earth on grassy banks. Rare or overlooked. First record: *Wheldon,* in *Macvicar's " Notes on British Hepaticæ,"* *Journ. Bot.,* 1907, p. 262.

E. 6. Near Caton, in fruit, 1899, *Wh.* (teste *Macvicar*).

Lophocolea cuspidata, Limpr.—Area, 1, 3, 6, 8. Range, 50–500 feet or higher.

Shaded banks, walls, and rocks. Rather rare? First record: *Wheldon, Journ. Bot.,* 1901, p. 299.

N. 1. Wall near Leck, *Wi.*
3. Near Leighton Beck, *Wi.*
E. 6. Heights Wood, near Clougha, *Wh.*
8. Near Leagram Hall, *Wh.*

Lophocolea heterophylla, Dum.—Area, 1, 2, 3. 4, 5, 6. 7, 8. Range, 0–600 feet or higher.

Tree trunks, rotting wood, damp rocks and stones, and on shaded banks amongst mosses. Very common. First record: *Wheldon, Journ. Bot.,* 1899, p. 517. " Stonyhurst and Hodder Valley."

Chiloscyphus polyanthos, Corda.—Area, 1, 4, 5, 6, 7, 8. Range, 25–1,000 feet or higher.

Bogs and marshes; sometimes in streams. Frequent, especially amongst the hills. First record: *Wilson, Journ. Bot.,* 1899, p. 517. " River Calder, near Garstang."

The var. *rivularis,* Nees., is abundant in streams and rills amongst the fells of East.

Chiloscyphos pallescens, Nees.—Area, 4. Range, 0–100 feet or higher.

Peat bogs and marshy places, attached to *Sphagnum* or decayed leaves and sticks. Apparently rare. First record: *Wheldon, M.E.C. Rep.,* 1907, p. 264.

W. 4. Near Pilling, 1898, *Wh.*

Saccogyna viticulosa, Dum.—Area, 1. Range, only seen at about 1,000 feet.

Shaded rocks and crumbling banks amongst the hills. Very rare. First record: The Authors, *Journ. Bot.,* 1906, p. 101.

N. 1. Limestone rocks in Ease Gill, May, 1905.

Associated with *Plagiochila spinulosa, Peltigera aphthosa,* and *Solorina saccata* in our only station.

Cephalozia bicuspidata, Dum.—Area, 1, 2, 3, 4, 5, 6, 7, 8. Range, 0–1,400 feet or higher.

Damp stones, rocks, and peaty banks, and amongst mosses on decaying wood. Common. First record: *Wheldon, Journ. Bot.,* 1899, p. 517. " Longridge."

Cephalozia Lammersiana, Spruce.—Area, 1, 3, 4, 7. Range, 20–1,000 feet or higher.

Peat bogs and damp banks by streams in shady places. Rare, although locally plentiful in some of the lowland mosses. First record: *Wheldon, Journ. Bot.,* 1899, p. 517.

N. 1. Ireby Fell, at 1,000 feet, *Wi.*
3. Heysham Moss.
W. 4. Pilling Moss, *Wh.* Cockerham Moss.
E. 7. Upper Grizedale, *Wi.*

Cephalozia connivens, Spruce.—Area, 1, 7, 8. Range, 200–1,200 feet.

Wet moors amongst mosses. Rare. First record: The Authors, *Journ. Bot.,* 1906, p. 101.

N. 1. Ireby Fell, *Wi.* Ease Gill.
E. 7. Nickey Nook, *Wi.*
8. Whitestone Clough, *Wh. & Wi.* Longridge Fell, *Wh.*

Cephalozia lunulæfolia, Dum.—Area, 2, 3. 4, 6, 8. Range, 25–1,700 feet.

On damp, turfy banks, rotting trunks in damp places, and creeping amongst *Sphagnum* and *Leucobryum* in bogs. Frequent. First record: *Wheldon. Journ. Bot.,* 1899, p 517. " Kemple End."

Cephalozia fluitans, Spruce.—Area, 1, 2, 4, 7, 8. Range, 25–2,000 feet.

Floating in wet bogs amongst *Sphagnum.* Rather rare, but locally abundant. First record: *Wheldon, Journ. Bot.,* 1899, p. 517.

N. 1. Greygarth Fell at 2,000 feet, *Wi*
2. Arkholme Moor, *Wi.*

W. 4. Cockerham Moss, *Wi.*
E. 7. Great Clough of Tarnbrook Fell.
8. Longridge Fell, 1898, *Wh.*

Cephaloziella byssacea, Warnst.—Area, 3, 4. 6, 7, 8. Range, 20–1,000 feet.

Usually creeping amongst mosses, on rocks, stones, and heathy ground. Frequent? First record: *Wheldon, Journ. Bot.,* 1899, p. 517.

N. 3. Heysham Moss.
W. 4. Cockerham Moss and on banks near Pilling, *Wi.*
E. 6. Caton, *Wh.* Udale.
·7. Barnacre, *Wi.* Damas Gill, *Wh.* Amongst *Andreæa falcata* in Hawthornthwaite Greave.
8. Longridge, 1896, *Wh.*

Odontoschisma denudatum, Dum.—Area, 7. Range, uncertain.

Bare, peaty soil in moist places on the fells. Very rare. First record: The present one.

E. 7. Tarnbrook Fell, 1900, *Wi.*

Odontoschisma Sphagni, Dum.—Area, 1, 3, 4, 6, 7. Range, 20–1,200 feet or higher.

Wet bogs amongst *Sphagnum.* Frequent, and particularly abundant on some of the lowland mosses. First record: The Authors, *Journ. Bot.,* 1899, p. 517.

N. 1. Ease Gill.
3. Heysham Moss, *Wi.*
W. 4. Cockerham Moss.
E. 6. Udale, 1899, and White Moss, Hindburn.
7. Great Clough of Tarnbrook Fell, *Wi.*

Kantia trichomanis, Gray.—Area, 1, 3, 4, 5, 6, 7, 8. Range, 0–1,950 feet.

Wet, clayey or peaty soil, damp woodland banks, and creeping amongst *Sphagnum* on moors. Common. First record: *Wheldon, Journ. Bot.,* 1899, p. 517. " Longridge."

Kantia Sprengelii, Pears.—Area, 1, 3, 6, 7, 8. Range, 80–2,000 feet.

Moorland bogs, often intermixed with mosses. Frequent in the hilly districts. First record: *Wheldon, Journ. Bot.,* 1901, p. 299.

N. 1. Summit of Greygarth Fell.
3. Warton Crag.
E. 6. Near Caton, *Wh.* Udale.
7. Railway cutting near Garstang, and in Upper Grizedale, *Wi.*
8. Longridge Fell, *Wh.*

Kantia submersa, Arnell.—Area, 3, 4. Range, 20–25 feet.

Very wet bogs amongst *Sphagnum.* Rare. First record: The Authors, *Journ. Bot.,* 1903, p. 17.

N. 3. Heysham Moss.
W. 4. Cockerham Moss, 1900.
The first account of this as a British species will be found in *Journ. Bot.* as cited above. It grows in the wettest parts of *Sphagnum* bogs and forms extensive patches, usually completely immersed. The leaves are very neatly and regularly bifarious, and the amphigastria have a deep and widely lunate sinus.

Kantia arguta, Lindb.—Area, 3, 6, 7. Range, 200–400 feet.

Shaded clayey or sandy banks, ditch sides, etc. Rare. First record: *Wheldon, Journ. Bot.,* 1902, p. 416.

N. 3. On sandy soil recently thrown out by rabbits on Warton Crag.
E. 6. Near Quernmore, 1900, *Wi.*
7. Calder Woods, near Garstang.

Bazzania trilobata, Gray.—Area, 1, 6, 7. Range, 1,000–1,600 feet.

Gritstone scars and stony slopes on the higher fells. Rather rare, but locally abundant. First record: The Authors, *Journ Bot.,* 1899, p. 517.

N. 1. Grit rocks on Greygarth Fell.
E. 6. Clougha, 1899, and Mallowdale Fell; Dale Gill, Hindburn, *Wi.*
7. Long Crag and Hell Crag, Over Wyresdale, *Wi.*
In almost every case accompanied by *Dicranum fuscescens, Mylia Taylori, Lepidozia Pearsoni* and *Scapania gracilis;* and frequently by *Sphenolobus minutus* and *Lepidozia reptans.*

Bazzania triangularis (Schleich.), Lindb.—Area, 1. Range, only seen at about 1,000 feet.

Shaded rocks amongst the higher hills. Rare. First record: The present one.

N. 1. Middle Ease Gill, May, 1905.

Lepidozia pinnata, Dum.—Area, 1, 6, 7. Range, 220–1,600 feet.

High gritstone scars, where it forms large masses in the hollows of loosely-piled blocks. Rare, but abundant where it occurs. First record: *Stabler, The Naturalist*, 1898, p. 232 (as *L. tumidula*, Tayl.).

    N. 1. Grit rocks on Greygarth Fell.
    E. 6. Clougha, 1881, *Stabler*. Still very abundant there !
       7. Long Crag, Over Wyresdale.

Lepidozia reptans, Dum.—Area, 1, 2, 3, 4, 6, 7, 8. Range, 20–1,950 feet.

On shaded banks, rocks, and rotting tree trunks, either in tufts or mixed with mosses. Frequent in East, much rarer elsewhere. First record: *Wheldon, Journ. Bot.*, 1899, p. 517. "Longridge."

    N. 1. Greygarth Fell, *Wi.* Middle Ease Gill.
      2. Abundant in Lords Lot Wood.
      3. Heysham Moss.
    W. 4. Rocks by the canal south-west of Galgate, *Wi.*
    E. 6, 7, 8. Frequent.

Lepidozia Pearsoni, Spruce.—Area, 1, 6, 7, 8. Range, 1,000–1,800 feet.

High stony, peaty slopes and gritstone scars on the higher fells. Frequent. First record: *Wheldon, Journ. Bot.*, 1901, p. 299.

    N. 1. Ease Gill and Greygarth Fell.
    E. 6. Clougha, Mallowdale, Haylot Fell, Udale, Botton Head Fell, and on the north-west side of Wardstone.
      7. Long Crag and Hell Crag, Over Wyresdale.
      8. Longridge Fell, 1899, *Wh.* Whitestone Clough.

Lepidozia trichoclados, Warnst.—Area, 6. Range, only seen at 1,300 feet.

Wet rocks and boggy, peaty ground. Rare. First record: The present one.

    E. 6. East side of Thrush Gill Fell, *Wi.*

Lepidozia setacea, Mitt.—Area, 1, 4, 6, 7, 8. Range, 25–1,800 feet.

Wet rocks, or in bogs amongst *Sphagnum* and other mosses. Rather rare. First record: *Jones & Wheldon, Journ. Bot.*, 1902, p. 415.

    N. 1. Upper Ease Gill.
    W. 4. Cockerham Moss, 1900.
    E. 6. Clougha, *Wh.*
      7. Summit plateau of Wardstone, *Wi.*
      8. Whitestone Clough.

Var. *sertularioides*, L.—Very wet bogs. Rare.

    W. 4. Cockerham Moss, August, 1902, *Jones & Wh.*

Blepharostoma trichophyllum, Dum.—Area, 1, 6. Rarge, 200–1,000 feet.

On damp, shaded mossy rocks. Rare. First record: *Wheldon, Journ. Bot.*, 1905, p. 96.

    N. 1. Gorge of the Greeta, near Wrayton.
    E. 6. Clougha, 1902, *Wh.*

Ptilidium ciliare, Hampe.—Area, 1, 5, 6, 7, 8. Range, 50–2,050 feet.

Damp, peaty ground and boggy heaths and moors, sometimes floating. Rather rare. First record: *Wheldon, Journ. Bot.*, 1899, p. 517.

    N. 1. Very fine on wet peat and in pools near the summit of Greygarth Fell.
    W. 5. Near Salwick, *H.B.*
    E. 6. Udale and Botton Head Fell, *Wh. & Wi.* Salter Fell, *Wi.*
      7. Near Scorton, *H.B.*
      8. North side of Longridge Fell, 1898, *Wh.* Near Dinkling Green.

Trichocolea tomentella, Dum.—Area, 1, 3, 6, 7. Range, 50–900 feet.

Damp woods and shaded, mossy banks. Rather rare. First record: *Wheldon, Journ. Bot.*, 1899, p. 517.

    N. 1. Ease Gill, near Leck, *Wi.*
      3. Woodwell, near Silverdale, *Wh.*
    E. 6. Dale Gill and near Botton, Hindburn.
      7. Upper Grizedale, *Wi.* Calder Valley, above Oakenclough.

Diplophyllum albicans, Dum.—Area, 1, 2, 6, 7, 8. Range, 200–2,050 feet.

Rocks and banks in woods on moorlands, and by streams. Very common on the siliceous formations, but rare on calcareous soils. First record: *Wheldon, Journ. Bot.*, 1899, p. 517. "Longridge."

Scapania compacta, Dum.—Area, 2. Range, 300–400 feet.

Clayey or sandy banks. Rare. First record: *Wilson, Journ. Bot.*, 1902, p. 416.

    N. 2. Arkholme Moor, 1902, and in Wash Dub Wood, *Wi.*

    C I

Scapania subalpina, Dum.—Area, 1. Range, uncertain; only seen at about 2,000 feet.

Probably commoner than is indicated by our only record. First record: The present one.

    N. 1. Near the summit of Greygarth Fell, July, 1899. Only this year determined for us by Mr. Macvicar, who refers it to the var. *undulifolia*, Nees.

Scapania aspera, Bernet.—Area, 1, 2, 3. Range, 0–1,000 feet.

Dry limestone rocks. Locally abundant. First record: *Wheldon, Journ. Bot.*, 1899, p. 517.

    N. 1. Ease Gill.
      2. Over Kellet and Kellet Seeds, *Wi.* Dalton Crag and near Carnforth.
      3. Silverdale, 1899, *Wh.* Yealand Storrs, Thrang Wood, etc., *Wi.* Warton Crag, *Wh.*

Scapania gracilis, Kaal.—Area, 1, 6, 7. Range, 500–1,800 feet.

Sandstone rocks amongst the hills. Frequent, ascending to near the summit of Wardstone. First record: *Wheldon, Journ. Bot.*, 1901. "Clougha" (as *Scapania resupinata*, Nees.).

It is a very variable species with us. From the var. *minor*, Pears., which is not unfrequent, it passes through various looser and compacter states to a very robust, densely tufted form (occurring on one or two of the higher gritstone scars, such as those of Wardstone and Clougha), which does not appear to have any varietal name. The tufts of this latter form are almost "corky" in texture when dry.

Scapania nemorosa, Dum.—Area, 1, 2, 3, 6, 7. Range, 200–1,500 feet.

Woods, and shaded banks and rocks. Rather rare. First record: *Wilson, Journ. Bot.*, 1901, p. 299.

    N. 1. Greygarth Fell and Middle Ease Gill.
      2. Kellet Park Wood.
      3. Near Warton, 1899, *Wi.*
    E. 6. Near Botton, Hindburn.
      7. Upper Grizedale, *Wi.*

Scapania purpurascens, Tayl.—Area, 2, 6, 7, 8. Range, 300–1,700 feet.

Wet rocks and stones by the sides of and in streams on the fells. Frequent in East. First record: *Wheldon, Journ. Bot.*, 1899, p. 517.

    N. 2. Gressingham Moor.
    E. 6. Clougha, Udale, Greenbank Fell, Dale Gill, etc.

    E. 7. Hawthornthwaite Fell, *H.B.* Marshaw Fell, *Wi.* Calder Valley above Oakenclough and Great Clough of Tarnbrook Fell, etc.
      8. Longridge Fell, 1896, *Wh.* Whitewell, *Fl. Prest. N.*, Burnslack Clough, *Wi.*

Scapania undulata, Dum.—Area, 1, 2, 6, 7, 8. Range, 300–2,050 feet.

Rocks and stones in or by moorland streams, more rarely in bogs. Common amongst the hills of North and East. First record: *Wheldon, Journ. Bot.*, 1899, p. 517.

Scapania uliginosa, Dum.—Area, 1. Range, seen only at about 900 feet.

Boggy moorlands. Rare. First record: The Authors, *Journ. Bot.*, 1902, p. 416.

    N. 1. Ireby Fell, 1901.

Mr. Macvicar is inclined to refer the Ireby plant to *S. obliqua* (*Arn.*), Schifn., but the material is not sufficient for a satisfactory discrimination between these two closely allied forms (see *Journ. Bot.*, 1907, p. 262).

Scapania irrigua, Dum.—Area, 1, 2, 3, 7. Range, 150–2,000 feet.

Wet sandy banks, and in bogs amongst *Sphagnum*. Rather rare. First record: The Authors, *Journ. Bot.*, 1902, p. 416.

    N. 1. Greygarth Fell and by the Cart Beck above Leck Hall.
      2. Quarry near Over Kellet, *J. W. Hartley, Wh. & Wi.*
      3. North side of Warton Crag.
    E. 7. Barnacre, near Garstang, 1900 ; Fairsnape Clough, *H.B.* Lower Bleasdale ; Greenside above Tarnbrook.

Scapania curta, Dum.—Area, 1, 3, 7. Range, 100–1,000 (?) feet.

Peaty banks and borders of footpaths on the hills. Very rare. First record: The Authors, *Journ. Bot.*, 1899, p. 517.

    N. 1. Ease Gill and by the road above Leck village.
      3. North side of Warton Crag, *Wi.*
    E. 7. Barnacre, *Wi.*

Radula complanata, Dum.—Area, 2, 3. Range, 0–300 feet.

Trunks and branches of trees. Frequent in parts of North ; not seen in East or West. First record: *Wheldon, Journ. Bot.*, 1899, p. 517.

    N. 2. Over Kellet and Nether Kellet, *Wi.* Near Carnforth, *Hartley, Wh. & Wi.* Hedgerows by the Lune near Caton and Aughton and near Whittington.
      3. Silverdale, 1898, *Wh.*

Madotheca lævigata, Dum.—Area, 3. Range, 0–300 feet.

Rocks, tree roots, and walls. Rare. First record : *Wheldon, Journ. Bot.,* 1899, p. 517.

N. 3. Near Silverdale, 1898, and Trowbarrow, *Wh.* Same neighbourhood, 1901, *Wi.*

Madotheca platyphylla, Dum.—Area, 2, 3, 8. Range, 0–750 feet.

Limestone rocks and walls, and more rarely on tree trunks. Common in the calcareous districts. First record : *Wheldon, Journ. Bot.,* 1899, p. 517. " Woodwell."

Not yet seen in District 1, but surely it occurs ?

Cololejeunea calcarea, Schiffn.—Area, 1. Range, 650–800 feet or higher.

Shaded, calcareous rocks. Very rare. First record · The Authors, *Journ. Bot.,* 1906, p. 101.

N. 1. Near the Witches Caves, Middle Ease Gill, on shady limestone scars ; locally abundant ; associated with *Metzgeria pubescens,* May, 1905.

Cololejeunea Rossettiana, Schiffn.—Area, 2, 3. Range, 100–400 feet.

Shaded limestone rocks. Rare. First record : *Wilson, Journ. Bot.,* 1906, p. 101.

N. 2. Limestone rocks in Leapers Wood, Kellet Seeds, March, 1905, *Wi.*
3. Limestone scar between Silverdale station and the village, *Wi.* Old mine shaft north side of Warton Crag.

Lejeunea cavifolia, Lindb.—Area, 1, 6, 8. Range, 100–2,050 feet.

Rocks, banks, and tree roots, often amongst mosses. Not common. First record : *Wheldon, Journ. Bot.,* 1899, p. 517 (as *L. serpyllifolia,* Dicks.).

N. 1. Ease Gill and Greygarth Fell, *Wi.* Near Leck.
E. 6. Caton, 1898, *Wh.* Udale.
8. Near Whitewell, *Wh.*

Microlejeunea ulicina, Evans.—Area, 1, 6, 8. Range, 150–600 (?) feet.

Trunks of trees. Rare. First record : The Authors, *Journ. Bot.,* 1905, p. 96.

N. 1. On hawthorns near Over Burrow.
E. 6. Clougha, *Stabler.*
8. On old trees, Gorge of the Hodder below Whitewell, March, 1903.

Marchesinia Mackaii, Gray.—Area, 2, 3. Range, 0–750 feet.

Shaded limestone rocks. Locally abundant. First record : *Stabler, The Naturalist,* 1898, p. 229.

N. 2. Over Kellet and Kellet Seeds, *Wi.* Dalton Crag.
3. Yealand, 1898, *Stabler.* Woodwell, Trowbarrow, and Thrang End near Silverdale, *Wh.* Deepdale, *Wi.*

Frullania Tamarisci, Dum.—Area, 1, 2, 3, 6. Range, 0–1,000 feet or higher.

Tree roots, banks, and rocks. Common in some parts of North, rare elsewhere. First record : *Wheldon, Journ. Bot.,* 1899, p. 517.

N. 1. Ease Gill and on Leck Fell scars at 1,000 feet.
2. Kellet Seeds, *Wi.*
3. Silverdale, 1898, *Wh.* Gatebarrow ; Middlebarrow, *Wi.*
E. 6. Udale.

Var· *cornubica,* Carr.

N. 3. Maritime rocks, Heysham, 1900, *Wi.*

Frullania fragilifolia, Tayl.—Area, 1, 2. Range, 400–700 feet.

Trunks of trees and rocks. Very rare. First record : The Authors, *Journ. Bot.,* 1899, p. 516.

N. 1. Ease Gill,
2. Dalton Crag.

Frullania dilatata, Dum.—Area, 1, 2, 3, 4, 5, 6, 7, 8. Range, 0–600 feet or higher.

Trunks of trees. Common in North ; rather rare in most parts of East and West. First record : *Wheldon, Journ. Bot.,* 1899, p. 516. " Silverdale and near Stonyhurst."

ANTHOCEROTALES.

Anthoceros lævis, L.—Area, 1, 2, 3. Range, 0–150 feet.

Bare earth in cultivated fields, especially amongst clover crops. Locally abundant. First record : The present one.

N. 1. Near Over Burrow.
2. Clover field near Over Kellet, October, 1906.
3. Fields near Yealand Redmayne, *Wi.*

LICHENES.

CŒNOGONIACEI.

Racodium rupestre, Per.—Area, 2, 6, 7, 8. Range, 200–1,720 feet.

Damp perpendicular or overhanging rocks in cloughs and hollows amongst the hills. Not unfrequent in East ; rare in the other divisions. Sterile.

N. 2. At 200 feet in Kellet Park Wood, near Capernwray.
E. 6. Dale Gill, Middle Gill, etc., Hindburndale ; Wolfhole Crag ; Clougha.
7. On the higher scars of both sides of Tarnbrook Fell and in the Great Clough ; also in Black Clough and other localities in Over Wyresdale. Calder Valley, above Oakenclough.
8. Burnslack Clough, etc.

This curious plant forms extensive blackish patches on the rocks, composed of interlacing filaments, usually sprinkled here and there with the grey thalline granules of some undeveloped and unrecognisable lichen.

COLLEMACEI.

Lichina confinis, Ag.—Area, 3. Range, 0.

On inter-tidal rocks on the coast of North. Rare.

N. 3. Limestone rocks between Jenny Brown's Point and Silverdale Cove ; grit rocks near Middleton and Heysham.

Synalissa symphorea, Nyl.—Area, 2. Range, ?

Rocks in upland districts, usually at no great distance from the sea. Very rare.

N. 2. Exposed limestone rocks on Dalton Crag.

British specimens have a less fruticulose appearance than continental ones, the branches being usually only nodulose. They occur as scattered black rosulate tufts affixed by a central point. All our examples are sterile.

Collema auriculatum, Hoffm.—Area, 1. Range, ?–900 feet.

Amongst mosses on the trunks of trees. Very rare. Not seen fruiting.

N. 1. Trees in Ease Gill, 1904.

A distinct species ; rigid, and with the lobes elegantly transversely wrinkled, by which it may be at once known from *C. flaccidum.*

Collema furvum, Ach.—Area, 3, 6. Range, 0–300 feet or higher.

On the mortar of old walls. Not common. Fruiting.

N. 3. Walls near Silverdale.
E. 6. Mortar of a bridge over the Conder near Clougha.

Collema flaccidum, Ach.—Area, 2. Range, ?

On mossy tree trunks and walls. Rare. Fruit rare.

N. 2. Tree trunk near Henridden, May, 1904. Mossy wall near Over Kellet ; a nearly smooth form with concave, almost urceolate apothecia (reaction with iodine negative).

Collema pulposum, Ach.—Area, 1, 3. Range, 0–1,000 feet or higher.

On banks or mud-capped walls in calcareous districts. Not common. Fruiting occasionally.

N. 1. Middle Ease Gill and Leck Fell.
3. Banks near Yealand and near Silverdale.

Collema glaucescens, Hoffm. (*C. limosum,* Ach.).—Area, 1. Range, only seen at about 1,000 feet.

On damp, bare limestone soil. Very rare. Fruiting.

N. 1. Middle Ease Gill.

This species occurs in South Lancashire at sea level, on glacial clay on the banks of the Mersey near Speke.

Collema ceranoides, Nyl.—Area, 3. Range, 0–400 feet or higher.

Upon earth-covered ledges of calcareous rocks. Rare. Not fruiting.

N. 3.　Heald Brow near Silverdale, March, 1906, and on Warton Crag.

Collema cheileum, Ach.—Area, 3, 8.　Range, 0–400 feet or higher.

On old mossy walls.　Rare, and sterile.

N. 3.　On an old wall near Heysham with *Tortula ambigua*.　Near Basil Point.

E. 8.　Wall near Barnsfold.

Collema granuliferum, Nyl.—Area, 1, 2, 3.　Range, 0–1,000 feet.

Upon rocks and walls.　Frequent in the calcareous districts, apparently absent elsewhere.　Not seen fruiting.

N. 1.　Near Ease Gill Kirk, ascending to 1,000 feet.

2.　Dalton Crag and near Borwick.

3.　Trowbarrow, Heald Brow, and other localities near Silverdale.

Collema melænum, Ach.—Area, 1, 2, 3, 8.　Range, 0–1,000 feet.

Calcareous rocks and walls.　Frequent locally.　Fruiting.

N. 1.　Middle Ease Gill, at 1,000 feet.

2.　Dalton Crag, near Henridden ; Over Kellet, and near Dunnald Mill Hole.

3.　Silverdale and Yealand ; mostly the *f. marginale*, Ach.　Warton Crag.

E. 8.　Near Dinkling Green.

Collema polycarpon, Koerb.—Area, 1, 2, 3.　Range, 0–1,000 feet or higher.

Calcareous rocks in hilly districts.　Not unfrequent in parts of North.　Fruiting abundantly.

N. 1.　On limestone rocks in Ease Gill, in fruit, 1904.

2.　Dalton Crag and on a wall near Capernwray.

3.　Near Silverdale Cove.

Collema multipartitum, Sm.—Area, 2, 3, 8.　Range, 0–800 feet or higher.

Calcareous rocks and walls.　Locally frequent, but not seen off the limestone.　Fruiting occasionally.

N. 2.　Near Over Kellet.

3.　Several localities near Silverdale.

E. 8.　Near Whitewell.

Collema isidioides, Nyl.—Area, 1, 2, 3.　Range, ?–1,000 feet.

Calcareous rocks.　Rare.　Sterile.

N. 1.　Ease Gill.

2.　Dalton Crag.

3.　Warton Crag, *J. A. Martindale*, *The Naturalist*, 1886, p. 320.　Still there !

Collemodium plicatile, Nyl.—Area, 3, 8.　Range, 0–600 feet.

Shaded, calcareous rocks.　Very rare.　Fruiting.

N. 3.　Rocks near Silverdale Cove.

E. 8.　Rocks in the Hodder Valley below Whitewell.

Collemodium fluviatile, Nyl.—Area, 8.　Range, ?

Rocks in the beds of rivers where occasionally submerged. Rare.　Sterile.

E. 8.　Stones in the river above Hodder Bridge.

Collemodium turgidum, Nyl.—Area, 1, 2.　Range, ?–950 feet.

On limestone rocks and banks and earth-capped walls.　Not common.　Fruiting occasionally.

N. 1.　Limestone rocks on Leck Fell at 950 feet.

2.　Near Capernwray ; fruiting.

Collemodium Schraderi, Nyl —Area, 2, 3.　Range, 0–800 feet or higher.

Exposed calcareous rocks.　Rare.　Sterile.

N. 2.　Dalton Crag, near Henridden.

3.　Heald Brow, near Silverdale, and sparingly on Warton Crag.

Leptogium subtile, Nyl.—Area, 3.　Range, unknown.

On the ground amongst mosses on bare, clayey, calcareous banks.　Rare.　Not seen fruiting.

N. 3.　Near Silverdale.

Forms thin, effuse, dark olive green patches, looking granular to the naked eye, but under the lens is like a miniature *L. lacerum* with digitate squamules.

Leptogium lacerum, Gray.—Area, 1, 2, 3.　Range, 0–1,000 feet or higher.

Mossy limestone banks, rocks, and walls.　Locally frequent. Fruit rare.

N. 1.　Middle Ease Gill at 1,000 feet (*f. fimbriatum*, Nyl.) and on Leck Fell.

2.　Amongst *Hypnum molluscum* on Dalton Crag near Henridden.

3.　Gatebarrow Wood and other localities near Silverdale ; Warton Crag.

Leptogium pulvinatum, Nyl.—Area, 2, 3.　Range, 0–800 feet or higher.

Exposed mossy limestone banks and rocks.　Frequent in North.　Fruit very rare.

N. 2.　Near Whittington ; Dalton Crag, and near Over Kellet.

3.　Gatebarrow Wood, Warton Crag, and near Carnforth.

Forms very dense rounded cushions of a mouse-coloured brown.

[*Leptogium lophæum*, Nyl.—Mr. Holmes suggested that some of our Silverdale specimens submitted to him might possibly belong to this sub-species.　We have not been able to see authentic specimens, and, in the absence of more definite confirmation, are unable to include it in our list at present.]

Leptogium scotinum, Fr.—Area, 1, 2, 3.　Range, 50–1,000 feet.

Mossy rocks and walls.　Rare.　Fruiting.

N. 1.　Ease Gill, 1901.

2.　Wall near Capernwray and on Dalton Crag.

Var. *sinuatum,* Malbr.—Mossy banks and earth-covered rocks and walls.　Not common.　Fruiting occasionally.

N. 1.　On mossy walls near Cowan Bridge and in Middle Ease Gill at 1,000 feet.

3.　Wall near Carnforth.

EPICONIODEI.

Sphinctrina turbinata, Fr.—Area, 1, 6.　Range, ?

Parasitic on species of *Pertusaria*.

N. 1.　On the thallus of *P. communis* on Dalton Crag and on *P. amara* near Kirkby Lonsdale.

E. 6.　On the thallus of *P. communis* in Hindburndale.

Calicium hyperellum, Ach.—Area, 1.　Range, 0–200 feet or higher.

On the trunks of oak trees.　Rare.　Fruiting.

N. 1.　On oaks between Hornby and Melling and near Wennington, September, 1903.　Near Nether Burrow.

The greenish-yellow thallus is probably more frequent than the above records indicate, but the curious fruit, resembling stout large-headed black pins, is infrequent, and is usually inconspicuous, occurring in deep crevices of the bark.

Calicium quercinum, Pers.—Area, 1, 2.　Range, ?

On the trunks of old trees.　Fruiting.

N. 1.　On oaks near Nether Burrow.

2.　Elm tree in Holme Area, near Kirkby Lonsdale.

Calicium trachelinum, Ach.—Area, 1, 2.　Range, ?

On old and decayed trunks of trees.　Fruiting.

N. 1.　On a dead oak near Tunstall, 1906.

2.　Holme Area near Kirkby Lonsdale, on both elm and oak.

The turbinate apothecia are dull-brown underneath and the thallus suffused with a greenish-yellow *Lepraria*.　The stipes vary much in length.

Calicium curtum, Turn. & Borr.—Area, 6.　Range, only seen at about 800 feet.

On wood and decayed trees.　Fruiting.

E. 6.　On a decorticated oak at Conder Head, sparingly, 1906.

Perhaps all the species of *Calicium* are more plentiful than the records indicate, but they are often sterile in this district and, therefore, unnameable.

Coniocybe pallida, Fr.—Area, 6.　Range, ?

On decaying tree trunks, near the ground.

E. 6.　Clougha, *R. Jacob* (teste *Mudd*.).

Trachylia tympanella, Fr.—Area, 6, 8.　Range, ?

On old posts and palings in hilly districts.　Rare.　Fruiting.

E. 6.　Crag Wood, near Clougha.

8.　Palings near Greystoneley.

The large black apothecia resemble those of a *Lecidea*, but have the surface covered with a thick, dark, sporal mass, which imparts an inky stain to the fingers.

Sphærophorus compressus, Ach.—Area, 6, 7.　Range, 1,200 feet or higher—700 feet.

Grit rocks on the fells.　Much rarer than either of the other species.　Sterile.

E. 6.　Salter Fell, Roeburndale ; Clougha Pike.

7.　Gavells Clough.

Sphærophorus coralloides, Pers.—Area, 1, 6, 7, 8.　Range, 2,000–700 feet.

Rocks and boulders of Silurian and Millstone Grit amongst the hills.　Frequent in North and East.　Fruit not seen.

N. 1. Leck Fell and near the summit of Greygarth Fell.
E. 6. Thrush Gill Fell, Hindburn ; Salter Fell, Roeburndale ; Clougha.
   7. Black Clough and Gavells Clough, Wyresdale. Summit of Wardstone at 1,820 feet.
   8. Wolf Fell.

Sphærophorus fragilis, Ach.—Area, 1, 6, 7, 8. Range, 1,820–700 feet.

In similar localities to the preceding species, but more frequent. Fruit not seen.

N. 1. Greygarth Fell at 1,550 feet.
E. 6. Many localities in Hindburn and Roeburndale, ascending to the summit of Botton Head Fell and Wolfhole Crag ; Clougha and Heights.
   7. Gavells Clough and Deer Clough. Summit of Wardstone at 1,820 feet.
   8. Wolf Fell.

CLADODEI.

Bæomyces rufus, DC.—Area, 1, 2, 6, 7, 8. Range, 200–1,200 feet or higher.

On sandy or gravelly banks and stones Frequent. Fruiting well.

N. 1. Ease Gill.
   2. Near Gressingham, September, 1903. Near Over Kellet.
E. 6. Middle Gill and Dale Gill, Hindburn ; Tatham Fell, near Wray.
   7. Barnacre, near Garstang ; Black Clough, Over Wyresdale.
   8. Near Dinkling Green.

Bæomyces roseus, Pers.—Area, 1, 6. Range, 700–1,600 feet or higher.

On damp, gravelly, turfy soil. Rare, though possibly overlooked, because rarely fruiting.

N. 1. By the Cant Beck above Leck Hall.
E. 6. Banks of the Little Moor Beck on Lythe Fell and in Dale Gill ; Wardstone.

The pale rose-coloured cephalodia are almost invariably present and help to distinguish it when barren.

Bæomyces æruginosus, DC.—Area, 2. Range, ?

On peaty banks and decayed mosses. Rare. Fruiting.

N. 2. Near Dalton Crag.

The apothecia are larger and much more shortly stipate than in the other species. It is a handsome species when fresh, the thallus then being green and the apothecia peach-blossom pink.

Stereocaulon coralloides, Fr.—Area, 1. Range, 500–2,050 feet.

On siliceous rocks amongst the hills. Rather rare. Sterile.

N. 1. Ease Gill, Leck Fell, and Greygarth Fell at 2,050 feet.

Stereocaulon evolutum, Graewe.—Area, 1, 6, 7, 8. Range, 340–2,050 feet.

On Millstone Grit rocks amongst the hills. Widespread, but not very abundant. Fruiting occasionally.

N. 1. Greygarth Fell, ascending to 2,050 feet.
E. 6. Clougha ; wall near Wennington at 340 feet.
   7. Deer Clough and Great Clough, Wyresdale ; walls in Bleasdale.
   8. Wall near Dinkling Green ; sparingly and stunted.

This is the plant which commonly passes as *S. paschale* in British Floras. Crombie states that the latter is one of the rarest British Lichens, and it has only been sparingly found among the Grampian Hills.

Stereocaulon denudatum, Floerke.—Area, 1, 7. Range, ?–2,050 feet.

In similar localities to the preceding, but less common. Fruit not seen.

N. 1. Greygarth Fell at 2,050 feet.
E. 7. At the head of Great Clough, Tarnbrook Fell.

Stereocaulon condensatum, Hoffm.—Area, 6. Range, ?–1,700 feet or higher.

Rocks on the higher fells. Very rare. Fruit not seen.

E. 6. High up on Wolfhole Crag.

Cladonia alcicornis, Floerke.—Area, 5. Range, about sea-level.

Mossy ground on the sandhills of the coast. Sterile.

W. 5. Sparingly near Lytham, 1899, *H. Wheldon*.

The specimens are small, but characteristic, with the marginal fibrils well developed ; these latter distinguish it from *C. luteoalba* when sterile.

Cladonia pyxidata, Fr.—Area, 1, 2, 3, 4, 5, 6, 7, 8. Range, 0–2,050 feet.

On banks, walls, rocks, and about the roots of trees. Common. Fruiting.

Var. *pocillum,* Fr.—Sandy banks and heaths.

N. 1. Leck Fell and in Ease Gill.
   2. Near Over Kellet ; Dalton Crag (*f. cervina*, Nyl.).

N. 3. Silverdale and Thrang End.
W. 5. Sandhills near St. Annes.
E. 7. Near Dolphinholme.

Var. *chlorophæa,* Floerke.—Mossy banks and walls.

W. 5. Near Fleetwood.
E. 6. Tatham Moor, Hindburn.
   7. Spreight Clough ; near Dolphinholme.

Cladonia pityrea, Floerke.—Area, 7. Range, ?

Heathy banks amongst the hills. Rare.

E. 7. Gully on Lea Fell, Wyresdale.

[*Cladonia Lamarckii,* Nyl.—A small *Cladonia* occurs creeping over *Trichostomum crispulum* near Ings Point, Carnforth, which is probably to be referred to this species. It has the very characteristic reaction K yellowish then ochraceous, C—, but, unfortunately, it has not yet been found with podetia developed.]

Cladonia fimbriata, Fr.—Area, 1, 2, 3, 4, 5, 6, 7, 8. Range, 0–?

Mossy banks and tree roots. Not uncommon. Fruit not seen.

N. 1. Near Cowan Bridge and between Leck and Burrow.
   2. Near Borwick, Dalton Crag, etc.
   3. Rail banks near Silverdale.
W. 4. Near Pilling and Preesall.
   5. Near Wrea Green.
E. 6. Lower Hindburndale.
   7. Parlick Pike.
   8. Near Whitewell.

Var. *conista,* Nyl.—As elsewhere in Britain, our specimens are sterile.

N. 2. Near Nether Kellet.
   3. Thrang End, near Silverdale.

Var. *subcornuta,* Nyl.

E. 6. Foxdale.
   7. Over Wyresdale.

Var. *radiata,* Nyl.

E. 7. Near Dolphinholme.

Cladonia gracilis, Hoffm.—Area, 1, 6. Range, 300–1,100 feet or higher.

Amongst mosses on rocky, peaty ground amongst the hills. Fruit not seen.

N. 1. Wall near Leck (*f. spinulifera*, Cromb.).
E. 6. Littledale Fell, where it sometimes passes into a densely squamulose form (*f. aspera*, Floerke), but usually has a few scyphiferous podetia mixed with the cornute ones, thus distinguishing it from *C. squamosa*.

Cladonia cervicornis, Schaer.—Area, 1, 6, 7, 8. Range, 500–2,050 feet.

On rocky, heathy ground amongst the hills. Frequent in East. Rarely fruiting.

N. 1. Greygarth Fell, ascending to 2,050 feet, and Lower Ease Gill.
E. 6. Salter Fell, Littledale Fell, Udale, and Clougha Pike.
   7. On the south-west side of Wardstone and Long Crag above Tarnbrook.
   8. Wolf Fell, on grit scars at 1,150 feet.

Cladonia sobolifera, Nyl.—Area, 1. Range, only seen at 1,750 feet.

On mossy gritstone scars high up on the fells. Not fruiting.

N. 1. Greygarth Fell ; a specimen resembling a less compact form of *C. cervicornis*, without reaction with KHO. ; probably belongs to this species.

Cladonia lepidota, Nyl.—Area, 1, 6, 7. Range, ?

On the earth amongst rocks on elevated moorlands. Rare. Fruiting.

N. 1. Lower Ease Gill.
E. 6. Clougha ; Foxdale.
   7. Wardstone.

Probably belonging to the *forma hypophylla*, Cromb., the fruit appearing to be almost sessile upon the squamules.

Cladonia furcata, Hoffm.—Area, 1, 2, 3, 4, 5, 6, 7, 8. Range, 0–2,000 feet.

Heathy and rocky ground and earth-covered walls. Frequent, especially amongst the hills. Fruit not seen.

The *f. spadicea*, Nyl., occurs on Greygarth Fell with the ordinary glaucous state.

Var. *spinosa,* Hook.

N. 3. Near Jenny Brown's Point, Silverdale.
E. 6. Conder Head and Clougha Pike.
   8. Wolf Fell.

Var. *corymbosa*, Nyl.

N. I.   Greygarth Fell   *H. Fisher* in *Victoria History of Lancashire.*

Var. *racemosa*, Mudd. (*C. racemosa*, Nyl.).

N. I.   Lower Ease Gill, *H. Fisher.* On Greygarth Fell the *forma recurva*, Flk., occurs at 550 feet and the *forma palamacea*, Nyl., at 2,050 feet.
E. 8.   Dewhurst Clough.

Cladonia furcatiformis, Nyl.?—Area, 3. Range, only seen at about 25 feet.

On mossy, peaty ground. Very rare. Sterile.

N. 3.   Heysham Moss ; somewhat doubtfully referred here.

Cladonia pungens, Floerke.—Area, 1, 2, 3, 4, 5. Range, 0–1,200 feet.

On the ground on heaths and stony places. Frequent. Sterile.

N. I.   Leck Fell.
2.   Dalton Crag and near Dunnald Mill Hole.
3.   Heysham Moss.
W. 4.   Grit rocks south-west of Galgate.
5.   Sandhills near St. Annes.

Forma foliosa, Floerke.

N. 3.   Castlebarrow and other localities near Silverdale.

Cladonia muricata, Crombie.—Area, 1. Range, ?–2,050 feet.

Damp, heathery ground in hilly districts. Rare. Sterile.

N. I.   Greygarth Fell at 2,050 feet.

[*Cladonia cenotea*, Schaer.—Var. *glauca*, Nyl. (*C. glauca*, Floerke). Area, 4. Range, only seen at 25 feet.

Amongst mosses on a peaty soil. Very rare. Sterile.

W. 4.   Cockerham Moss, November, 1906.

A slender plant with scyphoid axils ; repeatedly proliferous ; the subulate apices and branches glaucous and minutely squamulose. It is more slender than Hungarian specimens we possess, but otherwise very similar to them. We place the record in brackets pending confirmation.]

Cladonia squamosa, Hoffm.—Area, 1, 6, 7. Range, 150–2,050 feet.

On mossy, peaty ground in the hilly districts. Not common. Fruit not seen.

N. I.   Greygarth Fell and Ease Gill.   Turf-capped wall near Cowan Bridge.
E. 6.   Clougha ; Conder Head.
7.   Great Clough of Tarnbrook Fell and on Hell Crag.

It varies exceedingly in appearance, and is one of our most polymorphous species. The *forma subulata*, Schaer., occurs on Littledale Fell and in other localities. The *forma ventricosa*, Fr., is a handsome robust state, which we have only seen on Lythe Fell, near the falls of the Little Moor Beck. A plant occurs at 1,950 feet on Greygarth Fell and on Hell Crag which agrees exactly with French examples of *forma phyllopoda*, Wainio., which does not appear to be recorded as British.

Cladonia adspersa, Nyl.—Area, 8. Range, ?

Peaty, mossy banks in shady places.

E. 8.   North side of Longridge Fell, amongst *Mnium hornum.*

We have only recently received specimens enabling us to identify this species. It is probably, however, not common with us, as it is too striking a plant to be readily overlooked.

Cladonia subsquamosa, Nyl.—Area, 4. Range, only seen at about 100 feet.

On mossy walls. Rare. Not fruiting.

W. 4.   Wall near Ellel Grange.

Our specimen has rather the aspect of some of the species of the section *Erythrocarpœ*, but gives the rare yellow and then crimson reaction with KHO, which is our chief reason for referring it to this species. The podetia are generally simple or at most once furcate.

Cladonia cæspititia, Floerke.—Area, 1, 7, 8. Range, 300–1,520 feet or higher.

N. I.   Ease Gill near Leck, ascending to 1,520 feet on Greygarth Fell.
E. 7.   Black Clough and Marshaw Fell.
8.   Near Whitewell.

Cladonia delicata, Floerke.—Area, 1, 3. Range, ?

On rotting stumps of trees. Not common.

N. I.   Lower Ease Gill.
3.   Near Hawes Water, 1902.

Cladonia coccifera, Schaer.—Area, 1, 3, 4, 6, 7, 8. Range, 20–2,050 feet.

On sandy or peaty banks and heathery ground in hilly districts. Frequent in East and parts of North. Occasionally fruiting, but often sterile.

This beautiful and variable species is a frequent ornament of cloughs and fells, and ascends to 2,050 feet on Greygarth Fell. *Forma cornucopioides* has been noticed near Haylot and on Long Crag, Wyresdale. On Clougha and in Whitestone Clough occurs a form without scyphi and with very short podetia, rendering the apothecia almost sessile, which is probably var. *incrassata*, Fr.

D I

Cladonia digitata, Hoffm.—Area, 1, 2, 3, 4, 6, 7, 8. Range, 20–1,000 feet or higher.

Mossy tree trunks and heathery ground. Rarely fruiting.

N. I.   Middle Ease Gill at 1,000 feet.
2.   Dalton Crag.
3.   Heysham Moss.
W. 4.   Cockerham Moss.
E. 6.   Clougha and on old trees near Conder Head.
7.   Grizedale near Garstang ; Long Crag ; near Dolphinholme with *forma cerucha*, Nyl.
8.   Near Dinkling Green.

Cladonia luteoalba, nobis. Sp. Nov.—Area, 1, 6. Range, 1,100–1,550 feet.

Amongst mosses and lichens on high gritstone screes. Rare. Fruit very rare.

N. I.   Greygarth Fell at 1,550 feet, associated with *Cetraria aculeata* and *Cladonia furcata*, October, 1906.
E. 6.   Dale Gill, Hindburndale, at 1,100 feet.

This pretty and distinct-looking species was referred by eminent authorities, before the discovery of fruiting examples, to *C. alcicornis* and *C. sobolifera.* The fruit shows it to belong to the *Erythrocarpœ*, of which it must be regarded as a new species. Its characters are as follows :—Thallus macrophylline, lobes 5–10 mm. long, irregularly crenate, yellowish-green above (becoming blackish-green when old), pallido-sulphureous beneath, their apices, and sometimes their lateral margins, strongly incurved when dry, showing the pulverulent under surface, and rendering the leaflets concave. Podetia rare, short (3–5 mm.), cylindrical, from the surface of the leaflets, scyphi hardly dilated, bearing small marginal discrete scarlet apothecia. The yellowish underside is conspicuously displayed and at once attracts attention, as in *C. endiviæfolia* and *C. alcicornis.* The yellow colour of the under surface becomes more pronounced on applying KHO, and if this is followed immediately by CaCl₂ the colour is still further intensified. These reagents produce only indistinct reactions on the upper surface of the thallus. In addition to the localities named above, this species has been found in Westmorland (Dale Head, near Martindale, May, 1907, *Wi.*), on Silurian rocks at 1,600 feet, accompanied by *Cetraria aculeata* and *Platysma triste.*

Cladonia macilenta, Hoffm.—Area, 1, 2, 3, 4, 6, 7, 8. Range, 0–2,050 feet.

Mossy tree trunks, rotten stumps, and peaty banks. Very common in the hilly districts and extremely variable. Fruits occasionally.

Var. *coronata*, Nyl.—Mossy banks and tree roots.

N. I.   Greygarth Fell at 2,050 feet.
2.   Dalton Crag.
E. 7.   Near Dolphinholme.

Var. *ostreata*, Nyl.—Amongst mosses on decaying tree stumps and on bark. Rare.

N. 2.   Dalton Crag.
E. 7.   Over Wyresdale.
8.   Near Whitewell.

Var. *scabrosa*, Nyl.—On rotting stumps and tree roots.

W. 4.   Wall near Ellel Grange, with apothecia.
E. 6.   Lower Hindburn.

Cladonia bacillaris, Nyl.—Area, 6, 7. Range, ?

On bare, peaty soil on the fells of East. Fruiting.

E. 6.   Middle Gill, Hindburn ; Clougha.
7.   Grizedale, near Garstang, and Johnny Pye's Fell.

Cladonia Floerkeana, Fr. var. *trachypoda*, Nyl.—Area, 6, 7. Range, ?

Bare, peaty soil on the fells. Locally abundant, and fruiting freely.

E. 6.   Littledale Fell.
7.   Tarnbrook Fell and Grizedale Head.

The type is apparently absent from West Lancashire. The apothecia of our plant are very prone to lose much of their brilliant colouring on drying.

Cladina rangiferina, Nyl. — "Reindeer Moss." Area, 1. Range, only seen at about 2,000 feet.

Boggy alpine moorlands. Very rare. Sterile.

N. I.   Very sparingly amongst mosses on the summit of Greygarth Fell.

This rare species is readily distinguished from the next by its tomentose podetia and yellow reaction with KHO.

Cladina sylvatica, Nyl.—"Reindeer Moss." Area, 1, 2, 3, 4, 6, 7, 8. Range, 20–2,050 feet.

Heaths and moorlands, chiefly amongst the hills. Frequent, especially in East, but not seen fruiting.

Forma tenuis, Lamy.—Amongst mossy rocks on the fells.

E. 6.   Clougha and Wolfhole Crag.
7.   Wardstone.

Forma lacerata, Nyl.—Amongst sphagnum in bogs.

W. 4.   Cockerham Moss.

Var. *grandis,* Cromb.—Damp, elevated moorlands.

    N. 1.   Greygarth Fell at 2,050 feet.
    E. 7.   Marshaw Fell.

Var. *alpestris,* Nyl.—Damp moorlands.   Rare.

    W. 4.   Cockerham Moss.
    E. 7.   Wardstone.

Forma pumila, Leight.—This form occurs in very compact tufts, which have a spongy, greasy feel when fresh.

    E. 6.   Conder Head and Clougha.

Cladina uncialis, Nyl.—Area, 1, 2, 4, 6, 7, 8.   Range, 500–2,050 feet.

Amongst mosses on heaths and moors.   Frequent.   Fruit not seen.

Forma adunca, Cromb.—Amongst mosses on damp moors.   Rare.

    N. 1.   Near the summit of Greygarth Fell at 2,050 feet.

Forma bolacina, Cromb.—A shorter, pulvinate, and more branched state.

    E. 6.   Conder Head.

RAMALODEI.

Ramalina farinacea, Ach.—Area, 1, 2, 3, 4, 8.   Range, 0–400 feet or higher.

On the trunks of trees and more rarely on walls.   Rather frequent in North.   Not fruiting.

    N. 1.   Trees near Ireby, Thurland Castle, and Kirkby Lonsdale.   On walls and trees near Leck and Burrow.
       2.   Trees near Whittington, Capernwray, and Over Kellet.
       3.   Near Silverdale and Leighton Beck.   Trees in and near Grisdale Wood.
    W. 4.   Trees near Ellel Grange, very small and poor.
    E. 8.   Near Whitewell.

Ramalina fraxinea, Ach.—Area, 1.   Range, 100–200 feet.

On the trunks of old trees.   Very rare.   Not fruiting.

    N. 1.   On trees near Burrow.

This species, so common in many parts of England, appears to be singularly rare in Lancashire.

Ramalina fastigiata, Ach.—Area, 1, 2, 3.   Range, 0–300 feet.

On the trunks and branches of trees.   Occasionally in North. Fruiting.

    N. 1.   Lower Ease Gill; Thurland Castle, by the Lune below Kirkby Lonsdale, and on trees near Over Burrow.
       2.   Trees about Whittington.   Between Over Kellet and Capernwray.
       3.   Trees near Grisdale Wood, very sparingly.

Ramalina subfarinacea, Nyl.—Area, 3.   Range, 0–30 feet.

Rocks and walls near the sea coast.

    N. 3.   Basil Point, in the Heysham Peninsula, sparingly.

Occurs in small tufts, smaller than the next species and with pulverulent soredia, which resemble those of *R. farinacea,* but give a reddish reaction with KHO.

Ramalina scopulorum, Ach.—Area, 3.   Range, 0–30 feet.

Sandstone rocks by the sea.   Locally plentiful and fruiting.

    N. 3.   Maritime rocks near Middleton and Heysham, accompanied in both localities by the var. *incrassata,* Nyl.

Ramalina cuspidata, Nyl.—Area, 3.   Range, 0–30 feet.

Sandstone rocks by the sea.   Locally plentiful, but sterile.

    N. 3.   Maritime rocks near Heysham and Middleton, and with it in both localities the *forma minor,* Nyl.

Ramalina breviuscula, Nyl.—Area, 3.   Range, 0–30 feet.

Maritime sandstone rocks.   Rare.   Sterile.

    N. 3.   With the preceding near Heysham and Middleton.

This is the *R. polymorpha,* f. *depressa,* Cromb., of Leighton's Lichen Flora, Ed. 3., and in our paper on West Lancashire Lichens, Journ. Bot., 1904, p. 257, it was erroneously recorded as "*R. polymorpha,* Ach."

Usnea hirta, Hoffm.—Tree-beard Lichen.   Area, 1, 2, 3, 6, 7, 8.   Range, 100–700 feet.

On trees.   Rather rare; most frequent in District 1. Always sterile.

    N. 1.   Gorge of the Greeta; near Tunstall; near Burrow and Leck Hall.
       2.   Lords Lot Wood.
       3.   Near Silverdale.
    E. 6.   Wood near Pott Yeats, and in Hindburndale.
       7.   Between Marshaw and the Trough of Bowland.
       8.   Gorge of the Hodder below Whitewell.

Usnea plicata, Gray.—Area, 7.   Range, ?

Tree trunks in hilly districts.   Very rare.   Sterile.

    E. 7.   Trees near Lower Emmetts.

Usnea ceratina, Ach., var. *scabrosa,* Ach.—Area, 1, 6.   Range, 500–900 feet.

On trees and gritstone rocks in hilly districts.   Rare. Sterile.

    N. 1.   Lower Ease Gill.
    E. 6.   Near Clougha.   Hill north-west of Pott Yeats at 800 feet, abundant. Rocks near Conder Head.

Alectoria jubata, Nyl., var. *lanestris,* Ach.—Area, 7.   Range, 400–700 feet.

On the trunks of trees, generally of conifers.   Very rare and sterile.

    E. 7.   On Scots Pine between Marshaw and the Trough of Bowland.   On oak near Catshaw, very sparingly.

Alectoria bicolor, Nyl.—"Rock-hair."   Area, 6.   Range, 600–1,200 feet.

On mossy gritstone rocks in damp places amongst the hills. Rare and always without apothecia.

    E. 6.   Clougha.   Rocks near Conder Head.   Salter Fell, Roeburndale.   Thrushgill Fell, Hindburn.

Cetraria crispa, Nyl.—"Iceland Moss."   Area, 1.   Range, 1,990–2,050 feet.

On bare peaty ground and amongst mosses.   Rare, but locally plentiful.   Sterile.

    N. 1.   Summit ridge of Greygarth Fell.

Sometimes regarded as a sub-species or variety of *C. islandica* (which does not occur in West Lancashire), but considered to be a distinct species by Nylander.

Cetraria aculeata, Fr.—Area, 1, 2, 3, 4, 5, 6, 7, 8.   Range, 20–2,050 feet.

On sandy or stony heaths both in upland and lowland districts.   Common, especially amongst the hills of North and East.   Rare in West, where it occurs on the sandhills and mosses.

The *forma hispida,* Cromb., is almost equally frequent, and *forma acanthella,* Nyl., occurs in Deer Clough.

Platysma ulophyllum, Nyl.—Area, 6, 7.   Range, 500–700 feet.

On the trunks of conifers and old palings.   Apparently rare. Sterile.

    E. 6.   On palings, Lower Hindburn.
       7.   On Scots Pine between Marshaw and the Trough of Bowland.

Platysma diffusum, Nyl.—Area, 6.   Range unknown.

On palings or dead tree stumps.

    E. 6.   Clougha, *R. Jacob* in *Mudd's British Lichens,* 1861.

Platysma Fahlunense, Nyl.—Area, 6.   Range unknown.

On subalpine rocks and boulders.   Sterile.

    E. 6.   Clougha, *R. Jacob* in *Mudd's British Lichens,* 1861.
       7.   Long Crag, Wyresdale.

Platysma glaucum, Nyl.—Area, 1, 2, 3, 6, 7, 8.   Range, 0–2,050 feet.

Trees, walls and rocks.   Common amongst the hills; rare in the low country.   Apparently always sterile with us.

The *forma coralloidea,* Wallr., is not unfrequent, and passes by various gradations into the type.   The *forma fallax,* Nyl., occurs on trees near Pott Yeats.

Var. *tenuisectum,* Cromb.—Area, 6, 7.   Gritstone rocks and walls.   Less common than the type, but apparently connected with it by intermediate states.

    E. 6.   Clougha, *Martindale.*   North-west shoulder of Wardstone.
       7.   Marshaw Fell; Deer Clough; Hell Crag, and Long Crag.

[*Platysma juniperinum,* Nyl.—Area, 6.   Range, ?   "On earth and trees in woods."   Very rare if it really occurs.

    E. 6.   Clougha, Lancashire, *R. Jacob,* confirmed by *W. Mudd.* (*Manual of Brit. Lich.*).   But Crombie says, "Very doubtful in the North of England."]

PHYLLODEI.

Evernia prunastri, Ach.—Area, 1, 2, 3, 4, 6, 7, 8.   Range, 0–1,200 feet.

Trunks and branches of trees and more rarely on palings and walls.   Frequent and sometimes well developed in the hilly districts; much rarer and often very stunted elsewhere.   Apothecia only once seen, by the Lune below Kirkby Lonsdale.

Forma sorediata, Ach., has been observed near (1) Leck Hall, (7) Lower Emmetts, (8) By the Hodder below Whitewell.

Evernia furfuracea, Fr.—Area, 1, 2, 3, 6, 7, 8.   Range, 0–2,050 feet.

Tree trunks, rocks, and walls in hilly districts. Frequent; even abundant locally; rare in the low country. Apparently always sterile.

Forma scobicina, Nyl., occurs not unfrequently (as on Hell Crag, and near Clougha). *Forma ceratea.* Nyl., occurs on Greygarth Fell and Botton Head Fell, and probably elsewhere. We have not seen *forma nuda,* Cromb., within our district, although we have collected it in Westmorland.

Parmelia perlata, Ach.—Area, 1, 2, 3. Range, 100–400 feet.

Trunks of trees, rocks, and walls. Rare, and not luxuriant. Apothecia not seen.

> N. 1. Lower Ease Gill and by the Lune below Kirkby Lonsdale. Near Cowan Bridge, and Gorge of the Greeta. Elms near Tunstall.
> 2. Holme Area below Kirby Lonsdale Bridge.
> 3. Silverdale, but very scarce.

Parmelia ciliata, Nyl.—Area, 1.

On walls in subalpine districts. Rare.

> N. 1. Only once seen on a wall near Ireby.

Parmelia cetrarioides, Nyl.—Area, 6, 8. Range, ?

Tree trunks and on gritstone rocks. Very rare, and in small quantity where it occurs. Sterile.

> E. 6. Clougha.
> 8. Tree trunk near Chaigley.

Parmelia lævigata, Ach.—Area, 7, 8. Range, 300–800 feet or higher.

Rocks and tree trunks in upland districts. Rare. Sterile.

> E. 7. Marshaw Fell, on rocks.
> 8. Near Whitewell, on trees.

Parmelia scortea, Ach.—Area, 6. Range, only seen at about 80 feet.

Roots and trunks of trees near rivers. Rare. Sterile.

> E. 6. Trees by the Lune near Caton.

Parmelia saxatilis, Ach.—Area, 1, 2, 3, 4, 5, 6, 7, 8. Range, 0–2,080 feet.

Trees, walls, and rocks. Rare in the low country, common amongst the hills. Fruit not seen.

Forma furfuracea, Schaer.—On tree trunks and rocks.

> N. 1. Walls near Cowan Bridge.
> E. 6. Mallowdale Fell.
> 7. Long Crag.
> 8. Wood by the Hodder below Whitewell, on trees.

This form, when extreme, looks almost like a distinct species. The dense furfuraceous isidia sometimes almost obliterate the lobes of the thallus.

Parmelia sulcata, Tayl.—Area, 1, 3, 4, 5, 6, 7, 8. Range, 0–?

Tree trunks and occasionally on walls. Rather frequent, and commoner in the low country than the preceding.

The thallus is less rugulose than in the preceding species and never isidiose, by which, and its oblong white soredia, it may be readily distinguished. Near Mill Houses, Hindburn, there occurs a form with very narrow esorediate laciniæ, the abundant black rootlets, giving it a ciliate appearance quite different from the type, with which however it agrees in its chemical reactions. It may be the var. *lævis,* Nyl.

Parmelia omphalodes, Ach.—Area, 1, 2, 6, 7, 8. Range, ?–1,820 feet or higher.

Gritstone rocks on the fells. Abundant. Ascends to 1,820 feet on Wardstone.

Var. *panniformis,* Ach.—This variety, which forms a densely congested crust, and has a more dissected thallus, is not unfrequent on gritstone rocks in East.

Parmelia Borreri, Turn.—Area, 1, 2, 6. Range, 0–900 feet.

Tree trunks and, more rarely, on walls. Rare, and not seen in fruit.

> N. 1. Below Kirkby Lonsdale and near Wennington, on trees. On a wall near Cowan Bridge.
> 2. Near Aughton and Over Kellet.
> E. 6. Middle Gill, Hindburn.

Resembles in the field, *P. sulcata,* but is more rigid, with smaller punctiform soredia and different chemical reactions. It seems to gradually acquire a yellow sh tint in the herbarium.

Parmelia caperata, Ach.—Area, 1, 2, 3, 6, 7. Range, 25–600 feet.

On the trunks of trees, sometimes far above the ground; also, very rarely, on mossy walls. Locally plentiful in the north-east, becoming gradually rarer and more poorly developed towards the west and south. Always sterile.

> N. 1 & 2. General, and sometimes very fine on trees. On *walls* near Leck and Cowan Bridge.
> 3. Near Thrang End, and other places near Silverdale.

> W. 4. Trees near Ellel Grange, very small and starved.
> E. 6. Crook of Lune and near Caton and Wray. Lower Hindburn at 550 feet.
> 7. Damas Gill and Dolpinholme.

Parmelia conspersa, Ach.—Area, 1, 2, 6, 7. Range, 200–1,000 feet.

On Silurian and Millstone Grit rocks. Rather rare. Fruit uncommon.

> N. 1. Lower Ease Gill, and very fine and fruiting freely near Cowan Bridge. Fruiting on Ireby Fell at 600 feet.
> 2. Grit rocks east of Dunnald Mill Hole.
> E. 6. Gritstone wall between Clougha and Heights.
> 7. Damas Gill, on glaciated grit rocks.

Parmelia exasperata, Nyl.—Area, 1, 2, 6, 7. Range, 50–500 feet.

Trunks of trees. Not common, and often poorly developed.

> N. 1. Near Melling.
> 2. Trees near Aughton; between Carnforth and Over Kellet.
> E. 6. Trees near Clougha.
> 7. Near Dolphinholme.

The papillæ become abraded in herbaria, giving it a white-punctate appearance.

Parmelia subaurifera, Nyl.—Area, 2, 6, 8. Range, 150–500 feet.

Trunks of trees. Rare, and in small quantity.

> N. 2. Near Kirkby Lonsdale.
> E. 6. Trees in Lower Hindburndale.
> 8. Near Whitewell.

Parmelia prolixa, Nyl.—Area, 1, 6, 7. Range, 500–1,500 feet.

Rocks of Millstone Grit amongst the hills. Rare. Always sterile.

> N. 1. Ease Gill near Leck, and Greygarth Fell.
> E. 6. Foxdale Head; Clougha Scar.
> 7. Long Crag; Gavells Clough.

Parmelia fuliginosa, Nyl.—Area, 1, 2, 3, 4, 5, 6, 7, 8. Range, 0–800 feet or higher.

Rocks, walls, and trees. Common, especially on Millstone Grit walls in the more hilly districts. Fruit not seen.

Var. *lætevirens.*—Area, 1, 2, 3, 6, 7. Frequent on tree trunks and palings.

Parmelia lanata, Wallr.—Area, 7. Range, 1,400–1,500 feet.

On elevated gritstone scars. Rare. Not fruiting.

> E. 7. Hell Crag and Long Crag.

Parmelia tristis, Nyl.—Area, 7. Range, 1,450–1,500 feet.

In similar situations to the preceding. Rare. Sterile.

> E. 7. Hell Crag and Long Crag.

Occurs in small scattered tufts, resembling a small blackish fucus, and firmly adhering to the rock.

Parmelia physodes, Ach.—Area, 1, 2, 3, 4, 5, 6, 7, 8. Range, 0–2,080 feet.

Rocks, walls, palings, and tree trunks; also on heather branches on the moors. Very common except in West, where it is rather rare. Apothecia not seen.

Forma labrosa, Ach., is common, and the var. *tubulosa,* Mudd., occurs more rarely on rocks amongst the hills, as on Hell Crag.

Parmeliopsis ambigua, Nyl.—Area, 6. Range, ?

On roots of fir trees. Very rare.

> E. 6. Clougha, *R. Jacob* (in *Leighton's Lichen Flora*).

Peltidea aphthosa, Ach.—Area, 1, 3, 6. Range, 100–1,200 feet or higher.

Damp, shaded rocks and walls, especially on limestone. Rare.

> N. 1. Very fine in Ease Gill.
> 3. Near Bank Well, Silverdale.
> E. 6. Middle Gill, Hindburn.

When growing it is of a bright green colour, similar to the next species, and like it, changes to grey or cervine on drying.

Solorina saccata, Ach.—Area, 1, 2, 3. Range, 50–1,200 feet or higher.

Rocks, banks, and mud-capped walls in the limestone districts. Frequent in some parts of North. Fruits freely.

> N. 1. Limestone rocks in Ease Gill, ascending to 1,200 feet.
> 2. Dalton Crag and Kellet Seeds.
> 3. Eaves Wood; near Hawes Water; Heald Brow and other localities near Silverdale.

The *forma sorediosa,* Fries., occurs on Warton Crag with the type.

Peltigera canina, Hoffm —Dog's Liverwort. Area, 1, 2, 3, 4, 5, 6, 7, 8. Range, 0–2,050 feet.

Damp, mossy banks and rocks in sub-alpine districts. Frequent in North and East, and occurs in West on the sandhills near St. Annes, and near Pilling and Ellel Grange. Fruiting.

Peltigera rufescens, Hoffm.—Area, 1, 2, 3, 6, 8. Range, 0–1,500 feet or higher.

Mossy banks, old quarries, and mossy limestone walls. Rather rare; most frequent in the limestone districts. Fruiting.

> N. 1. Upper Ease Gill and Leck Fell scars.
>   2. Dalton Crag and near Henridden.
>   3. Heald Brow, Gatebarrow Wood, Cringlebarrow, and other localities near Silverdale.
> E. 8. Near Dinkling Green.

Var. *prætextata,* Floerke.

> N. 2. Near Over Kellet. Dalton Crag near Henridden.
> E. 6. Middle Gill, Hindburn.

Although resembling *P. canina* somewhat, this species is as a rule readily distinguished by its more adherent thallus, sub-glabrous and reddish lobes with crisped margins, and dark brown veins beneath. It has distinctly xerophilous tendencies.

Peltigera spuria, Leight.—Area, 4. Range, only seen at 25 feet.

On the ground amongst short grass. Very rare. Fruiting.

> W. 4. Cockerham Moss, 1900.

Very different in its erect habit from any form of *P. rufescens*, to which Crombie formerly referred it. It is more like a miniature form of *P. polydactyla*, but seems to be quite distinct from either.

Peltigera polydactyla, Hoffm.—Area, 1, 3, 4, 6, 7, 8. Range, 0–1,000 feet.

Grassy and mossy banks. Fruiting freely.

> N. 1. Middle Ease Gill.
>   2. Sandstone quarry between Over Kellet and Aughton.
>   3. Heald Brow Quarry, near Silverdale.
> W. 4. Wall near Ellel Grange.
> E. 6. Mill Houses, and near Hindburn Lower Bridge; Lower Salter, Roeburndale.
>   7. Grizedale at about 900 feet.
>   8. Near Dinkling Green.

Var. *hymenia,* Nyl.

> W. 4. Wall near Ellel Grange.

Peltigera horizontalis, Hoffm.—Area, 1, 2, 3, 6, 7, 8. Range, 0–1,300 feet or higher.

Mossy, shaded rocks amongst the hills. Not frequent. Fruit rare.

> N. 1. Middle Ease Gill.
>   2. Near Borwick; and Dalton Crag near Henridden.
>   3. Middlebarrow Wood near Silverdale. Near Yealand.
> E. 6. Mallowdale Fell and near Haylot.
>   7. Spreight Clough and Deer Clough.
>   8. Gorge of the Hodder below Whitewell.

Physcia parietina, De Not.—Area, 1, 2, 3, 4, 5, 6, 7, 8. Range, 0–2,050 feet.

Walls and roofs, rocks, and tree trunks. Very common, and generally fertile.

Forma cinerascens, Leight.

> N. 1. Bank of Lune near Over Burrow.
>   2. Over Kellet.
>   3. Near Heaton, and about Carnforth and Silverdale.
> W. 4. Near Nateby.

In a preliminary report on West Lancashire Lichens, we called attention to the remarkable prevalence of this lichen in the vicinity of cow byres, and on walls and gateways leading to fields in which cows are pastured. Further observation in West Lancashire and elsewhere, confirms our opinion that the presence of these animals in some way favourably influences the growth of this lichen.[*]

Physcia lychnea, Nyl.—Area, 8. Range, ?

Only rupestral in West Lancashire. Rare.

> N. 3. On Warton Crag.
> E. 8. Rocks near Whitewell.

Resembles a small, granulose, and elegantly divided state of the preceding species.

Physcia ciliaris, DC.—Area, 2. Range, ?

Trunks of trees in hedgerows. Very rare, and both sterile and depauperate.

> N. 2. Near Aughton, 1903.

This species is singularly rare in West Lancashire, as, according to Crombie, it is elsewhere in districts bordering the Irish sea.

Physcia pulverulenta, Nyl.—Area, 1, 2, 3, 6. Range, 0–200 feet or higher.

---

[*] Journal of Botany, 1904, p. 258.

Trunks of trees in hedgerows and fields; sometimes on stone walls. Frequent in North. Occasionally fruiting.

> N. 1. Thurland Castle and Tunstall.
>   2. Over Kellet.
>   3. Silverdale and Bolton-le-Sands.
> E. 6. Wall near Wray.

Physcia venusta, Nyl.—Area, 2. Range, 0–200 feet or higher.

Trunks of trees. Rare, but fruiting well where it occurs.

> N. 2. Between Carnforth and Over Kellet, and near Borwick.

Physcia pityrea, Nyl.—Area, 2, 3, 4. Range, 0–200 feet or higher.

Trunks of trees; rarely on walls. Apparently rare.

> N. 2. Between Carnforth and Over Kellet.
>   3. Silverdale. Wall near Heaton.
> W. 4. Nateby.

Physcia subdetersa, Nyl?—Area, 1, 2. Range, 100–200 feet.

On mossy tree trunks.

> N. 1. Willow tree by the Lune near Nether Burrow. On elms near Tunstall.
>   2. Holme Area below Kirkby Lonsdale.

We have some hesitation in recording this plant under the above name, as *P. subdetersa* is stated by Crombie to grow on rocks. The specimens have a yellowish medulla, becoming a rich deep yellow with Ca Cl_2, but showing no reaction with KHO. The margins and sometimes the surface of the thallus are copiously sprinkled with yellow soredia.

Physcia aquila, Nyl.—Area, 3. Range, 0–30 feet.

Sandstone rocks on the sea coast. Locally plentiful and bearing abundance of apothecia.

> N. 3. On rocks of Millstone Grit exposed to the sea near Heysham.

Its narrow convex laciniæ are very closely appressed to the rock; and it grows in situations where it must be often exposed to sea water.

Physcia stellaris, Nyl.—The type has not been observed.

Var. *leptalea,* Nyl.—Area, 2, 3, 5, 6. Range, 0–600 feet or higher.

Tree trunks by roadsides and in open woods. Frequent in some parts of North. Fruiting occasionally.

> N. 2. Near Henridden, and between Carnforth and Over Kellet.
>   3. Eaves Wood near Silverdale. Warton Crag. Near Sunderland Point.

> W. 5. Between Ansdell and Little Plumpton.
> E. 6. Trees near the Crook of Lune, Caton.

Physcia tenella, Nyl.—Area, 1, 2, 3, 4, 7, 8. Range, 0–1,000 feet or higher.

Trunks of trees, walls, and rocks. Abundant in North. Frequent in parts of East; rare in West. In fruit on a wall near Henridden, and near Greystoneley.

Physcia aipolia, Nyl.—Area, 1, 2. Range, ?

Trunks of trees. Very rare. Sterile.

> N. 1. Near Thurland Castle.
>   2. Near Gressingham.

Var. *cercidia,* Nyl.—In similar situations to the type, but rather more frequent.

> N. 1. Trees near Thurland Castle.
>   2. Near Whittington.

Physcia erosa, Leight.—Area, 2. Range, ?

Very rare, and sterile.

> N. 2. Near Henridden.

Physcia cæsia, Nyl.—Area, 1, 3, 8. Range, 0–2,060 feet.

Limestone rocks and walls. Not common. Not seen in fruit.

> N. 1. Wall on summit of Greygarth Fell at 2,060 feet.
>   3. Roof of a barn below Warton Crag.
>   8. Very fine on rocks, and more sparingly on walls near Dinkling Green.

Physcia adglutinata, Nyl.—Area, 2, 3. Range, ?

On trees and walls in shady places. Rare. Sterile.

> N. 2. Between Carnforth and Over Kellet, on trees.
>   3. Wall near Warton.

Gyrophora torrefacta, Cromb.—Area, 6, 7. Range, 700–1,800 feet.

On large gritstone blocks on the higher fells, mostly on their horizontal surfaces. Rather frequent in suitable localities in East. Fruiting rarely.

> E. 6. Botton Head Fell. Rocks about 300 yards north-west of the summit of Wardstone. Rocks north-west of Pott Yeats at 700 feet.

E. 7. Frequent on the crags above Tarnbrook, as Hell Crag and Long Crag; also in the Great Clough, and on the White side of Tarnbrook Fell. Rocks on the south-west side of Wardstone, ascending to the summit.

Some of the Wyresdale examples are very markedly and densely cribrose, having something of the appearance of *G. erosa*, but the under surface and chemical reactions indicate their true position under this species. The commoner form is usually found with these and occurs both with and without a few perforations in the thallus.

Gyrophora hyperborea, Ach.—Area, 6.

Gritstone rocks amongst the fells. Very rare. No recent record.

E. 6. Clougha, *R. Jacob*, (*Leighton's Lichen Flora*).

We have searched for this on several occasions without success; but the rocks on Clougha are extensive and it may still occur there.

Gyrophora polyphylla, Turn. & Bor.—Area, 1, 6, 7. Range, 600–1,800 feet.

On rocks and walls on the higher fells. Locally plentiful in North and East. Fruit not seen.

N. 1. On the county boundary wall on the summit of High Park, and in Lower Ease Gill. Very fine on Greygarth Fell at 1,520 feet.
E. 6. Salter Fell, Roeburndale; north-west side of Wardstone, and about Foxdale Head. Wall near Crag Wood at 600 feet, and abundant on rocks north-west of Pott Yeats at 700 feet.
7. Great Clough of Tarnbrook Fell; Long Crag; Grizedale Head, and south slopes of Wardstone.

Gyrophora flocculosa, Turn. & Bor.—Area, 6, 7. Range, 700–1,820 feet.

In similar situations to the preceding, but occurs much more sparingly. Fruit not seen.

E. 6. Clougha. Rocks by the Udale Beck. Conder Head.
7. Black side of Tarnbrook Fell; Gavells Clough. Wardstone summit.

[*Gyrophora vellea* (L.). Recorded from Clougha, Lancashire, by *R. Jacob*, with the mark ! by the Author, in *Mudd's Manual*. Probably the next species, which was the *Lichen velleus* of *Hudson's Flora Anglica*.]

Gyrophora polyrrhiza, Krb.—Area, 6. Range, only seen at about 1,180 feet.

On grit rocks, very sparingly. Sterile.

E. 6. Clougha, R. Jacob. (as *Gyrophora vellea* in Mudd's Manual). In small quantity on rocks above Foxdale Beck, July, 1906.

PLACODEI.

Pannaria brunnea, Nyl.—Area, 1. Range not ascertained.

On the ground in rocky, mossy situations amongst the hills. Fruiting.

N. 1. Lower Ease Gill.

Pannaria nigra, Nyl.—Area, 1, 2, 3, 6, 8. Range, 0–1,000 feet.

On calcareous rocks and walls. Common in North, and more sparingly in the limestone district of East. Fruiting.

N. 1. Wall above Leck Hall; Ease Gill; ascending to 1,000 feet on Leck Fell.
2. Over Kellet; Dalton Crag.
3. Frequent in the Silverdale district, and on Warton Crag, etc., near Carnforth.
E. 6. Salter, Roeburndale.
8. Whitewell; Greystoneley; Thornley; and near Dinkling Green.

Forms conspicuous black blotches on the grey limestone rocks. On close examination these will generally be found to be edged by a narrow bluish hypothalline line.

Coccocarpia plumbea, Nyl.—Area, 2. Range, 100–200 feet.

Trunks of trees. Very rare and poorly developed. Sterile.

N. 2. On trees by the Lune near Aughton, 1903. Near Arkholme.

Leproloma lanuginosa, Nyl.—Area, 1, 3, 6, 8. Range, 100–1,520 feet.

On shaded, mossy rocks amongst the hills. Very local. The fruit is unknown.

N. 1. Ascends in Ease Gill to 1,520 feet. Leck Fell.
3. Near Silverdale.
E. 6. Near Wennington.
8. Near Whitewell.

Lecanora crassa, Ach.—Area, 2, 3. Range, 0–870 feet or higher.

On calcareous rocks and banks, sometimes on walls. Locally abundant. Fruiting freely.

N. 2. Near Borwick; Dalton Crag; Kellet Seeds; and near Dunnald Mill Hole.
3. Plentiful in Silverdale and district. Also on Warton Crag.

Lecanora saxicola, Ach.—Area, 1, 2, 3, 8. Range, 0–500 feet or higher.

E 1

Rocks and walls. Somewhat rare, although widespread. Fruiting sparingly.

N. 1. Near Wrayton and Melling.
2. Borwick; and Highfield, Carnforth.
3. Silverdale.
E. 8. Rocks in the wood below Whitewell.

[*Lecanora fulgens*, Ach.—Reported from Clougha, Lancashire, by *Mr. R. Jacob* and confirmed by *Mudd*. Crombie records this as occurring on calcareous soils, and it seems unlikely to have occurred near Clougha.]

Lecanora murorum, Ach.—Area, 1, 2, 3. Range, 0–1,000 feet or higher.

Rocks and walls, especially in calcareous districts. Frequent in North. Fruiting freely.

Lecanora tegularis, Nyl.—Area, 3. Range, ?

Rocks and walls. Fruiting.

N. 3. Silverdale.

Lecanora callopisma, Ach.—Area, 3. Range, 0–400 feet or higher.

Calcareous rocks near the sea. Rare. Fruiting.

N. 3. Rocks on the shore, Silverdale. Warton Crag.

Lecanora sympagea, Nyl.—Area, 1, 2, 3, 7, 8. Range, 0–1,000 feet or higher.

Calcareous rocks and walls. Frequent in North. Fruiting.

N. 1. Ease Gill; Hornby.
2. Dalton Crag and near Borwick
3. Frequent in the Silverdale district.
E. 7. On the mortar of barns near Tarnbrook.
8. Near Whitewell.

Lecanora elegans, Ach.—Var. *tenuis*, Ach. (*Placodium elegans*, Link. Var. *discreta*, Schaer.). *Mudd. Man. Brit. Lich.*

On exposed rocks and stones.

E. 6. Clougha, Lancashire, *R. Jacob*, in *Mudd's British Lichens*, 1861.

Lecanora lobulata, Somm.—Area, 3. Range, 0–?

Maritime calcareous rocks. Very rare. Fruiting.

N. 3. Sparingly near Silverdale Cove. Wall near Basil Point.

Lecanora xantholyta, Nyl.—Area, 1, 2, 3, 8. Range, 0–900 feet or higher.

Shaded, calcareous rocks, best developed amongst the hills. Not unfrequent in North. The fruit is unknown.

N. 1. Very fine near Ease Gill Kirk.
2. Near Over Kellet and near Hebridden.
3. Gatebarrow Wood, Heald Brow and other places near Silverdale. Warton Crag.
E. 8. Sparingly near Whitewell.

Although frequently subeffigurate, its beautiful lemon yellow thallus never develops radiate marginal lobes, and is soft and leprose.

Lecanora laciniosa, Nyl.—Area, 2, 8. Range, 0–800 feet.

On mossy tree trunks or amongst moss on walls. Rare. Sterile.

N. 2. Between Carnforth and Over Kellet.
E. 8. A single patch amongst moss on a wall near Dinkling Green.

Very like a slender dissected form of *Physcia parietina*, but readily distinguished by the absence of any purple reaction with KHO.

Lecanora vitellina, Ach.—Area, 1, 2, 3, 4, 6, 7, 8. Range, 0–1,200 feet of higher.

Rocks, walls, and old palings. Frequent. Fruiting.

It ascends to 1,200 feet on Wolf Fell.

Lecanora medians, Nyl.—Area, 1. Range, ?

On perpendicular limestone rocks. Very rare. Sterile.

N. 1. Middle Ease Gill, plentiful in one locality only.

Resembles a pallid form of *L. murorum*, and without any reaction with KHO.

Lecanora citrina, Ach.—Area, 1, 2, 3, 4, 5, 6, 7. Range, 0–1,000 feet or higher.

Rocks and walls, especially abundant near villages. Common. Varies considerably in colour from greenish to bright citrine-yellow, according to the amount of moisture and degree of light to which it is subjected.

Lecanora erythrella, Nyl.—Area, 1. Range, only seen at about 1,000 feet.

Rocks in hilly districts. Very rare. Fruiting.

N. 1. Middle Ease Gill.

Lecanora ochracea, Nyl.—Area, 3. Range, 0–200 feet.

Maritime calcareous rocks. Locally abundant and fruiting freely.

> N. 3. Rocks on the shore near Silverdale, abundantly ; and more sparingly a little way inland. Very sparingly on a limestone wall near Grisdale Wood and on Warton Crag.

Lecanora ferruginea, Nyl.—Area, 1, 2, 3. Range, 0–700 feet or higher.

Siliceous rocks and walls. Rather rare. Fruiting.

> N. 1. Rocks in Ease Gill.
> 2. Walls near Over Kellet.
> 3. Silurian boulders near Grisdale plantation. Basil Point.
>
> All our specimens are apparently to be referred to the var. *festiva*, Nyl.

Lecanora pyracea, Nyl.—Area, 1, 7. Range, ?

On walls and rocks. Rare. Fruiting.

> N. 1. On a wall near Wrayton.
> E. 7. On a railway bridge, Barnacre, near Garstang.

Lecanora Turneriana, Nyl.—Area, 2, 3, 5. Range, 0–?

Calcareous rocks near the sea. Not common. Fruiting.

> N. 2. Dalton Crag.
> 3. Rocks on the coast near Silverdale.
> W. 5. Stones in the embankment near Fleetwood Docks.

Lecanora irrubata, Nyl.—Area, 1, 2, 3, 5. Range, 0–700 feet or higher.

Calcareous rocks and on the mortar of walls. Rare. Fruiting.

> N. 1. Rocks near Ease Gill Kirk.
> 2. Near Over Kellet.
> 3. Near Hawes Water ; Heald Brow and Cringlebarrow near Silverdale.
> W. 5. On mortar near Lytham.

Lecanora calva, Nyl.—Area, 1, 2, 3, 7. Range, 0–2,050 feet.

In similar situations to the preceding. Frequent in the limestone districts of North. Fruiting.

> N. 1. Greygarth Fell.
> 2. Dalton Crag.

> N. 3. With the preceding near Silverdale. Cringlebarrow and near Warton.
> E. 7. On mortar of a wall near Tarnbrook.
>
> The *forma incrustans*, Cromb., occurs on Dalton Crag and near Silverdale. Its apothecia are much smaller and immersed in the rock.

Lecanora candicans, Schaer.—Area, 2, 3, 8. Range, 0–800 feet or higher.

Calcareous rocks and walls. Rare. Fruiting.

> N. 2. Dalton Crag.
> 3. Heald Brow, Woodwell and other places near Silverdale. Rocks near Grisdale Wood.
> E. 8. Limestone rocks near Dinkling Green.

Lecanora exigua, Nyl.—Area, 1. Range unknown.

Hitherto only observed on walls. Rare or overlooked. Fruiting.

> N. 1. On the walls of a barn near Leck.

Lecanora Bischoffii, Nyl. — Area, 3. Range, only seen at about sea level.

Calcareous rocks on the sea coast. Rare. Fruiting.

> N. 3. Rocks near the seashore, Silverdale, the *forma immersa*, Cromb., in which the apothecia are sunk in the rock.

Lecanora galactina, Ach.—Area, 1, 2, 3, 4, 5, 6, 8. Range, 0–800 feet or higher.

Rocks and mortared walls. Frequent and usually fruiting well.

> N. 1. Ease Gill and near Leck. On mortar near Nether Burrow.
> 2. Borwick.
> 3. Limestone rocks near Silverdale and Warton. Near Torrisholme.
> W. 4 & 5. Stones in the Wyre embankment near Fleetwood on both sides of the river.
> E. 6. Rocks in the river bed at the Crook of Lune. Near Wray. Walls in Lower Hindburndale.
> 8. Gorge of the Hodder below Whitewell.
>
> The Nether Burrow specimens belong to the sub-species *L. urbana*, Nyl., having a white granulate thallus, pale pruinose apothecia and thicker paraphyses.

Lecanora subfusca, Nyl.—Area, 1, 5, 6, 7. Range, 0–500 feet or higher.

Trunks of trees. Rare. Fruiting.

> N. 1. Near Hornby.

Var. *campestris*, Nyl.—On walls and stones. Probably frequent.

> W. 5. On bricks near Lytham, and on stones by the Ribble near Preston.
> E. 6. Lower Hindburndale.
> 7. On a railway bridge, Barnacre near Garstang.
>
> The spores of this and other species allied to *L. subfusca*, often contain two large nuclei, giving them a sub-bilocular appearance. As this is not mentioned in our usual handbooks, young students are warned against erroneously regarding them as septate.

Lecanora allophana, Nyl.—Area, 1, 2, 3, 6, 7. Range, 0–600 feet.

Trunks of trees. Frequent. Fruiting.

> N. 1. Wood by the Lune near Kirkby Lonsdale. Near Leck.
> 2. Wash Dub Wood.
> 3. Grisdale plantation.
> E. 6. Crag Wood.
> 7. Near Abbeystead ; Barnacre, and Brock Valley near Garstang.

Lecanora epibryon, Ach.—Area, 1. Range, ?

On the ground on decaying moss. Very rare. Fruiting.

> N. 1. On moss near Wennington.
>
> The apothecia are much larger than in *L. subfusca*, and the habitat is peculiar ; but we have not had the name authoritatively confirmed, and only doubtfully admit a plant of such reputed rarity until further confirmation can be obtained.

Lecanora rugosa, Nyl.—Area, 1, 2, 3, 7. Range, 0–600 feet or higher.

Trunks of trees. Probably common in North. Fruiting.

> N. 1. Near Tunstall.
> 2. Near Dalton Houses. Over Kellet.
> 3. On trees between Silverdale and Yealand.
> E. 7. Spreight Clough.

Lecanora chlarona, Nyl.—Area, 1, 7. Range, not ascertained.

Trunks of trees. Not common. Fruiting.

> N. 1. On willow near Kirkby Lonsdale.
> E. 7. Near Abbeystead, Lea Fell, and Marshaw.

Lecanora atrynea, Nyl.—Area, 6. Range, not ascertained.

On sandstone rocks in hilly districts.

> E. 6. Dale Gill, Hindburn.

Lecanora gangaleoides, Nyl.—Area, 1, 3. Range, 0–1,200 feet or higher.

Rocks amongst the hills. Locally plentiful. Fruiting.

> N. 1. Lower Ease Gill.
> 3. Silurian boulder in a wall at Silverdale.

Lecanora angulosa, Ach.—Area, 1, 3, 7. Range not ascertained.

Trunks and branches of trees. Probably frequent in North. Fruiting.

> N. 1. On sycamores by the Lune near Over Burrow.
> 3. Trees between Silverdale and Yealand, with *Lecidea parasema*.
> E. 7. Near Dolphinholme.

Lecanora glaucoma, Ach.—Area, 1, 6. Range, 500–700 feet.

On rocks (generally Silurian) in hilly districts. Rare. Fruiting.

> N. 1. Ease Gill ; both the normal plant and a form with aggregato-conglomerate apothecia which is very similar to var. *Swartsii*, Nyl. Wall near Ireby. The *forma cinereopruinosa*, Leight., occurs on walls near Cowan Bridge.
> E. 6. Salter, Roeburndale.

Lecanora sulphurea, Ach.—Area, 1, 2, 3, 6. Range, 0–800 feet or higher.

Rocks and walls. Frequent. Fruiting sparingly.

> N. 1. Wall above Leck village, and near Cowan Bridge.
> 2. Dalton Crag on Silurian boulders.
> 3. Several localities near Silverdale and Carnforth, on Silurian boulders and stones in walls. Basil Point on gritstone wall.
> E. 6. Lower Hindburn.
>
> This species never seems to spread from the siliceous to the calcareous stones in the same wall. It is with us most abundant on Silurian rocks, and is very much rarer on Millstone Grit.

Lecanora varia, Ach.—Area, 1, 2, 3, 4, 5, 6, 7, 8. Range, 0–1,100 feet or higher.

Trunks and branches of trees and old wooden palings. Frequent. Fruiting.

> It ascends to 1,100 feet in Dale Gill, Hindburn, on willow. This and the next two species are sometimes hardly separable and appear to pass into each other.

Lecanora conizæa, Nyl.—Area, 6, 7, 8. Range, 0–800 feet.

Tree trunks and old palings. Rare ? Fruiting.

E. 6. Palings near Crag Wood.
7. Barnacre near Garstang. Over Wyresdale.
8. Near Chaigley, and in the wood below Whitewell.

Lecanora conizæoides, Nyl.—Area, 1, 3, 4, 6, 7, 8. Range, 0–600 feet or higher.

Trunks of trees. Frequent. Fruiting.

N. 1. Near Cowan Bridge.
3. Near Silverdale. On juniper bushes on Warton Crag.
W. 4. Trees between Skippool and Shard Bridge.
E. 6. Trees, Crag Wood.
7. Brock Valley and near Garstang. Near Dolphinholme on furze. On Scots pine in the Trough of Bowland.
8. Near Whitewell.

Lecanora expallens, Ach., var. *lutescens*, Nyl.—Area, 1, 2, 6, 8. Range, 0–400 feet or higher.

Trunks of trees. Frequent, but rarely fruiting.

N. 1. Trees by the Lune below Kirkby Lonsdale, and near Wennington.
2. Trees by the Lune opposite Caton, and Holme Area below Kirkby Lonsdale.
E. 6. Trees at the Crook of Lune near Caton.
8. Gorge of the Hodder below Whitewell.

Lecanora polytropa, Schaer.—Area, 1, 3. Range, 0–1,400 feet or higher.

Rocks in hilly districts. Fruiting.

N. 1. Amongst *Lecidea geographica* in Lower Ease Gill, a form with an almost evanescent granulose thallus, probably var. *acrustacea*, (Schaer.) of *Mudd's Manual*. Upper Ease Gill, very fine. Wall near Leck.
3. Silurian rocks in a wall near Silverdale.

A diffuse sorediose thallus, which very rarely fruits, is common in North, ascending to the summit wall of Greygarth Fell, and is probably a state of this species. It occurs also in Hindburndale.

Lecanora intricata, Nyl.—Area, 1. Range, ?

Rocks and walls in hilly districts. Rare. Fruiting.

N. 1. Lower Ease Gill.

Lecanora atra, Ach.—Area, 1, 2, 3, 4, 6. Range, 0–1,000 feet or higher.

Rocks and walls; more rarely on trees. Frequent. Fruiting freely.

N. 1. Ease Gill. Thurland Castle, etc.
2. Walls near Borwick.
3. Wall near Carnforth. Walls between Morecambe and Heysham, and near Middleton.
W. 4. Bridge near Saltcoats.
E. 6. Trees near Caton, and walls near Wray.

Lecanora badia, Ach.—Area, 1, 6. Range, not ascertained.

Rocks, walls, and stones. Perhaps more common than the records indicate.

N. 1. Lower Ease Gill, and near Leck.
E. 6. Lythe Fell.

Lecanora ventosa, Ach.—Area, 1, 8. Range, 500–1,200 feet or higher.

On Silurian or more rarely on Millstone Grit rocks in hilly districts. Locally abundant and fruiting freely.

N. 1. Silurian rocks and walls in Lower Ease Gill, and on High Park.
E. 8. Millstone Grit rocks on Wolf Fell, less finely developed, but fruiting.

In our specimens the yellow colour of the thallus is little evident, but it is thick, verrucose-rugose and greenish-cinereous. The blood-red apothecia render it a striking species.

Lecanora tartarea, Ach.—Area, 1 6, 7, 8. Range, ?–2,000 feet.

Rocks and tree trunks. Frequent in the more elevated districts of North and East, but far from being so luxuriant and well-developed as it is in parts of the neighbouring counties of Westmorland and Yorkshire.

Lecanora subtartarea, Nyl.—Area, 1, 2. Range, ?

Trees and rocks. Rather rare. Sterile.

N. 1. Trees near Burrow ; and by the Leck Beck near Cowan Bridge.
2. Between Over Kellet and Capernwray.

Lecanora parella, Ach.—Area, 1, 2, 3, 4, 6, 7, 8. Range, 0–2,050 feet.

Rocks, walls, and trunks of trees. Very common, especially in the hilly parts of the district. The *forma porinoides*, Cromb., occurs sparingly in Lower Ease Gill. Fruiting.

Var. *Turneri*, Nyl.

N. 1. Trees near Wennington.
E. 6. Near Mill Houses.

Lecanora pallescens, Nyl.—Area, 2. Range, ?

Tree trunks. Apparently rare.

N. 2. Trees near Dalton Hall, and near Henridden.

Appears to be too near *L. parella* to be entitled to full specific rank, and hardly separable except by its chemical reaction.

Lecanora calcarea, Somm.—Area, 1, 2, 3, 8. Range, 0–1,200 feet or higher.

Limestone rocks and walls. Common. Fruiting.

N. 1. Middle Ease Gill and Leck Fell.
2. Near Over Kellet.
3. Rocks on the shore, Silverdale. Cringlebarrow. Warton Crag, etc.

Var. *contorta*, Nyl.

N. 3. Near Silverdale. Cringlebarrow.
E. 8. Near Dinkling Green.

Rarely more than five or six of the large sub-globular spores are fully developed, and they are often irregularly angular through mutual pressure, though normally nearly spherical. They are usually hyaline, but are occasionally rendered sub-opaque by being densely filled with finely granular protoplasm.

Lecanora epulotica, Nyl.—Area, 3. Range, only seen at about 300 feet.

Limestone rocks. Rare. Fruiting.

N. 3. Cringlebarrow.

Looks intermediate between *L. calcarea* and *L. lacustris*. Differs from the first named in its somewhat more elongate spores, which in our examples are always eight in number, and in its pale apothecia circumcised from the thallus, which in our plant is whitish.

Lecanora gibbosa, Nyl.—Area, 1. Range, ?–2,050 feet.

Rocks and walls in hilly districts. Rare. Fruiting.

N. 1. Wall near the summit of Greygarth Fell, and in Middle Ease Gill.

Lecanora Dicksonii, Nyl.—Area, 1, 7. Range, 1,200–2,060 feet.

On siliceous rocks and stones in mountainous districts. Locally plentiful. Fruiting.

N. 1. Grit rocks and boulders in Upper Ease Gill and on Greygarth Fell.
E. 7. Great Clough of Tarnbrook Fell.

[*Lecanora percæna*, Ach. (*L. glaucocarpa*, Whlnb.).—Lancashire, R. Jacob.

Probably where Mr. Jacob's other West Lancashire specimens were collected and therefore in our district, but the exact locality s not stated. All the specimens we have seen have the reaction K Cl red, and belong to the next species.]

Lecanora fuscata, Nyl.—Area, 1, 7, 8. Range, 250–1,100 feet or higher.

Siliceous rocks in hilly districts. Probably frequent and often overlooked. Fruiting.

N. 1. Walls near Cowan Bridge and Leck, and rocks at High Park, Ease Gill.
E. 7. Gritstone rocks on Threaphaw Fell at 1,100 feet.
8. With *Lecidea mollis* on Wolf Fell.

Lecanora pruinosa, Nyl.—Area, 1, 8. Range, ?

Calcareous rocks and the mortar of walls. Rare? Fruiting.

N. 1. On a mortared wall near Wennington.
E. 8. Limestone rocks near Dinkling Green.

Lecanora simplex, Nyl.—Area, 5. Range, ?

On the sandstone copings of walls. Rare. Fruiting.

W. 5. On a wall near Lytham, a form with gyrophoroid apothecia which may be *forma complicata*, Cromb.

The absence of thallus and the congregated apothecia in which the disk is little visible owing to their involute margins, cause this lichen to closely resemble an Opegrapha. Some of its forms have been described as varieties of *O. Persoonii*, but the internal structure is totally different.

Pertusaria globulifera, Nyl.—Area, 6. Range, only seen at about 100 feet.

Trunks of trees. Rare.

E. 6. On a tree by the Crook of Lune.

Pertusaria amara, Nyl.—Area, 1, 2, 3, 4, 6, 7, 8. Range, 0–800 feet or higher.

On trunks of trees. Common in North and East ; rare, if at all, in West.

This is much the most abundant of the *Pertusariæ* in West Lancashire, and is readily recognised by its very bitter taste when chewed, resembling that of quinine.

Pertusaria lactea, Nyl.—Area 1, 6, 7. Range, 100–1,000 feet or higher.

On gritstone rocks in hilly districts. Rare and always sterile.

N. 1.   Near Cowan Bridge, and Hornby.
E. 6.   Hindburndale, and near Caton.
    7.   Black Clough, Marshaw Fell.

Pertusaria communis, DC.—Area, 1, 2, 3, 4, 6, 7, 8. Range, 0–700 feet.

Trunks of trees. Frequent, especially in North. Fruiting.

Pertusaria dealbata, Nyl.—Area, 1, 3, 6, 7. Range, 0–1,820 feet or higher.

Rocks of Millstone Grit, etc., on the higher fells. Locally plentiful. Apothecia not seen.

N. 1.   Ease Gill, and ascending to 1,520 feet on Greygarth Fell.
    3.   Grit rocks near Heysham.
E. 6.   Lythe Fell and Middle Gill, Hindburn ; Salter Fell, and Clougha Pike.
    7.   Great Clough of Tarnbrook Fell, with a form approaching *forma corallina*, Cromb. Wardstone at 1,820 feet. Black Clough.

Pertusaria pustulata, Nyl.—Area, 1. Range, only seen at about 140 feet.

Trunks of trees. Apparently very rare. Fruiting.

N. 1.   Trees by the Lune near Over Burrow.

Not unlike *P. communis*, but with smaller spores, different reactions, and more numerous ostioles in the verrucæ.

Pertusaria Wulfenii, DC.—Area, 6, 8. Range, ?

Tree trunks in upland districts. Rare. Fruiting.

E. 6.   Middle Gill and below Wolfhole Crag.

Var. *glabrescens*, Nyl.

E. 8.   Tree trunks in the wood below Whitewell.

Pertusaria leioplaca, Schaer.—Area, 1, 2, 7, 8. Range, 0–700 feet.

On the smooth bark of trees. Frequent in North. Fruiting.

N. 1.   Trees near Wennington and Nether Burrow.
    2.   Near Nether Kellet.
E. 7.   Gully on Lee Fell.
    8.   Gorge of the Hodder below Whitewell.

Forma hexaspora, Nyl.

N. 2.   Kellet Seeds.

The Nether Kellet specimens are peculiar in having one to three ostiola, and varying with one, three or four spores in the asci of a single apothecium. The thallus gives a distinct yellow reaction with KHO.

Phlyctis agelæa, Koerb.—Area, 1. Range, not ascertained.

On the roots and lower parts of the trunks of trees. Rare.

N. 1.   Trees near Wennington.

Phlyctis argena, Koerb.—Area, 2, 3, 8. Range, 0–400 feet or higher.

Tree roots and trunks. Rare. Fruiting.

N. 2.   Near Whittington.
    3.   Trees near Silverdale.
E. 8.   Whitewell.

Thelotrema lepadinum, Ach.—Area, 8. Range, ?

Trunks of trees. Rare. Fruiting.

E. 8.   Trees in the wood below Whitewell.

Urceolaria scruposa, Ach.—Area, 1, 2, 3, 6, 7, 8. Range, 200–1,000 feet.

Rocks and walls. Rather uncommon. Fruiting.

N. 1.   Between Leck and Burrow, and in Ease Gill at 1,000 feet.
    2.   Dalton Crag.
    3.   Warton Crag.
E. 6.   Clougha.
    7.   Black Clough.
    8.   Whitestone Clough.

Urceolaria bryophila, Nyl.—Area, 8. Range, ?

On mosses and *Cladoniæ*. Rare or overlooked? Fruiting.

E. 8.   On *Cladonia pyxidata* and moss in Whitestone Clough.

Lecidea lurida (Swartz.).—Area, 1, 2, 3, 8. Range, 100–1,100 feet.

On earth among calcareous rocks. Frequent in limestone districts. Fruiting.

N. 1.   Leck Fell and Ease Gill, ascending to 1,100 feet.
    2.   Dalton Crag.

N. 3.   Warton Crag.
E. 8.   Encrinitic limestone near Dinkling Green.

Looks at first glance like one of the *Phyllodei* owing to its well-developed thalline squamules.

Lecidea Friesii, Ach. (*Psora caradocensis*, Mudd non Leight.). —Area, 6. Range, only seen at about 800 feet.

On decorticated tree trunks. Rare.

E. 6.   On an old oak tree near Conder Head, May, 1906.

Lecidea ostreata (Hoffm.).—Area, 2.

Tree trunks. Rare. Sterile.

N. 2.   Trees near Aughton.

Lecidea dispansa, Nyl.—Area, 6. Range, only seen at about 900 feet.

On water-worn sandstone blocks in streams. Fruiting.

E. 6.   In the Foxdale Beck.

Lecidea crustulata (Ach.).—Area, 1, 2, 6, 7, 8. Range, 500–2,050 feet.

Sandstone rocks and stones. Probably frequent in East. Fruiting.

N. 1.   Summit of Greygarth Fell.
    2.   On an erratic Silurian boulder on Dalton Crag.
E. 6.   Dale Gill, Hindburn, and summit plateau of Wardstone. Clougha Quarry.
    7.   Deer Clough. Near Abbeystead.
    8.   Fairsnape Fell, on small pebbles.

Very partial to small loose stones, and is no doubt frequently overlooked owing to its resemblance to a young state of *L. contigua*, which abounds in similar localities.

Lecidea lucida, Ach.—Area, 1, 2, 3, 6, 7, 8. Range, 100–600 feet or higher.

Shaded sandstone rocks and walls. Local. Fruit not seen.

N. 1.   Near Cowan Bridge. Near Melling.
    2.   Near Docker. Whittington Moor.
    3.   On Silurian boulders in a wall near Silverdale.
E. 6.   Near Mill Houses, and several places in Lower Hindburndale.
    7.   Near Abbeystead. Bleasdale.
    8.   Sparingly near Whitewell.

A sterile thallus occurs on rotten trunks on Warton Crag which is probably a state of this species, *forma theoietea* (Ach.).

Lecidea flexuosa (Fries.).—Area, 6. Range, ?

On rails and decaying stumps of trees. Rare? Fruiting.

E. 6.   Lower Hindburn at 550 feet.

Lecidea decolorans, Floerke.—Area, 1, 4, 6, 7, 8. Range, 25–2,060 feet.

On bare, peaty ground, especially in the hilly districts. Common on some parts of the fells. Fruiting freely.

An exceedingly variable lichen, the colour of the apothecia especially being very inconstant. When they are blackish or greenish it is very inconspicuous, but when, as is frequently the case they are of a bright flesh colour, it at once arrests attention, especially in wet weather. After the heather has been burnt this lichen is often the first vegetation to appear, covering large surfaces with its grey thallus. The *forma aporetica*, Koerb., occurs on Longridge Fell and probably elsewhere.

Lecidea querna, Ach.—Area, 2. Range, only seen at about 100 feet.

On tree trunks. Rare, or overlooked when sterile.

N. 2.   On oaks near Capernwray, sparingly, in fruit.

Lecidea sanguinaria (L.).—Area, 1, 6, 7. Range, 500–2,050 feet.

Sandstone rocks, and less commonly on tree trunks, in hilly districts. Locally abundant and fruiting well.

N. 1.   Greygarth Fell, ascending to the summit wall.
E. 6.   Rocks and trees on Littledale Fell, and in Foxdale. Conder Head. Walls near the source of Rowton Brook. Dale Gill, Hindburndale.
    7.   White side of Tarnbrook Fell, and Long Crag, etc., in Over Wyresdale.

Lecidea enteroleuca, Ach.—Area, 1. Range, ?

Rocks and walls. Rare. Fruiting.

N. 1.   Rocks in Ease Gill. Wall near Cowan Bridge.

Lecidea parasema (Ach.).—Area, 1, 2, 3, 6, 7. Range, 0–600 feet.

Tree trunks and old palings, more rarely on rocks and walls. Frequent in North, rare elsewhere. Fruiting.

N. 1.   Near Wennington.
    2.   Wash Dub Wood. Trees near Dalton Crag and Whittington. Over Kellet. Trees by the Lune between Caton and Halton.
    3.   Between Silverdale and Yealand.
E. 6.   On walls in Lower Hindburn.
    7.   Spreight Clough.

Var. *tabescens,* Leight.—On the bark of young trees.

N. 1. On sycamores by the Lune near Over Burrow.
   3. Near Hawes Water, and Thrang End. Near Yealand.

Var. *elæochroma,* Ach.

N. 1. Trees in the Gorge of the Greeta.
   2. Dalton Crag.

Lecidea uliginosa (Schrad.).—Area, 1, 6, 7. Range, 700–1,700 feet or higher.

Moist, peaty ground on heaths. Rare or overlooked.

N. 1. Leck Fell.
E. 6. By the Little Moor Beck, Hindburn. Foxdale.
   7. Wardstone Breast.

Lecidea neglecta, Nyl.—Area, 2.

Not seen properly developed.

N. 2. A lichen collected at Borwick, and seen elsewhere in the district creeping over mosses, is referred by Mr. Martindale to *Lepraria lobificans,* Nyl., which is generally supposed to be an imperfect state of *L. neglecta,* Nyl.

Lecidea mollis (Wahl.).—Area, 8. Range, only seen at about 1,200 feet.

On exposed gritstone rocks. Fruiting.

E. 8. Wolf Fell.

The epithecium in section is rich reddish-brown. Iodine gives blue then vinous-violet.

Lecidea coarctata (Sm.).—Area, 1, 2, 6, 8. Range, 0–1,200 feet or higher.

On shaded sandstone rocks and walls. Apparently rare. Fruiting.

N. 1. On walls near Nether Burrow, the *forma involuta,* (Tayl.).
   2. Loose stones near Dalton Crag.
E. 6. Hindburn. Foxdale.
   8. Wolf Fell (*forma glebulosa,* Sm.)

Lecidea Kochiana, Hepp.—Area, 1. Range, only seen at 1,550 feet.

On exposed grit rocks in hilly districts. Fruiting.

N. 1. Greygarth Fell.

The thallus resembles that of some forms of *L. rivulosa,* but the apothecia are sunk so as to be even with the surface, without margin, and with a smooth disk.

Lecidea segregans, Nyl.—Area, 1. Range, only seen at 350 feet.

On Silurian boulders and stones. Very rare. Fruiting.

N. 1. On a wall built of water-worn stones near Leck.

Somewhat doubtfully referred to this species by Mr. E. M. Holmes. Leighton does not give its behaviour with reagents. In our specimens the reactions are K-(K Cl) blood-red, Cl alone, faint reddish coloration of the apices of the verrucæ (? where abraded). Fruit sparingly produced in distant groups of small aggregato-confluent apothecia.

[*Lecidea Mullensis,* Stirt.—A lichen occurs on the summit of Wardstone, and on high, exposed rocks elsewhere, which Mr. Holmes thought to be very like this species. It is very darkly cinereous, and the areolate-diffract thallus has a columnar appearance. Unfortunately, we have been unable to find good fruiting specimens, so that for the present it must remain doubtful.]

Lecidea lapicida, Fr.—Area, 1, 2. Range, 200–1,550 feet or higher.

Exposed rocks and stones amongst the hills. Fruiting.

N. 1. Greygarth Fell at 1,550 feet.
   2. Near Gressingham.

Lecidea lithophila, Ach.—Area, 6. Range, ?

Rocks and stones amongst the hills. Fruiting.

E. 6. Clougha, April, 1904.

Lecidea plana, Lahm.—Area, 1, 6, 8. Range, ?–2,050 feet.

Rocks and walls amongst the hills. Rare. Fruiting.

N. 1. Ease Gill, and on the summit of Greygarth Fell.
E. 6. On gritstone rocks, Clougha.
   8. Wolf Fell.

Lecidea rivulosa, Ach.—Area, 1, 6, 7, 8. Range, 200–2,050 feet.

Siliceous rocks and walls amongst the hills. Frequent. Fruiting.

N. 1. On Coniston Grit rocks, Ease Gill, and on Yoredale Grit rocks on Greygarth Fell up to 2,050 feet.
E. 6. Salter Fell. Mallowdale Fell. Thrushgill Fell.
   7. Wardstone at 1,800 feet. Stones by the Wyre below Abbeystead.
   8. Wolf Fell, and near Whitewell and Dinkling Green.

F I

Lecidea fusco-atra, Ach.—Area, 1, 6. Range, ?

Rocks and stones. Rare. Fruiting.

N. 1. Lower Ease Gill, the *forma fumosa,* Ach.
E. 6. Hindburndale.

Lecidea lactea (Floerke).—Area, 1. Range, 500–900 feet.

On Silurian rocks in hilly districts. Very rare. Fruiting.

N. 1. Sparingly on Silurian rocks in Lower Ease Gill.

Lecidea sub-Kochiana, Nyl.—Area, 1. Range, between 1,000 and 2,050 feet.

On siliceous rocks in mountainous districts. Fruiting.

N. 1. Greygarth Fell, ascending to the summit ridge.

Lecidea contigua, Fr.—Area, 1, 2, 3, 4, 6, 7, 8. Range, 0–2,050 feet.

Rocks and stones of Millstone Grit ; more rarely on the Silurian or limestone. Common on the hills of North and East ; rarer elsewhere. Fruiting.

This species varies greatly in the characters of the thallus and size of the apothecia. The following forms have been noticed.

Forma limitata, Leight.—Near Over Kellet in North and frequent in the cloughs of East.

Forma steriza, Ach.

N. 1. Near Ease Gill Kirk.
E. 7. Grizedale, near Garstang.

Forma platycarpa, Fr.

E. 6. Water-washed rocks, Clougha.
   7. Gavells Clough and Black Clough.

Forma leprosa, Leight.

N. 1. Near Gressingham.

Forma calcarea, Fr.

N. 3. Silverdale.

Forma hydrophila, Fr.

E. 7. Submerged in stream, Great Clough of Tarnbrook Fell, and below Abbeystead.

Forma nobilis, Fr.

E. 6. Foxdale.

Lecidea confluens (Web.).—Area, 1, 3, 7. Range, 0–2,050 feet.

Walls, principally in mountainous districts. Fruiting.

N. 1. Wall on the summit of Greygarth Fell. Lower Ease Gill.
   3. Heysham.
E. 7. Great Clough of Tarnbrook Fell.

Lecidea tenebrans, Nyl.—Area, 1. Range, 1,200–2,050 feet.

Rocks and walls in mountainous districts. Local. Fruiting.

N. 1. Upper Ease Gill and Greygarth Fell.

Lecidea calcivora (Ehrh.).—Area, 1, 2, 3, 7, 8. Range, 0–1,200 feet or higher.

Limestone rocks and walls. Common in parts of North. Fruiting.

N. 1. Ease Gill.
   2. Dalton Crag and Over Kellet.
   3. Near Silverdale Station ; Heald Brow ; Gatebarrow Wood, Cringlebarrow, and other localities near Silverdale. Warton Crag.
E. 7. On a limestone boulder on Stake House Fell.
   8. Near Dinkling Green.

This, like several species of *Lecanora* and *Verrucaria,* eats deeply into the limestone, giving it a characteristic pitted appearance.

Lecidea Metzleri (Koerb.).—Area, 3. Range, 0–300 feet or higher.

Limestone rocks near the sea shore. Rare. Fruiting.

N. 3. Silverdale. Cringlebarrow.

The apothecia are occasionally somewhat imbedded, but never deeply as in the preceding species, and the thallus is frequently of a darker grey, although sometimes it is sub-evanescent.

Lecidea canescens (Dicks.).—Area 1, 2, 3, 4. Range, 0–400 feet.

On trees and walls, especially near villages. Frequent in North. Fruit not seen.

N. 1. Near Cowan Bridge.
   2. Borwick and Nether Kellet.
   3. Thrang End ; Silverdale ; and near Heysham and Heaton.
W. 4. Near Nateby, and elsewhere, but small and ill-developed.

Lecidea stellulata, Tayl.—Area, 1. Range, 600–2,050 feet.

Silurian rocks in mountainous districts. Rare. Fruiting.

N. 1. With *Lecidea geographica*, High Park, Lower Ease Gill. Summit of Greygarth Fell.

Lecidea disciformis, Fr. Area, 1.

On smooth bark of trees. Apparently rare. Fruiting.

N. 1. Near Wennington.
E. 6. Wood near Mill Houses.

Lecidea myriocarpa (DC.).—Area, 1, 2, 3, 4, 5, 6, 7, 8. Range, 0–500 feet or higher.

Trunks of trees and on rocks. The corticole state is common, especially in North; saxicole forms are rare. Fruiting.

Forma quercina, Rabh.—Trees near Nether Burrow.

Forma leprosa, Leight.

N. 1. On sandstone rocks in the Gorge of the Greeta.

Lecidea hyperiza, Stirton.—Area, 1. Range, ?

On sycamore trees. Rare. Fruiting.

N. 1. Trees in a wood by the Lune below Kirkby Lonsdale.

We have seen no specimen of Stirton's plant, but our specimens agree well with the description in Leighton's *Lichen Flora*. The reaction of the hymenium with iodine is very intense. The thallus is here and there limited by a thin serpentine black hypothallus.

Lecidea chalybeia, Borr.—Area, 2. Range, only seen at about 700 feet.

On sandstone rocks and stones.

N. 2. On loose erratic stones on Dalton Crag.

Lecidea lutosa (Mont.).—Area, 8. Range, only seen at about 1,200 feet.

Rocks in hilly districts. Rare.

E. 8. Wolf Fell.

Lecidea Templetoni, Tayl.—Area, 1. Range, only seen at 950 feet.

Creeping over mosses in hilly districts.

N. 1. On *Neckera crispa* on Leck Fell.

Lecidea cœruleo-nigricans (Lightf.).—Area, 1, 2, 3. Range, 0–1,000 feet or higher.

On rocks, walls, and earthy banks in the limestone districts. Locally abundant and fruiting freely.

N. 1. Near Leck, ascending on Leck Fell Scars to 1,000 feet.
2. Dalton Crag, and between Carnforth and Borwick.
3. Near Silverdale Station and the Cove; Eaves Wood; Gatebarrow Wood. Near Yealand. Warton Crag.

Lecidea lenticularis, Ach.—Area, 1, 3. Range, ?

Rocks and walls. Not common. Fruiting.

N. 1. Sandstone quarry near Wennington.
3. Silverdale.

Lecidea tricolor (With.).—Area, 7. Range, ?

Trunks of trees. Probably more frequent than the only record indicates. Fruiting.

E. 7. Claughton, near Garstang.

Lecidea albo-atra (Hoffm.).—Area, 1, 2, 6, 7. Range, ?

Trunks of trees. Rare or overlooked. Fruiting.

E. 6. Trees by the river, Crook of Lune.
7. On oaks, Barnacre, near Garstang.

Forma ambigua (Ach.).

N. 2. On stones near Gressingham.

Forma margaritacea (Ach.).

N. 1. Gorge of the Greeta, on rocks.

Lecidea Œderi (Web.).—Area, 1. Range, 500–900 feet.

On rocks in mountainous districts. Rare. Fruiting.

N. 1. On Silurian rocks in Ease Gill, May, 1904.

Lecidea aromatica (Sm.).—Area, 1, 2, 3, 4, 8. Range, 0–800 feet or higher.

On crumbling rocks, walls, and stony banks. Fruiting.

N. 1. Near Melling.
2. Dalton Crag.
3. Silverdale, and Warton Crag.
W. 4. Saltcoat.
E. 8. Limestone rocks near Dinkling Green.

Lecidea exanthematica (Sm.).—Area, 1, 2, 3. Range, 0–1,000 feet or higher.

Calcareous rocks. Frequent in parts of North. Fruiting.

N. 1. Ease Gill.
2. Dalton Crag.
3. Warton Crag. Near Hawes Water, The Cove, and other localities about Silverdale.

Lecidea umbrina, Ach.—Area, 1, 3. Range, 0–1,000 feet or higher.

On stones and rocks. Rare.

N. 1. Lower Ease Gill. Higher up, near the Dry Foss, the var. *pelidniza,* Nyl., occurs.
3. Stones (Silurian) in a wall near Silverdale.

Lecidea milliaria, Fr.—Area, 3. Range, only seen at about 300 feet.

On bare earth. Rare. Fruiting.

N. 3. Mud-capped wall on Warton Crag.

Lecidea sabuletorum, Floerk.—Area, 1, 2, 3. Range, ?

On decaying mosses on rocks and banks. Rare. Fruiting.

N. 1. Near Wennington.
2. Near Carnforth, and on Dalton Crag.
3. Near Yealand.

Lecidea endoleuca, Nyl.—Area, 1, 4. Range, 0–200 feet or higher.

On the bark of trees. Apparently not common. Fruiting.

N. 1. Trees by the Lune below Kirkby Lonsdale. Near Thurland Castle.
W. 4. Trees near Pilling.

Lecidea muscorum (Sw.).—Area, 1, 3. Range, ?

On earthy banks amongst mosses. Rare or overlooked. Fruiting.

N. 1. Lower Ease Gill.
3. On *Frullania Tamarisci* and *Trichostomum,* Castlebarrow, Silverdale.

Lecidea effusa (Sm.).—Area, 1, 4, 8.

On shrubs and trees. Probably frequent, but rarely fruiting.

W. 4. With apothecia on stumps near the shore south of Glasson Dock.

Var. *inundata* (Fr.).—On stones in streams. Rare.

N. 1. Stones at the Dry Foss, Ease Gill.

Var. *fuscella* (Fr.).—On trees.

E. 8. On elms on the north side of Longridge Fell.

Lecidea geographica (L.).—Map Lichen. Area, 1, 6, 7, 8. Range, 200–2,050 feet.

Finest on Silurian rocks, on which it forms a striking object. Usually much more poorly developed on sandstone. Common in the upland districts of North and East.

Lecidea viridi-atra, Ach.—Area, 1, 7, 8. Range, 300–2,050 feet.

Exposed siliceous rocks in mountainous districts. Frequent in East.

N. 1. Greygarth Fell summit.
E. 6. Clougha.
7. Wardstone. White side of Tarnbrook Fell.
8. Wolf Fell, at 1,150 feet.

Lecidea petræa (Wulf.).—Area, 7. Range, ?

On sandstone rocks and stones. Fruiting.

E. 7. Bed of the Wyre below Abbeystead, with *f. fuscescens,* Leight.

[**Lecidea geminata** (Flot.).—Lancashire. *Mr. R. Jacob, teste Mudd.* Exact locality not stated.]

Lecidea obscurata (Ach.).—Area, 1, 7. Range, ?

On rocks and stones. Fruiting.

N. 1. Ease Gill (*f. ferrata,* Nyl.).
E. 7. Grit rocks below Long Crag, Wyresdale, in insulated patches amongst *Lecidea contigua.*

Lecidea lavata, Fr.—Area, 8. Range, only seen at about 1,000 feet.

On water-washed gritstone rocks. Apparently rare. Fruiting.

E. 8. Stones in the Burnslack Beck.

A distinct-looking species, with large apothecia, resembling externally some forms of *L. contigua* more than the plants under which it has been occasionally placed as a variety, such as *L. atro-alba* and *L. petræum.* The elevato-adnate apothecia, thick margin, and murali-locular spores will always distinguish it. The thalline reactions—not given by Leighton—are K— C—.

Lecidea concentrica (Dav.).—Area, 1, 2, 6, 7, 8. Range, 100–2,000 feet.

Rocks and walls, both limestone and sandstone. Frequent in parts of North and East. Fruiting.

N. 1. Near Wennington and Cantsfield. Cowan Bridge and Upper Ease Gill and ascending to near the summit of Greygarth Fell.
2. Near Arkholme and Borwick.
E. 7. Gavells Clough.
8. Gorge of the Hodder below Whitewell.

Var. *impressula*, Leight.

E. 6. On walls near Quernmore.

Var. *coarctata*, Leight. Rocks by moorland streams. Fruiting.

N. 1. Stones in the bed of the Cant Beck, Leck Fell, at 800 feet.
E. 6. Foxdale Beck. Little Moor Beck and Lower Hindburn.

The last variety has the appearance of a distinct species. It spreads extensively, forming large irregular patches very different from the small orbicular thallus of the type.

Lecidea rimosa (Dicks.).—Area, 1. Range, 1,900–2,070 feet.

Mountain rocks. Very local. Fruiting.

N. 1. Limestone rocks and wall on the summit of Greygarth Fell.

Lecidea cupularis (Ehrh.).—Area, 1, 3, 6, 8. Range, 0–1,200 feet or higher.

On rocks, usually limestone ; more rarely on sandstone. Fruiting.

N. 1. Near the Dry Foss, etc., in Ease Gill.
3. Rocks near Hawes Water and on Warton Crag.
E. 6. On sandstone rocks in Dale Gill, Hindburndale, at 1,200 feet.
8. Limestone rocks near Dinkling Green.

Lecidea truncigena, Ach.—Area, 3. Range, only seen at 100 feet.

Mossy tree trunks. Rare or overlooked. Fruiting.

N. 3. With *Orthotrichum stramineum* and *Verrucaria gemmata* on a tree near Silverdale in very small quantity.

Lecidea parasitica (Floerk.).—Area, 1. Range, ?

On the thallus of *Lecanora parella*. Not common.

N. 1. On *L. parella*, Ease Gill.

Opegrapha herpetica, Ach.—Area, 3, 7, 8. Range, 0–500 feet or higher.

On trees. Frequent in North.

N. 3. Tree branches below Grisdale Wood.
E. 7. Near Abbeystead reservoir.
8. On elm on the north side of Longridge Fell.

Opegrapha atra, Pers.—Area, 2, 5, 7. Range, 200–700 feet.

On the branches of trees and bushes. Frequent, but not abundant.

N. 2. Whittington (*f. nigrita*, Leight.). Dalton Crag on hazels (*f. parallela*, Leight.).
E. 6. Near Wray.
7. Near Dolphinholme. On ash in the Gorge of the Wyre.

Opegrapha Turneri, Leight.—Area, 2. Range, ?

On hazel and elder branches in hedgerows. Rare.

N. 2. Hedgerows near Over Kellet.

Opegrapha saxicola, Ach.—Area, 3. Range, 0–50 feet.

Maritime rocks. Rare.

N. 3. Gritstone rocks near the sea shore, Heysham.

Var. *Chevallieri*, Leight.

N. 3. Calcareous rocks on the shore, Silverdale Cove.

Opegrapha varia, Pers.—Area, 1, 3, 6, 7. Range, 0–700 feet or higher.

On trunks and twigs of trees. Probably frequent.

Forma pulicaris, Lightf.

N. 1. On oaks near Nether Burrow.
3. Silverdale.
E. 6. Foxdale. On birch, Lower Hindburndale, at 500 feet.

Forma notha, Ach.

E. 7. On oak bark, Barnacre, near Garstang.

Opegrapha vulgata, Ach.—Area, 1, 2, 3, 7, 8. Range, 0–400 feet or higher.

Trunks and branches of trees. Probably frequent.

N. 1. Near Kirkby Lonsdale.
2. Near Borwick.
3. Near Hawes Water, Silverdale.
E. 7. Wood near Abbeystead reservoir.
8. Gorge of the Hodder below Whitewell.

Arthonia lapidicola, Tayl.—Area, 7. Range, only seen at about 1,800 feet.

Exposed rocks in mountainous districts. Fruiting.

E. 7. Near the summit of Wardstone. First detected by Mr. E. M. Holmes amongst specimens of *Lecidea* submitted to him by the Authors.

Arthonia punctiformis, Ach.—Area, 2. Range, ?

On the smooth bark of trees, especially on young branches. Rare. Fruiting.

N. 2. Wash Dub Wood, on young hazels.

Arthonia epipasta, Ach.—Area, 2. Range, ?

Branches of trees. Very rare. Fruiting.

N. 2. Wash Dub Wood, sparingly.

Arthonia Swartziana, Ach.—Area, 1, 2, 3. Range, 0–200 feet.

Trunks of trees. Frequent in North ; rare elsewhere. Fruiting.

N. 1. Near Nether Burrow.
2. Wash Dub Wood, and near Carnforth.
3. Grisdale near Warton ; and near Yealand.

Arthonia astroidea, Ach.—Area, 2, 3. Range, 0–200 feet.

On the smooth bark of trees. Fruiting.

N. 2. Wash Dub Wood.
3. Grisdale Plantation, near Warton.

Arthonia pruinosa, Ach.—Area, 7. Range, ?

Trunks of trees.

E. 7. Near Abbeystead.

Stigmatidium crassum, Dub.—Area, 2. Range, ?

On the trunks of trees. Apparently very rare. Fruiting.

N. 2. On oak near Borwick. Near Bolton-le-Sands.

Graphis elegans (Sm.).—Area, 1, 2, 3, 6, 7, 8. Range, 0–1,100 feet.

On the trunks of holly, oak, hazel, mountain ash, and birch. Frequent amongst the hills. Fruiting.

N. 1. Near Wennington.
2. Near Dalton.
3. Warton Crag. Cringlebarrow.
E. 6. Hindburn ; Roeburndale ; Foxdale ; Clougha, etc.
7. Abbeystead, Lea, Tarnbrook and Marshaw. At 1,100 feet on mountain ash in Black Clough.
8. Near Whitewell.

Forma stellata (Leight.).

E. 6. On birch, Tatham Moor, Hindburn.

Graphis scripta, Ach.—Area, 1, 2, 3, 6, 7, 8. Range, 0–1,000 feet.

Tree trunks and branches. Frequent and variable. Fruiting.

N. 1. Gorge of the Greeta (*f. tremulans* and *f. recta*, Humb.). Near Whittington (*f. recta*, Humb.), on hazels at 1,000 feet in Middle Ease Gill.
2. Dalton Crag, on hazels. Kel et Seeds.
3. Near Silverdale ; and also *f. recta*, Humb. Cringlebarrow.
E. 6. Near Lower Salter, Roeburndale. Tatham Moor (*f. minuta*, Leight.).
7. Wyresdale, frequent.
8. Wood below Whitewell (*f. flexuosa*, Leight.).

Var. *serpentina*, Ach.

N. 3. Near Silverdale.

Graphis inusta, Ach.—Area, 2, 6. Range, ?

Tree trunks in woods and shady places. Rare. Fruiting.

N. 2. High Field near Kellet Seeds.
E. 6. Trees in Lower Roeburndale.

Graphis sophistica, Nyl.—Area, 1, 3, 8. Range, 0–1,000 feet.

Trunks and branches of trees. Infrequent. Fruiting.

N. 1. Ease Gill.
3. Silverdale. Warton Crag. Grisdale Plantation near Warton (*f. divaricata*, Leight.).
E. 8. Gorge of the Hodder below Whitewell.

Var. *pulverulenta* (Sm.).

E. 8. Wood below Whitewell.

PYRENODEI.

Normandina pulchella, Bor.—Area, 1. Range, ?

Mossy tree trunks. Very rare. Sterile.

N. 1. Middle Ease Gill, May, 1905. On elm near Nether Burrow.

This might easily be passed over as a small *Cladonia* when not well developed. The upturned and inflexed margin gives the squamules a peculiarly elegant appearance.

Endocarpon miniatum (L.)—Area, 1, 2, 3, 8. Range, 0–1,000 feet or higher.

Dry limestone rocks   Frequent in parts of North ; rare in East. Apothecia frequent.

N. 1. Ease Gill, especially fine near the Kirk, and on Leck Fell.
    2. Near Borwick and near Bolton-le-Sands.
    3. Several localities near Silverdale. Warton Crag.
E. 8. Gorge of the Hodder below Whitewell.

Var. *complicatum* (Sw.).—In damper situations.

N. 1. Leck Fell, and Ease Gill.
    2. Near Over Kellet.
    3. Near Hawes Water Tarn.

Endocarpon rufescens, Ach.—Area, 1, 2, 3. Range, 0–1,200 feet.

On earth amongst calcareous rocks. Not common. Fruiting.

N. 1. Leck Fell.
    2. Dalton Crag. Over Kellet.
    3. Rocks near Hawes Water. Cringlebarrow. Heald Brow. Warton Crag.
Very different in habit and appearance from the preceding, more adnate, and sometimes of a bright chestnut-red colour. It always occurs in small cushions and never in very widely extended patches.

[*Endocarpon fluviatile* (DC.).

Occurs on the Westmorland side of the Leck Beck in Ease Gill. We have not yet succeeded in finding it on our side, but further search may prove it to be a West Lancashire species.]

Verrucaria halophila, Nyl.—Area, 3. Range, 0–25 feet.

Maritime sandstone rocks. Rare. Fruiting.

N. 3. Rocks on the seashore, Heysham, in fruit.

[*Verrucaria epigæa* (Pers.). — Recorded from Lancashire by *Mr. R. Jacob* in *Mudd's Manual*. Locality not stated.]

Verrucaria Dufourei, DC.—Area, 1, 3. Range, 0–1,000 feet or higher.

Calcareous rocks and walls. Rare ? Fruiting.

N. 1. Ease Gill.
    3. Silverdale ; Warton.

Verrucaria lævata, Ach.—Area, 8. Range, ?

Stones in the beds of rivers. Not common. Fruiting.

E. 8. In the Hodder below Whitewell.

Verrucaria margacea, Whlnb. — Area, 2, 4, 6, 7. Range, 25–1,000 feet or higher.

Rocks and stones in the beds of streams. Fruiting.

Var. *æthiobola*, Whlnb.

N. 2. On submerged stones in Wash Dub Brook. Stream near Borwick.
W. 4. In the Conder below Galgate.
E. 6. Lythe Fell and Dale Gill, Hindburndale.
    7. In the stream flowing from Windy Clough, Clougha Pike.

Verrucaria mutabilis, Borr.—Area, 1, 2, 3. Range, ?

On scattered stones and pebbles. Fruiting.

N. 1. Near Gressingham.
    2. Small stones near Dalton Crag.
    3. Cringlebarrow.

Verrucaria maura, Whlnb.—Area, 3. Range, 0–25 feet.

Rocks on the sea coast. Rare.

N. 3. Rocks south of Heysham.

Verrucaria mauroides, Schaer.—Area, 1, 2. Range, ?

Rocks and walls amongst the hills.

N. 1. Lower Ease Gill.
    2. Dalton Crag.

Verrucaria viridula (Schrad.).—Area, 1, 3, 6. Range, only seen at about 300 feet.

Rocks and stones.

N. 1. Near Nether Burrow.
    3. On stones on Cringlebarrow.
E. 6. Foxdale.

Verrucaria nigrescens (Pers.). — Area, 1, 2, 3, 7. Range, 0–2,050 feet.

Calcareous rocks. Frequent in North. Fruiting.

N. 1. Greygarth Fell. Near Leck village.
    2. Dalton Crag.
    3. Eaves Wood, near Silverdale. Cringlebarrow and Warton Crag.
E. 7. On an erratic limestone boulder at 850 feet, Stake House Fell.

Verrucaria glaucina, Ach. — Area, 1, 2, 3. Range, 0–800 feet.

Calcareous rocks, especially near the sea. Fruiting sparingly.

N. 1. Wall near Tunstall.
    2. Dalton Crag.
    3. Rocks on the shore, Silverdale. Cringlebarrow.

Verrucaria rupestris, Schrad.—Area, 1. Range, ?–2,050 feet.

Calcareous rocks and walls. Fruiting.

N. 1. Ease Gill and near the summit of Greygarth Fell. Wall near Leck. Near Burrow a form with tartareo-farinose thallus, which is perhaps the var. *muralis*, Ach.
E. 8. Mortared wall near Dinkling Green (var. *muralis*, Ach.).

Verrucaria integra, Nyl.—Area, 6. Range, only seen at 200 feet.

E. 6. On a wall near Wray.
This species closely resembles the preceding, but has much larger spores.

Verrucaria calciseda, DC.—Area, 1, 2, 3, 7, 8. Range, 0–1,000 feet or higher.

Limestone rocks and walls. Frequent in North ; rare in East.

N. 1. Ease Gill, ascending to 1,000 feet.
    2. Dalton Crag.
    3. Frequent in the Silverdale district. Cringlebarrow and Warton Crag.
E. 7. On an erratic limestone boulder at 850 feet on Stake House Fell.
    8. Near Whitewell and Dinkling Green.

Verrucaria conoidea, Fries.—Area, 1, 2, 3, 8. Range, 0–1,200 feet or higher.

Limestone rocks and walls. Frequent in North. Fruiting.

N. 1. Near Ease Gill Kirk.
    2. Over Kellet.
    3. Silverdale Cove, Cringlebarrow, Warton Crag.
E. 8. Gorge of the Hodder below Whitewell.

Verrucaria immersa, Leight.—Area, 1. Range, ?–2,030 feet.

Limestone rocks and walls. Fruiting.

N. 1. Ease Gill. Wall on summit of Greygarth Fell at 2,030 feet.
E. 8. Near Dinkling Green.
The spores as well as the hymenium become rich brownish yellow with iodine without any previous blue colouration. The large colourless spores readily divide at the median line.

Verrucaria saxicola, Mass.—Area, 3. Range, unknown.

Limestone rocks, associated with *Lecanora calva, L. ochracea,* and *Pannaria nigra*. Fruiting.

N. 3. Warton Crag.
Forms greyish macular patches, with numerous very minute apothecia. Readily overlooked, and, no doubt, more frequent in the limestone districts than the single record indicates.

Verrucaria gemmata, Ach.—Area, 2, 3. Range, ?

On trees. Rare. Fruiting.

N. 2. Oak tree near Whittington. Dalton Crag and near Capernwray.
    3. Near Silverdale with *Lecidea truncigena*.

Verrucaria epidermidis, Ach.—Area, 1, 2, 3, 6. Range, 0–1,000 feet.

On the smooth bark of trees. Not infrequent, but easily overlooked. Fruiting.

N. 1. On hazel, Middle Ease Gill, at 1,000 feet ; also, at the same altitude, the var. *fallax*, Nyl., on mountain ash.
    2. Between Carnforth and Over Kellet.
    3. Trees near Grisdale Wood.
E. 6. Hindburndale.

Verrucaria chlorotica, Ach.—Area, 1, 3. Range, ?

On trees and rocks. Fruiting.

N. 1 Near Cowan Bridge and Nether Burrow.
    3. Near Silverdale.

Var. *carpinea*, Schaer. (*Pyrenula fusiformis*, Hepp.).

N. 3. Trees on the North side of Warton Crag.

Verrucaria linearis, Leight.—Area, 1.

Calcareous rocks in sub-alpine districts. Fruiting.

N. 1.   Ease Gill, with *Lecidea calcivora*.

We have not been able to compare this with a specimen of Leighton's plant, but although very near to *V. chlorotica*, and agreeing with it in its reaction with iodine, it differs in colour and in the linear obtuse spores.

Verrucaria nitida (Weig.).—Area, 3, 8.   Range, 0–400 feet or higher.

Trunks of trees. Rare. Fruiting.

N. 3.   Near Silverdale.

E. 8.   Gorge of the Hodder near Whitewell.

Verrucaria umbrina (Wahlb.).—Area, 6.

Stones and rocks in mountain streams. Fruiting.

E. 6.   Stones by the Foxdale Beck.

Verrucaria subpyrenophora, Leight.—Area, 6. Range, only seen at about 1,000 feet.

Damp sub-alpine sandstone rocks. Rare. Fruiting.

E. 6.   On broken gritstone rocks near the stream in Dale Gill, Hindourndale, 1907.

Paler than *V. umbrina*, with broader spores and eight spores in each ascus. The paraphyses are distinct and very slender, and the spores so dark when old as to render the muriform divisions invisible.

Verrucaria terebrata (Mudd), Leight. (*Sphæromphale terebrata*, Mudd.).—Area, 1. Range, ?–2,050 feet.

Limestone rocks and walls in hilly districts. Rare. Fruiting.

N. 1.   Wall on the summit of Greygarth Fell.

Microthelia ventosicola, Mudd.—Area, 1. Range, 500–1,200 feet.

On the thallus of *Lecanora ventosa*. Rare. Fruiting.

N. 1.   On the above species on Silurian rocks near Lower Ease Gill, May, 1904.

INDEX I.

GEOGRAPHICAL NAMES.

INDEX II.

ORDERS, GENERA AND ENGLISH NAMES.

English names in italics ; Local names in quotation marks ; Orders in small capitals

ADDITIONS.

During the period in which this work has been passing through the press the following additional items of information have been obtained.

METEOROLOGY.

The rainfall for the first seven months of 1907 on the summit of Fairsnape Fell is given below for comparison with the corresponding period of 1906, as described on p. 65.

Jan.	Feb.	Mar.	April	May	June	July
4·75	5·44	8·32	3·27	5·40	13·82	6·60

It will be observed that notwithstanding the exceptionally heavy fall for June the total rainfall for the seven months is slightly lower than that for the same period in 1906, viz., 47·60 inches as against 48·26 inches.

THE FLORA.

The following new records require insertion, those species not included in the preceding pages being indicated by an asterisk.

Ranunculus peltatus, Schrank.—Add district 7 ; Canal near Garstang, *Wi.*

Fumaria officinalis, L.—Add district 6 ; Fields near Halton.

Fumaria purpurea, Pugs.—Add district 2 ; Near Bolton-le-Sands. Also district 6 ; Fields near Halton.

Draba verna, L.—Add district 2 ; Limestone quarry near Nether Kellet, *Wi.*

Silene noctiflora, L.—Add district 2 ; Abundant in a cultivated field near Nether Kellet. Should probably rank as a colonist.

Alsine rubra, Crantz.—Add district 5 ; Sandy bank near Rossall, *Wi.*

Sagina maritima, Don.—Add district 5 ; Near Fleetwood Docks, *Wh.*

Spiræa Filipendula, L.—Add district 2 ; Near Nether Kellet.

Rubus plicatus, W. & N.—Add district 2 ; By ditches and in a swamp near Nether Kellet.

Rubus gratus, Focke.—Add district 6 ; Near Lancaster Moor.

Alchemilla arvensis, Scop.—Add district 6 ; Fields near Halton.

Rosa canina, L., var. *urbica* (Leman).—Add district 6 ; Near Lancaster.

*Vicia tetrasperma, Moench.— Four-seeded Tare. English type. Colonist. Area, 6. Range, only seen at about 50 feet. Cultivated fields. Rare. First record : The present one.

E. 6. Turnip field near Halton.

Circæa alpina, L., var. *intermedia*, Ehrh.—Add district 2 ; By the Lune near Aughton, *Wi.*

Pimpinella major, Huds.—Add district 4 ; Roadside near Cabus, *Wi.*

Sherardia arvensis, L.—Add district 6 ; Fields near Halton.

Montia fontana, L.—Add district 2 ; Between Over Kellet and Aughton, *Wi.*

Callitriche vernalis, Koch.—Add district 3 ; Ditches near Bare.

Epilobium obscurum, Schreb.—Add district 3 ; Ditches near Bare.

Lysimachia vulgaris, L.—Add district 6 ; Swamp near Halton.

Myosotis sylvatica, Hoffm.—Add district 2 ; Woods by the Lune near Aughton, *Wi.*

Echium vulgare, L.—Add district 3 ; Near Heysham Dock, *P. F. Lee.*

*Euphrasia scotica, Wetts.—Area, 6. Range, only seen at 800 feet.

Damp sandy-peaty ground on moors. Rare. First record : The present one.

E. 6. Tatham Moor, Hindburndale, 1907, *Wi.*

The following species occurred with or near it :—*Calluna vulgaris, Oligotrichum incurvum* and *Hypnum Patientiæ.*

Acorus calamus, L.—Add district 3 ; Canal near Hest Bank.

Lemna gibba, L.—Add districts 2, 6 and 7 ; Canal about Skerton, Lancaster, and Garstang.

Lemna polyrhiza, L.—Add districts 2, 3, 6, 7 ; Canal at Skerton, Hest Bank, Lancaster and Garstang.

*Potamogeton lucens, L.—Great Pond-weed. English type. Native. Area, 3. Range, only seen at about 20 feet. P. July–August.

Deep water in lakes. Very rare. First record : The present one.

N. 3. Hawes Water near Silverdale, 1907, *Wi.*

Potamogeton densus, L.—Add district 3 ; Canal near Bolton-le-Sands, and ditches near Bare.

*Carex axillaris, Good.—Axillary-clustered Sedge. English type. Native. Area, 5. Range, only seen at about 75 feet. P. June–July.

Marshy edges of canals and streams. Rare. First record : *John Moss, Journ. Bot.,* 1905, p. 95.

W. 5. Canal near Brock ! *J. Moss,* who sent us fresh specimens through *Rev. P. J. Hornby.*

This plant is said to be a hybrid between *Carex vulpina* and *Carex remota.* If it is really a hybrid these seem to be the most probable parents in our district. Mr. Hornby having indicated the locality, we have seen the plant in situ, and both the supposed parents are found in the vicinity.

Carex acuta, L.—Add district 4 ; Ditches near Cockerham, *Wi.*

Carex hirta, L., var. *hirtæformis.*—Add district 3 ; Marsh near Bare.

Phleum arenarium, L.—First record : J. Sidebotham, in Lowe's " *British Grasses,*" vol. i., 1865. " Fleetwood."

Calamagrostis epigejos, Roth.—Add district 5 ; Bank near Rossall Point, *Wi.*

Ammophila arenaria, Link.—First record : E. J. Lowe, " *British Grasses,*" vol. i., 1865. " Lytham."

Glyceria plicata, Fries.—Add district 4 ; Ditches near Cockerham, *Wi.*

*Serrafalcus racemosus, Parl.—Smooth Brome Grass. British type. Native. Area, 2, 5. Range, 0–200 feet or higher. B. June–July.

Fields, grassy banks, and waste ground. Probably overlooked, and may be more frequent than the records indicate. First record : The present one.

N. 2. Quarry and field near Nether Kellet.

W. 5. Waste ground near Fleetwood Docks, *Wh.*

Athyrium Filix-fæmina, Roth.—First record : E. J. Lowe, " *Our Native Ferns,*" vol. 2, p. 16, 1865. " Eccleston."

Asplenium Ruta-muraria, L.—First record : E. J. Lowe, *loc. cit.,* p. 225. " Near Chaigeley."

Phyllitis Scolopendrium, Newm.—First record : E. J. Lowe, *loc. cit.,* p. 231, 1865. " Chaigeley."

Polystichum aculeatum, H. Schott.—First record : E. J. Lowe, " *Our Native Ferns,*" vol. i., p. 198, 1865. " Chaigeley Manor, Morecambe Bay and Preston."

Polystichum angulare, Presl.—First record : E. J. Lowe, *loc. cit.,* p. 75, 1865. " Somewhat abundant at Chaigeley in Lancashire."

*Var. *dubium*, Wollast.—Near Preston, J. Stansfield, *loc. cit.,* p. 86.

Lastrea Filix-mas, Presl.—First record : E. J. Lowe, *loc. cit.,* p. 246. " Morecambe and Lancaster."

Var. *Borreri*, Newm.—Chaigeley Manor, E. J. Lowe.

Lastrea aristata, Britt. & Rendle.—First record : E. J. Lowe, *loc. cit.,* p. 286. " Longridge Fell."

*Var. *dumetorum*, Moore.—Longridge Fell and Chaigeley Manor, E. J. Lowe.

---

*Var. *tanacetifolia*, Moore.—Abundant between Clitheroe and Preston, and between Preston and Lancaster, E. J. Lowe.

Chara aspera, Willd.—Add district 2 ; In the canal near Bolton-le-Sands.

Tortula intermedia, Berk.—Add district 6 ; Wall near Halton.

Physcomitrium pyriforme, Brid.—Add district 3 ; By muddy ditches near Bare.

Hypnum aduncum, Hedw., var. *intermedium*, Schimp.—Add district 6 ; Swamp near Halton.

Hypnum cordifolium, Hedw.—Add district 6 ; Swamp near Halton.

Riccia glauca, L.—Add district 6 ; Field near Halton.

Mr. Charles Bailey, F.L.S., has this year observed the following additional alien plants on the sandhills near St. Annes.

*Delphinium Ajacis, L.	Melilotus indica, All.
*Adonis æstivalis, L.	*Orlaya grandiflora, Hoffm.
*Rapistrum Linneanum, Bois.	*Bupleurum protractum, Link.
*Raphanus sativus, L.	B. rotundifolium, L.
*Myagrum perfoliatum, L.	*Caucalis daucoides, L.
*Neslia paniculata, Desv.	*Centaurea melitensis, L.
*Eruca sativa, Lamk.	Lithospermum arvense, L.
*Sisymbrium Columnæ, Jacq.	*Amsinckia lycopsoides, Lehm.
var. stenocarpum, R. & F.	*A. angustifolia, Lehm.
*Lepidium virginicum, L.	Anagallis cœrulea, Lamk.
Silene noctiflora, L.	Chenopodium murale, L.
*Saponaria Vaccaria, L.	*C. hybridum, L.
*Malva parviflora, L.	*Phalaris brachystachus, Link.
*M. nicæencis, All.	*Bromus tectorum, L.
*Trigonella corniculata, L.	*Secale cereale, L.
*T. polycerata, L.	*Lolium multiflorum, L.

THE DRY FOSS, EASE GILL, NEAR LECK,
Showing Scar Limestone Rocks, with Mountain Ash.

NAVEL POT, LECK FELL.

(Greygarth Fell in the Distance)

Showing hanging vegetation on vertical limestone sides.
Actæa spicata grows in this pot-hole on inaccessible rock-ledges.

LIMESTONE PAVEMENT, GATEBARROW WOOD.
Yew and Ash in foreground.

VIEW IN THE GREAT CLOUGH OF TARNBROOK FELL,
OVER WYRESDALE.

SUMMIT PLATEAU OF FAIRSNAPE FELL, AT 1,600ft.,

Shewing denudation of peat, with islands of Cotton-grass and Heather.

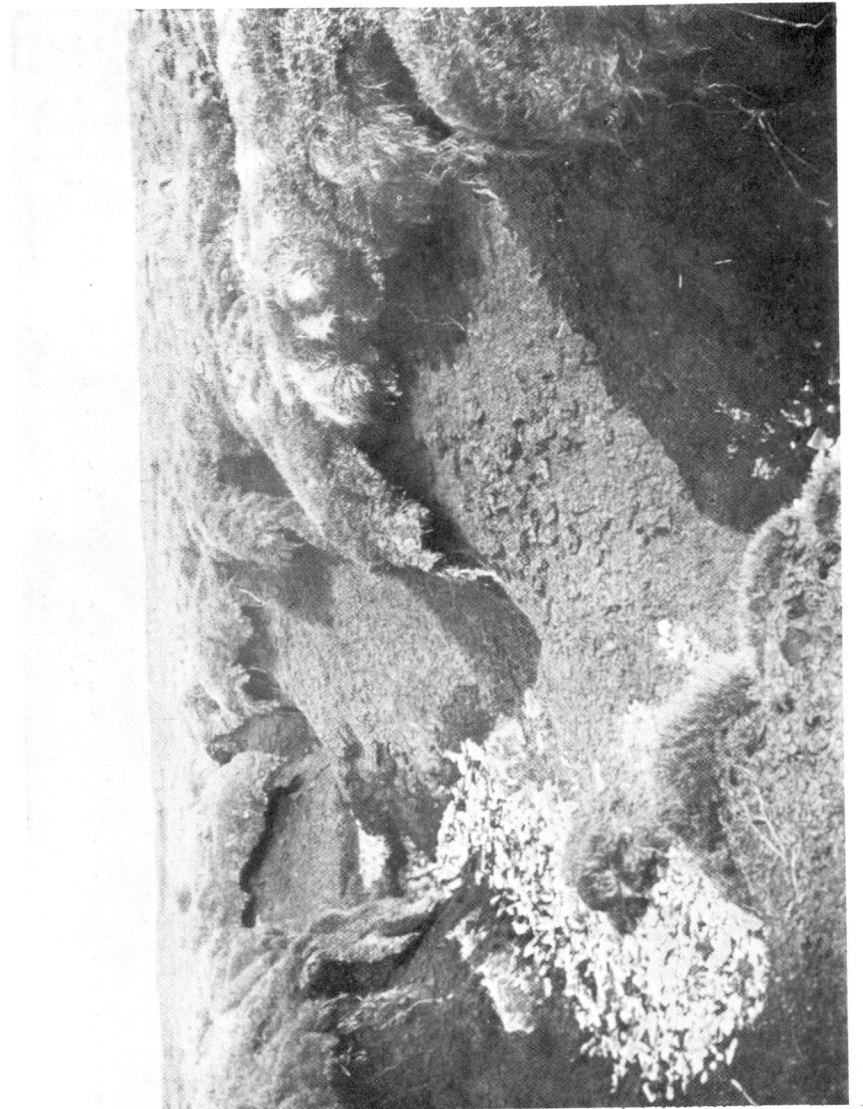

ONE OF THE NUMEROUS INTRICATE GULLIES ON FAIRSNAPE FELL.

Showing deep peat, capped with Cotton-grass, Bilberry and Heather.   Broken Millstone Grit in the dry bed of the gully.

VIEW LOOKING ACROSS A POT-HOLE CALLED THE RUMBLING HOLE
OR FAIRY'S WORKSHOP, LECK FELL.

Mountain Ash on the left ; Ling above and Dog's Mercury, etc., below on the right.

LIMESTONE PAVEMENT WITH JUNIPER THICKETS,
NORTH SIDE OF WARTON CRAG.

LIMESTONE CREVICES, DALTON CRAG,
Showing *Polygonatum officinale* and *Lastrea rigida*.

HAWES WATER, Near SILVERDALE,

From the slopes of Trowbarrow; Junipers in foreground.

A LIMESTONE ROCK-CREVICE, DALTON CRAG,
Filled with *Polypodium Robertianum* and *Lastrea rigida*.

VIEW IN LOWER EASE GILL, NEAR LECK.
A Glen in the Upper Silurian (Coniston Grit) Rocks.

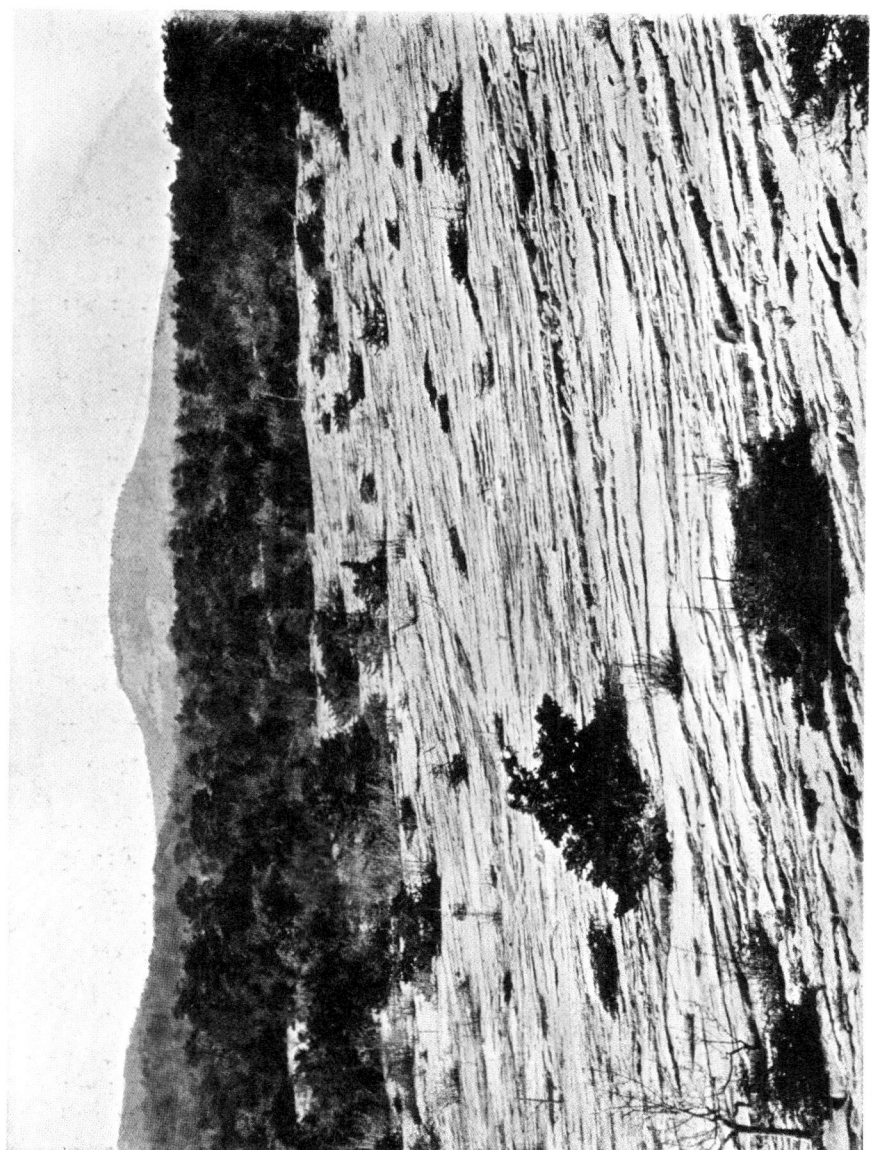

CREVICED LIMESTONE PAVEMENT OR "CLINTS."
GATEBARROW WOOD.

HELL CRAG, TARNBROOK FELL,
Looking east below the Millstone Grit Cliffs.

WEST LANCASHIRE VICE COUNTY No 60

REFERENCE

NORTH
1 LECK
2 KELLET
3 SILVERDALE AND HEYSHAM

WEST
4 PILLING
5 KIRKHAM

EAST
6 WRAY AND QUERNMORE
7 WYRESDALE AND BLEASDALE
8 LONGRIDGE

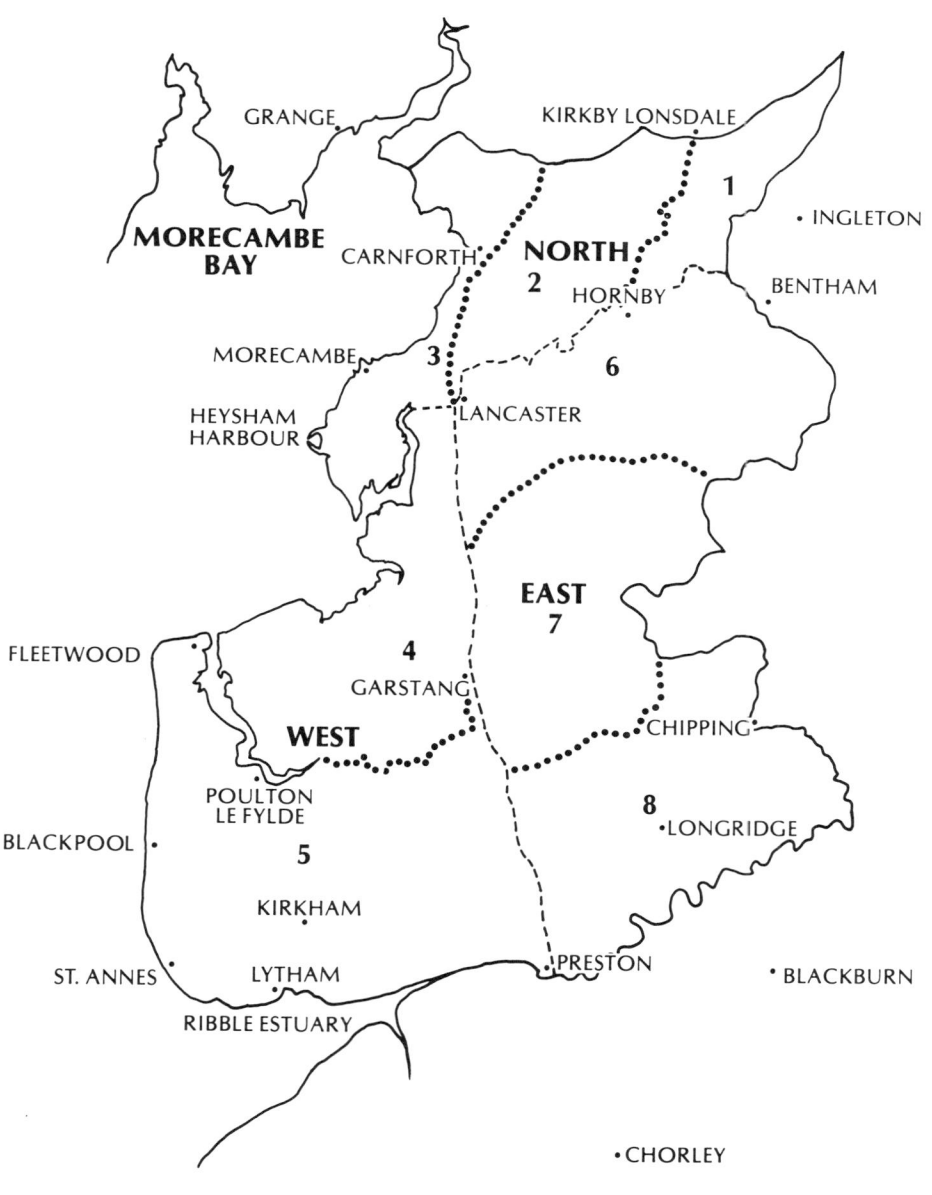